D0215340

Chemical and Energy Process Engineering

Chemical and Energy Process Engineering

Sigurd Skogestad

CRC Press is an imprint of the
Taylor & Francis Group, an **informa** business

CRC Press
Taylor & Francis Group
6000 Broken Sound Parkway NW, Suite 300
Boca Raton, FL 33487-2742

Printed in the United States of America on acid-free paper
10 9 8 7 6 5 4 3 2 1

International Standard Book Number-13: 978-1-4200-8755-0 (Hardcover)

Library of Congress Cataloging-in-Publication Data

Skogestad, Sigurd.
Chemical and energy process engineering / Sigurd Skogestad.
p. cm.
Includes index.
ISBN 978-1-4200-8755-0 (hardback : alk. paper)
1. Chemical engineering. 2. Power resources. 3. Power (Mechanics) I. Title.

TP155.S555 2008
660--dc22 2008023563

Visit the Taylor & Francis Web site at
http://www.taylorandfrancis.com

and the CRC Press Web site at
http://www.crcpress.com

Preface

In everyday life we use and surround ourselves with products: cars, gasoline, plastic bags, glue, telephones, clothing, computers, lamps, airplanes, makeup, fishing rods and toilet paper. To produce these products we need raw materials. But this is not enough. We also need a "**process**",[1] and in this book we concentrate on this process (or the path) from raw materials (feedstock) to products (including energy).

The development of a new process is demanding and exciting. Especially important are the choice of reaction conditions (pressure, temperature, degree of conversion) and the method for separation. The final process must be optimized to be competitive. Unfortunately, it is not often that an engineer gets a chance to participate in the development of a completely new process, but it is also very interesting to analyze and understand existing processes (which we focus on in this book).

This book uses three basic principles:

1. Mass is conserved (mass balance = material balance)
2. Energy is conserved (energy balance = the first law of thermodynamics),
3. Any system will proceed towards a more probable state (with more disorder) – and left to itself it will end up in a state of *equilibrium* (the second law of thermodynamics)

In this book, we will mainly apply these principles at the *macro scale*. This gives simple equations and the most essential information with relatively little effort. The same principles also apply if we look at the details on the micro scale, but rather than simple algebraic equations we then often end up with partial differential equations which are difficult to work with analytically.

The goal of this book is primarily to give a foundation for industrial process engineering calculations. But the principles are general and also apply for biological processes. I have attempted to write a book that is intellectually stimulating by deriving most of the formulas from the three basic principles mentioned above. This also promotes insight and understanding. Emphasis is placed on presenting analytical methods which give physical insight and which can be used for "calculations by hand." Even though the calculations are often done with computers, it is imperative that one has the physical insight to be able to validate the results by performing simple calculations by hand.

To make analytical calculations possible and give insight, it is often necessary to simplify. Many industrially important processes take place in a gaseous phase. In

[1] In dictionaries, **process** has many definitions. One from Webster is the following: "Process. A continuous action, operation or series of changes taking place in a definite manner."

addition to assuming ideal gas, we simplify our calculations significantly by assuming *perfect mixing*. This allows us to not look at the detailed flow pattern at micro scale.

This book has two main target audiences:

1. Students who need a basis in process engineering.
2. Practicing engineers and students at a more advanced level who need a reference book for practical calculations.

These targets are partly contradictory, but on the other hand a student who has invested time and money in a book should expect that it will prove useful later in his or her career. A suggestion to the students: Do not sell this book! The book contains many examples to illustrate the use of theory on actual problems. The emphasis is on obtaining numbers that can actually be used in the real world, so much of this book will useful in your career as a process engineer!

A few words on using this book

Chapter 1 summarizes the most important notation, definitions and conversion factors. You should browse through this chapter, which is primarily meant to be used for later reference.

The reader is assumed to have some previous knowledge of physics, chemistry and thermodynamics. Most of this is summarized in Appendix A. I recommend you start by browsing through this material, possibly reading some parts carefully, since it forms the basis for later chapters.

The focus of this book is on the general principles of material and energy balance. This begins in Chapter 2 where we introduce the general balance principle and mass balances which are the most important tool for process engineers. This continues in Chapter 3 with chemical reaction systems and in Chapter 4 with energy balances and so on.

This book was originally written to be used in a process engineering course for second year chemical and petroleum engineering students at NTNU in Trondheim. The material has been taught in approximately the following order: We start with topics on basic thermodynamics from Appendix A (about two weeks) and then cover the following chapters: 2, 3, 4, 5, 6, 7, 8, and finally (if we have time) Chapter 9. The course has also included a field trip to a plant and an associated required project where the students utilize their knowledge on a specific chemical process.

This book is suitable also for students and engineers with other backgrounds, such as energy engineering, process engineering, mechanical engineering or control engineering. The book can also be used for courses in more advanced subjects, in particular the "bonus" Chapter 11 on process dynamics, and Chapters 7 and 8 on entropy and thermal power.

Use the index. To find specific information, such as conversion factors, definitions and data, you should use the index at the end of the book.

Further reading

Many books can be recommended for further reading. This applies in particular to thermodynamics where the following texts are recommended, depending on your background:

J.M. Smith and H.C. van Ness, *Introduction to chemical engineering thermodynamics*, McGraw-Hill, 6th Edition, New York, 1996.

M.M. Abbot and H.C. van Ness, *Thermodynamics with chemical applications*, Schaum's Outline Series, McGraw-Hill, New York, 1989.

M.J. Moran and H.N. Shapiro, *Fundamentals of engineering thermodynamics*, Wiley, 6th Edition, 2007.

K. Denbigh, *The principles of chemical equilibrium*, Cambridge Press, 4th Edition, 1981.

D.R. Gaskell, *Introduction to the thermodynamics of materials*, Taylor & Francis, 4th Edition, New York, 2003.

There are also several good textbooks that focus on material balance calculations for chemical process engineering, and these may be useful for gaining additional insight or finding additional problems:

R.W. Felder and R.W. Rousseau, *Elementary principles of chemical processes*, Wiley, 3rd Edition, 2000.

D.M. Himmelblau and J.B. Riggs, *Basic principles and calculations in chemical engineering*, Prentice Hall, 7th Edition, 2004.

T.M. Duncan and J.A. Reimer, *Chemical engineering design and analysis. An introduction*, Cambridge University Press, 1998.

It may also be interesting and rewarding to return to the classic:

Olaf A. Hougen and Kenneth M. Watson, *Chemical process principles.*, Wiley. Part 1: Material and energy balances (1943). Part 2: Thermodynamics (1947). Part 3: Kinetics and catalysis (1947).

Book home page

The book has its own web page: `http://www.nt.ntnu.no/users/skoge/`. Here you will find:

- Comments on and corrections to the book
- Complete solutions to starred exercises
- Additional exercises
- The MATLAB files used in the examples

Good luck with the reading of the book!

Thanks

There are many I would like to thank – for discussions, comments, corrections, inspiration, patience, translation, love, support .... The list is very long and if I listed them all I am afraid I would forget someone. Anyway, thank you all – you know who you are.

Epilogue

An epilogue belongs of course at the end of a text, but I'm placing it here because it is probably better to read it before you start.

The balance principle is the key to process engineering calculations. We start by considering a small part of the world (usually indicated by a dotted line in figures) which we call our "system." The outside is the surroundings. Process streams may enter and exit our system, and energy can be provided in the form of heat or work. We can quite easily formulate mass balances based on "counting up" what goes in and out of our system.

Energy balances are usually more difficult to formulate, mainly because there are so many forms of energy, and because it makes use of *thermodynamics* with all its associated identities and variable transformations. The most important form of energy for us is *internal energy* (U). This is the energy of the molecules which includes chemical bonding energy, weaker forces between the molecules, as well as the thermal kinetic energy related to the movement of the molecules. In addition to its internal energy, a stream also contributes an associated flow (pV) work. By introducing the enthalpy $H = U + pV$ we no longer need to worry about this flow work (see also page 99):

Enthalpy is the sum of a stream's internal energy and flow work

The energy balance (the first law of thermodynamics) for a steady-state (stationary; static) open flow process (process with inlet and outlet streams) is

$$H_{\text{out}} - H_{\text{in}} = Q + W_n \quad [J/s] \tag{1}$$

where H is the stream enthalpy, Q is the heat supplied and W_n is the supplied non-flow work, (n stands for "non-flow"). That is, W_n is the overall work W minus the flow work (pV work) already included in H.

Note that H is the enthalpy of the *stream* and not of the system itself – the system is unchanged for a steady-state process. Also note that no assumption of constant pressure is made.[2]

The notion of the state of the system is also extremely important. If we consider a system (or a process stream) in internal equilibrium, then the state of the system is uniquely determined by specifying two variables in addition to the composition, for example enthalpy H and pressure p. The value of all other state variables, for example temperature T, entropy S are then given. Note that work W and heat Q deal with the transfer between systems and are not state variables.

An important consequence of the state notion is that one can evaluate changes in a real process by considering an imaginary process operating between the same start and end states. One can for example compute changes in enthalpy (which is a state variable) for an open system by considering an imaginary reversible process between the same conditions in a closed system.

[2] It is easy to confuse the energy balance for a steady-state *open* system in (1) with the energy balance for a *closed system* with constant pressure. Both are often written in the form $\Delta H = Q + W$, but ΔH has different meanings in the two cases – it is change in *stream* enthalpy for an open system and the change in *system* enthalpy for a closed system. You are now warned!

Of equal importance is the concept of entropy and the second law of thermodynamics. In short, the entropy ("degree of disorder") of a system is a state function and the second law of thermodynamics says the total entropy of the universe (system plus its surroundings) must always increase. At the equilibrium state the total entropy reaches its maximum. This results in simple and practical results, including equilibrium constants for chemical reaction, phase equilibrium relationships, and the Carnot factor $\eta = 1 - T_C/T_H$ for the maximum fraction of heat that can be transferred to work.

In summary, thermodynamics is a very useful tool – it is amazing how much practical knowledge can be obtained from the first and second laws of thermodynamics and the idea that internal energy and entropy are state variables.

To sum up: The basic theory of mass and energy balances is quite simple and, together with the phase and chemical equilibrium theory and some knowledge about rates, it constitutes the basis for what a process engineer needs to know.

A word of caution: even though the basic theory is simple, a fair amount of experience and insight is needed to put it to use. The best method of achieving this is by solving exercises, which is highly recommended!

About the author

Sigurd Skogestad was born in Norway in 1955 and is a professor of chemical engineering at the Norwegian University of Science and Technology (NTNU) in Trondheim. After obtaining his M.Sc. degree in chemical engineering at NTNU in 1978, he worked for four years in the industry at Norsk Hydro's Research Center in Porsgrunn, gaining valuable experience in the areas of process design and simulation. Moving to the US in 1983, he received a Ph.D. degree from the California Institute of Technology in 1987. Since his return to Norway in 1987, he has been a professor at NTNU, except for visiting Professorships at the University of California at Berkeley in 1994/95, and at the University of California at Santa Barbara in 2001/02.

The author of more than 140 international journal publications and 200 conference publications, he is the principal author with Ian Postlethwaite of a widely used textbook on *Multivariable Feedback Control* (Wiley, 1996, 2005). The goal of his research is to develop simple yet rigorous methods to solve problems of engineering significance. Research interests include the use of feedback as a tool to reduce uncertainty, change the system dynamics, and generally make the system more robust and well-behaved (including self-optimizing control). Other interests include plantwide control, interactions between process design and control, and distillation column design, control and dynamics.

Contents

1 Notation, concepts and numbers . . . 1
1.1 Notation . . . 1
1.2 Always check the units! . . . 7
1.3 Some conversion factors . . . 8
1.4 Some important numbers . . . 15
1.5 Some important concepts . . . 18
1.6 Unit operations . . . 21
1.7 Batch and continuous process . . . 27
1.8 A little about economy . . . 29
1.9 Some fun and useful energy exercises . . . 31
1.10 Global energy consumption . . . 37

2 Derivation of balance equations . . . 39
2.1 The balance principle . . . 39
2.2 The balance equation . . . 42
2.3 Mass balances without accumulation . . . 47
2.4 Recycle . . . 55
2.5 Systematic formulation and solution of mass balances . . . 58
2.6 Use of spreadsheet program . . . 59
2.7 Examples of recycle without reaction . . . 62
2.8 Flash calculations . . . 64
2.9 Summary: Procedure for deriving balance equations . . . 66
2.10 Degrees of freedom and solvability . . . 66
2.11 Simulation versus design . . . 75
2.12 Summary . . . 76

3 Mass balances with reaction . . . 77
3.1 Introduction . . . 77
3.2 The component balance . . . 77
3.3 Steady-state component balance . . . 78
3.4 Conversion and extent of reaction . . . 79
3.5 Selectivity and yield . . . 82
3.6 Reaction and recycle . . . 85
3.7 Atomic balances . . . 86
3.8 Independent reactions and matrix formulation . . . 88

3.9 Reaction with chemical equilibrium 91
3.10 Summary 94

4 The energy balance 95
4.1 The general energy balance (open system) 95
4.2 Energy forms 96
4.3 Work forms 98
4.4 Alternative formulations of the energy balance 100
4.5 Calculation of enthalpy 105
4.6 Energy balance for mixing processes 107
4.7 Valve: Isenthalpic pressure relief 114
4.8 Real fluids: Thermodynamic state diagrams 115
4.9 Energy balance with chemical reaction 118
4.10 Energy balance with kinetic and potential energy 125
4.11 Summary of energy balance 128

5 Heat exchange 129
5.1 Introduction 129
5.2 Calculation (design) of heat exchangers 131
5.3 Simulation of heat exchangers 139

6 Compression and expansion 143
6.1 Introduction 143
6.2 Compression (increase of pressure) 144
6.3 Expansion in turbine 144
6.4 Reversible shaft work 145
6.5 Reversible shaft work for ideal gas 148
6.6 Actual work and examples 149
6.7 Pump work 154
6.8 Compression and expansion of real gases 155

7 Entropy and equilibrium 161
7.1 The laws of thermodynamics 161
7.2 Calculation of entropy 163
7.3 Equilibrium 173
7.4 Introduction to vapor/liquid equilibrium 179
7.5 Flash calculations 189

8 Work from heat 197
8.1 Thermodynamics 197
8.2 Heat engine and the first law 198
8.3 Heat engine and the second law 199
8.4 Reverse heat engine: Refrigeration and heat pump 203
8.5 Efficiency 209
8.6 Ideal work and exergy 212
8.7 Gas power plant 225
8.8 Summary 236

9 Mechanical energy balance . . . 237
9.1 The "regular" energy balance . . . 237
9.2 Mechanical energy . . . 238
9.3 Reversible shaft work and friction . . . 238
9.4 The mechanical energy balance . . . 239
9.5 Compressible flow in pipe (gases) . . . 247
9.6 A remark on friction . . . 249
9.7 Summary . . . 250

10 Chemical reaction engineering . . . 253
10.1 Reaction kinetics . . . 253
10.2 Reactor calculations and reactor design . . . 260

11 Process dynamics . . . 273
11.1 Introduction . . . 273
11.2 Modeling: Dynamic balances . . . 274
11.3 Dynamic analysis and time response . . . 284
11.4 Linearization . . . 301
11.5 Dynamic simulation with examples . . . 303
11.6 Process control . . . 322
11.7 Summary . . . 325

A Some thermodynamics and physical chemistry . . . 327
A.1 Concept of mol . . . 327
A.2 Balancing chemical reactions . . . 328
A.3 Thermodynamic concepts . . . 329
A.4 Thermodynamic diagrams . . . 333
A.5 Equations of state . . . 334
A.6 Work, heat and energy . . . 343
A.7 Volume change work for closed system . . . 346
A.8 Internal energy . . . 347
A.9 Enthalpy . . . 348
A.10 Heat capacity . . . 349
A.11 Adiabatic reversible expansion of ideal gas . . . 350
A.12 Pressure independence of U and H for ideal gas: Joule's experiment . 353
A.13 Calculation of enthalpy . . . 354
A.14 Thermochemistry . . . 357
A.15 Alternative reference states . . . 364

B More thermodynamics: Entropy and equilibrium . . . 369
B.1 Entropy and the second law of thermodynamics . . . 369
B.2 Definition of entropy . . . 370
B.3 Carnot cycle for ideal gas . . . 373
B.4 Calculation of the system's entropy . . . 377
B.5 Mixtures (variable composition) . . . 381
B.6 Equilibrium . . . 383
B.7 The fundamental equation of thermodynamics and total differentials . 389

C Differential balances: Examples . . . 393
C.1 Emptying of gas tank . . . 393
C.2 Logarithmic mean temperature difference . . . 394
C.3 Batch (Rayleigh) distillation . . . 396

D Summary of the whole book . . . 399

E Additional problems . . . 405
E.1 Test exam . . . 405
E.2 Solution . . . 407
E.3 Some more exercises . . . 411

F Data . . . 415

G Solutions to starred exercises . . . 423

Index . . . 425

1

Notation, concepts and numbers

Collected here is a bit of everything that could have been placed in an Appendix, but I find them so important that I chose to place them first. The chapter starts with an overview of the notation and a discussion on the choice of basis. Engineers need numbers for practical calculations, and we provide an overview of conversion factors and some important numbers. After defining some important concepts, we next provide an overview of some important unit operations. We look at the difference between batch and continuous processes and give an introduction to economic analysis. Finally, you find some fun energy exercises which will make you familiar with the use of numbers and conversion between units.

1.1 Notation

Symbol	Name	SI unit
C_p	heat capacity, constant pressure	[J/K]
C_V	heat capacity, constant volume	[J/K]
c_p, c_V	specific heat capacity	[J/K kg]
c	concentration	[mol/m^3]
E	(total) energy	[J]
H	enthalpy	[J]
M	molar mass	[kg/mol]
m	mass	[kg]
n	number of mols	[mol]
Q	supplied heat	[J]
p	pressure	[N/m^2] = [Pa]
S	entropy	[J/K]
T	temperature	[K]
t	time	[s]
U	internal energy	[J]
V	volume	[m^3]
v	velocity	[m/s]
W	supplied work	[J]
x_i	mole fraction of component i	[mol i/mol total]
ρ	(mass) density	[kg/m^3]

The most important symbols are given in the table. They follow the international conventions (IUPAC and ISO). The following applies:

- Units are in the SI system which has seven basic units: m (meter), kg (kilogram), s (second), K (Kelvin), mol, A (ampere) and cd (candela). This book uses the first five of these. In addition, we use the derived SI units Newton (N), Pascal (Pa), Joule (J) and Watt (W):
  $N = kg \cdot m \cdot s^{-2}$
  $Pa = N \cdot m^{-2} = kg \cdot m^{-1} \cdot s^{-2}$
  $J = N \cdot m = kg \cdot m^2 \cdot s^{-2}$
  $W = J/s = kg \cdot m^2 \cdot s^{-3}$
  Conversion factors to some non-SI units are given on page 8.
- SI-notation is also used for prefixes:
  E (exa) $= 10^{18}$
  P (peta) $= 10^{15}$
  T (tera) $= 10^{12}$
  G (giga) $= 10^9$
  M (mega) $= 10^6$
  k (kilo) $= 10^3$
  h (hecto) $= 10^2$
  c (centi) $= 10^{-2}$
  m (milli) $= 10^{-3}$
  μ (micro)$= 10^{-6}$
  n (nano) $= 10^{-9}$
  p (pico) $= 10^{-12}$
- In English (American) literature one finds some non-SI numbers and prefixes, including
  billion $= 10^9$
  trillion $= 10^{12}$
  M $= 10^3$ (because M is the roman number 1000)
  MM $= 10^6$ (yes, this is really a strange one!)
- To indicate small fractions, one often uses the non-SI terms
  ppm $= 10^{-6}$ (parts per million)
  ppb $= 10^{-9}$ (parts per billion)
- E or e is commonly used to indicate exponentials with base 10, e.g., E-4 $= 10^{-4}$.
- The **universal gas constant** is
  $R = 8.314510$ J/K$\cdot$ mol (SI units)
  The value of R in some other units is
  $R = 1.987$ cal/mol$\cdot$ K
  $R = 1.987$ Btu/lb $\cdot$ R (here, the latter R stands for degrees Rankine) [1]
  $R = 82.06$ cm^3 atm/mol$\cdot$ K
  $R = 0.08206$ l atm/mol$\cdot$ K
- The standard value for the acceleration of gravity is (IUPAC)
  $g = 9.80665$ m/s^2
  However, we often set $g = 10$ m/s^2 as it varies around the earth anyway.

[1] Unfortunately, the number of symbols and letters is limited, so the same letter is sometimes used in different meanings. For example, W is the symbol for work, but W is also the unit Watt for power (1 W = 1 J/s); R is the universal gas constant, but R is also the unit for degrees Rankine.

- **Standard temperature and pressure (STP).** According to IUPAC, the standard temperature is 273.15 K (0°C) and the standard pressure is
  $p^{\ominus} = 1$ bar $= 10^5$ N/m^2.
  However, most thermodynamic data are given at the **standard ambient temperature**
  $T_0 = 298.15$ K
  and we actually rarely use the "standard temperature" 273.15 K in this book. Also, note that until 1982 the standard pressure was set as
  1 atm = 1.013250 bar.
- Superscript $^{\ominus}$ (or ^{o}) is used generally to indicate standard or reference states. For a gas at temperature T, the standard state is a (hypothetical) state as ideal gas at $p^{\ominus} = 1$ bar and temperature T.
- Standard enthalpy of formation $\Delta_f H^{\ominus}$ and other thermodynamic quantities are given at 1 bar and 298.15 K (25 °C), unless otherwise stated.
- Superscript * often indicates pure components.
- Superscript $'$ often indicates ideal gas.
- Subscript $_0$ often indicates initial state at time t_0, or feed stream to a reactor, or standard temperature T_0.
- Subscript $_f$ often indicates final state at time t_f.

Note

- We use lowercase letter for mass m [kg] and number of moles n [mol]. However, these are exceptions, because we use capital letters for most other **extensive** quantities (which are variables that depend on the size of the system), for example V, H, S, U, Q and W.
- The following are examples of **intensive** quantities (which are point variables that do not depend on the size of the system): c, M, p, T, x and ρ.
- We follow the IUPAC-convention and let Q indicate *supplied* heat and W indicate *supplied* work (to the system from the surroundings). For work, the opposite older convention, where W is work performed by the system (on the surroundings), is still commonly used, especially in mechanical engineering literature, i.e., W gets the opposite sign.
- In SI units, the **molar mass** M is in [kg/mol], for example, water has $M = 18.015 \cdot 10^{-3}$ kg/mol. Usually, we give molar mass in [g/mol], which is not a standard SI unit, for example $M = 18.015$ g/mol for water. A closely related quantity is the relative molar mass (**molecular weight**) M_r [dimensionless], which has the same numerical value as M [g/mol] but without the units. For example, water has M=18.015 g/mol and M_r=18.015.
- Exact relationship between mass, number of moles and molar mass:

$$m[\text{kg}] = n[\text{mol}] \cdot M[\text{kg/mol}] \tag{1.1}$$

  (which may be viewed as the definition of molar mass M).
- We use brackets with the meaning "with unit," for example [kg] means "with unit kg." Brackets are used sometimes as "extra-information" in the equations. For

example, we can write the ideal gas law as

$$p[\mathrm{N/m^2}] = \frac{n[\mathrm{mol}] \cdot \mathrm{R[J/mol\ K]} \cdot \mathrm{T[K]}}{\mathrm{V[m^3]}}$$

However, if we introduce numbers in the equations then units *should* be included. Thus, the units are now necessary information rather than extra information, and we may use "common" parenthesis rather than brackets. For example, the ideal gas law with numbers,

$$p = \frac{1\mathrm{mol} \cdot 8.31(\mathrm{J/mol\ K}) \cdot 298.15\mathrm{K}}{25 \cdot 10^{-3}\mathrm{m}^3} = 99055\ \mathrm{N/m^2} = 0.991\ \mathrm{bar}$$

- **Mole fraction**. Consider a mixture with c components (molecules), where we have n_i mol of component i. The total number of moles n (sometimes denominated n_{tot}) is

  $$n = n_1 + n_2 + \ldots + n_i + \ldots + n_c = \sum_{i=1}^{c} n_i = \sum_i n_i \quad [\mathrm{mol}]$$

  The mole fraction x_i [mol i/mol total] is defined as

  $$x_i \triangleq \frac{n_i}{n} \tag{1.2}$$

  and since $n = \sum_i n_i$ we always have that the sum of the mole fractions is 1,

  $$\sum_i x_i = 1 \tag{1.3}$$

  Mass and volume fractions are defined in the same way. For mole, mass or volume fractions, it is common to use ppm (parts per million $= 10^{-6}$) and ppb (parts per billion $= 10^{-9}$), which are not SI units.

1.1.1 Choice of basis – Consistent notation

Most quantities can be given with different basis. For example, the quantity of a material can be given on a mole basis (chemist), weight basis (mechanical engineer) or volume basis (sales people). To avoid uncertainty about the basis, we can introduce special symbols, and we choose to use subscript m for mole basis (e.g., H_m [J/mol] for molar enthalpy) and lowercase letter for mass basis (e.g., h [J/kg] for specific enthalpy). We will not introduce any special symbol for volume basis.

Molar quantity (mole basis). An extensive quantity X can be divided by the number of moles n [mol], to get the molar quantity, which is here indicated with subscript m, i.e.,

$$X_m = X/n$$

which is an intensive quantity. Examples are molar heat capacity, molar enthalpy and molar volume:

$$\begin{aligned} C_{p,m} &= C_p/n \quad [\mathrm{J/K\ mol}] \\ C_{V,m} &= C_V/n \quad [\mathrm{J/K\ mol}] \\ H_m &= H/n \quad [\mathrm{J/mol}] \\ V_m &= V/n \quad [\mathrm{m^3/mol}] \end{aligned}$$

The subscript m is omitted sometimes when it is obvious from the context that it is related to a molar quantity. For example, the standard enthalpy of formation $\Delta_f H_m^\ominus$ [J/mol] is written as $\Delta_f H^\ominus$ [J/mol].

Specific quantities (mass basis). An extensive quantity X can be divided by the mass m [kg], to get the specific quantity, which is here indicated by lowercase letter, i.e.,

$$x = X/m$$

which is an intensive quantity. Examples are specific heat capacity, specific enthalpy and specific volume:

$$\begin{aligned} c_p &= C_p/m \quad [\text{J/K kg}] \\ c_V &= C_V/m \quad [\text{J/K kg}] \\ h &= H/m \quad [\text{J/kg}] \\ v &= V/m \quad [\text{m}^3/\text{kg}] \end{aligned}$$

Note that the density ρ [kg/m^3] is the inverse of the specific volume, i.e.,

$$\rho = \frac{m}{V} = \frac{1}{v} \quad [\text{kg/m}^3]$$

- Some exact relations follow from the definitions above. For example, we have the relation between molar volume V_m and density ρ,

$$V_m \ [\text{m}^3/\text{mol}] = \frac{M \ [\text{kg/mol}]}{\rho \ [\text{kg/m}^3]}$$

  where M is the molar mass, and the relation between molar and specific heat capacity

$$C_{p,m} \ [\text{J/K mol}] = \text{c}_\text{p} \ [\text{J/kg K}] \cdot \text{M} \quad [\text{kg/mol}]$$

Rates and dot notation. For a continuous process, the flow rate (per unit of time) is often indicated by the use of dot notation,[2] $\dot{X}$. Examples are

- molar rate (molar flow): $\dot{n}$ [mol/s]
- mass rate (mass flow): $\dot{m}$ [kg/s]
- volumetric rate (volumetric flow): $\dot{V}$ [m^3/s]
- enthalpy rate (enthalpy flow): $\dot{H}$ [J/s]

Example 1.1 *In this example we illustrate the use of consistent notation. Consider a continuous process where water is heated from 10 °C to 25 °C by the use of an electric heater. The mass flow of water is*

$$\dot{m} = 2 \ kg/s$$

[2] Some, including control engineers, use the dot notation to mean time derivative, $\dot{x} \equiv \frac{dx}{dt}$ (change of a variable per unit of time), which is not the same as a rate variable. To avoid confusion, the whole dot notation can be omitted by using special symbols for flow rates, and the following is common: Molar flow $F \equiv n$ [mol/s], mass flow $w \equiv m$ [kg/s] and volumetric flow $q \equiv \dot{V}$ [m^3/s]. We use this later in Chapter 11 on process dynamics where we deal with both time derivatives and rates and, to avoid confusion, we omit the use of dot notation. However, in the rest of the book we use dots to mean rates, as defined above.

(a) Calculate the molar [mol/s] and volumetric [m^3/s] flows and the supplied power [W=J/s]. (b) Calculate the specific [J/kg] and molar heat [J/mol] supplied. (c) How much water is heated up in 1 hour (in kg, mol and m^3) and how much heat is supplied in this period [J]. (d) What is the cost per hour if the energy cost is 0.2 \$/kW h.

Use the following data for water (liquid): Molar mass $M = 18 \cdot 10^{-3}$ kg/mol; density $\rho = 1000$ kg/m^3; specific heat capacity $c_p = 4.18 \cdot 10^3$ J/K kg.

Solution. *(a) Molar and volumetric flows*

$$\dot{n} = \dot{m}/M = (2kg/s)/(18 \cdot 10^{-3} kg/mol) = 111 \ mol/s$$

$$\dot{V} = \dot{m}/\rho = (2kg/s)/(1000kg/m^3) = 0.002 \ m^3/s$$

For the energy balance (more about this in Chapter 4), the supplied power (heat per unit of time) is

$$\dot{Q} = \dot{m}c_p(T_{\text{out}} - T_{\text{in}}) = 2 \ kg/s \cdot 4.184 \cdot 10^3 J/kg \ K \cdot (25K - 10K) = 125.5 \cdot 10^3 \ J/s = 125.5 \ kW$$

(b) Supplied heat per kg (specific heat)

$$q = \frac{\dot{Q}}{\dot{m}} = \frac{125.4 \cdot 10^3 J/s}{2kg/s} = 62.8 \ kJ/kg$$

Supplied heat per mol (molar heat)

$$Q_m = \frac{\dot{Q}}{\dot{n}} = 1131 \ J/mol$$

(c) In the time period $\Delta t = 3600$ s (1 hour), we heat up the following amount

$$m = \dot{m}\Delta t = 2(kg/s) \cdot 3600s = 7200 \ kg$$

$$n = \dot{n}\Delta t = 400 \cdot 10^3 \ mol = 400 \ kmol$$

$$V = \dot{V}\Delta t = 7.2 \ m^3$$

and the supplied heat in the period is

$$Q = \dot{Q}\Delta t = 125.4 \cdot 10^3 \text{J/s} \cdot 3600\text{s} = 451.44 \cdot 10^6 J = 451.44MJ$$

(d) The cost for 1 hour is

$$125.4\text{kW} \cdot 1\text{h} \cdot 0.2 \ \$/\text{kWh} = 25.1 \ \$$$

The disadvantage with the consistent notation introduced above is the large number of equations and symbols. For example, consider the energy balance from the example. With *consistent notation* it can be written in the following equivalent forms

$$\begin{aligned}
\text{Mass basis}: \quad & Q \ [\text{J}] = & m \ c_p(T_{\text{out}} - T_{\text{in}}) \\
\text{Mole basis}: \quad & Q \ [\text{J}] = & n \ C_{p,m}(T_{\text{out}} - T_{\text{in}}) \\
\text{Mass rate basis}: \quad & \dot{Q} \ [\text{J/s}] = & \dot{m} \ c_p(T_{\text{out}} - T_{\text{in}}) \\
\text{Mole rate basis}: \quad & \dot{Q} \ [\text{J/s}] = & \dot{n} \ C_{p,m}(T_{\text{out}} - T_{\text{in}}) \\
\text{Specific mass basis}: \quad & q \ [\text{J/kg}] = & c_p(T_{\text{out}} - T_{\text{in}}) \\
\text{Specific molar basis}: \quad & Q_m \ [\text{J/mol}] = & C_{p,m}(T_{\text{out}} - T_{\text{in}})
\end{aligned} \tag{1.4}$$

1.1.2 Simplified notation

An alternative approach, which we frequently use in this book, is to write the equations using "mass basis symbols" (1.4) for all choices of basis. The advantage is fewer special cases, which makes it easier to focus on the contents of the equation. As an example, with the *simplified notation*, the energy balance is in all cases written in the form

$$Q = mC_p(T_{\text{out}} - T_{\text{in}}) \quad [\text{J; J/s; J/kg; or} \quad \text{J/mol}] \tag{1.5}$$

with the *same* symbols (Q, m and C_P) for all choices of basis. The particular **choice of basis** for m, or, equivalently, the choice of units, may be indicated by use of brackets, for example:

$$\begin{aligned}
\text{Mass basis:} \quad & Q\,[\text{J}] = & m[\text{kg}] \cdot \text{C}_\text{p}[\text{J/kg K}] \cdot (\text{T}_\text{out}[\text{K}] - \text{T}_\text{in}[\text{K}]) \\
\text{Mole basis:} \quad & Q\,[\text{J}] = & m[\text{mol}] \cdot \text{C}_\text{p}[\text{J/mol K}] \cdot (\text{T}_\text{out}[\text{K}] - \text{T}_\text{in}[\text{K}]) \\
\text{Mass rate basis:} \quad & Q\,[\text{J/s}] = & m[\text{kg/s}] \cdot \text{C}_\text{p}[\text{J/kg K}] \cdot (\text{T}_\text{out}[\text{K}] - \text{T}_\text{in}[\text{K}]) \\
\text{Mole rate basis:} \quad & Q\,[\text{J/s}] = & m[\text{mol/s}] \cdot \text{C}_\text{p}[\text{J/mol K}] \cdot (\text{T}_\text{out}[\text{K}] - \text{T}_\text{in}[\text{K}]) \\
\text{Spec. mass basis:} \quad & Q\,[\text{J/kg}] = & C_p[\text{J/kg K}] \cdot (\text{T}_\text{out}[\text{K}] - \text{T}_\text{in}[\text{K}]) \\
\text{Spec. molar basis:} \quad & Q\,[\text{J/mol}] = & C_p[\text{J/mol K}] \cdot (\text{T}_\text{out}[\text{K}] - \text{T}_\text{in}[\text{K}])
\end{aligned}$$

(the latter two cases correspond to setting $m = 1$ kg and $m = 1$ mol, respectively). We have then just one symbol for heat (Q) instead of many (Q, $\dot{Q}$, q and Q_m). Similarly, the symbol m is used for both mass [kg] and mass flow [kg/s], and in some cases even for [mol] and [mol/s]. Furthermore, we use C_p to mean also molar and specific heat capacity (e.g., we write for water $C_p = 4.18$ kJ/K kg and $C_p = 75.3$ J/K mol instead of $c_p = 4.18$ kJ/K kg and $C_{p,m} = Mc_p = 75.3$ J/K mol).

Nevertheless, there will be cases where we follow the *consistent* notation and introduce dots (for rates), lowercase letters (for specific quantities) or subscript m (for molar quantities).

1.2 Always check the units!

If SI units are used for all numbers in an equation, then the answer is always in SI units. One can then get away with omitting the units from the equations, which is frequently done in this book to save space (or actually because the author was lazy). *That is, if you find a number without units then it is understood that it is in SI units.* For example, the molar volume for ideal gas at 0 ^{o}C and 1 atm is equal to

$$V_m = \frac{RT}{p} = \frac{8.3145 \cdot 273.15}{1.01325 \cdot 10^5} = 22.414 \quad \text{l/mol}$$

(l/mol here means liters per mol). Here, there are several things that are "not good":

1. It is mathematically wrong to omit units in intermediate calculations, but as mentioned it is understood that we use SI units if nothing else is said, i.e., we have in reality

$$V_m = \frac{8.3145 \cdot 273.15}{1.01325 \cdot 10^5} \frac{\text{m}^3}{\text{mol}} = 22.414 \cdot 10^{-3} \frac{\text{m}^3}{\text{mol}} = 22.414 \quad \frac{\text{l}}{\text{mol}}$$

2. We should insert units for each number to make sure that we actually end up with the expected SI unit. This gives a very useful check that the equations are dimensionally correct. That is, it is recommended you write

$$V_m = \frac{8.3145 \frac{\text{J}}{\text{K mol}} \cdot 273.15 \text{ K}}{1.01325 \cdot 10^5 \frac{\text{N}}{\text{m}^2}} = 22.414 \cdot 10^{-3} \frac{\text{J}}{\text{K mol}} \cdot K \cdot \frac{\text{m}^2}{\text{N}}$$

Here we easily see that the [K]s cancel, and that [N] is cancelled because [J] = [Nm], i.e., we get

$$\frac{\text{J}}{\text{K mol}} \cdot \text{K} \cdot \frac{\text{m}^2}{\text{N}} = \frac{\text{m}^3}{\text{mol}}$$

which is indeed the SI unit for molar volume. As 10^{-3} m^3 = 1 l, we then get as expected that $V_m = 22.414$ l/mol.

1.3 Some conversion factors

Unfortunately, the world has not yet completely converted to metric (SI) units, and especially in the US the transition is slow. In spite of many efforts, most people in the US are still using the old "British" units, or *U.S. customory units* as they are officially known. In any case, non SI-units are still in use and are also found in older literature, so an engineer needs to know the most important units and conversion factors.

Time. The standard SI unit for time (t) is s (second). In addition, the following derived SI units are common:

1 min = 60 s.

1 h (hour, hr) = 60 min = 3600 s.

1 d (day) = 24 h = 86400 s.

1 y (year) is usually taken as 365 days = 8760 h.
Sometimes the symbol a (annum) is used, i.e., 1 y = 1 a.

The operating time for a continuous plant is often assumed to be 8000 h per year, which corresponds to an uptime (availability) of 91%.

Mass. The standard SI unit for mass (m) is kg (kilogram).

In English (American) literature one finds the unit

1 lb (pound) = 1 lb$_m$ (pound mass) = 0.45359237 kg (exact).

A common derived SI unit is

1 t (ton, tonne) = 1000 kg (exact).

However, note that in English literature one can find the terms (short) ton = 2000 lb = 907.185 kg (US) and (long) ton = 2240 lb = 1016.047 kg (British). The term tonne is sometimes used to clearly indicate that it is a metric ton (1000 kg).

Mole. In SI units, a mole (= 1 mol) is the number of molecules (or atoms) in 12 grams of carbon-12 = $N_A = 6.02214 \cdot 10^{23}$ (Avogadro's number); see page 327 for details.

In English literature one can also find the rather **strange unit** "pound mole" (lb-mol) defined as

$$1 \text{ lb-mol} = 0.453593 \text{ kmol} = 453.593 \text{ mol} = 453.593 \text{ g-mol}$$

The reason for introducing these strange units is to make the molar mass (molar mass) equal in SI and English units. For example, the molar mass of water is M = 18.015 g/mol = 18.015 kg/kmol = 18.015 lb/lb-mol. Note also that the unit 1 g-mol ("gram mole") found in English literature means 1 mol (the standard SI unit), and similarly 1 kg-mol means 1 kmol.

Length. The standard SI unit for length is m (meter). A derived unit is Ångstrøm:

$$1 \text{ Å} = 10^{-10} \text{ m} = 0.1 \text{ nm}.$$

The commonly used British/American *foot* (ft) was defined in 1960 to be exactly 0.3048 m.[3] In addition, we have the British/American length units *inch*, *yard* and *mile*, and we have:

$$\begin{aligned} 1 \text{ inch} &= 1'' = \frac{1}{12} \text{ ft} = 0.0254 \text{ m} \quad \text{(exact)} \\ 1 \text{ ft} &= 1' = 0.3048 \text{ m} \quad \text{(exact)} \\ 1 \text{ yard} &= 3 \text{ ft} = 0.9144 \text{ m} \quad \text{(exact)} \\ 1 \text{ mile} &= 1760 \text{ yards} = 1609.344 \text{ m} \quad \text{(exact)} \end{aligned}$$

At sea, one uses the nautical mile = 1852 m (exact).

Exercise 1.1* *What is a speed of 100 mph (miles per hour) in SI units [m/s]?*

Volume. The standard SI unit for volume (V) is m^3, but for smaller volumes it is common to use the SI-derived unit l (liter) and we have:

$$1 \text{ l} = 10^{-3} \text{ m}^3 \text{ (exact)}.$$

In US literature one often finds the units cubic foot and gallon:

$$1 \text{ cu.ft. (ft}^3) = 28.317 \text{ l}$$

$$1 \text{ gal (US)} = 231 \text{ inch}^3 \text{ (exact)} = 3.78541 \text{ l}$$

The latter should not be confused with the somewhat less common British "imperial" gallon,

$$1 \text{ gal (Imperial)} = 4.54609 \text{ l (exact)}$$

The oil industry uses barrel (bbl):

$$1 \text{ bbl} = 42 \text{ gal (US) (exact)} = 158.99 \text{ l} = 0.15899 \text{ m}^3.$$

By the way, my "favorite unit" ☺ is *acre-foot* which is used in the US to indicate volumes for water storage and oil reservoirs. 1 acre = 43560 ft^2 = 4046.9 m^2 and we have 1 acre-foot = 43560 ft^3 = 1233.5 m^3.

[3] A British/US foot (ft) is defined as 0.3048 m (exact). For comparison, the old Norwegian foot was 0.3137 m, whereas the Swedes, poor guys, had a foot that was only 0.2969 m.

For gas volumes, see page 14.

Exercise 1.2 *On food products in the US, you may find the mysterious units* **ounce (oz).** *and* **fluid ounce (fl.oz).**. *These are defined as 1 oz. = (1/16) lb and 1 fl.oz. = (1/128) gal (US). Show that 1 oz = 28.35 g and 1 fl.oz. = 29.57 ml. (Thus, for water, which has a density of 1 ml/g, 1 fl.oz. is slightly more than 1 oz.)*

Temperature. The standard SI unit for temperature is K. A derived SI unit is Celsius [C or ^{o}C] and we have (exact):

$$T[\mathrm{K}] = \mathrm{t}[^o\mathrm{C}] + 273.15$$

In English literature one finds the unit Fahrenheit [F or ^{o}F] and its "absolute" counterpart Rankine [R]. We have (exact)

$$\begin{aligned} t[^o\mathrm{C}] &= \frac{t[^o\mathrm{F}] - 32}{1.8} \\ T[\mathrm{K}] &= T[\mathrm{R}]/1.8 \end{aligned}$$

In thermodynamic equations (e.g., the ideal gas law), one must *always* use "absolute" temperature T (i.e., in K or R). This is why I often use the lowercase letter (t) for temperature when it is given in degrees Celsius or Fahrenheit (we must then live with the small problem that t can also mean time). On the other hand, note that for temperature *differences* (changes) the units K or ^{o}C are the same (and R or ^{o}F are the same); for example, for specific heat (C_p) we have 1 J/kg K = 1 J/kg ^{o}C, and for a temperature difference (ΔT) we have 1 K = 1 ^{o}C.

Force. The standard SI unit for force (F) is N (Newton). 1 N = 1 kg m s^{-2}).

From the old "metric" cgs-system (where cm and g were used instead of m and kg, and which was replaced by the SI-system in 1960) we have the "small" unit dyn, which is no longer in common use,

1 dyn = 10^{-5} N.

Then, there is the more strange American/British unit

1 lb_f (*pound force*)= 4.4482216 N.

This is equal to the gravitational force that a mass of 1 lb (0.454 kg) has at the earth's surface. This is a rather stupid unit which makes it necessary to introduce the **mystical factor** g_c in all equations with force. Newton's 2nd law is normally written $F = ma$ where a is the acceleration. No "factor" is needed in this equation. However, if force (F) is given in units lb_f (*pound force)* we must write $F = \frac{ma}{g_c}$ where g_c is the acceleration of gravity used when defining lb_f. In American/British units the acceleration is in ft/s^2, and we get

$$g_c = 32.1740 \frac{\mathrm{lb}_m \cdot \mathrm{ft}}{\mathrm{lb}_f \cdot \mathrm{s}^2}$$

Note that in SI units, we have, as expected, $g_c = 32.1740 \frac{0.453593 kg \cdot 0.3048 m}{4.4482216 N \cdot s^2} = 1$ (dimensionless), and the factor g_c drops out.

Pressure. The standard SI unit for pressure (p) is Pa ($=$ N m^{-2} $=$ kg m^{-1} s^{-2}), but 1 Pa is a rather small pressure, so we often use the derived SI unit

$$1 \text{ bar} = 10^5 \text{ Pa (exact)}.$$

The conversions to bar from some non-SI units are:

$$\begin{aligned} p[\text{bar}] &= p[\text{atm}] \cdot 1.013250 \\ p[\text{bar}] &= p[\text{mmHg}]/750.061 = \text{p[torr]}/750.061 = \text{p[torr]} \cdot 133.322 \cdot 10^{-5} \\ p[\text{bar}] &= p[\text{psi}]/14.5038 \end{aligned}$$

Here the US/British unit for pressure [psi] is the same as [lb_f in^{-2}] (pound-force per square inch). Also note that

$$1 \text{ atm} = 1.01325 \text{ bar} = 760.0 \text{ mm Hg} = 14.696 \text{ psi}.$$

In older literature (and in old process design manuals) one may find "technical atmosphere" ([kp/cm^2] =[at]):

$$p \text{ [bar]} = p \text{ [at]} \cdot 0.9807$$

In every day life (e.g., tire pressure) and in industry, one often uses "gauge pressure" which is the pressure difference to the atmosphere,

$$\text{gauge pressure} + \text{atmospheric pressure} = \text{absolute pressure}$$

In English literature, gauge pressure is indicated with g (for gauge) and German literature with o (for overpressure). We have then, with the standard atmospheric pressure set at 1 atm,

$$p \text{ [bar]} = p \text{ [barg; baro]} + 1.01325 \text{ bar}$$

$$p \text{ [bar]} = \frac{p[\text{psig}]}{14.5038} + 1.01325 \text{ bar}$$

In some cases, the letter a is used to explicitly show that one means the real ("absolute") pressure, i.e.,

$$1 \text{ bara} = 1 \text{ bar}.$$

$$1 \text{ psia} = 1 \text{ psi}.$$

Example. *If the pressure is given as 0.400 barg, then the real (absolute) pressure is 0.400 bar + 1.013 bar = 1.413 bar.*

Example. *The pressure in car and bicycle tires is given often in psig (pounds per square inch gauge). A tire pressure of 30 psig corresponds to an absolute pressure of (30 / 14.504) bar + 1.013 bar = 3.082 bar.*

Example. *If the pressure is given as 0.400 ata, then this is probably the absolute pressure in technical atmosphere, i.e., the real (absolute) pressure is* $0.4 \cdot 0.9807 = 0.392$ *bar.*

Example. *If the pressure is given as 0.400 ato, then this is probably gauge pressure in technical atmosphere, i.e., the real (absolute) pressure is* $0.4 \cdot 0.9807 + 1.013 = 1.405$ *bar.*

Energy and power. The standard SI unit for energy is J ($=$ N m $=$ kg m^2 s^{-2}).

The conversion factor between the old unit *calorie* and Joule is

$$1 \text{ cal} = 4.184 \text{ J (exact)}.$$[4]

Note that the "common calorie" still used in everyday language and on many food labels is actually 1 kcal.

For electrical energy, it is common to use kilowatt-hour,

$$1 \text{ kWh} = 3600 \text{ kJ} = 3.6 \text{ MJ (exact)}$$

In English literature one finds the unit Btu (British thermal unit),

$$1 \text{ Btu} = 1.05505 \text{ kJ}.$$[5]

Another (rather stupid) unit used for mechanical energy in English literature is

$$1 \text{ ft} \cdot \text{lb}_f = 1.35582 \text{ J}.$$

With energy in ft $\cdot$ lb_f, the mystical factor g_c (see page 10) appears also in the energy equations, for example, the kinetic energy becomes $E_K = \frac{mv^2}{g_c}$. Of course, if we use SI units, $E_K = mv^2$.

For really small energies, at the molecular level, the unit **electronvolt** is common

$$1 \text{ eV} = 1.60218 \cdot 10^{-19} \text{ J}$$

On the other end of the scale, we have the unit **ton of oil equivalent** (toe), which is used as a measure of the energy content in fuels. The unit is the (lower) heat of combustion for 1 ton of crude oil, but this value of course depends on the source of the oil. IEA/OECD has more precisely defined

$$1 \text{ toe} = 41.868 \text{ GJ}$$

For **power** (energy per unit of time), the standard SI unit is Watt [W],

$$1 \text{ W} = 1 \text{ J/s}$$

The unit W (and also kW, MW, GW, TW) is widely adopted, even in English (American) literature, but some people use the British unit

$$1 \text{ Btu/hr} = 0.2931 \text{ W}$$

The old unit horsepower (hp) is still quite common,

$$1 \text{ hp (US)} = 550 \text{ ft} \cdot \text{lb}_f/\text{s} = 745.7 \text{ W} = 0.7457 \text{ kW}.$$

This American/British unit is also known as a "mechanical horsepower."

Apparently, the horses varied in strength around the world, and a Norwegian (and German) horse was apparently somewhat weaker, since a Norwegian *hestekraft* (also known as a "metric horsepower") was only 735.5 W (rarely used today). This is the power needed to lift 75 kg up 1 meter in a period of 1 second (735.5 W = 75 m $\cdot$ kg_f/s = 75 kg $\cdot$ 9.80665 m/s^2$\cdot$ 1 m / 1 s); you would need to lift about 76 kg to get 1 hp (US). What about a real horse? The peak

[4] The conversion factor between (thermochemical) calorie and Joule is 4.184 (exact), but also other "calories" have been in use, including the *International Table (IT)*-calorie from 1956 with conversion factor 4.1868, and the "at 15 ^{o}C"-calorie from 1950 with conversion factor 4.1855. The problem dates back to 1800's when it was not clear that heat and work had the same unit, and different units were used for heat (calorie) and work (Joule).

[5] The unit calorie was originally defined as the amount of heat necessary to raise the temperature of 1 gram of water by 1 ^{o}C. Analogously, the unit Btu was defined as the amount of heat necessary to raise the temperature of 1 lb of water by 1 ^{o}F. For water we then have that $C_p(l) \approx 1$ Btu/lb F$\approx$ 1 cal/g K = 4.184 kJ/kg K = 4.184 cal/g C.

power over a few seconds has been measured to be as high as 15 hp, but over longer periods an average horse gives somewhat less than 1 hp.

For some more fun (and even useful) facts about energy and power, see Section 1.9 (page 31).

Exercise 1.3* *In their design book, Seider et al.* (Process design principles, *Wiley, 1999) state that a heuristic for an air cooler is that the fan power is 5 hp/(MMBtu/hr). What is the expected fan power (kW) for an air cooler that removes 1 MW of heat?*

Density. In SI units, density ($\rho = m/V$) is given in kg/m^3.

To convert from English units use

$$\rho \text{ [kg/m}^3] = \rho \text{ [lb/ft}^3] \cdot 16.018$$

A related concept is **specific gravity** (spgr), which is the normalized (dimensionless) density defined by

$$\text{Specific gravity (spgr)} = \frac{\rho[\text{kg/m}^3]}{\rho_{\text{ref}}[\text{kg/m}^3]} = \frac{\rho[\text{lb/ft}^3]}{\rho_{\text{ref}}[\text{lb/ft}^3]}$$

Note that the specific gravity is the same in SI and American units. The reference for liquids and solids is normally water (liquid) at 4°C, where $\rho_{\text{ref}} = 1000$ kg/m^3 = 62.43 lb/ft^3. For gases, the reference is normally air.

Specific gravity of liquid. To be precise, when using specific gravity, the data should also refer to the temperatures of the substance and reference. Thus, for solids and liquid the notation

$$\text{spgr} = 1.264^{25/4}$$

means the following: The density of the substance at 25°C is 1.264 times higher than the density of the reference substance (water) at 4°C. For cases where no temperatures are given, one can assume ambient temperature (20°C) for the substance and 4°C for water, i.e., $\rho_{\text{ref}} = 1000$ kg/m^3.

API gravity. In the petroleum industry, they have many strange units, and yet another one is the so-called American Petroleum Institute (API) gravity (or ^{o}API). To convert to the more "normal" specific gravity, use

$$\text{spgr}^{60F/60F} = \frac{141.5}{^o\text{API}+131.5}$$

where we note that the densities of both the substance (petroleum fluid) and the reference (water) are at 60°F (= 15.6^{o}C). Fortunately, the density of water is almost constant from 0°C to 16°C, so one can still set the reference density as 1000 kg/m^3. Note that a "heavy" component with a high density has a small API gravity and *vice versa.* Also note that spgr = 1.0 corresponds to API gravity = 10.

Classification of crude oil. Crude oil is classified as light, medium or heavy, according to its measured API gravity. A light crude oil is defined as having an API gravity higher than 31.1 °API (density less than 870 kg/m^3). A medium oil is defined as having an API gravity between 22.3 °API and 31.1 °API (density between 870 kg/m^3 and 920 kg/m^3). A heavy oil is defined as having an API gravity below 22.3 °API (density over 920 kg/m^3).

Specific gravity of gases. For gases the reference substance is normally air. Then at 25°C and 1 atm we have $\rho_{\text{ref}} = 1.184$ kg/m^3 (see exercise). Often, the densities of gas and air are given at the same conditions, and if we assume ideal gas then the specific gravity is equal to the relative molar mass,

$$\text{spgr (gas)} = \frac{\text{density gas } (T,p)}{\text{density air } (T,p)} = \frac{M[\text{g/mol}]}{M_{\text{air}}[\text{g/mol}]} \quad \text{(assuming deal gas)}$$

where M is the molar mass of the gas and $M_{\text{air}} = 28.97$ g/mol.

Exercise 1.4 *Show that the density of air at 25° C and 1 atm is 1.184 kg/m³ when you assume ideal gas.*

Gas volumes. The amount (quantity) of a material is specified by giving its mass (m) or total mol (n). Historically, it has been common to use volume as a measure of quantity, because it is easy to understand and measure. The problem is that the volume depends also on temperature and pressure. This has made it necessary to introduce so called "standard" or "normal" states for volume – both for gas and liquid. Here, we consider gases, where the ideal gas is used as the basis.

For an ideal gas the volume is

$$V = n\,\frac{RT}{p} = nV_m$$

where

$$V_m = \frac{RT}{p}$$

is the molar volume for ideal gas. Thus, at a given ("standard" or "normal") T and p, *the "standard" or "normal" volume V is a direct measure of the total number of moles n.*

- At standard temperature and pressure (STP), $T = 273.15$K (0°C) and $p = 1$ bar, the molar volume of an ideal gas is $V_m = 22.711$ l/mol (this value is listed as a "fundamental constant" by IUPAC).
- At $T = 298.15$K (25°C) and $p = 1$ bar, the molar volume of an ideal gas is $V_m = 8.3145 \cdot 298.15/10^5 = 24.790 \cdot 10^{-3}$ m³/mol = 24.79 l/mol.
- The older unit "**normal cubic meter**" [Nm³] is defined as the theoretical volume the substance would have as ideal gas at 0 °C and 1 atm. In this state

$$V_m = \frac{8.3145 \text{ J/K mol} \cdot 273.15 \text{ K}}{1.01325 \cdot 10^5 \text{ N/m}^2} = 22.414 \cdot 10^{-3} \text{ m}^3/\text{mol} = 22.414 \text{ l/mol}$$

  or $1/V_m = 1/(22.414 \cdot 10^{-3}\text{m}^3/\text{mol}) = 44.615$ mol/m³. We have then that

  1 Nm³ is 44.615 mol.

- A "**standard cubic meter**" [Sm³] was originally defined (US) at 60°F (15.56 °C) and 0 psig (1 atm). In this state $V_m = 8.3145 \cdot 288.706/(1.01325 \cdot 10^5) = 23.690 \cdot 10^{-3}$ m³/mol = 23.690 l/mol or $1/V_m = 42.211$ mol/m³. Thus,

  1 Sm³ (US) is 42.211 mol .

- The newer "*ISO-standard cubic meter*" is defined at 15 °C and 1 atm. In this state $V_m = 8.3145 \cdot 288.15/(1.01325 \cdot 10^5) = 23.645 \cdot 10^{-3}$ m^3/mol = 23.645 l/mol, or $1/V_m = 42.292$ mol/m^3 (which is 0.2% higher than the US-value). Thus,

  1 Sm^3 (ISO) is 42.292 mol .

Production rates. Many strange units, most of them non-SI, are in common use in the process industry to denote production volumes. *Some examples are (in order of increasing strangeness, according the the author's view):*

1 TPD (ton per day)= 1 t/d = 1000 kg/d = 11.574 $\cdot 10^{-3}$ kg/s

1 TPA (ton per annum) = 1 t/y = 2.740 kg/d = 31.71 $\cdot 10^{-6}$ kg/s

1 KTPA (kton per annum) = 1000 t/y = 2740 kg/d = 31.71$\cdot 10^{-3}$ kg/s

1 MTPA (million ton per annum)= 10^6 t/y = 2740 t/d = 31.71 kg/s

1 MTA (metric ton per annum) = 1 t/y = 31.71$\cdot 10^{-6}$ kg/s

1 gpm (US gallons per minute) = 63.09$\cdot 10^{-6}$ m^3/s

1 BPD (barrels per day) = 1 bbl/d = 0.15899 m^3/d = 1.8401$\cdot 10^{-6}$ m^3/s

1 MIGD (million imperial gallons per day) = 4546 m^3/d = 0.05262 m^3/s

1 MM lbs/hr = 10^6 lb/h = 453.6 t/h = 126.00 kg/s

Finally, a common unit (US) for large flows of gas:

1 BCFD (billion cubic feet per day) = 10^9 Sft^3/d = 327.74 Sm^3/s

Note here that 1 Sm^3(US) is 42.211 mol (see page 14), so 1 BFCD is 13.834 kmol/s.

This list is pretty confusing; the letter M can mean mega (million) or metric or the roman numeral for 1000! The combination MM means million – please do not ask me why this unit still is popular in English literature...

Note that in a typical processing plant, 10 kg/s is a "large" flow, 1 kg/s is a "medium" flow and 0.1 kg/s is a "small" flow; see the illustration in Figure 1.9.

Exercise 1.5 *Another common unit (US) for gas flow is standard cubic meters per hour, [scmh]=[SCMH] = [Sm^3 (US)/h]. Show that 10^3 scmh is 11.725 mol/s.*

1.4 Some important numbers

As an engineer, it is very important to have an idea about order of magnitudes and to be able to make quick estimates. It is a great advantage to know by heart some numbers. For example, one should know the molar mass (molecular weight) of the most important elements and components (e.g., the molar mass of water is about $M = 18$ g/mol) and that the gas constant $R = 8.31$ J/mol K. Below you find some other numbers that are important to know. References to sources of physical data are given in page 415.

Air. The **normal pressure** ("atmospheric pressure") at the surface of the earth is defined as 1 atm = 1.013250 bar (where 1 bar is 10^5 N/m^2 = $10^5 Pa$). In most cases, the *composition* of (dry) [6] air can be set to 21 mol% O_2 and 79% N_2. If you want to be more accurate you can use

78.1% N_2, 0.9% Ar and 21% O_2.[7]

In addition, air contains a varying amount of water. In *saturated*[8] air, the content of water is about 0.6mol% at 0 ^{o}C, approx. 3% at 25 ^{o}C and 100% at 100 ^{o}C (because the vapor[9] pressure of water is ca. 0.006 atm, 0.03 atm and 1 atm at the three temperatures).

The *molar mass* M for (dry) air is 28.97 g/mol (or approximately 29 g/mol). The *heat capacity* of air is $C_p = 29$ J/mol K (= 1.0 kJ/K kg) and one then has $\gamma = C_p/C_V \approx 1.4$. The *density* of air at 1 bar and 25 ^{o}C assuming ideal gas is

$$\rho^{\ominus}(g) = \frac{p^{\ominus} M}{RT} = \frac{1.013 \cdot 10^5 \cdot 29 \cdot 10^{-3}}{8.31 \cdot 298.15} = 1.19 \text{ kg/m}^3$$

Water. The *molar mass* for water is M = 18.015 g/mol. You should know some physical constants for water (here given at 25 ^{o}C):

$$\begin{aligned}
\text{Liquid}: \quad & \rho(l) = 1000 \ \frac{\text{kg}}{\text{m}^3} \\
& C_p(l) = 4.18 \ \frac{\text{kJ}}{\text{kg K}} = 75.4 \frac{\text{J}}{\text{mol K}} \\
\text{Gas}: \quad & \rho^{\ominus}(g) = \frac{p^{\ominus} M}{RT} = 0.74 \ \frac{\text{kg}}{\text{m}^3}; \\
& C_p(g) = 1.87 \ \frac{\text{kJ}}{\text{kg K}} = 33.6 \ \frac{\text{J}}{\text{mol K}} \\
\text{Heat of vaporization}: \quad & \Delta_{\text{vap}} H = 2444 \text{ kJ/kg} = 44.03 \text{ kJ/mol}
\end{aligned}$$

Note that the **heat capacity** of water is $C_p(l, H_2O)$ = 1 cal/g K =1 kcal/kg K because the original definition of one calorie was that it is the heat necessary to raise the temperature of 1 g water by 1°C (1K).[10]

At water's normal boiling point[11] of 100 ^{o}C (T_b = 373.15K) the heat of vaporization is 2257 kJ/kg = 40.66 kJ/mol, which is somewhat less than at 25°C; see Exercise A.5 (page 359) for details. Note that the heat of vaporization of water is very large – evaporating water requires the same energy as heating it up 584K.Water vapor is also known as **steam**.

[6] By **dry basis** we mean that any water in the gas has been removed before calculating the composition.

[7] The air's CO_2 content, which is of so much concern because of global warming, is only about 0.04% (400 ppm), but it is steadily increasing.

[8] By **saturated** we generally mean that the state is in equilibrium with another phase. Specifically, saturated air is in equilibrium with water (liquid), i.e., the air contains the maximum quantity of water and a further increase gives condensation.

[9] **Vapor** means "saturated gas," that is, vapor is gas in equilibrium with a liquid phase

[10] Note that for temperature *differences* (changes), it does not matter if we use K or ^{o}C; for example, for specific heat (C_p), which is the energy required to raise (change) the temperature by 1 degree, we have 1 J/kg K = 1 J/kg ^{o}C.

[11] The **normal boiling point** (T_b) is the point where the liquid starts boiling at 1 atm, i.e., the temperature where the vapor pressure is 1 atm.

Other components. The *heat capacity* of liquid and gas for some components is given in the form of nomograms in the Appendix (pages 421 and 422). We notice that the heat capacity for most liquid is around 2 kJ/kg K, and that water stands out with a much higher value of 4.18 kJ/kg K.[12]

The *heat of vaporization* for **hydrocarbons** is typically about 400 kJ/kg (i.e. much lower than for water which has a value of 2444 kJ/kg). For hydrocarbons C_nH_m with $n > 4$, the (higher) **heat of combustion** (gross heating value) (to H_2O(l)) is about 48000 kJ/kg, which we note is about 100 times larger than their heat of vaporization. See page 362 for more about combustion reactions. The heat of combustion per kg is somewhat larger for smaller hydrocarbons, and methane has a (higher) heat of combustion of 55501 kJ/kg. This is called the "higher" heat of combustion because H_2O (l) is the product rather than H_2O (g). To indicate the energy content as a fuel, it is common to use the *lower* heat of combustion (also known as the net heating value) with water as a gas product, which is typically about 3500 kJ/kg lower because the heat released by condensing water is not included. The lower heat of combustion is 50014 kJ/kg for methane (g), 44736 kJ/kg for n-hexane (l), $\approx$ 42000 kJ/kg for heating oil, $\approx$ 29300 kJ/kg for coal and $\approx$ 17300 kJ/kg for wood. Note that the energy unit "ton of oil equivalent" (toe) is defined as the lower heating value (to H_2O (g)) for 1 ton of crude oil, or more precisely, 1 toe = 41.868 GJ.

Heat capacity for ideal gas. For an ideal gas, the heat capacity C_p is a function of temperature only and we also have that (we here omit the subscript m for molar quantity, i.e., C_p means $C_{p,m}$)

$$\boxed{C_p - C_V = R} = 8.3145 \text{ J/mol K} \tag{1.6}$$

For ideal gases, the heat capacity can be estimated from statistical thermodynamics by adding the contributions from

- the molecule's movements (translation, rotation and vibration),
- the electronic contribution at extremely high temperatures.

Isolated, each atom in a molecule has 3 degrees of freedom for motion, and the total number of motional degrees of freedom is therefore $3N_a$, where N_a is the number of atoms in the molecule. For an ideal gas, each "active" degree of freedom gives a contribution of $\frac{1}{2}R$ [J/K,mol] to the heat capacity C_V. At sufficiently high temperatures (where all degrees of freedom are "active") we therefore have that the contribution from motion is $C_V = \frac{3N_a}{2}R$ or

$$\text{molecular motion at high temperature}: \quad C_p = \left(\frac{3N_a}{2} + 1\right) R$$

Of the $3N_a$ degrees of freedom for motion, 3 degrees of freedom are for translation, 3 for rotation (2 for linear molecules) and the rest for vibration.

[12] It is typical that water stands out. This is mainly because of hydrogen bondings in its liquid phase. For example, the boiling point at 100°C is much higher than one should expect for such a small molecule.

The translational degrees of freedom are "active" already at "low" temperatures, the rotational contributions become active at "medium " temperatures, while the contributions from vibration become active only at "high" temperatures. In addition, at "extremely high" temperatures there contributions from the electronic degrees of freedom. We then have:

- For a mono-atomic gas ($N_a = 1$, for example He or Ar), there is just translation and the contribution to the heat capacity from motion is $C_p = (3/2 + 1)R = 2.5R = 20.79$ J/mol K.
- For a diatomic molecule ($N_a = 2$, for example N_2), there are $3N_a = 6$ degrees of freedom for motion. At "medium" temperatures, only the three degrees of freedom from translation and the two from rotation are active and we have $C_p = (5/2 + 1)R = 3.5R = 29.1$ J/mol,K. At "higher" temperatures, the contribution from vibration becomes active and the total contribution from motion is $C_p = (6/2+1)R = 4R = 33.2$ J/mol,K. For example, C_p for N_2 (ideal gas) is 29.12 [J/mol,K] at 298 K ("medium" temperature) and increase to 32.69 J/mol,K at 1000 K ("high" temperature). At "extremely high" temperatures, it increases further (e.g. to 37.05 J/mol,K at 3000 K) because of additional contributions from the electronic degrees of freedom
- For *larger molecules*, all the three rotation's degrees of freedom are active at "medium" temperatures and the contribution from motion is $C_p = (6/2 + 1)R = 4R = 33.2$ J/mol K. At "higher" temperatures, the contribution from the remaining $N_a - 6$ vibration's degrees of freedom become active (especially for large molecules) and the value for C_p is higher than 33.2 J/mol K. For example, C_p (ideal gas) at 25 ^oC (298 K) is equal to
  33.63 J/mol K for H_2O
  35.89 J/mol K for NH_3
  35.52 J/mol K for CH_4
  73.76 J/mol K for C_3H_8

Heat capacities for ideal gases are important for practical calculations (see page 342), and empirical expressions are used in practice to capture the temperature dependency (see page 355), for example $C_p(T) = A + BT + CT^2 + DT^3$. Note that the heat capacity for an *ideal gas* always increases with temperature (but for a real gas the pressure dependency makes the heat capacity for saturated vapor (real gas) approach zero at the critical point).

1.5 Some important concepts

In this book, we focus on the *process* (course of events, "the path") that takes us from one state (feed) to another state (product). We then need to define more precisely the system and the surroundings where the process takes place. In the following, we list some definitions and important concepts. Most of them are from the field of thermodynamics and more details are given in Appendix A.

Process – The course of events from an initial state to a final state. In our processes, this happens by transport, transfer or transformation of mass and/or energy, and

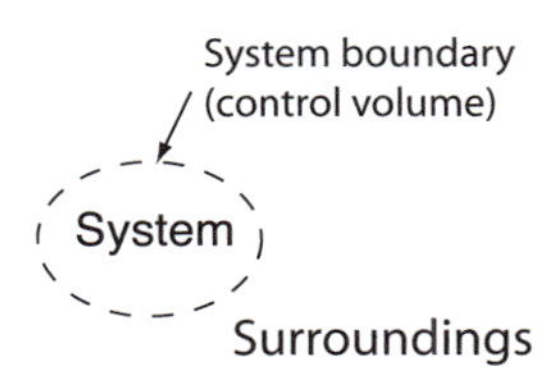

Figure 1.1: System and surroundings

the initial state is the feed (inflow) and the final state is the product (outflow).

System – The part of the world that we choose to analyze. The outside is called the **surroundings** – which can also be viewed as a system.

System boundary (control volume) – The spacial boundary between the system and its surroundings, for example, the outside of a piece of equipment.

Closed system – System that does not exchange mass with its surroundings.

Isolated system – System without any interaction with the surroundings, that is, a closed system without exchange of heat or work with the surroundings.

Example of an isolated system: A perfect thermos bottle (vacuum flask).

Open system – System with exchange of mass with the surroundings.

Adiabatic system (process) – System (process) without exchange of heat (with the surroundings), i.e., $Q = 0$.

State – Quantitative characterization of the system at a given time.

State variable (function) – A property (variable) of the system that depends only on the system's (equilibrium) state at a given time, and not on the path the system followed to get to this state.

Examples of thermodynamic state variables are mass (m), enthalpy (H), entropy (S), pressure (p), temperature (T), volume (V), density (ρ), composition (x_i, c_i), etc.

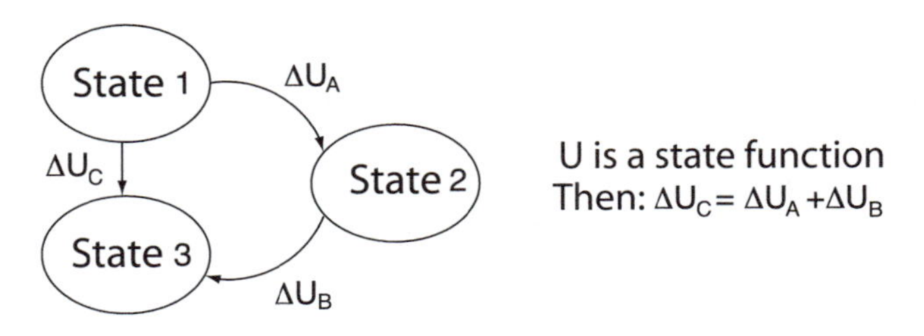

Figure 1.2: Example of state variable (function)

In addition to the composition, at least two state variables must be specified to determine the system's (thermodynamic) state – the values of the remaining state quantities are then a function of these (see also page 332). In our systems, the following "base variables" are often chosen : Quantity (mass) of all components (alternatively, total mass and composition), pressure p and temperature T.

- The term "**equation of state**" is normally used for a *constitutive equation*[13] that gives a relationship between a fluid's pressure, volume and temperature. It can, for example, be used to calculate the pressure p given the systems volume, temperature and composition. The simplest equation of state is the **ideal gas law**

$$pV = nRT \tag{1.7}$$

where the gas constant $R = 8.3145$ J/K mol is an universal constant, independent of composition. The ideal gas law applies well at low pressure and at high temperature (see figure page 336). For real gases, more complicated equations are used with parameters that depend on composition, for example, the Redlich-Kwong's equation of state (see page 340).

Extensive variable – A physical quantity whose value depends on the quantity of matter in the system, for example mass, volume, energy and enthalpy.

Intensive variable – A physical quantity ("point variable") which is independent of the quantity of matter, for example pressure, temperature, density, concentration and molar enthalpy.

Stream – Usually, a mass that flows, for example in a pipe, but in this book it is generalized to include any quantity (bulk) of material [kg; kg/s]. A stream is defined by its quantity (for example kg or kg/s) and state. The state is typically defined by specifying the following intensive variables: composition, pressure, temperature and phase distribution (instead of the last two, one can specify the enthalpy).

Reversible process – Hypothetical process where the driving forces are always zero, such that the process is always in equilibrium (both internally and with its surroundings).

(Such a process is not feasible in practice, for example because it would take an infinite time, but a hypothetical reversible process between two given states can be used to obtain the change in state variables for a real process).

Batch process – Process where the supply of feed and/or removal of products *is not* continuous. This includes any process which takes place in a closed system. *Example:* cooking of rice in a pot.

A **semi-batch** process is a batch process with a continuous feed or product stream (but not both). *Example:* Batch distillation where the feed is supplied as "a batch," but the product is withdrawn continuously.

[13] A **constitutive equation** is a relation between physical quantities that is specific to a material or substance, and does not follow directly from physical laws.

Continuous process – Process with continuous feed (inlet stream) and continuous removal of product (outlet stream). This is a special case of an open system. For our processes, a continuous process is the same as a **flow process**.

Steady-state process – Continuous process where all the variables are independent of time (Note! This does *not* mean that "nothing happens" because the feed (inlet) stream is different from the product (outlet) stream. *Example:* Mixing of cold and hot water in a shower process.

Non-steady state or dynamic process – Process where the variables vary with time, for example, cooking of rice in a pot. Any batch or semi-batch process is a dynamic process.

Constant conditions

- *isotherm, isothermal*: constant temperature
- *isobar, isobaric*: constant pressure
- *isochor*: constant volume
- *isenthalpic*: constant enthalpy
- *isentropic:* constant entropy

1.6 Unit operations

The development of chemical engineering as a separate field from applied chemistry around 1900 is closely related to the concept of unit operations. It was found that certain basic operations in a process were similar in a paper mill, a diary, a chemical plant and in a refinery, and this simplified considerably the exchange of knowledge and made it possible to educate generalists rather than specialists for each industry. Some unit operations and/or their associated apparatus/unit are (here listed alphabetically):

Absorption (Figure 1.4a). Absorption is a process where a gas is dissolved in a liquid. In an **absorber** or absorption column the gas and liquid are contacted in a vertical column (see column) in a countercurrent fashion. The objective is to remove one or several "heavy" components from the gas by transferring them to (absorbing them into) the liquid. There are two feed streams (a gas entering from the bottom and a liquid entering from the top) and two product streams (gas and liquid exiting at the other end). In a stripping column (stripper), the reverse process takes place.

Boiler Heat exchanger where evaporation takes place on the cold side.

Coalescer Unit that separates emulsions, for example, separation of oil from water. Basically, it works as a reversed emulsifier.

Column (tower) (Figure 1.4a). Vertical apparatus where a light phase (usually a gas, but it can be a liquid) and a heavy phase (liquid) are contacted in a countercurrent fashion. The light phase flows upwards in the column and the heavy phase flows downwards. Examples are absorption columns and strippers;

a distillation column consists of two **column sections** with the feed entering between them.

The objective is to exchange one or several components between the two phases. It is important to achieve good contact between the phases, with a large contact area and good mixing. To achieve this, one uses as "internals" either plates (trays) or packings. There are two main kinds of **packings**: random packings (e.g., Rashig-rings) and the more expensive structured packings. **Trays** are usually used in large columns. Packings are used in columns where a small pressure drop is required, e.g., vacuum columns.

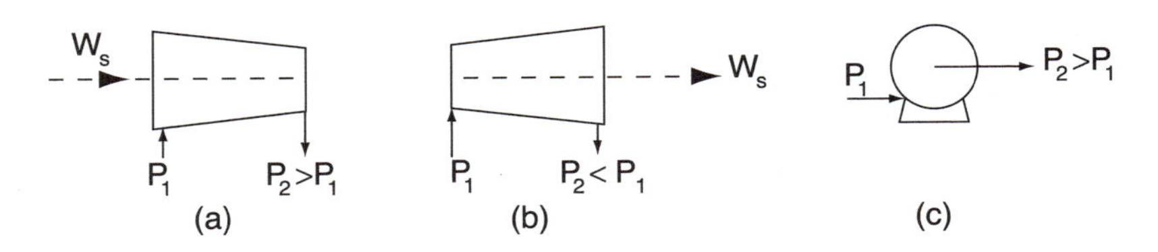

Figure 1.3: (a) Compressor. (b) Turbine. (c) Pump

Compressor (Figure 1.3a). Unit that increases pressure of a gas-stream by supplying mechanical work (shaft work, W_s). The compressor efficiency η is the fraction of supplied work that gives a "useful" increase in pressure. More precisely, $W_s^{\text{rev}} = \eta W_s$, where W_s^{rev} is the work required for a reversible process with the same pressure increase. The remaining "lost work" $(1 - \eta)W_s$ increases the gas temperature by friction loss. The opposite of a compressor is a turbine (see this).

Condenser Heat exchanger where condensation takes place on the hot side.

Crystallizer Apparatus where solid crystals are precipitated from a (super) saturated liquid.

Cyclone Unit that separates particles or fluids by making use of their density difference. A **hydrocyclone** separates solids from liquids or separates liquids of different density.

Decanter Unit where two phases (liquid or solid) are separated by gravity.

Demister A unit that removes entrained liquid droplets from a gas stream.

Distillation (Figure 1.4b). Process where a feed F is separated into a "light" product (distillate D) and a "heavy" product (bottom product B) by utilizing the difference in volatility (vapor pressure; boiling point) of the components. Distillation is the most common unit operation for separating liquid mixtures. In the laboratory, the simple **flashing** (boiling, evaporation) of a liquid feed followed by condensation of the vapor is sometimes referred to as distillation (see single-stage batch (Rayleigh) distillation on page 396). However, in industry, distillation refers to a unit with a trayed or packed column section (see column) and where some of the condensed overhead product is returned to the column

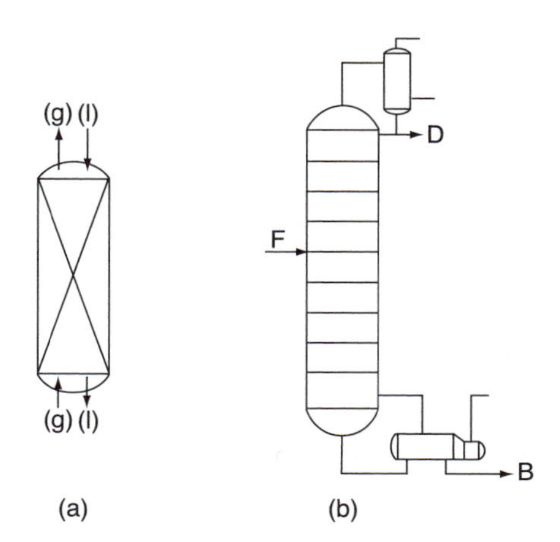

Figure 1.4: (a) Absorber/stripper (column section). (b) Distillation column

as liquid **reflux**. The countercurrent flow of vapor and liquid in the column results in successive evaporation and condensation which improves the degree of separation. The number of such successive steps in the column is called the *number of theoretical stages* (N).

In a **batch distillation** column, the feed F is charged to the bottom of the column as a batch, and the distillate product D is usually removed continuously from the top. For a multicomponent feed, one may obtain a distillate product for each component in the order of their boiling points. The remaining liquid (residue) in the bottom at the end of the batch is the "heavy" product B. The difference between batch and continuous distillation is illustrated in Figure 1.8.

A **continuous distillation** column has continuous feed and product withdrawal, also of the bottom product, see Figure 1.4b. It usually consists of the following sub-units: Two trayed or packed column sections with the feed entering between them, two heat exchangers (reboiler and condenser), two holdup vessels (condenser drum and reboiler sump) and a splitter for the reflux. The upward vapor stream (boilup) in the column is generated by supplying heat Q_H at high temperature in the **reboiler** which evaporates parts of the liquid; the remaining is withdrawn continuously as the bottom product (B). The downward liquid stream (**reflux**) is generated by the removal of about the same quantity of energy Q_C at a lower temperature in the *condenser* at the top of the column, and returning some of the condensed liquid as reflux; the remaining withdrawn as the top (distillate) product (D). In the top section of the column (above the feed point), the heavy components condense and are absorbed into liquid. In the bottom section, the light components evaporate and are thereby removed (stripped) from the liquid and transferred to the gas. As a result, the light components are concentrated in the top and the heavy components in the bottom of the distillation column.

Dryer Heat exchanger with wet solid on the cold side, causing some of the liquid (usually water) to evaporate.

Evaporator Heat exchanger with boiling liquid on the cold side, causing some or all of the liquid to evaporate.

Extraction Exchange of components between two liquid streams. This can take place in a vertical or horizontal extraction column where the two liquid streams flow countercurrently.

Fan or blower A simple compressor (see this), but where the main purpose is the transfer of gas and not increase of pressure.

Filter; filtration unit Unit that separates particles from liquid or gas. Some of the liquid or gas in the feed passes through a porous medium (e.g., a filter cloth) to form the *filtrate*, whereas the remaining particles and some retained liquid form the *filter cake*.

Fittings Parts used to connect pipes; see Table 9.1 (page 246) for details.

Flash; flash drum Unit where the feed stream is "flashed" to a lower pressure, thereby causing a partial evaporation, resulting in a liquid and vapor product. The liquid and vapor products are usually assumed to be in equilibrium.

Flotation Process for separating a suspended phase (solids, liquid, particles) from a liquid by using surfactants and wetting agents.

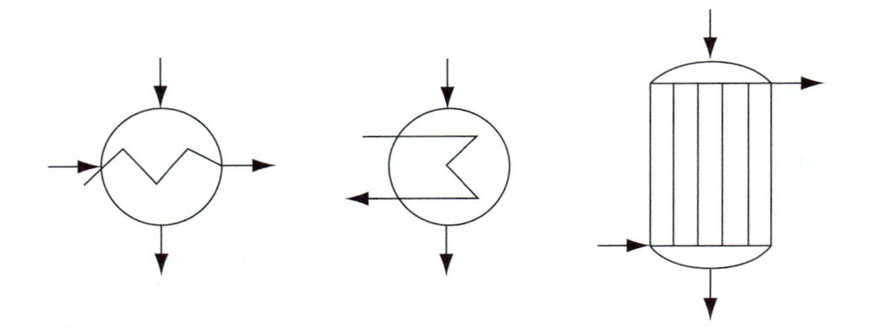

Figure 1.5: Alternative flow sheet symbols for heat exchanger

Heat exchanger (Figure 1.5). Unit where heat is transferred from a hot stream to a cold stream through a solid barrier (wall); see Chapter 5 for more details.

Membrane unit Unit that by *permeation* exchanges mass between two streams (gas or liquid) through a film (the membrane). A membrane can be viewed as a "filtration of molecules," rather than the normal filtration of particles that takes place in a filter.

Mixer (1) Unit where two or more streams are combined to a product stream. (2) Unit where a stream is treated to get a more uniform product (e.g., blender).

Pump (see Figure 1.3c). Simple "compressor" used to increase the pressure of a liquid stream by supplying mechanical work.

Quencher Unit where a hot feed stream (usually gas) is brought into direct contact with a cold stream (often cold water) to be rapidly cooled. A quencher is often used following a reactor to stop the reaction.

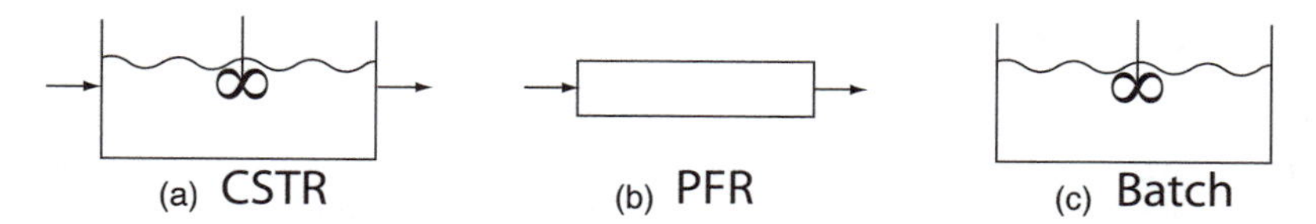

Figure 1.6: Reactor types

Reactor (see Figure 1.6). Unit where a chemical conversion takes place. Some types of reactors: (a) Continuously stirred tank reactor (CSTR), (b) plug flow reactor (PFR) and (c) batch reactor.

Scrubber An absorption column designed to remove undesirable particles, drops or components from a gas stream. In most cases, water is used as the liquid.

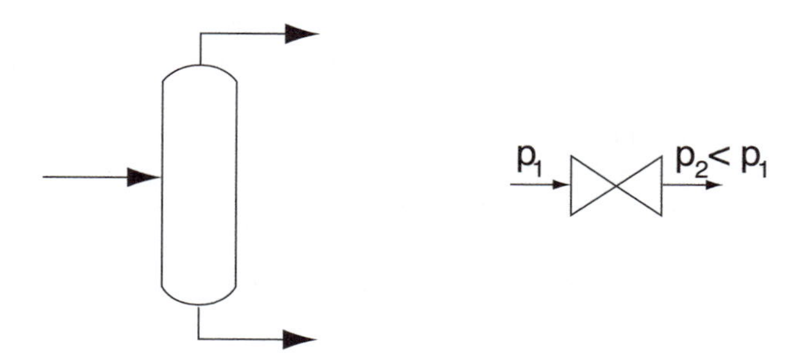

Figure 1.7: Simplified flowsheet symbols for (a) Separator. (b) Valve

Separator (Figure 1.7a). Common term for a wide range of unit operations (typically tanks or "drums") that separate a feed stream into two (or more) outlet streams with different compositions. Examples: gas/liquid-separator (flash tank), liquid/liquid-separator, etc. (If the outlet streams have equal compositions, the unit is called a **splitter**).

Settler See decanter.

Stripping (Figure 1.4a). Stripping is a process where dissolved gas comes out of a liquid (opposite of absorption). A **stripper** is a vertical column with two feed streams (gas and liquid) and two product streams (gas and liquid), where the objective is to remove (strip off) a "light" component from the liquid and transfer it to the gas. In an absorber, the reverse process takes place.

Turbine (expander) (Figure 1.3b). Rotating machine (engine) that extracts mechanical work (W_s) from a fluid flow by using the pressure head. Early turbine examples are windmills and water wheels. A gas turbine (expander) is the opposite of a compressor, and liquid (water) turbine is the opposite of a pump.

In an ideal (reversible) turbine, there is no friction and all of potential pressure head is converted to mechanical work. The efficiency η indicates the fraction of the ideal (reversible) work that is extracted, that is, $W_s = W_s^{\mathrm{rev}}\eta$. The remaining lost work increases the fluid temperature by friction loss.

Valve (Figure 1.7b). A valve is a device that regulates the flow of substances (gases, liquids, slurries) by partially obstructing its passageways, resulting in a pressure drop. In a **control valve**, the flow can be adjusted by changing the valve position (z). The **valve equation** gives the dependency of flow on valve position and pressure drop. A typical valve equation for liquid flow is

$$q = \underbrace{C_d f(z) A}_{C_v} \sqrt{\Delta p/\rho} \tag{1.8}$$

where q [m^3/s] is the volumetric flowrate, C_d (dimensionless in SI units) is the valve constant (relative capacity coefficient), z is the relative valve position (0 is fully closed and 1 is fully open), $f(z)$ is the valve characteristic (e.g., $f(z) = z$ for a linear valve), A [m^2] is the cross sectional area of the valve (at its inlet or outlet), $\Delta p = p_1 - p_2$ [N/m^2] is the pressure drop over the valve, and ρ [kg/m^3] is the fluid density. The mass flowrate is m [kg/s] $= \rho q$ and the flow velocity is v [m/s] $= q/A$ (at the valve inlet or outlet). A typical value for a control valve is $C_d \approx 1$ (see Example 9.2, page 244). $C_v = C_d f(z) A$ [m^2] is the **valve coefficient** (capacity coefficient), which depends on the valve opening. Note that the valve coefficient C_v' provided by the valve manufacturer, usually is the flow in gallons per minute (gpm) of cold water when the valve pressure drop is 1 psi, and to convert to SI units this value needs to be divided by 41625.

Exercise 1.6* *Prove that the expression for converting the manufacturer's valve coefficient C_v' to SI units is* $C_v[m^2] = C_v'(\text{manufacturer})/41625$.

A **choke (throttle) valve** is a valve where the primary objective is to reduce the pressure rather than to regulate flow.

A **Joule-Thompson** valve is a valve where the primary objective is to reduce the temperature of a non-ideal gas, by making use of the fact it requires energy to lower the pressure because of the attractive forces between the gas molecules (except at very high pressures).

And more... In particular, there are many units that handle solids and particles:

- agitator
- blender
- classifier
- conveyor belt
- centrifuge
- crusher, grinder, mill (for solids size reduction)

- electrostatic precipitator
- electrostatic filter
- lifter
- fluidizer
- homogenisator, emulsifier, disperser
- prilling tower
- screw (pump for solid material)
- scrubber, spray tower, venturi scrubber
- sedimenter
- sieve

Mechanical engineers are good at many of these operations (but mechanical engineers become noticeably nervous and start flickering if an operation involves chemical reactions and especially if it involves moles...).

1.7 Batch and continuous process

1. CONTINUOUS:

Separation
C (product) D
Reactant
B
Feed
A
Heat
exchanger
Heat
Reactor
A+B→C+D
A+B

2. BATCH (Same apparatus used for many operations):

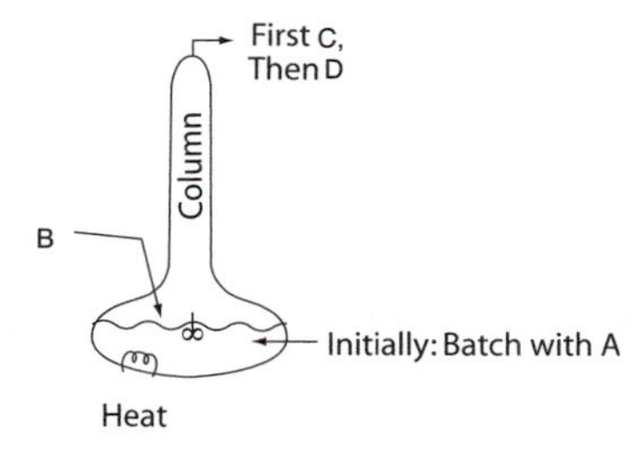

1. Heat up
2. Reaction with addition of B (A+B→C+D)
3. Distillation (Product C)
4. Distillation (Product D)
5. Distillation residue: Unreacted A and B to the next batch

Figure 1.8: Continuous and batch process with the same feed and product

The difference between a batch and continuous process is illustrated in Figure 1.8. In a continuous process, there is a separate unit (heat exchanger, reactor, column, etc.)

for each step of the process and the units are arranged sequentially with a continuous flow between them. In addition, there is usually a continuous recycle in order to improve yield and avoid losses. In a batch process, which is often a direct scale up of the lab-scale process, the same unit is used for several steps of the process, and the subprocesses take place sequentially in time rather than in space.

1.7.1 Typical production rates for a continuous process

A continuous process is usually operated 24 hours a day. We typically assume 8000 operating hours per year, which is about 90% of the total of 8760 hours.

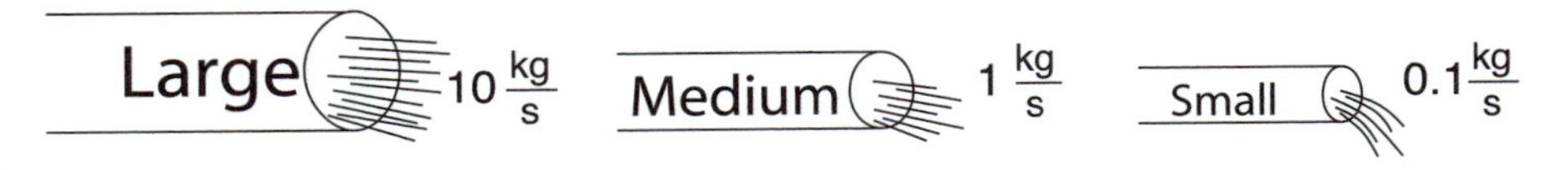

Figure 1.9: Large, medium and small production rates

- A **small** production rate for a continuous process is about 1000 ton/year $\approx$ 3 ton/d (d = day) $\approx$ 0.1 ton/h (h = hour) $\approx$ 2 kg/min $\approx$ 0.03 kg/s.
- For a **large** plant for bulk chemicals, the production rate is typically about 250 times larger, i.e., $\sim$ 10 kg/s.
- For giant plants (for example, an oil refinery), the production rate can be a factor 10 larger than this, say 100 kg/s.
- In summary, the production rate may vary from about 0.03 kg/s (very small plant) to 0.1 kg/s (small plant), 1 kg/s (medium plant), 10 kg/s (large plant) and 100 kg/s (giant plant).
- *Rule of thumb:* A "**medium**" stream (see Figure 1.9) is around 1 kg/s.

What about molar flows? A typical molar mass is $M = 30$ g/mol, which means that the molar production rate typically varies from 1 mol/s (very small plant) to 3000 mol/s (giant plant).

Note that you must multiply by a factor of 3.6 to go from kg/s to ton/h (t/h), and to go from mol/s to kmol/h.

1.7.2 Typical production rate in a batch process

In batch production, it is often practical to start a new batch every day. Let us assume that we have 250 days of production per year and there is 1 batch per day. Then we have

- A yearly production below 500 ton/year (where batch production is usually better; see above) corresponds to a batch smaller than 2 ton.
- A yearly production above 5000 ton/year (where continuous production is usually better) corresponds to a batch of more than 20 ton.

This means that continuous operation should be considered if the batch size exceeds about 2 ton; see also Example 1.2 (page 30).

1.7.3 Batch versus continuous

Advantages with continuous operation. Continuous production usually has a substantially better energy efficiency than batch production, because one can use heat integration where an energy demand in one part of the process is matched by an energy excess in another part. It is also easier to automate a continuous process such that it needs fewer operators. In short, the operation costs are lower than with batch production. In addition, there are some processes that are less suitable for batch production. This applies especially to gas phase processes with a short residence time in the reactor, which includes many of the most important industrial processes.

Disadvantages with continuous operation. Compared to a batch plant, a continuous plant is more complicated and usually requires a larger investment for the same capacity. For small production rates, a batch process is therefore more profitable. Batch plants are also significantly more flexible as a standard apparatus can be used to produce a large range of products. A continuous plant, on the other hand, is designed for a specific feed and product. Typically, it takes at least 2-3 years to build a new continuous process – which may exceed the market life time for some special products. Thus, batch production is the only viable alternative in many cases.

Choice. Until about 1900, nearly all production was batch, and in the period from about 1900 to 1980 the most important task for a process engineer was to replace old-fashioned batch plants with modern continuous plants. Most of the training of process engineer was (and is) therefore directed to continuous processes. But more recently, it has become clearer that batch processes are preferable when great flexibility is desired and also for small production rates. In general, continuous processes are better for larger production volumes and we have

- Processes with a yearly production above $\sim$ 5000 ton/year (0.17 kg/s) are usually continuous.
- Processes with production rates less than $\sim$ 500 ton/year (0.017 kg/s) are operated batchwise.

For production rates between about 500 and 5000 ton/year, a closer evaluation should be made.

1.8 A little about economy

The **investment cost** I [in \$ or any other currency] for a unit or process section depends on the production rate m [kg/s]. The following relationship is often used,

$$I = c_0 + c_1 m^q \quad (\text{typically} : q \approx 0.67) \tag{1.9}$$

Here, the scaling exponent q is almost always smaller than 1, that is, we have an *economy of scale* advantage because the investment increases less than proportional to the production. The value $q \approx 0.67$ is physically reasonable, because the production capacity m increases proportional to the volume of the apparatus, while the investment I typically increases proportional to the area of the apparatus. If L [m] represents a typical length of the apparatus, then $m \propto L^3$ and $I \propto L^2$ and it follows that $I \propto m^{2/3}$ (the symbol $\propto$ here means "proportional to").

One often estimates the investment I only for the main equipment, that is, one does not include the cost of valves, smaller pumps, measurement devices, automatization (control), installation, etc. To obtain the total investment I, the cost of the main equipment is then multiplied by an experience factor ("Lang-factor"). The Lang-factor for a continuous process is typically around 5.

To compare alternative processes for producing a given amount of product, one often considers the **total annualized cost (TAC)** which is the sum of the production costs P [$/y] and the capital (investments) costs. For the capital costs, we here use a very simple approach, often used in practice: Assume that the plant will run for T years and use linear (constant) depreciation of the investment I over this period. The capital cost is then I/T [$/y] and the total annualized cost is

$$\mathrm{TAC} = P + I/T \quad [\$/\mathrm{y}] \tag{1.10}$$

A typical depreciation period is $T = 10$ years. The process is profitable if TAC is less than the income from selling the products. The expression in (1.9) for I can be used both for continuous and batch plants, but the values of c_0 and c_1 are usually smaller for a batch plant. On the other hand, the operational costs P are usually substantially larger in a batch plant, especially for operators, energy and raw materials (because of larger losses). We next illustrate the ideas for economical analysis with a simple example.

Example 1.2 Economic analysis of batch versus continuous production. *For a* **continuous** *(c for* continuous*) production plant which produces a material X, the total investment cost is assumed to be*

$$I_c[\$] = 10 \cdot 10^5 + 0.5 \cdot 10^5 m^{0.7}$$

where m [t/y] is the yearly production. Note that scaling exponent is assumed to be 0.7. I_c is here the total investment cost, i.e., the Lang-factor has already been included. We note that it costs one million dollars to build a plant with zero production capacity. The **production cost** *P_c for the continuous plant, without the capital cost, is assumed to be*

$$P_c[\$/y] = 2 \cdot 10^6 + 100 \cdot m[t/y]$$

Here, a constant term of $2 \cdot 10^6$ $/y is included to account for fixed costs like wages for supporting staff and insurance, while the variable term included cost for energy, operators and raw materials.

From (1.10), the total annualized cost (TAC) is $\mathrm{TAC} = P_c + I_c/T \quad [\$/\mathrm{y}]$, *where we choose a depreciation period of $T = 10$ years. The cost per ton product for our example is then $C_c = P_c/m + I_c/(10m)$ or*

$$C_c[\$/t] = 2 \cdot 10^6/m + 100 + 10^5/m + 0.5 \cdot 10^4 m^{-0.3} = 100 + 2.1 \cdot 10^6/m + 0.5 \cdot 10^4 m^{-0.3} \tag{1.11}$$

where m is in [t/y]. For example, with a production of $m = 10000$ t/y, the production cost in $ per ton is:

$$C_c[\$/y] = 100 + 210 + 315 = 625$$

For a **batch** *(b for* batch*) production plant to produce the same material X, we assume in this example that the investment cost is 1/10 of the cost for the continuous plant, i.e., $I_b = I_c/10$, whereas the variable cost for energy, raw materials and operators are 10 times higher (but with no fixed costs, since the batch plant can be used also to produce other products)*

$$P_b[\$/y] = 1000 \cdot m[t/y]$$

The total production cost (again with linear depreciation over 10 years) is $P_b + I_b/10$[\$/y] and the cost per kg becomes

$$C_b[\$/t] = 1000 + 10^4/m + 0.5 \cdot 10^3 m^{-0.3} \tag{1.12}$$

where m is in [t/y]. For example, with a production of m = 10 t/y, the production cost becomes, in dollar per ton:

$$C_b[\$/t] = 1000 + 1000 + 250 = 2250$$

The table below compares the production cost [\$/t] of continuous and batch production for some values of capacity m [t/y]:

	$m[t/y]$	1	10	100	1000	10000	100000	1000000
Continuous :	$C_c[\$/t]$	2100000	213000	22400	2830	625	279	181
Batch :	$C_b[\$/t]$	11500	2250	1225	1072	1032	1016	1008

For this particular example, the costs for batch and continuous production cross at about 4000 t/y (however, the numbers can be completely different in another case).

1.9 Some fun and useful energy exercises

Here you find some exercises to practice how to convert between different units. The exercises also give you insight into typical sizes and numbers, especially related to energy and energy usage. For this reason, I have provided the answer to the exercises, but it is strongly recommended that you solve them yourself. Please include units in all calculations to check that the formula you use is consistent. If you miss information to solve a problem, then this can be found earlier in this chapter (use the index!).

Exercise 1.7* *One day, the energy price for electric power is 0.1 \$/kWh. What is this in \$/GJ? (***Answer.** *27.8 \$/GJ.)*

Exercise 1.8* *One day, the price of crude oil with density 0.8 kg/l is 1 \$/l (= 3.79 \$/gal = 159 \$/bbl). What does this correspond to in \$/GJ (with water as gas combustion product)? (***Answer.** *27.8 \$/GJ (assuming the lower heating value is 45 MJ/kg).)*

Thus, the energy price is the same if we pay 0.1 \$/kWh (electricity) and 1 \$/l (oil). This is because the energy contents if 1 kWh (=3600 kJ) is the 1/10 of the energy contents in 1 l oil (= 36000 kJ = 1 l · 0.8 kg/l · 45e3 kJ/kg). We derive from this the rule:

- **The price (in *any* currency) of 1 kWh electricity should be at least 1/10 the price of 1 liter oil.**

The reason we say "at least" is because electricity is a more valuable energy source than oil. First, it gives no losses when used for heating, whereas some energy will be lost in the flue gas when we burn oil for heating. Second, and more importantly, it can easily be converted to work in an electro motor with almost 100% efficiency, whereas less than about 50% (and in many cases much less) of the heating value in oil can be converted to work. You can read more about the thermodynamic background for extracting work from heat in Chapter 8.

Exercise 1.9* *Given the above energy costs, what is the cost of taking a bath, that is, how much does it cost to heat up 100 l of water from 10 °C to 50 °C with the use of electricity or oil?*

Exercise 1.10* *How much oil [kg] is needed to heat up 100 l of water from 10 °C to 50 °C?*

Exercise 1.11* *The oil price is one day $100/barrel. What is this per liter and per gallon?*

Exercise 1.12* *The price of natural gas varies a lot depending of the location. In October 2005, the gas price varied from $0.75/million Btu in Saudi-Arabia to $8/million Btu in Western Europe and $14/million Btu in the United States. What does this correspond to in $/GJ?* (**Answer.** *0.71 $/GJ (Saudi), 7.63 $/GJ (Europe), 13.25 $/GJ (USA)*)

Exercise 1.13* *What does a natural gas price of $8/million Btu (Western Europe Nov. 2005) correspond to in $/Sm^3$?*

Exercise 1.14* *The natural gas production in the Norwegian part of the North Sea was in 2001 about* $53 \cdot 10^9$ *Sm^3 (which was about 2% of world's production of natural gas). What does this correspond to in kg/s and what is the combustion value in GW (assume that the natural gas has molar mass of 18 g/mol and that its heating value (lower heat of combustion) is 42 MJ/Sm^3)?*

Exercise 1.15* *In addition to the* $53 \cdot 10^9$ *Sm^3 mentioned above, an additional* $34 \cdot 10^9$ *Sm^3 was reinjected into the ground as pressure support for oil extraction and to avoid flaring. (a) What is the potential value of the reinjected gas, with a gas price of 8 $/million Btu (Europe, Nov. 2005)? (b) What is its potential value for producing electric energy, if we assume a 50% efficiency for converting it to electric energy and the electricity price is 0.1 $/kWh.*

Exercise 1.16* *How much CO_2 [Sm^3] is created when we burn 1 l gasoline (assume that the density for gasoline is 0.7 kg/l and that the mole ratio C:H in gasoline is 1:2)?*

Exercise 1.17* *Assume that there are 600 million cars in the world, which annually on average are driven 20000 km with a gasoline consumption of 0.1 l/km. Assume that gasoline has a density of 0.7 g/l and that the lower heat of combustion is 45000 kJ/kg. What is the consumption of gasoline (in l/y and kg/s) and energy (GW) for all these cars?* (**Answer.** *1.2 E12 l/y, 26636 kg/s, 1.2 E12 J/s = 1.2 TW).*

Exercise 1.18* The world's yearly energy consumption *(2008) corresponds to about 12000 million ton of oil equivalents (toe). What does this correspond to in EJ/y, in kWh/y, and in TW? What is the average energy consumption [kW] per person? (Data: 1 toe corresponds to 42 GJ and there are 6 billion people).* (**Answer.** *504 EJ/y,* $1.4 \cdot 10^{14}$ *kWh/y, 16 TW, 2.7 kW/person.)*

In terms of money, with an energy price of 27.8 $/GJ (see Exercises 1.7 and 1.8), the cost for a global energy consumption of 504 EJ/y = 504^10^9 GJ/y, is $14000 \cdot 10^9$ $/y or 14000 billion US dollars per year. This corresponds to about 2300 US dollars per year for every person on the earth.

To solve the next exercise, you need to make use of the following **rule of thumb**:

- **The heating value of 1 Sm^3 natural gas is about the same as 1 kg oil ($\approx$ 1.2 l oil).**

- This rule implies that **the price (in *any* currency) of 1 l oil should be at least the price of 1 Sm^3 natural gas** (they have about the same energy contents but we say "at least" because oil is more valuable, e.g., it is easier to transport).

To justify the rule, recall first that the lower heating value for oil is about 42 MJ/kg (as 1 toe= 41.868 GJ, see page 12). **Natural gas** consists mainly of methane, but may also contain significant amounts of ethane and propane and also inerts such as CO_2 and N_2. Methane has a higher (gross) heating value (GHV) of 37.7 MJ/Sm^3 and a lower heating value of 33.9 MJ/sm^3. Ethane has a higher (gross) heating value (GHV) of 66.0 MJ/Sm^3 and a lower heating value of 60.4 MJ/sm^3. From this it is clear that the heating value of *natural gas* depends strongly on its composition, so the above rule of thumb (1 toe =1 Sm^3 natural gas) is quite rough. Also, note that the natural gas sold to the customers[14] ("**sales gas**") usually has a lower heating value than the original natural gas, and a typical value is to set 1 toe = 1.14 Sm^3 sales gas.

Exercise 1.19 *The world's largest natural gas field is offshore Qatar in the Arabian Gulf.* It is given that the total reserves in this field are 900 TCF. How many years of the world's current energy consumption does this correspond to?*

Exercise 1.20 * Solar cells. *Taking into account that some light is reflected, the earth receives from the sun about 120000 TW which is about 7500 times more than the current worldwide energy consumption (16 TW in 2008; see Exercise 1.18). Thus, if we could make effective use of the solar energy, there would no need to worry about the world's energy supply or global warming. The most direct way to utilize the energy from the sun is by solar photavoltaic cells.*

The average solar radiation to the earth is about 240 W/m^2, but not all this reaches the surface. Assume that the average value in the Sahara is 220 W/m^2. How large of an area must one cover with solar cells to meet the world's energy demand if the efficiency in a solar cell for conversion to electricity is 15%?

Exercise 1.21 * Electricity price with solar cell. *Let us be a bit conservative, and assume that the radiation from the sun is 100 W/m^2 on average. What is the resulting energy price (in \$/kWh) if we assume 15% efficiency, a price of solar cells of 1000 \$/m^2 and linear depreciation over 10 years?*

Comment: *The electricity production from solar cells is currently (2008) approx. 9000 MW = 0.09 TW, which is only 0.06% of the global energy consumption of 16 TW. The electricity production from solar cells increases by about 50% per year, and as the cost to produce solar cells is reduced (here the process engineer will play an important role!), the resulting energy price may eventually reach 0.2 \$/kWh or lower. The oil company Shell produced in 2001 a study that predicted that solar cells around 2050 would be the planet's most important source of electricity; however, at present, most experts in the energy field find this prediction unlikely.*

[14] The **sales gas** to customers should typically have a gross heating value (GHV) between 39.5 and 43.6 GJ/Sm^3 – and the lower heating value (LHV) is then about 4 GJ/Sm^3 lower. The GHV specification is to assure that the customers' gas burners will function properly without need for adjusting the air to gas ratio. Thus, natural gas usually needs to be processed, by removing ethane and heavier components, before it is sold.

Human energy consumption

Exercise 1.22* *What is the average human energy consumption in Watt [=J/s] given the data below for the daily energy consumption?*

Minimum to survive	900 kcal/d
Student, teacher, office worker	1900 kcal/d
Lumberman, athlete	4500 kcal/d

Exercise 1.23* *How much tortilla chips per day must a student eat to meet her energy demand (given: tortilla chips has an energy content (heat of combustion) of 2150kJ/100g) ?*

Exercise 1.24* *The following formula is given for the maximum work a person can perform over a period P:*

$$\dot{W}[\mathrm{W}] = \frac{2500}{\ln \mathrm{P[s]}}$$

What is the maximal work over a period of (a) 2 seconds and (b) 1 hour?

Comment: *Note that* work W *and* energy consumption E *are two different things. Only a fraction of a person's energy consumption from burning food is converted into work in the muscles (maybe 10-30%).*

Exercise 1.25* *What is the average* work W *(in Watt) when a weight lifter lifts a weight of 200 kg up 1 m in a period of 1 second?*

Exercise 1.26* *What is the average* work W *(in Watt) when a person that weighs 80 kg climbs up a 500 m high mountain in a period of one hour?*

Exercise 1.27* *The* **human heart** *can be viewed as a pump. From (6.14), the pump work is $\dot{W}_s = \dot{V}\Delta p$ where $\dot{W}_s$ [W] is the work per unit of time (power), $\dot{V}$ [m^3/s] is the volumetric flow and Δp [N/m^2] is the pressure increase. Calculate the work that a human heart performs given the following data:*

- *The blood flow for a person at rest is 5 l/min and it goes to both the lungs and the body (via the two heart chambers).*
- *The pressure increase for the blood that goes to the lungs is 25 mmHg.*
- *The pressure increase for the blood that goes to the body is 130 mmHg.*

Next, a few exercises on the **heat loss Q for humans**. With a constant body temperature, the steady-state energy balance (4.13) applies,

$$E = Q + W \quad [\mathrm{J; J/s}]$$

- $E = \Delta H_{\mathrm{combustion}}$ – net energy intake from food (released by burning in cells)
- W – work performed by body on surroundings
- Q – heat from body to surroundings (net cooling)

(Note: To keep the numbers positive, we have reversed the definitions (signs) for E, Q and W compared to the standard sign convention used elsewhere in this book.)

In most cases, at least over a longer time period, the work W is insignificant compared to the heat Q (for example, the net work W is zero if you climb up and down a mountain – even if it does not feel like it!). Thus, in practice all the energy E ends up as heat Q:

$$E \approx Q$$

Some typical data for the required energy E for various activities:

sleeping	70 W
sitting	100 W
walking	250 W
jogging	500 W

To remove the corresponding amount of heat Q, we consider three cooling mechanisms:

1. Heat loss by breathing (Q_1)
2. Heat loss by heat exchange through the skin (Q_2)
3. Heat loss by sweating (Q_3)

The total heat loss is

$$Q = Q_1 + Q_2 + Q_3$$

In the first exercise, you will find that the breathing heat loss Q_1 is insignificant in most cases.

Exercise 1.28* *A person breathes 20 l/min and the outgoing air is 20 °C warmer than the incoming. What is the heat loss by breathing when we neglect the change in the air's water contents ($Q_1 = mc_P\Delta T$)?*

If the air temperature is not too high and you take it easy, the heat loss Q_2 through the skin provides sufficient cooling, as found in the next exercise.

Exercise 1.29* *As discussed in more detail in Chapter 5 on heat exchange, the heat Q_2 [W] transferred from the body to the air through the skin is, see equation (5.1),*

$$Q_2 = UA\Delta T \quad [\mathrm{J/s}]$$

where $\Delta T = T_{\mathrm{body}} - T_{\mathrm{air}}$ [K or C] is the temperature difference, A [m^2] is the surface area of the body (skin area) and U [W/m^2 K] is the overall heat transfer coefficient. For humans, U may vary from about 2 W/m^2 K (well dressed and gentle breeze) to 20 W/m^2 K (swimming in water). The skin area typically varies from 1.6 m^2 (female) to 1.8 m^2 (male). In this exercise, we assume $U = 5$ W/m^2 K (a typical value) and $A = 2$ m^2, such that $UA = 10$ W/K. The normal body temperature (T_{body}) is 37.0°C.

Problem: At which air temperature will you start sweating when (a) $Q = 100$ W (sitting), (b) $Q = 250$ W (walking) and (c) $Q = 500$ W (jogging).

If the temperature is sufficiently high, **sweating** is the only means to keep the body cool. Cooling is then provided by evaporation of water according to the equation,

$$Q_3 = m \cdot \Delta_{\mathrm{vap}} H \quad [\mathrm{J; J/s}]$$

where m [kg; kg/s] is the amount of water evaporated and $\Delta_{\mathrm{vap}}H$ [J/kg] is the heat of evaporation for water, which is 2444 kJ/kg at 25°C (see page 16). In the next exercise, you will find that even for moderate energy consumptions, the amount of water needed for sweating may be several liters per day.

Exercise 1.30 * *A person stays in a hot climate such that the heat loss due to the difference between the body's and air's temperatures is negligible. How much water must the person evaporate/sweat/perspire in a day (24 hours) if the energy consumption is 100 W?*

Hiking the Grand Canyon in Arizona (US) is a great experience, but one needs to be a bit careful because, as opposed to climbing a mountain, the return hike is the hard part. The elevation difference from the rim at the top to the river at the bottom is about 1360 m, and as you start descending into the canyon you see warning signs:

> **Warning.** DO NOT attempt to hike from the canyon rim to the river and back in one day. Each year hikers suffer serious illness or death from exhaustion.

Another sign tells the sad story of a young lady who died in the canyon:

> Margaret L. Bradley was a 24-year-old athlete and medical student who finished the 2004 Boston Marathon in a little more than three hours. On July 2, 1984, she died in the Grand Canyon of dehydration. Margaret and a companion left the South Rim mid-morning for what they thought was a 15 mile (24 km) day hike. They failed to carry a map, so they were unaware that the proposed hike was actually 27 miles (43 km). The predicted high temperatures for the inner canyon was 105°F (41°C). Each carried inadequate food and water. Margaret had 1.5 liters of water, 2 energy bars and 1 apple. They were unprepared for the extreme heat, excessive distance and lack of water. By mid-afternoon, the hottest part of the day, the two ran out of water and were severely heat stressed. Margaret's companion could no longer continue. Thinking it would be best to go for help, they decided to separate - their final mistake. Her companion, after resting out the heat of the day, made it out alive. Two days after beginning their trip, park rangers found Margaret's body.

While hiking up the canyon the heat production Q is at least 600 W. You can imagine what happens if you leave a kettle of water on the stove at 600 W – the water disappears fast, and this is what happened to Margaret Bradley. If you run out of water, you should rest and use as little energy as possible, like the companion who survived (and like most people in hot climates do).

Exercise 1.31 * *A sign at the top of the Grand Canyon says that for every hour of hiking in the canyon, you should drink 0.5 to 1 liter of water. Prove this statement by computing the energy (in Watt) required to evaporate this amount of water.*

In spite of these warnings, my daughter Hanne and I did hike from the South Rim to the river and back in one day. Hiking down the Kaibab Trail and up the Bright Angel Trail, we started early at 6 am and returned 12 hours later, including detours to the Phantom Ranch and the Plateau Point. This was on June 8, 2008, and the temperature at the bottom of the canyon was about 37°C. Actually, the hike was not very hard, but we did drink a lot of water.

Exercise 1.32 * *Hanne was wearing a pulse monitor, and when we got back, it estimated that she had used 7000 calories during the hike. How much water does this correspond to if all of the cooling was due to sweating?*

1.10 Global energy consumption

We end this chapter by looking into the world's future. A high standard of living is closely related to the use of energy. Table 1.1 shows the global energy consumption by source for the period 2000 to 2050, as predicted in 2008 by the Shell oil company. Energy use is predicted to double in the period 2000 to 2050. Fossil fuels (oil, gas and coal) are expected to remain the main energy source. In any case, providing the world with sufficient and sustainable energy is a major challenge and will require significant efforts and technology developments by the world's process engineers.

Table 1.1: Estimated global energy consumption by primary source [EJ/y = 10^{18}J per year]. *Source: Shell Energy scenarios (2008).*

	2000	**2010**	**2020**	**2030**	**2040**	**2050**
Oil	147	176-177	186-191	179-192	160-187	141-157
Natural gas	88	109-110	133-139	134-142	124-135	108-122
Coal	97	137-144	172-199	186-210	202-246	208-263
Nuclear	28	30-31	30-34	34-36	38-41	43-50
Biomass	44	48-50	52-59	59-92	54-106	57-131
Solar	0	0-1	2-7	22-26	42-62	74-94
Wind	0	1-2	9-9	17-18	27-28	36-39
Other renewables (incl. hydroelectric)	13	18-19	28-29	38-40	50-51	62-65
Total	417	524-531	628-650	692-734	738-815	769-880

2

Derivation of balance equations

The balance principle is the process engineer's most important tool. First we deal with balances in general. Then we formulate a more detailed procedure for deriving balance equations with an emphasis on mass balances. Recycling improves the effectiveness of a process, but the resulting balance equations become coupled and more difficult to solve. Finally, we look at the degrees of freedom and the solvability of the equations.

2.1 The balance principle

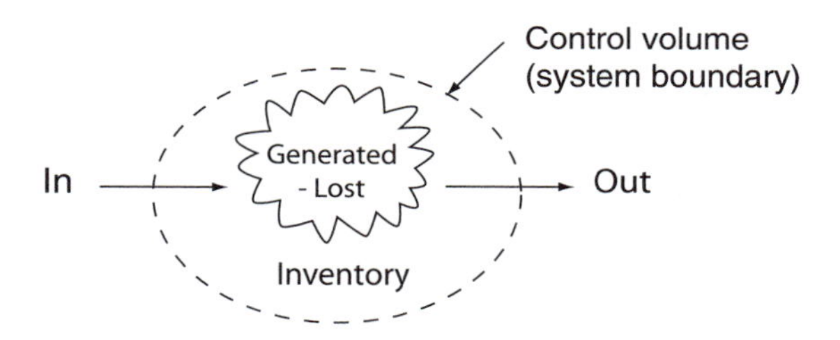

Figure 2.1: The balance principle. The dashed line defines the system boundary.

Consider a "quantity" (mass, energy, etc.) that may be balanced (counted). The general **balance equation** for a system constrained by a boundary (control volume) over a given time period is (see Figure 2.1)

$$\underbrace{\text{Change in inventory}}_{\text{accumulated within system}} = \underbrace{\text{In} - \text{Out}}_{\text{through system's boundary}} + \underbrace{\text{Generated} - \text{Lost}}_{\text{internally in system}} \tag{2.1}$$

or with symbols

$$\Delta B = B_{\text{in}} - B_{\text{out}} + B_{\text{generated}} - B_{\text{lost}} \tag{2.2}$$

- **Inventory** B: Quantity located within the system's boundary at a given time.
- **Change in inventory**, ΔB: Accumulated over the time period.
- **In**: Supplied to the system over the time period.
- **Out**: Removed from the system over the time period.
- **Generated** (formation term): Formed in the system over the time period.
- **Lost** (loss term): Lost in the system over the time period.

Balance equations can only be formulated for extensive variables, that is, for variables that depend on the system's quantity (mass).

Examples of quantities that can be balanced: total mass, component mass, number of moles (total or component), energy, momentum, population, money.

Examples of quantities that cannot be balanced: volume (imagine mixing gases with different pressures), concentration, temperature, pressure (the latter three are not even extensive variables) and enthalpy (it is energy and not enthalpy that should be balanced).

Example 2.1 Bank account. *The following balance applies for a bank account over a given time period, e.g., a year:*

$$\text{Change in inventory} = \text{Deposited} - \text{Withdrawn} + \text{Interest} - \text{Fees}$$

Here, "interest" is a formation term (generated) while "fees" is a loss term. Knowledge of this basic principle of accounting should guarantee process engineers a bright future as bank managers.

Example 2.2 Balance over the number of students in a class room. *Let B denote the number of students. We consider a lecture (the time period) and assume there are no child births (generated) or deaths (loss), so that $B_{\text{generated}} = B_{\text{lost}} = 0$. Furthermore, we assume that the auditorium is emptied after each lecture, so that the "change in inventory" is zero. The balance equation is then $0 = B_{\text{in}} - B_{\text{out}}$, or $B_{\text{out}} = B_{\text{in}}$, or*

$$\text{Number of students out} = \text{Number of students in}$$

Conserved quantities

A conserved quantity is by definition a quantity for which we don't have terms for generation or loss,

$$\text{Generated} = 0; \quad \text{Lost} = 0$$

The balance equation for a conserved quantity then assumes the simple form (**"conservation principle"** or **"conservation law"**):

$$\text{Change of inventory} = \text{In} - \text{Out} \tag{2.3}$$

$$\Delta B = B_{\text{in}} - B_{\text{out}} \tag{2.4}$$

In this book, we will focus on two conserved quantities

1. Total mass m [kg]

2. Energy E [J]

A third important conserved quantity is

3. Momentum (mv) [N]. The balance equation for momentum is

$$\frac{d(mv)}{dt} = \sum_i F_i \tag{2.5}$$

where $\sum_i F_i$ is the sum of all forces acting on the system, including gravitation and friction. Equation (2.5) is also known as **Newton's second law**. The momentum balance is the core of fluid mechanics and is used to find the relationship between pressure and flow. We won't explicitly use the momentum balance in this book, but we do consider the "mechanical energy balance" in Chapter 9 which provides similar information.

Comments:

- *The atomic mass of each element* is also a conserved quantity, but the atomic balance is *indirectly included* in the stoichiometry of the reaction equations, i.e., they are "used up" when we balance the reactions; see Appendix A.2 (page 329). For this reason we will usually not explicitly formulate atomic balances.
- For systems without chemical reaction, the *component mass* ([kg A] or [mol A]) is also a conserved quantity, and also the total number of moles [mol].
- The fact that mass is a conserved quantity is fairly obvious and has been known for a long time.
- On the other hand, it is much less obvious that energy is a conserved quantity, and this was only established around 1850 with the *first law of thermodynamics.*
- But isn't energy generated in exothermic chemical reactions? The answer is **no**. What happens is that some of the molecule's internal energy is *converted* from chemical bonding energy to thermal energy, but the amount of energy is constant.
- Strictly speaking, mass is a conserved quantity only for systems without nuclear reactions and at velocities far from the speed of light. A nuclear reactor involves a mass change and converts it to energy according to Einstein's famous formula $\Delta E = \Delta mc^2$. It is therefore assumed in this book that we consider systems without nuclear reactions and at velocities far from the speed of light $c = 3 \cdot 10^8$ m/s.

Contributions from mass flows

In our systems, the terms in "In - Out" are often split in two contributions: (1) "bulk" transport with streams (for flow processes) and (2) transfer by other means (for example, heat through a wall), that is

$$\underbrace{\text{In} - \text{Out}}_{\text{Through the system's boundary}} = \underbrace{\text{In} - \text{Out}}_{\text{with streams (bulk transport)}} + \underbrace{\text{In} - \text{Out}}_{\text{other means (through wall)}}$$

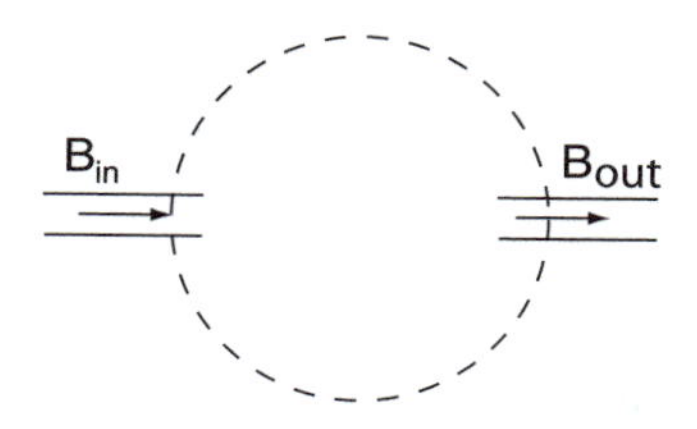

Figure 2.2: Contribution to "In - Out" from mass flow (stream)

2.2 The balance equation

Let us summarize the use of balance equations in more mathematical detail, and provide some examples. Introduce

B – Inventory of balanced quantity (within the control volume at a given time)

We can make balances over a given time period or per unit of time:

1. **Balance over a given time period (for batch process).** We consider the time period from start (at time t_0) to end (at the *final* time t_f) of the process. Over this time period

$$\Delta t = t_f - t_0$$

the change in inventory is

$$\Delta B = B_f - B_0 \tag{2.6}$$

and from (2.1) we derive the **general balance equation for B over the period Δt:**

$$\boxed{\Delta B = B_{\rm in} - B_{\rm out} + B_{\rm generated} - B_{\rm lost}} \quad [\text{kg, \$, mol A, persons, } \ldots] \tag{2.7}$$

We often use this equation for *batch* processes where we consider the time from filling of reactants $B_{\rm in}$ (at t_0) to removal of products $B_{\rm out}$ (at t_f).

2. **Balance per unit of time (rate form for continuous process).** This is used for processes with a continuous supply ($\dot{B}_{\rm in}$) and removal ($\dot{B}_{\rm out}$). We use dot variables to clearly indicate that these are rates (but later we usually omit the dots to simplify notation). From (2.7) we can derive the **general (dynamic) balance equation for B at any time t:**

$$\boxed{\frac{dB}{dt} = \dot{B}_{\rm in} - \dot{B}_{\rm out} + \dot{B}_{\rm generated} - \dot{B}_{\rm lost}} \quad [\text{kg/s, \$/s, mol A/s, persons/s, } \ldots] \tag{2.8}$$

Here, dB/dt denotes the rate of change of the inventory (accumulation) of the quantity B.

Proof of (2.8): We consider the time period from time t_0 to time $t_0 + \Delta t$, where Δt is small (for example, 1 second). We assume that the feed rates, etc. are constant over this short time period, that is, we have $B_{\rm in} = \dot{B}_{\rm in}\Delta t$, etc. From (2.7) we then get for the time period Δt:

$$\Delta B = \underbrace{\dot{B}_{\rm in}\Delta t}_{B_{\rm in}} - \underbrace{\dot{B}_{\rm out}\Delta t}_{B_{\rm out}} + \underbrace{\dot{B}_{\rm generated}\Delta t}_{B_{\rm generated}} - \underbrace{\dot{B}_{\rm lost}\Delta t}_{B_{\rm lost}}$$

We divide both sides by Δt and then let $\Delta t \to 0$. The definition of derivative, $dB/dt \triangleq \lim_{\Delta t \to} \Delta B/\Delta t$, then gives (2.8). □

Important. (2.7) and (2.8) are the same, but (2.7) is over a time period Δt, whereas (2.8) is the rate at a given time t.

Note that we in $B_{\rm in}$ (or $\dot{B}_{\rm in}$) include *all* inflows (and correspondingly all outflows in $B_{\rm out}$). With several inflows we have $B_{\rm in} = B_1 + B_2 + \cdots = \sum_i B_i$, where B_1, B_2, etc. denote the individual inflows (for example, separate streams).

Special case A: Conserved quantity (mass and energy)

For a conserved quantity (total mass, energy or momentum), there are no terms for "generated" or "lost," that is, $B_{\rm generated} = 0$ and $B_{\rm lost} = 0$, and the general balance equation becomes

$$\text{Over time period } \Delta t: \qquad \Delta B = B_{\rm in} - B_{\rm out} \ [\text{kg, J, } \ldots] \tag{2.9}$$

$$\text{At time } t: \qquad dB/dt = \dot{B}_{\rm in} - \dot{B}_{\rm out} \ [\text{kg/s, J/s, } \ldots] \tag{2.10}$$

Special case B: No accumulation

In most of this book (except Chapter 11) we consider processes where there is no accumulation over the time period, that is, the inventory within the system does not change. For such a system the following applies:

1. For a **batch process** where the tank is emptied between each batch, we have $\Delta B = 0$ over the batch period Δt and the balance equation becomes

$$\text{Over time period } \Delta t: \qquad B_{\rm out} = B_{\rm in} + B_{\rm generated} - B_{\rm lost} \quad [\text{kg, J}, \ldots] \tag{2.11}$$

2. For a **continuous process** without accumulation (= *steady-state process*) there is no time variation, i.e., $dB/dt = 0$, and the balance equation becomes

$$\text{At time } t: \qquad \dot{B}_{\rm out} = \dot{B}_{\rm in} + \dot{B}_{\rm generated} - \dot{B}_{\rm lost} \quad [\text{kg/s, J/s}, \ldots] \tag{2.12}$$

   which is the most common case studied in this book.

   Comment. In most continuous processes, steady state (without accumulation) is never fully attained because it takes time to reach this state, and also because there are always changes (*disturbances*) in feed streams and changes in the process equipment. Nevertheless, the idealized assumption of a steady-state process is used for calculation and design of most process plants.

Note that equations (2.11) and (2.12) are identical if we omit the dots on B. In this book, we usually omit the dot notation, and we can use the same equation for both batch and continuous processes.

Combined special case A and B: Total mass balance without accumulation

Since *total mass* is a conserved quantity, the mass balance (2.9) and (2.10) without accumulation becomes $0 = m_{\rm in} - m_{\rm out}$, or

$$\boxed{m_{\rm in} = m_{\rm out}} \quad [\text{kg; kg/s}] \tag{2.13}$$

(2.13) applies to both (1) batch and (2) continuous processes at steady state, and is probably the process engineer's most useful tool. (2.13) always holds for total mass, and if no chemical reactions takes place it also holds for component mass and component moles.

The next section (page 47) features many uses for the steady state mass balance (2.13). However, before getting to this, we consider two examples where we need the dynamic mass balance (2.10). First, we consider filling a bathtub with water and next a membrane unit (or actually an artificial kidney).

Example 2.3 Process with accumulation: Filling a bathtub. *Let m [kg] be the mass of water in a bathtub. Mass is a conserved quantity and from (2.10) the dynamic mass balance for a bathtub at time t becomes:*

$$\frac{dm}{dt} = \dot{m}_{\mathrm{in}} - \dot{m}_{\mathrm{out}} \quad [\mathrm{kg/s}] \tag{2.14}$$

During filling with the plug in, the outflow is zero, $\dot{m}_{\mathrm{out}} = 0$, and the mass balance becomes $\frac{dm}{dt} = \dot{m}_{\mathrm{in}}$. The filling increases the inventory m of mass ($dm/dt > 0$), that is, we have a "dynamic process" where mass is accumulated. This increase continues until we turn off the water (then $\dot{m}_{\mathrm{out}} = 0$ and also $\dot{m}_{\mathrm{in}} = 0$) or the bathtub flows over. In both cases, we get

$$\dot{m}_{\mathrm{out}} = \dot{m}_{\mathrm{in}} \tag{2.15}$$

At this point, we have no further accumulation in the system ($dm/dt = 0$), and (2.15) represents the mass balance of a continuous steady-state process.

Example 2.4 Mass balances for an artificial kidney (membrane unit). *In this example, we formulate mass balances for urea (which is the most important component in urine) during cleansing (dialysis) of a patient's blood with an artificial kidney (dialysis machine, which for a process engineer is a membrane unit). This is a continuous process. For the membrane unit itself, we will assume that there is no accumulation (steady state mass balance), while for the patient, we consider accumulation (dynamic mass balance).*

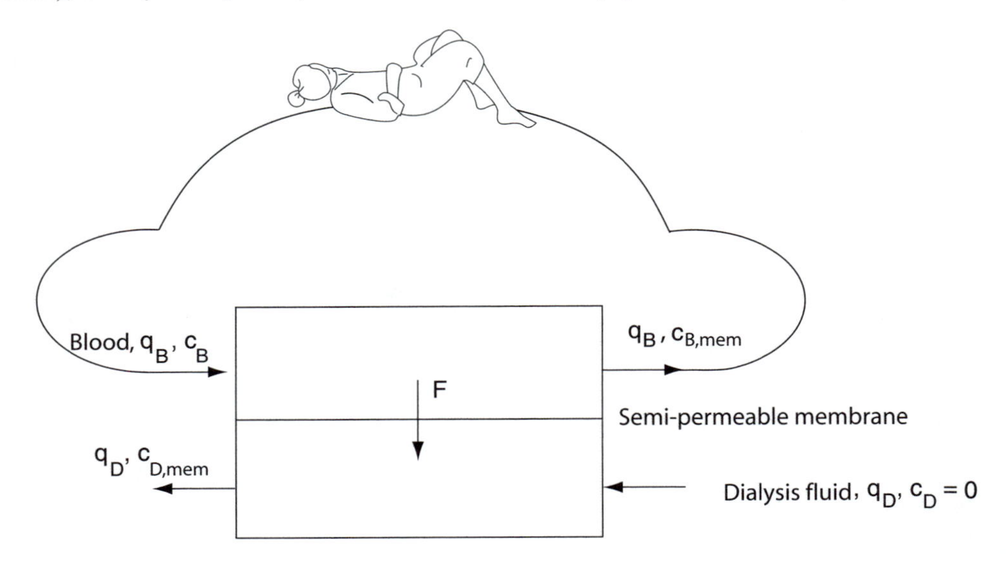

Figure 2.3: Dialysis using a membrane unit (artificial kidney). Notation: c [mol/m^3 = mmol/l] = concentration of urea. B = blood. D = dialysis fluid. mem = leaving membrane unit.

More details on the process: *During dialysis, urea is transferred from the blood (B) through a semi-permeable membrane to the dialysis liquid (D), as schematically shown in Figure 2.3. Here, q [m^3/min] is the volumetric flow and c [mol/m^3] is the concentration of urea. The dialysis liquid contains some of the same components as in blood (so that the blood won't be drained for important salts), but we assume that it contain no urea, that is, $c_D = 0$. We have countercurrent flow in the membrane unit (with blood and dialysis fluid entering at opposite ends of the membrane unit as shown in the figure).*

For simplicity, we here assume that we have a perfect membrane unit, *where the urea concentration in the exiting blood is the same as in the entering dialysis fluid, i.e.,*

$$c_{B,\text{mem}} = c_D = 0$$

This means that the returning blood contains no urea (we will relax this assumption in Exercise 2.2).

Data. *Urea is present in all the body fluids, not only in the blood. In this exercise we assume that we have 50 liters of body fluid which initially ($t = 0$) contains 50 mmol/l of urea. The membrane unit is connected for 240 minutes, with constant flows of 0.3 l/min blood and 0.5 l/min dialysis fluid. (This means that 72 liters of blood go through the membrane unit during the four hours the patient is connected.) We thus have that*

$$c_{B,0} = 50 \text{ mmol/l}, \quad c_D = 0 \text{ mmol/l}$$

$$q_B = 0.3 \text{ l/min}, \quad q_D = 0.5 \text{ l/min}$$

$$V = 50 \text{ l}, \quad t = 240 \text{ min}$$

Problem. *How do the concentrations of urea in the blood (c_B) and in the exiting dialysis fluid ($c_{D,\text{mem}}$) change with time?*

Mass balance patient (dynamic, i.e., with accumulation). *We start by formulating the urea mass balance for the patient. The boundary (control volume) then includes the patient, but not the membrane unit. We assume that urea is evenly distributed in all of the $V = 50l$ body fluids. The concentration of urea in the body $c_B(t)$ will decrease with time t when the patient is connected to the membrane. This can be derived more exactly by setting up the* dynamic mass balance *for the amount of urea n [mol] in the body. Using $B = n$ in (2.8) gives*

$$dn/dt = \dot{n}_{in} - \dot{n}_{out} \quad \text{[mol/s]}$$

To simplify, we have here neglected the generation and conversion of urea by chemical reactions in the body, that is, $\dot{n}_{\text{generated}} = 0$ and $\dot{n}_{\text{loss}} = 0$. We further have for the patient

$$\begin{aligned} n &= Vc_B \text{ [mol]} \\ \dot{n}_{out} &= q_B c_B \text{ [mol/s]} \\ \dot{n}_{in} &= q_B c_{B,\text{mem}} \text{ [mol/s]} \end{aligned}$$

and we get with the assumption of constant volume V of body fluids:

$$V dc_B/dt = q_B c_{B,\text{mem}} - q_B c_B$$

With the assumption $c_{B,\text{mem}} = 0$ (perfect membrane unit) this gives the differential equation

$$V dc_B/dt = -q_B c_B$$

Separation of variables gives $dc_B/c_B = -(q_B/V)dt$, which upon integration gives

$$c_B(t) = c_{B0} \cdot e^{-q_B t/V}$$

With the given data, we have that $(q_B t/V) = 0.3 \text{ l/min} \cdot 240 \text{ min}/50 \text{ l} = 1.44$ and since $e^{-1.44} = 0.237$ we get

$$c_B(240\ min) = 0.237 c_{B0} = 0.237 \cdot 50 \text{ mmol/l} = 11.8 \text{ mmol/l}$$

That is, during a period of four hours, the amount of urea in the blood is reduced to 23.7% of its initial amount. This result was derived for an ideal membrane, but actual membranes give similar values.

Mass balance membrane units (steady state, that is, without accumulation). *Let us now formulate the urea mass balance for the membrane unit. The boundary (control volume) now includes the membrane unit, but not the patient. The mass of blood and dialysis fluid (and thereby of urea) within the actual membrane unit is relatively small, and we will therefore neglect the accumulation of urea in the membrane unit. We then have $dn/dt \approx 0$ where n here is the amount of urea accumulated in the membrane unit. (We couldn't make this assumption for the patient because the amount of body fluids is very large). The membrane has two sides and we set up the steady state urea balance (in=out) for each side of the membrane unit:*

$$\text{Blood side}: \quad q_B c_B = qBc_{B,\text{mem}} + F \quad [\text{mol/s}]$$

$$\text{Dialysis fluid side}: \quad q_D\ c_D + F = q_D\ c_{D,\text{mem}} \quad [\text{mol/s}]$$

The two equations are coupled by the amount F transferred through the membrane from the blood to the dialysis fluid. We can eliminate F by adding the equations (or alternatively by setting up a total mass balance over the membrane unit) and derive

$$\text{Total (in = out)}: \quad q_B c_B + q_D c_D = q_B c_{B,\text{mem}} + q_D c_{D,\text{mem}} \quad [\text{mol/s}] \tag{2.16}$$

These mass balance equations are independent of how the actual membrane functions, that is, they apply for both cocurrent and countercurrent flows. Equation (2.16) can be used to calculate the urea concentration $c_{D,\text{mem}}$ of the existing dialysis fluid. In our case, with a "perfect membrane unit" and countercurrent flow, (2.16) gives that the concentration of urea in the exiting dialysis fluid is initially ($t = 0$)

$$c_{D,\text{mem}} = \frac{q_B c_B + q_D c_D - qBc_{B,\text{mem}}}{q_D} = \frac{0.3 \cdot 50 + 0.5 \cdot 0 - 0.3 \cdot 0}{0.5} = 30 \text{ mmol/l}$$

and at the end of the treatment ($t = 240$ min) it decreases to $c_{D,\text{mem}} = (0.3 \cdot 11.8/0.5) = 7.0$ mmol/l.

Remark 1 Similarity of membrane unit and heat exchanger. *The membrane unit is almost identical to the shell-and-tube heat exchanger in Figure 5.2 (page 130), except that the tubes are semi-permeable fibers where mass is transferred rather than heat. The blood flows through the fibers (tubes) while the dialysis fluid flows countercurrently on the outside; this is the same as in a heat exchanger.*

Remark 2 Perfect membrane unit. *For the "perfect membrane unit" assumption ($c_{B,\text{mem}} = 0$) to be true the following must hold:*

- *The flows are countercurrent (as shown in the figure)*
- *The membrane is sufficiently effective (for example, with a very large membrane contact area A), so that we approximately have zero driving force at the end of the membrane where the blood leaves, that is, $\Delta c = c_{B,\text{mem}} - c_D = 0$.*
- *The flow of dialysis fluid q_D [m^3/s] is larger than the flow of blood q_B [m^3/s].*

The need for the last assumption can be shown by setting $c_D = 0$ and $c_{B,\text{mem}} = 0$ into the total mass balance (2.16) for urea. We find that $c_B/c_{D,\text{mem}} = q_D/q_B$ and since $c_B > c_{D,\text{mem}}$ (urea is transferred from the blood to the dialysis fluid) we must then have that $q_D > q_B$.

Next are two follow-up exercises for the membrane example. It is well known from heat transfer that countercurrent flow is more effective than cocurrent and in the first exercise this is confirmed for mass transfer in a membrane.

Exercise 2.1* Perfect membrane unit with cocurrent flow. *By mistake the patient is connected to the membrane unit so that the flow is cocurrent, that is, the two flows enter at* *the same end. With a perfect membrane unit we can then assume* $c_{B,\mathrm{mem}} = c_{D,\mathrm{mem}}$ *at the exiting end. Assume also here* $c_D = 0$ *(no urea in entering dialysis fluid).*

(i) Use the total urea mass balance for the membrane unit to derive an expression for $c_{B,\mathrm{mem}}/c_B = (1-\epsilon)$*, where* ϵ *is the membrane unit's "efficiency" (which was 1 for ideal membrane with countercurrent flow). Note that* ϵ *should only depend on* q_B *and* q_D*.*

(ii) Put this into the dynamic urea mass balance for the patient, and use the given data to calculate the concentration of urea in the blood (c_B*) at the end of the treatment (t=240 min). (You will find that the urea concentration is about double of that found with countercurrent flow).*

The next exercise should be done after you have read Chapter 5 on heat exchange.

Exercise 2.2 Mass balance for a real membrane unit. *Here, we use a more detailed model of a countercurrent membrane unit. The mass transfer though the membrane is assumed to be described by* **Fick's law***:*

$$F = kA\Delta\bar{c} \quad [\mathrm{mol/min}]$$

where F [mol/min] is the amount of urea transferred, k [m/min] is the mass transfer coefficient (or permeance) which is assumed constant, A [m^2*] is the area of the membrane, and* $\Delta\bar{c} = c_B - c_D$ *[mol/*m^3*] is the average (mean) concentration difference across the membrane.*

Above we assumed a "perfect membrane unit" (or more precisely, we assumed that the mass transfer is very effective so $N \triangleq kA/q_D \gg 1$*) where the local concentration difference was 0 at the end where blood exits (*$\Delta c = c_{B,\mathrm{mem}} - c_D = 0$*), but here we will not make this assumption. Otherwise, we use the same data as given above, including* $c_D = 0$ *(no urea in entering dialysis fluid).*

(a) **Simplified model.** *In order to simplify the calculations, we first assume perfect mixing on each side of the membrane (there is then no difference between cocurrent and countercurrent flows), that is, we have* $\Delta c = c_{B,\mathrm{mem}} - c_{D,\mathrm{mem}}$ *along the whole membrane unit.*

(i) Eliminate $c_{D,\mathrm{mem}}$ *from the mass balance equations and use this to derive an expression for* $c_{B,\mathrm{mem}}/c_B = (1-\epsilon)$*, where the membrane efficiency* ϵ *is a function of* kA, q_B, q_D*.*

(ii) Compute the efficiency ϵ *for the two cases* $N = 1$ *and* $N = \infty$*.*

(b) **Countercurrent flow membrane model.** *The simplified model gives valuable insights, but it is too pessimistic because the actual average* Δc *is larger.*

(i) Compare the equations for heat exchange and mass transfer in a membrane, and explain why the correct expression for the average $\Delta\bar{c}$ *in the membrane unit is the logarithmic mean concentration difference,*

$$\Delta\bar{c}_{\mathrm{lm}} = \frac{\Delta c_1 - \Delta c_2}{\ln(\Delta c_1/\Delta c_2)}$$

where for countercurrent flow the local concentration differences at the two ends are $\Delta c_1 = c_{B,\mathrm{mem}} - c_D$ *and* $\Delta c_2 = c_B - c_{D,\mathrm{mem}}$*.*

(ii) Find the concentration c_B *at* $t = 240$ *min for the two cases* $N = 1$ *and* $N = \infty$*. Also find the efficiency* ϵ *for the two cases.*

2.3 Mass balances without accumulation

The simplest balance, but also the most important one in engineering practice, is the total mass balance for a process without accumulation, see equation (2.13) and

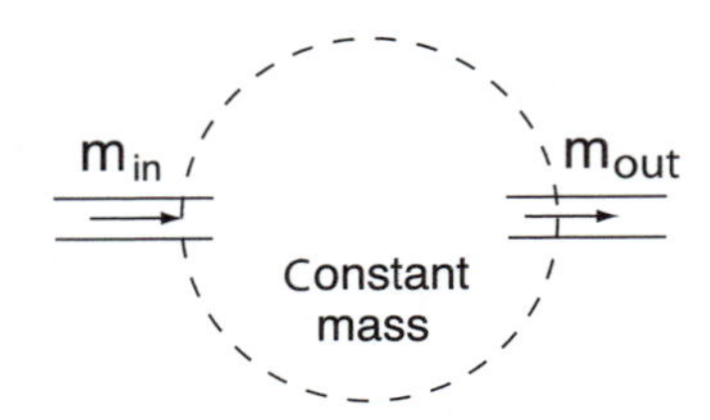

Figure 2.4: Mass balance without accumulation

Figure 2.4:

$$m_{\text{in}} = m_{\text{out}} \quad [\text{kg; kg/s}]$$

This balance applies to the following two cases:

1. **Batch** process with no accumulation; $m_{\text{in}} = m_{\text{out}}$ [kg] (over period Δt)
2. Steady-state **continuous** process; $m_{\text{in}} = m_{\text{out}}$ [kg/s]

Note that $m_{\text{in}} = m_1 + m_2 + \cdots$ can consist of several inflows, and correspondingly $m_{\text{out}} = m_I + m_{II} + \cdots$ can consist of several outflows. By writing $m = \rho V$, where V [m^3] is the volume and ρ [kg/m^3] the density, the total mass balance becomes $\rho_{\text{in}} V_{\text{in}} = \rho_{\text{out}} V_{\text{out}}$. For the special case of *constant density*, that is, $\rho_{\text{in}} = \rho_{\text{out}}$, the total mass balance can then be written as a "volume balance" (the quotation marks are absolutely necessary as volume is <u>not</u> a conserved quantity):

$$V_{\text{in}} = V_{\text{out}} \quad [\text{m}^3; \text{m}^3/\text{s}] \qquad \text{(only for constant density!)}$$

The mass balance (2.13) applies also to component mass for cases *without reaction* (here written for an arbitrary component A):

$$m_{A,\text{in}} = m_{A,\text{out}} \quad [\text{kgA; kgA/s; molA; molA/s}] \qquad \text{(no reaction!)} \tag{2.17}$$

You may think it is trivial to set up such balances, but in practice it is not always so simple unless you work *systematically*. The following procedure is recommended (for more details see Table 2.2, page 67):

1. Obtain an overview of the problem by drawing a flow sheet
2. Summarize the stream data – either in a table or directly on the flow sheet
3. Define the control volume(s) (usually one for each unit)
4. Formulate balances (for total mass, moles, energy, etc.) for each control volume
5. Solve the equations

When solving the equations by hand, you need to start the calculations at the right point in the process, for example, by choosing a "clever" **basis**. This usually corresponds to specifying a flow rate, for example, at the reactor feed. The idea is to have sufficient information at some point to start the calculations. Alternatively (and this always works!) you can use the general (systematic) approach: Formulate all the equations (mass balances, etc.) and solve them simultaneously, for example, using MATLAB or a spreadsheet program.

We now consider some examples of mass balance without accumulation. First, we consider mixing processes (which are very important industrially), and then proceed in the next section to more complicated processes with recycle.

Example 2.5 Continuous mixing process. *10 kg/s of a 5% (meaning 5 weight%) ethanol solution is produced by mixing two streams from two tanks, that contain 1% and 41% ethanol, respectively (the rest is water, all on a weight basis). How much is needed from each tank?*

Solution. *The flow sheet with stream data is shown in Figure 2.5(a). E denotes ethanol. We lack data for the two feed flows which we denote m_1 and m_2 [kg/s] (if desirable, you can use dot notation, $\dot{m}_1$ and $\dot{m}_2$, to explicitly show that these are flow rates, but here we omit the dots and instead include the units in the equations).*

The steady-state total mass balance ("In=Out") gives

$$m_1 \text{ [kg/s]} + m_2 \text{ [kg/s]} = 10 \text{ kg/s}$$

The component balance for ethanol [kg E/s] gives:

$$0.01 \text{ kg E/kg} \cdot m_1 \text{ [kg/s]} + 0.41 \text{ kg E/kg} \cdot m_2 \text{ [kg/s]} = 0.05\text{kg E/kg} \cdot 10\text{kg/s}$$

Thus, there are two equations with two unknowns. We find $m_1 = 9$ kg/s and $m_2 = 1$ kg/s.

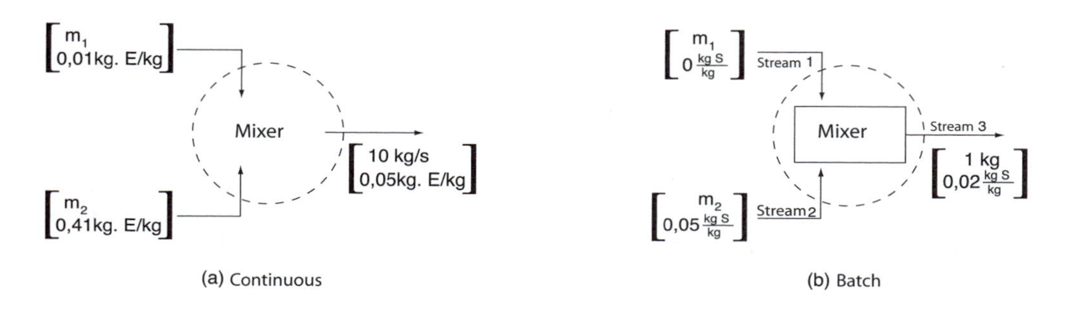

Figure 2.5: Mixing of streams in (a) continuous process (Example 2.5) and (b) batch process (Example 2.6)

The following example is very similar, but instead of a continuous process with a rate balance [kg/s], we consider a batch process where we balance [kg] over a certain period of time.

Example 2.6 Batch mixing process (beaker). *From clean water and an aqueous solution with 5 % S (NaCl), we want to produce 1 kg of 2% NaCl (everything is in weight%). How much is needed of each reactant?*

Solution. *The flow sheet is shown in Figure 2.5(b). The two unknown feeds are denoted m_1 and m_2 [kg]. The total mass balance "In=Out" gives*

$$m_1 + m_2 = 1 \quad [kg]$$

The component balance for S gives:

$$0 + 0.05 m_2 = 0.02 \cdot 1 \quad [kg\ S]$$

and we find $m_2 = 0.4$ kg (water) and $m_1 = 0.6$ kg (S).

The next example illustrates that volume is not a conserved quantity unless one assumes constant density, and illustrates the choice of basis and scaling.

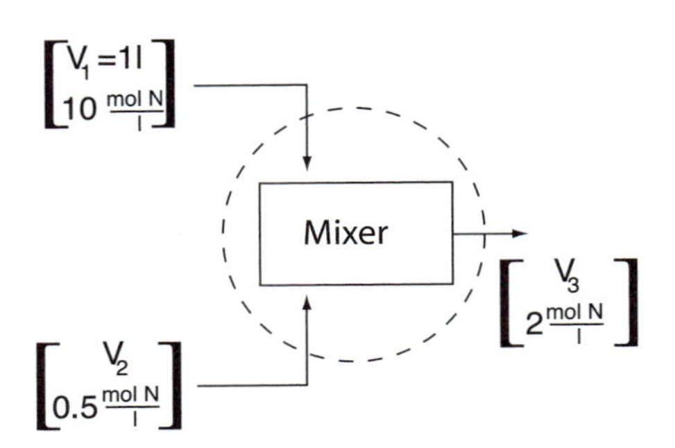

Figure 2.6: Another mixing example

Example 2.7 Another mixing example. *A stream of 2 m^3/h with 10 mol/l NaOH (stream 1) is to be diluted with a solution of 0.5 mol/l NaOH (stream 2) to produce a product with 2 mol/l NaOH (stream 3). How much of stream 2 must be added?*

Solution. *The flow sheet is shown in Figure 2.6. We choose 1 l of stream 1 as a basis (even though it is specified that the amount is 2 m^3/h; we will "fix this up" later). The total mass balance is $m_1 + m_2 = m_3$ [kg] and by introducing the density $\rho = m/V$ [kg/m^3] we get*

$$\rho_1 V_1 + \rho_2 V_2 = \rho_3 V_3$$

where we have $V_1 = 1$ l with our assumed basis. The densities are not given so we do not have enough information to solve this exercise. We therefore **assume that the density is the same for all streams**, *that is, $\rho_1 = \rho_2 = \rho_3$ (this is of course not entirely correct, but it is common to neglect mixing volumes). The mass balance then gives the "volume balance"*

$$\underbrace{V_1}_{=1l} + V_2 = V_3 \quad [l]$$

The mass balance for NaOH on molar basis gives (no assumption of equal density is needed here)

$$10 \cdot 1 + 0.5 V_2 = 2 V_3 \quad [mol\ NaOH]$$

and by combining the two equations we find $V_2 = 5.33l$ which is the amount that is needed for 1 l of stream 1. To consume 2 m^3/h of stream 1 we need to scale everything by the factor 2 [$(m^3/h)/l$], that is, we need 10.66 m^3/h of stream 2.

Example 2.8 Mixing of gasoline. *98 octane gasoline with 1 weight% aromatics is to be produced by mixing three fractions. Fraction 1 is 90 octane with 0.2% aromatics. Fraction 2 is 105 octane with 0.2% aromatics. Fraction 3 is 95 octane with 0.5% aromatics. How much is needed of each fraction?*

For simplicity, assume that the octane number p of the blend (outstream) is the mass average of the octane numbers p_i of the streams that are mixed (instreams):

$$p = \frac{\Sigma_i m_i p_i}{\Sigma_i m_i} \quad \text{(linear mixing of property } p \text{ on mass basis)} \tag{2.18}$$

Solution. *Write balances ("In=Out") for total mass, aromatics and octane:*

$$m_1 + m_2 + m_3 = m_{\text{out}} \text{ [kg]}$$

$$0.2 m_1 + 2 m_2 + 1.5 m_3 = 1 \cdot m_{\text{out}} \text{ [kg aromatics]}$$

$$90m_1 + 105m_2 + 95m_3 = 98 \cdot m_{\text{out}} \text{ [octane number]}$$

We use $m_{\text{out}} = 1$ kg total mass as a basis, and then have three linear equations with three unknowns. This is easily solved, for example, using MATLAB:

```
A = [1 1 1; 0.2 2 0.5; 90 105 95]; b=[1; 1; 98];
m=inv(A)*b
```

This gives the mass fractions (m): 0.1111, 0.3556 and 0.5333.

Example 2.9 Distillation of methanol.

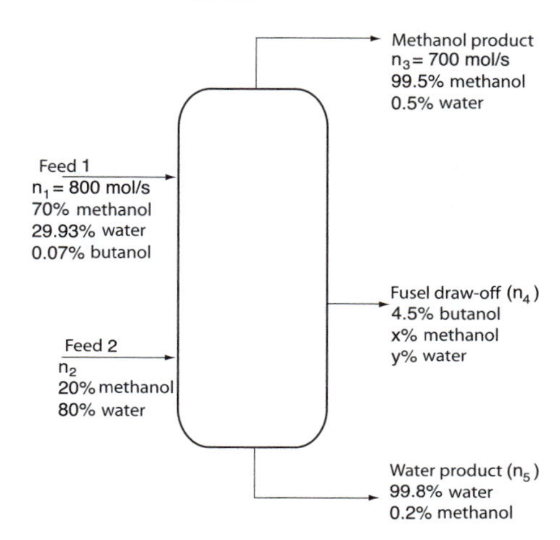

Figure 2.7: Distillation column with two feeds and side stream

In a continuous methanol plant there is a distillation column where methanol is removed as a top product and water as a bottom product. The column has two feeds. There is some butanol in one of the feeds and the butanol is removed as a "fusel oil" in the middle of the column. (Distillation is based on differences in volatility, and one would expect that butanol, which boils at a higher temperature than water, would be removed at the bottom of the column. This does not happen in this case because water and butanol in the liquid phase "do not like each other," and butanol is therefore "pushed up" the column). More information is given in Figure 2.7 (all numbers are in mol-%). Steady-state operation is assumed. We want to determine the amount of feed 2 (n_2).

(a) How many independent mass balances can be formulated for the column?

(b) How much additional information do you need to solve the problem?

(c) Is the system solvable if it is specified that the fusel contains 2% methanol (if yes, find the answer)?

(d) Is the system solvable if instead it is specified that the amount of fusel is n_4=20 mol/s (if yes, find the answer)?

Solution. *(a) We can formulate 3 independent mass balances:*

$$\text{Total}: \quad 800 + n_2 = 700 + n_4 + n_5 \quad \text{[mol/s]}$$

$$\text{Methanol}: \quad 0.7 \cdot 800 + 0.2n_2 = 0.995 \cdot 700 + xn_4 + 0.002 \cdot n_5 \quad \text{[molM/s]}$$

$$\text{Butanol}: \quad 0.0007 \cdot 800 = 0.045 \cdot n_4 \quad \text{[molB/s]}$$

Note that we have converted the percentages to mole fractions, so x represents the mole fraction of methanol in the side stream. Also note that $x + y + 0.045 = 1$ so the water fraction y in the side stream can be found when we know x.

(b) In summary, we have 3 independent equations with 4 unknowns: n_2, n_4, n_5, x. *Thus, we need 1 extra piece of information (e.g., 1 additional equation, or the value of 1 of the unknowns).*

(c) We are given $x = 0.02$*, and since* x *is an unknown the problem should now in principle be solvable (the only potential problem is that the solution may be non-physical, e.g., negative flows). Let us proceed with the solution. The amount of fusel (*n_2*) can be obtained from the butanol balance:* $n_4 = 0.0007 \cdot 800/0.045$ *mol/s = 12.44 mol/s. Substituting this into the two other balances gives:*

$$800 + n_2 = 700 + 12.44 + n_5 \quad [\text{mol/s}]$$

$$560 + 0.2n_2 = 696.5 + 0.249 + 0.002 \cdot n_5 \quad [\text{molM/s}]$$

We have 2 equations in 2 unknowns. The last equation (the methanol balance) gives $n_2 = 683.7 + 0.01n_5$ *which inserted into the total balance gives* $n_5 =$ *779.1 mol/s and we find* n_2*=691.5 mol/s (which is a physical solution, so it is OK).*

(d) The problem is not solvable, because the specification $n_4 = 20$ *mol/s is not consistent with the previous information (the butanol balance). In addition, we are missing a piece of information to find the distribution between methanol and water.*

Comment. *Here, we solved the equations by hand. A more general approach is to add any additional information as extra equations, and then solve the resulting equation set numerically. For example, in part (c) we can add the equation* $x = 0.02$ *and then get 4 equations in 4 unknowns. However, note that the equation set is not linear in this case because of the term* xn_4 *in the mass balance for methanol, so a non-linear equation solver would be needed. In part (d), we would not find any solution, and most likely the equation solver would issue an error message such as "singular matrix."*

Here are some exercises that you can solve yourself. State clearly any additional assumptions!

Exercise 2.3 Mass balance distillation.* *A 2000 kmol/h feed stream F with 60 mol% methanol and 40 mol% water is separated in a distillation column and gives two products: A "light" methanol product (distillate D) which contains 2 mol% water, and a "heavy" water product (bottom's B) which contains 5 mol% methanol.*

(a) Draw a flow sheet and formulate two mass balances.

(b) Calculate the amount of methanol product.

Exercise 2.4 What is the best way to rinse a beaker?* *We have a 0.5 l beaker where there is a residue of 10 ml water solution with some salt S, with concentration 100 g S/l. We want to rinse the beaker using pure water. After each rinsing there is a residue of 10 ml.*

(a) What is the concentration in the beaker after two rinsings with 200 ml water each?

(b) What is the concentration in the beaker after four rinsings with 50 ml water each?

(c) What is the best way to rinse a beaker if we want to use as little water as possible?

Exercise 2.5* Absorption column. *The absorption process in Figure 2.8 uses oil (stream 1) to remove benzene from polluted air (stream 2). (a) How many independent mass balances can be set up? (b) Calculate the amount of oil (stream 1).*

Exercise 2.6 *An intermediate product stream has the following composition on weight basis: 23%* HNO_3 *and 57%* H_2SO_4 *(the rest is water). However, this does not match the desired composition of the sales product: 27%* HNO_3 *and 60%* H_2SO_4*. Thus, we need to mix the intermediate with concentrated* HNO_3 *(90%) and concentrated* H_2SO_4 *(93%). How much of the intermediate is used to make 1000 kg product?*

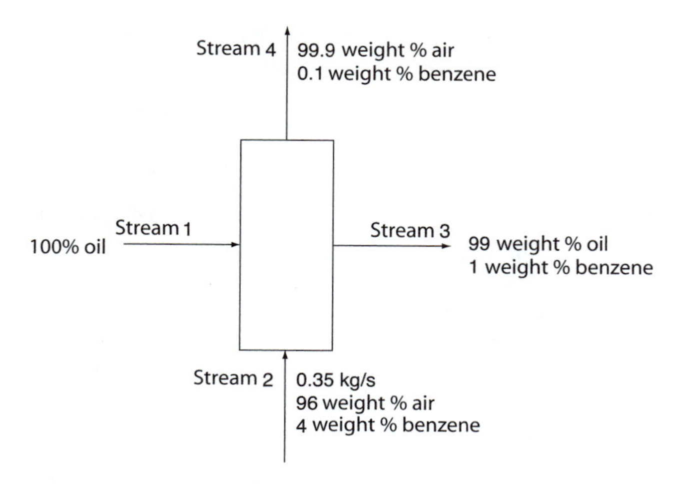

Figure 2.8: Absorption column

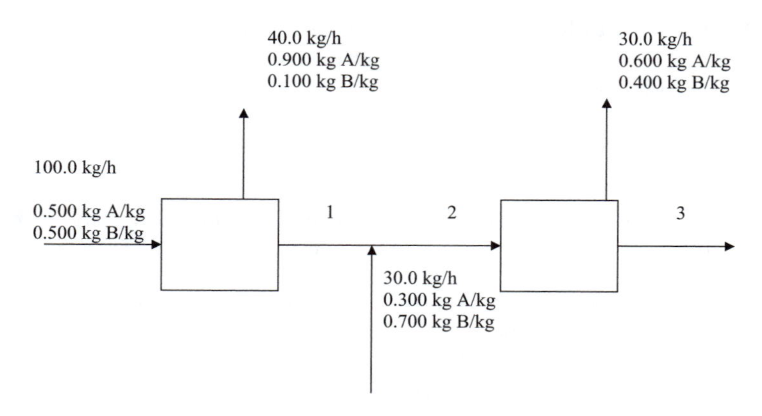

Figure 2.9: Flowsheet of simple example process

Exercise 2.7 *The continuous process in Figure 2.9 consists of a separator, a mixer (shown very simplified as an arrow that is added to stream 1) and another separator. Find the amounts and compositions of streams 1, 2 and 3 when steady-state conditions are assumed.*

Example 2.10 Linear interpolation. *In Perry's Handbook (1984), the following data are found for the density ρ [g/ml] of ethanol-water mixtures at 20 $^\circ C$ (weight% ethanol):*

$$0.99823(0\%), 0.98187(10\%), 0.96864(20\%), 0.95382(30\%)$$

$$0.93518(40\%), 0.91384(50\%), 0.89113(60\%), 0.86766(70\%)$$

$$0.84344(80\%), 0.81797(90\%), 0.80424(95\%), 0.78934(100\%)$$

Problem: *Determine the density of a mixture with 63 weight% ethanol using linear interpolation.*

Solution. *Linear interpolation corresponds to plotting the given values for y (here ρ) as a function of x (here a weight fraction) and extending straight lines between neighboring points. With two neighboring points (x_1, y_1) and (x_2, y_2) the value of y is then for a given x given by:*

$$y = (1-\alpha)y_1 + \alpha y_2; \quad \alpha = \frac{x - x_1}{x_2 - x_1} \tag{2.19}$$

The density of a mixture with 63 weight% ethanol is then estimated by a linear interpolation to be $0.7 \cdot 0.89113 + 0.3 \cdot 0.86766 = 0.88409$ *g/ml.*

Comment 1: *Draw a figure with ρ as a function of weight fraction and you will see that this is correct. Physically, this means that one can produce a mixture with 63 weight% by mixing* 0.7 *weight fraction of a 60%-mixture and 0.3 weight fraction of a 70%-mixture. This is also known as the lever rule.*

Comment 2: *Linear interpolation is accurate as long as the data points lie close to each other. If we only had data for 0% and 100% then we would by linear interpolation find the density* $0.37 \cdot 0.99823 + 0.63 \cdot 0.78934 = 0.8666$ *g/ml which is quite different from the "correct" value of 0.88409 g/ml found above.*

Exercise 2.8 Density of ideal mixture. *Show that if you neglect the mixing volume (ideal mixture), then the density of a mixture can be calculated using the following formula ("mixing rule for density of ideal mixture")*

$$\rho_{\rm id} = v_1\rho_1 + v_2\rho_2 = \frac{1}{\frac{w_1}{\rho_1} + \frac{w_2}{\rho_2}} \tag{2.20}$$

where v_1 and v_2 are volume fractions (before mixing), w_1 and w_2 are mass fractions (for mass, before or after mixing is the same), and ρ_1 and ρ_2 are the density of the pure liquids.

This implies that for an ideal mixture it is not correct to use linear interpolation with mass fractions for density, as done in Example 2.5. Rather, one should use volume fractions, or eq. (2.20), as illustrated next.

Example 2.10 (page 54) continued. *Here, we use the more correct interpolation formula (2.20) to estimate the density of a mixture with 63 weight% ethanol*

$$\rho = \frac{1}{\frac{w_1}{\rho_1} + \frac{w_2}{\rho_2}} = \frac{1}{\frac{0.7}{0.89113} + \frac{0.3}{0.86766}} = 0.88396 \text{ g/ml}$$

which is slightly lower than the value 0.88409 g/ml found in Example 2.10.

Exercise 2.9* **Mixing volume for ethanol-water.** *Use the density data from Example 2.10 to determine the volume when you mix 1 l 96-volume% ethanol and 1 l water at 20 °C. You will find that there is a "loss," but before you do this, answer subtasks (a)-(d):*

(a) What is the mass composition of 96 volume% ethanol? (Note that volume% always refers to volumes before mixing!)

(b) Use the given density data to calculate the mass of water and ethanol in 1 l 96 volume% ethanol.

(c) We will now look at a mixture of 1 l 96 volume% ethanol and 1 l water. Calculate the composition of the mixture in weight% and volume-%.

(d) Use the given density data and eq. (2.20) to determine the density of your mixture.

(e) Finally: What is the volume of your mixture?

Solution. *(a) 94.99 weight% ethanol. (b) 40.25 g water and 763.98 g ethanol. (c) 42.39 weight% ethanol and 48.20 volume% ethanol (the answer is somewhat more than 48 volume-% because volume is not a conserved quantity, and the original 1 l of 96% therefore corresponds to more than 1 l of pure water + ethanol). (d)* $1/((w_1/\rho_1)+(w_2/\rho_2)) = 1/((0.239/0.91384)+(0.761/0.93518)) = 0.92999$ g/ml. *(e) And finally: mixing reduces the volume from 2 l to 1.938 l – this is because water and ethanol "like each other."*

2.4 Recycle

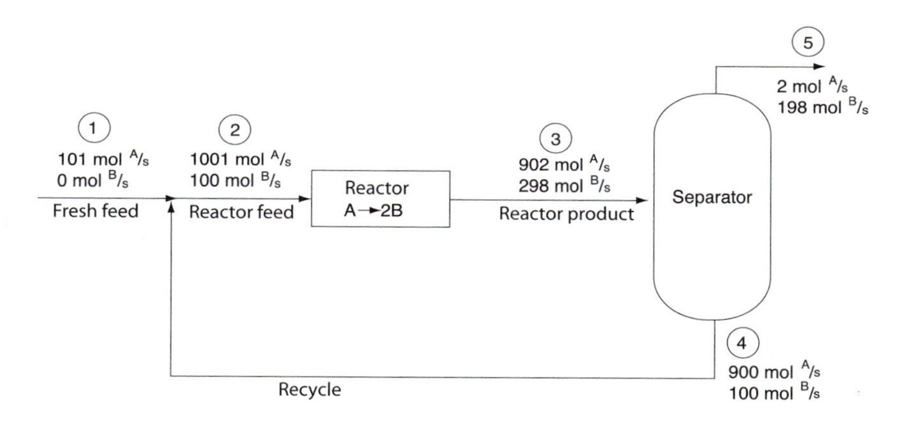

Figure 2.10: Flowsheet for process with reactor, separation and recycle

Recycle is a very effective and common method to

1. increase the yield of a process (financial gain)
2. avoid emissions (environmental gain)

Especially, it is common with chemical reactions to recycle unreacted reactants to increase yield. An example can be seen in Figure 2.10 where unreacted A is recycled. Note the difference between the "fresh" feed (1) to the process as a whole and feed (2) to the reactor. The conversion of reactant A in the reactor is only 9.9% (per pass), but because of the recycle the conversion of A in the process as a whole is 98.0%. (The reasons for the low conversion per pass can be many: Maybe component B causes "coking" in the reactor if its concentration is too high; maybe there is a reverse reaction that limits the conversion; maybe the reaction is exothermic and there is a maximum temperature allowed.)

Exercise 2.10* **Check of solution.** *Check that the solution given in the flow sheet in Figure 2.10 satisfies the mass balances.*

In practice, it may take some time to "build up" the amount and composition of the recycle stream to its steady state value. Here, we go "directly" to the steady state, that is, we assume that the amount of recycle has reached the value where it no longer changes.

With recycle one must make sure that all components that cannot react (surplus reactants and inerts) have a "way out" – otherwise they will accumulate in the process and the recycle stream will go to infinity. Because of this one must often remove or purge a fraction of the recycle stream to avoid accumulation of inerts (see Example 2.11 and Figure 2.11, page 56).

Let us now consider in detail a similar recycle example, where a purge stream is needed to avoid accumulation of inerts.

Example 2.11 Reactor with recycle and purge. *Consider the process in Figure 2.11.*

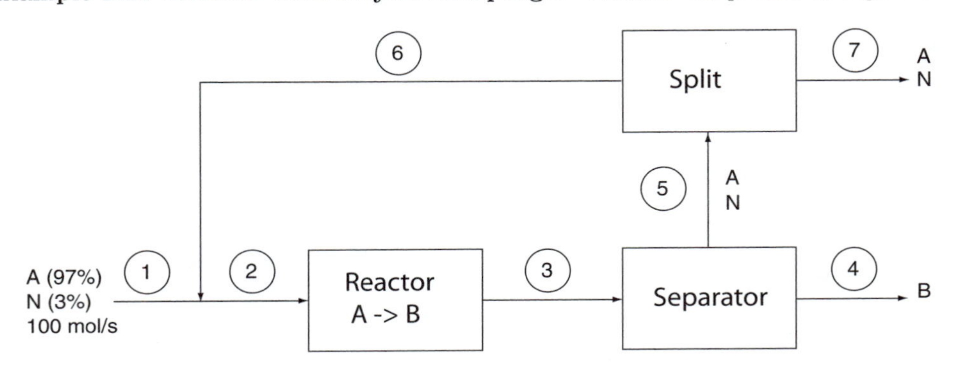

Figure 2.11: Continuous process with reaction, separation, recycle and purge (used also in EXCEL spreadsheet example).

The feed is 100 mol/s and contains components A (97%) and N (3%), where N is an inert component. In the reactor, A is converted to B according to the reaction

$$A \rightarrow B$$

The reaction is reversible and goes to equilibrium. Therefore, at the exit of the reactor (stream 3) it is given that

$$K = \frac{c_B}{c_A} = \frac{n_{B,3}}{n_{A,3}} = 0.3911$$

Unreacted A is separated from B and is recycled to the reactor. Let us initially assume ideal separation, where all of components A and N go in stream 5, and all of B goes to the product (stream 4) (we will later look at the case where a fraction x of the B from the reactor goes to stream 5, but so far it is assumed that $x = 0$). To avoid accumulation of the inert in the recycle system, we need to introduce a purge (stream 7). However, the purge gives an undesirable loss of reactant A, which can be reduced by increasing the recycle.

Question. *Find the amount of recycle that results in a loss of A of 9 mol/s.*

Without recycle. *To motivate the use of recycle, let us first look at the case without recycle, that is,* $n_6 = 0$. *There are three components, so we can formulate three independent component (mass) balances for the reactor*

$$\begin{aligned} n_{A,3} &= n_{A,2} - \xi = 97 - \xi && \text{[mol A/s]} \\ n_{B,3} &= n_{B,2} + \xi = \xi && \text{[mol B/s]} \\ n_{N,3} &= n_{N,2} = 3 && \text{[mol N/s]} \end{aligned}$$

Here, the "extent of reaction" ξ *[mol/s] is the amount converted according to the reaction* $A \rightarrow B$ *(which from the stoichiometry of the reaction* $A \rightarrow B$ *equals the amount of A reacted and the amount of B formed). Furthermore,* $n_{B,3}/n_{A,3} = K = 0.3911$ *(equilibrium). Inserting from the mass balance for A and B this gives* $\frac{\xi}{97-\xi} = 0.3911$ *and we find* $\xi = 27.27$ [mol/s]. *The mass balances then give for the reactor product (stream 3)*

$$\begin{aligned} n_{A,3} &= 69.73 && \text{[mol A/s]} \\ n_{B,3} &= 27.27 && \text{[mol B/s]} \\ n_{N,3} &= 3 && \text{[mol N/s]} \end{aligned}$$

With an ideal separator we then find that the product (stream 4) is 27.27 mol B/s, and there is a loss of 69.73 mol A/s in stream 7 (= stream 5 in this case without recycle). This is much larger than the loss of 9 mol/s that we are aiming for, so we clearly need to introduce recycle.

With recycle. *We recycle a fraction f of stream 5 to the reactor (stream 6). We want to find the value of f that results in a loss of A of 9 mol/s. The solution (stream data) is summed up in the table below. All numbers are in mol/s.*

	1	*2*	*3*	*4*	5	6	*7*
	Fresh feed	Rx.feed	Rx.prod	Prod.		Recycle	Purge
A	*97*	*313*	*225*	*0*	*225*	*216*	*9*
B	*0*	*0*	*88*	*88*	*0*	*0*	*0*
N	*3*	*75*	*75*	*0*	*75*	*72*	*3*
Total	*100*	*388*	*388*	*88*	*300*	*288*	*12*

The numbers in this table were obtained by solving the mass balance "by hand" as follows:

1. *The fresh feed (stream 1) is known and we also have information about the exit streams (streams 4 and 7), so we start by formulating balances around the whole process:*
2. *Stream 7 (purge) can be set up directly: it contains 9 mol/s A (given loss) and 3 mol/s N (from a balance for N over the whole process).*
3. *Stream 4 (product) is also easy to find: From a balance for A over the whole process, it follows that* $97 - 9 = 88$ *mol/s A is converted, and from the stoichiometry this must be equal to the amount of B in stream 4.*
4. *...... which is equal to the amount of B in stream 3.*
5. *From the equilibrium, the amount of A in stream 3 is equal to* $88/0.3911 = 225$ *mol/s*
6. *.... which is equal to the amount of A in stream 5.*
7. *The compositions of streams 5 and 7 are equal, so that the amount of inert (N) in stream 5 is* $\frac{3}{9} \cdot 225 = 75$ *mol/s.*
8. *We can now calculate stream 6 (=stream 5 - stream 7),*
9. *and finally, stream 2 (=stream 1 + stream 6), and we can fill in the rest of the table.*

Comment. *The amount of recycle (stream 6) is 288 mol/s which is 2.88 times the amount of the fresh feed (stream 1). This means that the recycle fraction is* $f = 288/300 = 0.96$. *We note that by recycling unreacted A, we have been able to reduce the loss of A from 69.73 mol/s to 9 mol/s. This loss can be further reduced by*

1. *Increasing the recycle fraction f. The loss of A can in theory be reduced to 0 by letting $f \rightarrow 1$, but the recycle flow will then go to infinity.*
2. *Introducing a new separation process which separates A from N.*

The optimal choice of process and process conditions is in general determined by economics.

2.5 Systematic formulation and solution of mass balances

In more complicated cases it is difficult to solve the mass balances "by hand" as done in the example above. Instead, we must be systematic and formulate all the equations (balances) before solving them. This is illustrated next.

Example 2.11 continued. Detailed equations for reactor with recycle. *We want to formulate all the equations for the process in Figure 2.11. This involves balances for the three components over each unit (we do* not *include the total mass balance since this is the sum of the three component balances). For the separator, we assume that a fraction x of B from the reactor goes to stream 5, and assume ideal separation for the other two components. We further assume that the recycle fraction is f.*

The mass balance equations are: Mixing point. *3 mass balances (stream 2 = stream 1 + stream 6):*

$$n_{A,2} = n_{A,1} + n_{A,6}$$

$$n_{B,2} = n_{B,1} + n_{B,6}$$

$$n_{N,2} = n_{N,1} + n_{N,6}$$

Reactor. *3 mass balances (stream 3 = stream 2 + generated by reaction)*

$$n_{A,3} = n_{A,2} - \xi$$

$$n_{B,3} = n_{B,2} + \xi$$

$$n_{N,3} = n_{N,2}$$

Separator. *3 mass balances (stream 4 + stream 5 = stream 3):*

$$n_{A,4} + n_{A,5} = n_{A,3}$$

$$n_{B,4} + n_{B,5} = n_{B,3}$$

$$n_{N,4} + n_{N,5} = n_{N,3}$$

Splitter. *3 mass balances (stream 5 + stream 7 = stream 5):*

$$n_{A,6} + n_{A,7} = n_{A,5}$$

$$n_{B,6} + n_{B,7} = n_{B,5}$$

$$n_{N,6} + n_{N,7} = n_{N,5}$$

We have also some additional information that must now be stated in equation form. First, for the reactor, we have the equilibrium relationship:

$$K = n_{B,3}/n_{A,3}$$

For the **separator**, *there are 3 specifications for the separation of each component:*

$$n_{A,5} = f_A n_{A,3}; \quad f_A = 1$$

$$n_{B,5} = f_B n_{B,3}; \quad f_B = x$$

$$n_{N,5} = f_N n_{N,3}; \quad f_N = 1$$

For the **splitter**, *it is given that the 3 components split the same way (with recycle fraction f):*

$$n_{A,6} = f \cdot n_{A,5}$$

$$n_{B,6} = f \cdot n_{B,5}$$

$$n_{N,6} = f \cdot n_{N,5}$$

In all there are 19 equations. How many unknowns are there? Stream 1 ($n_{A,1}$, $n_{A,2}$, $n_{A,3}$) is given and we further assume that the values of K (the equilibrium constant), f (the recycle fraction) and x (the separation factor for B) are given. Thus, we have 19 unknowns: 6 unknown streams (stream 2 to stream 7) each with 3 unknown amounts, and in addition the unknown extent of reaction ξ. In summary, we have 19 equations and 19 unknowns, so if we assume that the 19 equations are independent (which they are), the system of equations has a unique solution.

Comment: In our "hand calculations" (page 57), f was not given, so we had 20 unknowns. But on the other hand, we specified $n_{A,7} = 9$ mol/s, so that we had 20 equations.

In general there are three ways of solving such problems:

1. *Simultaneously:* Solve all the equations as one large set of equations. This is the most general method, but it requires that we have an equation solver, for example we may use MATLAB. Today, most commercial process simulators use this method.
2. *Sequentially:* Start with the feed and proceed sequentially through the process. If we encounter any unknown streams, then we make a guess of their value, for example, by setting them equal to zero. We may then iterate until we achieve convergence where the answer no longer changes. This method is commonly used when using a spreadsheet (see below).
3. *By hand:* It is here very important to find a "smart" point to begin the calculations. This can, for example, involve making balances over the whole process (as we did in the example above, page 56-58) or assuming a basis somewhere. Often, it is smart to begin with the reactor, for example, by choosing a basis of 100 mol/s reactor feed. If this basis is not consistent with other given information, we can later scale all the streams.

2.6 Use of spreadsheet program

Here, we solve the equations in Example 2.11 (continued) sequentially using a spreadsheet program. The example also provides an introduction to Excel.

Example 2.11, further continued. Spreadsheet solution (and an introduction to Excel), see Table 2.1.

For the given example we use the following sequential solution procedure:

1. *We start with stream 1 (given).*
2. *We assume, initially, that the recycle (stream 6) is 0 mol/s (this is correct for the case without recycle, but not for the case with recycle).*
3. *We now know stream 2 and with the assumption of equilibrium at the exit of the reactor we can determine the extent of reaction ξ and the reactor product (stream 3).*

Table 2.1: Excel spreadsheet for mass balance: Without recycle

	A	B	C	D	E	F	G	H
1			exercise w/ SPREADSHEET					
2		Data						
3	Feed	100						
4	Composition	0,97	A					
5		0	B					
6		0,03	N					
7	equilibrium cont	0,3911						
8	x (fract B str5)	0						
9	f (recycle. fract.)	0,96						
10								
11								
12		Stream 1	Stream 2	Stream 3	Stream 4	Stream 5	Stream 6	Stream 7
13	Amount A	97	97	69,72899	0	69,72899	66,93983	2,789159
14	Amount B	0	0	27,27101	27,27101	0	0	0
15	Amount N	3	3	3	0	3	2,88	0,12
16	Total	100	0	100	27,27101	72,72899	69,81983	2,909159
17								
18	Extent of reaction			27,27101				

4. *With the given data for the separator, we can calculate streams 4 and 5. We will consider, initially, the case with ideal separation ($x = 0$).*
5. *With the given recycle fraction f, we find stream 7 and a "new" stream 6.*
6. *We then return to step 3 and find a new stream 2; from this we calculate a new stream 3, etc. The iterations continue until it converges to the steady state solution, that is, until n_6 no longer changes (this can take many iterations, especially if f is close to 1).*

This procedure is used in the Excel spreadsheet described in the following.

START OF EXCEL MINI-COURSE

See Table 2.1.

1. Write the title in cell C1 (For spreadsheet novices: Click first on cell C1 with the mouse. This will highlight the border around the cell. Enter the text (do not be concerned if the text continues outside the cell) and press ''Enter.''
2. Enter the process specifications: Begin by entering ''Data'' in cell B2, and continue filling data into the cells in rows 3 to 9. Note that in my version of Excel, a comma is used as a decimal point (see Table 2.1), that is, I enter 0,97 instead of the ''normal'' 0.97 (the decimal point standard can be changed by going to Tools, Options, International).
3. Fill in the text in cells A13-A16 and in cells B12-H12 (Stream 1 to 7). We now formulate the mass balances.
4. Begin by calculating the amount of A in stream 1 (cell B13). You can do this by entering ''=B3*B4'' in cell B13, that is, by entering cell addresses. But as all veteran spreadsheet users know, it is quicker and simpler to click on the individual cells you are referring to instead of entering cell addresses. So click instead on cell B13 and enter ''=''. This tells the spreadsheet that you are about to enter a formula. Then choose B3 (total amount of feed) with the mouse (the letters ''B3'' will now appear in cell B13). Enter the symbol for multiplication, ''*''. Then choose cell B4 (fraction A) with the mouse. The formula is now complete; finish by pressing ''enter.''
   The number 97 should now appear in cell B13. Do the same with components B and N in stream 1 (cells B14 and B15).
5. Now, we want to enter the total amount of stream 1 in cell B16. It is of course equal to B3, but do not use this. Instead, we want to set cell B16 equal to the sum of the three cells above (the reason for this is that we will later copy this to all the other total amounts). This can be done by entering = in cell B16. The text SUM will now appear above cell A1. Click on this. The program will now propose to add the three cells above -- reply OK to this.

   A. Without recycle.
6. Go to stream 6 and fill in 0's for the amounts of A, B and N in stream 6 -- we assume

for now that we have no recycle.

7. The formula for the total amount of stream 6 in cell G16 can be copied from cell B16 by using ''Copy'' (ctrl-C) and ''Paste'' (ctrl-V). You might as well continue doing this for the total amount also of the other five streams.
8. Stream 2 is the sum of stream 1 and stream 6. You can first go to A in stream 2 (cell C13) and enter = and then select A in stream 1 (cell B13), enter +, select A in stream 6 (cell G13), and press enter. The resulting formula can be copied also for components B and N in the two cells below (for example by using ctrl-c and then ctrl-v twice). 97, 0, 3, 100 should now appear for stream 2.
9. We are now ready to calculate stream 3, which is the reactor product. We know the feed (stream 2) and the equilibrium constant K, so in principle this should be fine. We wish to use the extent of reaction ξ as an (auxiliary) internal variable. We have from the mass balance over the reactor that $n_{A,3} = n_{A,2} - \xi$; $n_{B,3} = n_{B,2} + \xi$. At the exit of the reactor we have equilibrium, $n_{B,3}/n_{A,3} = K$. By combining these two equations we are able to derive $\xi = (K \cdot n_{A,2} - n_{B,2})/(1+K)$. With the given cell addresses this is equal to (B7*C13 - C14)/(1+B7). Enter this into the auxiliary cell D18. After pressing ''enter'' the number 27.27101 should appear.
10. You can now calculate stream 3: $n_{A,3} = n_{A,2} - \xi$, $n_{B,3} = n_{B,2} + \xi$, $n_{N,3} = n_{N,2}$. Enter these formulas by using the calculated value of ξ in cell D18 (for example ''=C13-D18'' in cell D13). The numbers 69.73, 27.27, 3, 100 should now appear for stream 3.
11. Stream 5 contains: all A from stream 3 (''=D13'' in cell F13) , the fraction x of B from stream 3 (=D14*B8 in cell F14) and all N from stream 3. Since x=0, 69.73, 0, 3, 72.73 should appear for stream 5.
12. Stream 4 = Stream 3 - Stream 5 (''=D13-F13'' in cell E13; and a copy of this in the two cells below). It should appear 0, 27,27, 0, 27,27 in stream 4.

B. Let us now continue with recycle included.

13. Stream 6 (recycle) is a fraction f of stream 5 (''=F13*B9'' in cell G13). Excel will display an error message about a ''circular reference.'' This is because stream 6 affects itself via the calculation of stream 2 etc. To fix this click on Tools on the top menu and then Options and then Calculation. Choose manual, check iteration and set Maximum iterations to 1.
14. Now, try calculating stream 6 again (''=F13*B9'' in cell G13). You should not see any error message. Continue by filling in the corresponding equations for components B and N. It should appear 66.94, 0, 2.88, 69.82 for stream 6.
15. Stream 7 = stream 5 - stream 6 (''=F13-G13'' in cell H13 etc.). It should appear 2.79, 0, 0.12, 2.91 for stream 7.
Your spreadsheet should now be as shown in Table 2.1.
16. You can now continue calculating by pressing the F9 key. Each time you press F9, the new value for stream 2 will be inserted and a new calculation will be carried out. Keep F9 pressed until the calculation converges (with x=0 and f=0.96 we should end up with a total of 12.00 [mol/s] in stream 7 (purge) and 288.00 [mol/s] in stream 6 (recycle); **see the table on page 57.**

We will now make use the main strength of a spreadsheet: How easy it is to change the data and recalculate. Enter for example x=0.2 (20% of B goes in stream 5). By keeping F9 pressed, you will find that the amount purged (stream 7) is 14.75 [mol/s] and the amount recycled (stream 6) is 354.03 [mol/s].

END OF EXCEL MINI-COURSE

Solve the following on your own, using the spreadsheet from the previous Example:

Exercise 2.11* *(a) What is the purge and recycle if $x = 0.5$ and $f = 0.96$?*
(b) What is the purge and recycle if $x = 0.5$ and $f = 0.99$?
(c) What is the purge and recycle if the feed contains 99% A and 1% N ($x = 0, f = 0.96$)?

Solve the following on your own, using hand calculations ($x = 0$):

Exercise 2.12* *(a) What is the recycle (stream 6) if the loss of A is only 1 mol/s? (b) What is the recycle (stream 6) if the feed does not contain any inert (that is, the feed is 100 mol A/s) and we desire 0 loss of A?*

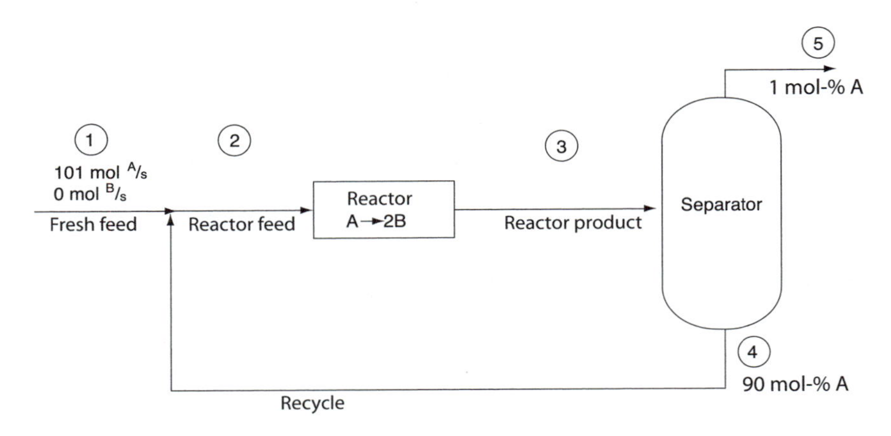

Figure 2.12: Flow sheet for exercise process with reactor, separation and recycle

Exercise 2.13* *Consider the process in Figure 2.12 where the "heavy" recycle stream (stream 4) is 2000 mol/s. What is the reactor conversion for each pass ($X_A = (n_{A,2} - n_{A,3})/n_{A,2}$) ?*

Exercise 2.14 *Consider the process in Figure 2.12 where the reactor conversion for each pass is 3% (i.e., $X_A = (n_{A,2} - n_{A,3})/n_{A,2} = 0.03$) ? What is the recycle?*

Exercise 2.15 *Consider the process in Figure 2.12 where we have equilibrium at the exit of reactor with $K_x = x_B^2/x_A = 0.1$. What is the recycle flow?*

You may not get very excited about the reactions $A \rightarrow 2B$ or $A \rightarrow B$ used in the above examples, but this avoids confusing the issue. The approach for real reactions is very similar; and we will take a closer look at balances for systems with chemical reactions in Chapter 3.

2.7 Examples of recycle without reaction

Let us now take a look at an example of recycle without reaction.

Example 2.12 Continuous crystallization process with recycle. *5000 kg/h of an aqueous solution with 20% (all numbers are in weight%) of a potassium salt (abbreviated K) is mixed with a recycle stream. This is sent to an evaporator where water is removed so that the stream now contains 35% K. This stream is sent to crystallization/filtration. The filtrate (liquid) with 30% K is recycled. The filter cake (product) consists of mostly solid crystals (K), but 1/26 of the product is remaining liquid (filtrate). Determine all the stream amounts.*

Solution. *The flow-sheet is sketched in Figure 2.13 where we have introduced the (unknown) mass flows m_i [kg/h] for five streams. We have not introduced a symbol for the combined feed to the evaporator because this is strictly speaking unnecessary if we choose the control volume around the evaporator including the mixing point as shown in the figure.*

Note that in the flow sheet we have split the product (filter cake) in two imaginary streams: Crystals (m_4) and filtrate (m_5). It is specified that the product consists of 1/26 liquid, that is, we have the following extra piece of information (which is not shown on the flow sheet)

$$m_5 = (m_4 + m_5)/26 \quad \text{or} \quad m_4 = 25m_5 \tag{2.21}$$

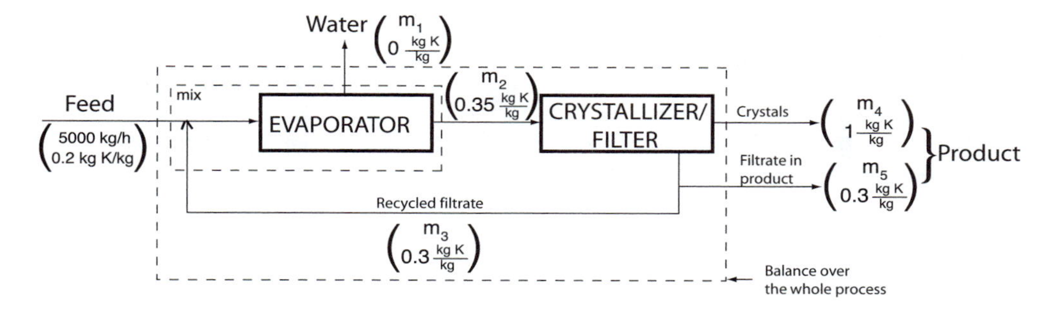

Figure 2.13: Continuous crystallization process with recycle

We have two units and can formulate balances around each of these. We choose the two control volumes as the whole process and the evaporator including the mixing point (we could have replaced one of these with the crystallizer; try this). For each of the two control volumes we can set up a total mass balance and a component mass balance for K.

Total balance and component balance for K ("Out = In") over the whole process:

$$5000 = m_1 + m_4 + m_5 \quad [kg/h] \tag{2.22}$$

$$0.20 \cdot 5000 = 1 \cdot m_4 + 0.30 m_5 \quad [kgK/h] \tag{2.23}$$

Total balance and component balance for K over the evaporator including the mixing point:

$$5000 + m_3 = m_1 + m_2 \quad [kg/h] \tag{2.24}$$

$$0.20 \cdot 5000 + 0.30 \cdot m_3 = 0 + 0.35 m_2 \quad [kgK/h] \tag{2.25}$$

This gives 5 equations with 5 unknowns. The five equations can be solved simultaneously.

Alternatively, for hand calculations, we should start at a point where we have sufficient information. If it is difficult to begin, it is often smart to choose a basis for a stream in the recycle loop that has a known composition (and then rescale all the streams at the end to match the given feed stream). For example, we could choose $m_2 = 100$ kg as a basis, and start applying balances from this point.

However, in our case, we can avoid this by starting with the K-balance over the whole process; from (2.21) we have $m_4 = 25 m_5$ which inserted into the total balance for K (2.23) gives $m_5 = 39.53$ kg/h and $m_4 = 988.14$ kg K/h. The total balance over the whole process (2.22) then gives $m_1 = 3972.3$ kg/h. We are then left with two equations, (2.24) and (2.25), in two unknowns, m_2 and m_3. The solution is $m_2 = 13834$ kg/h and $m_3 = 12806$ kg/h. In summary, we have then:

	Feed	Stream 1	Stream 2	Stream 3	Stream 4	Stream 5
m [kg/h]	5000	3972	13834	12806	988	40
K [weight %]	20%	0%	35%	30%	100%	30%

Comments.

1. *We should always check if the solution is correct. For example, we can check the water balance over the whole process*

$$0.8 \cdot 5000 = m_1 + 0.7 m_5 \quad [kg/h]$$

(which is correct) and the water balance over the crystallizer

$$0.65 m_2 = 0.7(m_3 + m_5) \quad [kg/h]$$

(which also is correct).

Extra exercise: *Check the total balance and K-balance over the crystallizer.*

2. *The feed to the evaporator (external feed + stream 3) is 17806 kg/h and contains 27.2 % K (show this!).*
3. *The recycle stream is large. This is because relatively little K is removed in the crystallizer (the amount of K in the liquid is only reduced from 35% to 30%).*

Try to do the following on your own:

Exercise 2.16* Another recycle problem. *2 kg/s of an aqueous solution with 10 weight% of a salt (S) is mixed with a recycle stream. The combined stream is sent to an evaporator where pure water is removed as gas so that the remaining liquid contains 45% S. This is then crystallized and sent to a filter where pure salt (S) is removed while the filtrate (liquid), which contains 20 % S, is recycled.*

(a) Sketch a flow sheet and formulate the mass balances. Find the amount of recycle.

(b) Assume now that the feed in addition contains 1 weight% of another salt (T). The salt T always remains in the water solution (liquid phase), so we purge (drain) 0.04 kg/s of the recycled stream in order to avoid accumulation of T. Otherwise, all the data given above is the same (e.g., the feed is 2 kg/s and contains 10% S, etc.) What is fraction of T in the purge? Calculate the amount recycled.

Batch processes with recycle

Recycle can also be used with batch processes. For example, some residue can be recycled to the next batch, and the assumption of "steady state" means that this has been done an infinite number of times so that the composition and amount of the residue has reached its steady state value.

Example 2.13 Batch process with recycle. *A factory produces a medicine by taking 1 g of active component A and mixing it with water to obtain 1000 l. The product is then drained, but there will always be a rest of 100 l which is "recycled" to the next batch. The first batch has a composition of 1 mg/l; the second batch 1.1 mg/l; the third batch 1.11 mg/l; the fourth batch 1.111 mg/l and so on. (Exercise: show this!)*

Comment: *The composition will in this case quickly reach a steady state value even though we never reach the steady-state exactly. We can check the calculations by formulating a steady state mass balance with the assumption that nothing is accumulated. We can make a total balance for component A at steady state: The product of 900 l in each batch must contain 1 g A, that is, the composition is 1 g/900 l = 1.111 mg/l – which is correct.*

2.8 Flash calculations

Flash calculations are used for mass balance problems with equilibrium between the vapor and liquid phases. This topic is discussed in more detail in Section 7.5 (page 189), but the material can be covered at this point, if desired. A short summary is given here. We write the vapor-liquid equilibrium (VLE) for a multicomponent mixture on K-value form,

$$y_i = K_i x_i$$

where y_i and x_i are the mole fractions of component i in the vapor and liquid phases, respectively. The "K-values" K_i depend on temperature T, pressure p and composition (x_i and y_i), but in many cases the dependency on composition can be neglected. For example, this is the case for ideal mixtures which follow **Raoult's law**:

$$K_i = p_i^{\text{sat}}(T)/p$$

where the **saturation pressure** $p_i^{\text{sat}}(T)$ is the equilibrium vapor pressure of pure component i at temperature T. A simple (and important) flash is to specify p and T (pT-flash). It is simple because the K_i's are then constant for an ideal mixture. It is important, because a pT-flash corresponds to the common process where a partially evaporated feed stream (with flowrate F and mole fractions z_i) is separated into a liquid (L) and vapor product (V); see Figure 2.14. For each of the N_c components we

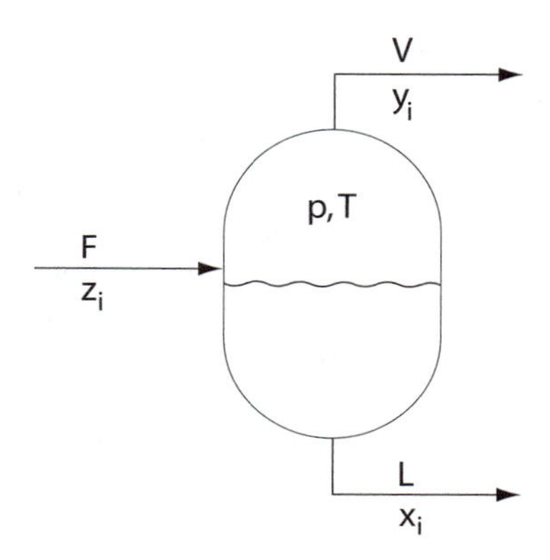

Figure 2.14: Flash tank

can write a material balance

$$Fz_i = Lx_i + Vy_i \quad [\text{mol i; mol i/s}] \tag{2.26}$$

Similarly, the total mass balance is

$$F = L + V \quad [\text{mol; mol/s}] \tag{2.27}$$

Substituting $y_i = K_i x_i$ into (2.26) gives $Fz_i = Lx_i + VK_i x_i$, and solving with respect to x_i gives $x_i = (Fz_i/(L + VK_i)$. Introducing $L = F - L$ (the total mass balance) gives

$$x_i = \frac{z_i}{1 + \frac{V}{F}(K_i - 1)}$$

Here, we cannot directly calculate x_i because the vapor split V/F is not known. To find V/F we may use the fact that the liquid mole fractions must sum up to 1, $\Sigma_{i=1}^{N_c} x_i = 1$, or alternatively that the vapor mole fractions must sum up to 1, $\Sigma_i y_i = \Sigma_i K_i x_i = 1$. However, it has been found that an even better choice is to use the combination $\Sigma_i(y_i - x_i) = 0$, because it results in an equation with good numerical properties; this is the so-called **Rachford-Rice** flash equation,

$$\Sigma_i \frac{z_i(K_i - 1)}{1 + \frac{V}{F}(K_i - 1)} = 0 \tag{2.28}$$

which for given K_i's is a monotonic function in V/F and is easy to solve numerically. A physical solution must satisfy $0 \leq V/F \leq 1$. After having obtained V/F from (2.28), we can find x_i and y_i. For more details on the pT-flash, including examples and exercises, see page 193. Bubble point and dew point calculations are discussed on page 189 and onwards.

2.9 Summary: Procedure for deriving balance equations

In Table 2.2 (page 67) we present a systematic procedure for deriving balance equations. Step 1 is to obtain an overview of the process by making a flow sheet for the process. On the flow sheet, the streams are represented by lines while the equipment (unit operations) are represented by boxes or are given special symbols (some are shown in Chapter 1.6). In order to gain an overview one should write, for each stream, the total amount, composition, temperature and pressure. For gases, the composition is usually given on a mole basis (which is equivalent to volume basis for ideal gas). For liquid processes, also a mass basis is common. In some cases where the amount of water varies, the composition is given on a pure water (dry) basis. For example, this is common when giving the composition of air. Note that we show the nominal values (normal operation) for pressure, temperature and flow on the flow sheet. Do not confuse this with mechanical constraints, for example for maximum pressure or temperature, which are sometimes indicated on some units.

After having drawn the flow sheet, enter stream data and quantify any other information in equation form. The next step is to formulate the balance equations, which is usually elementary accounting. However, it is important to work systematically, especially on larger and more complicated processes.

To quantify flowrates, an initial **basis** for a selected stream is often chosen, for example 1 kg or 100 mol/s. If necessary, we can later **rescale** (up or down) all the streams to the desired quantity. Mass, energy, volumes, etc. (all extensive variables) will scale with the same factor provided the efficiencies of the units remain constant.

2.10 Degrees of freedom and solvability

In principle, it is easy to formulate and solve the mass and energy balance equations, but in practice it is not always so easy to tell whether we have enough information to solve the problem, or if we possibly have duplicated some pieces of information. In this section, we try to get some insight into these issues.

In general, the model equations for a process are formulated based on the following information:

1. **Mass balances** (1 for each of the N_c components for each control volume).
2. **Energy balance** (1 for each control volume).
3. **Stoichiometry of chemical reactions** (which we use in the component mass balances).

Table 2.2: Procedure for deriving balance equations

---

1. **Make a simple flow sheet of the process including all streams and unit blocks.** (This give an excellent overview!)

2. **Choose a basis (if necessary).** This means that one specifies the amount of one stream, for example, the feed rate if is not given. A "clever" choice of basis can often simplify the calculations, and we can later rescale the flows (see item 10).

3. **Enter the stream data and other given information on the flow sheet and provide symbols for the unknown variables.** In general, a stream with N_c components is specified by giving the amounts of each component plus two specifications, that is, in total $N_c + 2$ independent pieces of information. Typically, we specify temperature and pressure, that is,

$$\text{Stream data} = \begin{bmatrix} \text{total amount} \\ \text{composition} \\ \text{temperature} \\ \text{pressure} \end{bmatrix}$$

   It is then simple to identify missing data.

   - We have here chosen to specify the total amount and composition, but we could instead specify the amounts of each of the N_c components.
   - Enthalpy is often specified instead of temperature because, in general, it is a more unambiguous specification (see page 332).
   - If we only are interested in mass balances, there is usually no need to specify temperature and pressure.

4. **Quantify other given information not shown on the flow sheet.** This can be data for chemical reactions (for example conversion, equilibrium constant or reaction rate) and data for separation units.

5. **Use consistent units.** For the mass balance this means choosing mass or molar basis. To convert between various specification, one typically needs data for densities and molar masses.

6. **Check if the problem is solvable.** You should not go too far here, but a quick analysis (see page 70) is recommended before you start defining control volumes and formulating balances.

7. **Define control volumes.** One can usually make control volumes around each unit, but there are other possibilities:

   - Mixing points are often combined with the following ("downstream") unit.
   - Splitters are often combined with the previous ("upstream") unit.
   - Total balances are often formulated for the whole process, which then replaces the balances above one of the single units.

8. **Formulate balances** for total mass, component mass, energy, etc. for each control volume. Be careful to avoid extra dependent equations (which can be derived from other equations and hence do not contain extra information); for example, the total mass balance is equal to the sum of all component balances.

9. **Solve the equations.** Start by checking whether the equation system can be solved, that is, make sure that the number of **independent** equations = number of unknowns (see page 66).

10. **If necessary, scale the solution to the desired throughput.** This is done by applying the same scale factor on all extensive variables (flows, heat duties, etc.), which assumes that the efficiencies of the process units are independent of scaling.

---

4. **Process specifications** (for example given feed composition, given split fractions, given composition of streams, equilibrium constants, reactor temperature, product purities, etc.).
5. **Physical constraints and definitions** (for example that mole fractions sum up to 1, and that flow rates are positive).
6. **Physical quantities and laws** (constitutive equations; for example ideal gas law).

How many independent equations can we formulate? For mass balances, the following applies:

- For a process with n units (blocks) and N_c components, we can formulate $n \cdot N_c$ independent mass balances. This also applies for cases with chemical reactions.

Furthermore we can formulate

- 1 energy balance for each unit (typically used to determine temperature) (see Chapter 4).
- 1 force balance/mechanical energy balance for each unit (typically used to determine pressure) (see Chapter 9).

Do not fool yourself by including "extra" balances that do not contain any new information (but of course they can be used to check the calculations):

- The balance for total mass is the sum of the balances for the N_c components.
- "The overall (total) balance" over the whole process (for a component or total mass) is the sum of the corresponding balances for all the n units.

Example 2.12 continued. *Let us return to the recycle example on page 62. Here we have $N_c = 2$ components (water and potassium salt) and $n = 3$ units (mixer, evaporators and crystallizer/filtration), so we can formulate $n \cdot N_c = 6$ independent mass balances. However, in the example we combined the mixer and evaporator. Then, the evaporator feed is not a variable and we need only 4 balances.*

Do we have enough information to find a solution? This is an important question which is not always easy to answer. Here we will describe two ways: the mathematical "equation-counting" method and a "faster" method based on physical insight.

2.10.1 The equation-counting method

In principle, one can determine whether a problem is solvable by counting the number of independent equations and subtracting the number of unknown variables. The result is the number of degrees of freedom

$$n_{\text{DOF}} = n_{\text{unknowns}} - n_{\text{independent equations}}$$

and we have that

1. If $n_{\text{DOF}} = 0$ there are no remaining degrees of freedom. That is, we have as many equations as we have unknowns and the problem is solvable (provided the equations are truly independent).

2. If $n_{\text{DOF}} > 0$ then there are more unknowns than independent relations between them. We then need n_{DOF} additional pieces of information in order to find a unique solution to the problem, otherwise the problem is under-specified and has an infinite number of solutions.
3. If $n_{\text{DOF}} < 0$ then there are more independent equations than unknowns, that is, the system is over-specified. This is most likely because the equations are not really independent, that is, we have somehow used the same information twice. But there are also cases where the information or the requirements are inconsistent, so that the system is unsolvable.

Example 2.11 (page 56) continued. *There are three components (A, B and N) and by setting up mass balances for the four units (mixing point, reactor, separator and splitter) we derive 12 mass balance equations. We have, in addition, 1 equilibrium equation, 3 equations for the given split fractions in the ideal separator and 3 equations for the splitter, that is, in all 19 equations. With the given feed stream there are 20 unknowns (the amount of A, B and N in the 6 streams, the extent of reaction for one reaction and, in addition, the split fraction). We thus find that* $n_{\text{DOF}} = n_{\text{unknown}} - n_{\text{indep. equat.}} = 20 - 19 = 1$, *that is, we lack 1 piece of information. Thus, if we for example specify the loss of A, the equation system is in principle solvable (which we can verify by solving the equations).*

However, in practice, it may be difficult to determine the number of degrees of freedom using this approach. First of all, the method can be somewhat time-consuming. It is also easy to make mistakes. For example, even with $n_{\text{DOF}} = 0$, it might well be the case that we have dependent equations (with the same information) so that the system is actually not solvable. The following alternative simplified method may therefore be helpful.

2.10.2 Simplified "quick" physical analysis

An alternative way of determining whether the system is solvable is to use Table 2.3 (page 70). Here insight about various process units is used to check whether we have enough information to solve the problem.

Example 2.11 (page 56) continued (quick analysis of degrees of freedom). *We want to determine whether the problem is solvable by using the method in Table 2.3. We know the feed stream. No further information is needed for the mixing point. In the reactor one independent reaction takes place, and we have given the equilibrium constant (so this is OK!). There is given sufficient information for the separator to calculate how all the components split into the two products. For the splitter we need one piece of information and we have given the amount A lost in the purge. From this quick analysis we conclude that the problem is indeed solvable.*

Example 2.14 Paper machine with fiber recycle. *A simplified sketch of a paper machine with 7 units (blocks) is shown in Figure 2.15. Here, a block without a symbol indicates a simple stream split or mix, while a block with a line is a separator between the water-rich phase (lower stream) and the fiber-rich phase (upper stream). In the paper machine itself, large amounts of water are required, and large recycles are necessary to avoid loss of fiber to the drain (stream J).*

Assume that each stream has two "components:" water and fiber. We then need to keep track of two variables in each stream, which we here choose as the flow (kg total/s) and fiber fraction c_i *[kg fiber / kg total]. Assume a steady state situation.*

Table 2.3: Quick analysis of whether a problem is solvable

In addition to the equations for mass and energy balances for each process unit, we need the following pieces of information to solve the problem (and compute all stream data):

- **Feed streams**: We must know "everything" – that is, the amount of each component and, if necessary, the temperature, pressure and phase distribution.
- **Mixer**: Do not need more data (as mass and energy balances give everything).
- **Reactor**: Need one piece of information for each *independent* reaction, for example, a given conversion, extent of reaction or equilibrium constant of each independent reaction. (See page 88 to determine the number of independent reactions.)
- **Splitter** with two outlet streams with the same composition as the feed: Need one piece of information (the split factor f).
- **Separator** (distillation, flash, crystallization, etc.) with two outlet streams with different compositions: Need N_c pieces of information where N_c is the total number of components, for example the split fraction f_i for each component. If we have a separator with inflow (0) and two outlet streams (1 and 2), we then have for any component i

$$m_{i,1} = f_i m_{i,0}; \quad m_{i,2} = (1 - f_i) m_{i,0} \quad [\text{kg; kg/s; mol; mol/s}]$$

- **Heat exchanger**: Need one piece of information in order to determine the amount of heat transferred.
- **Compressor/turbine/pump**: Need one piece of information to determine the supplied/performed work (in addition to the efficiency of the equipment).

A **quick analysis** then is:

> *If we lack one or more of these pieces of information, then they must be replaced by the same number of* other *independent pieces of information.* For example, the composition of a product stream can be given instead of the composition of a feed stream, or instead of information about splits fractions in a separator.

Comments:

1. **Pressure.** In addition to what is listed above we need information about the pressure in all units where this is necessary for the calculations.
2. **Dynamics.** For dynamic computations, where time enters as a variable, we also need to know the initial state of the system, as expressed by the holdup (inventories) of all extensive variables (inventory of all components, masses and energy) in all units. This is not necessary for steady state calculations where all inventories are assumed constant.

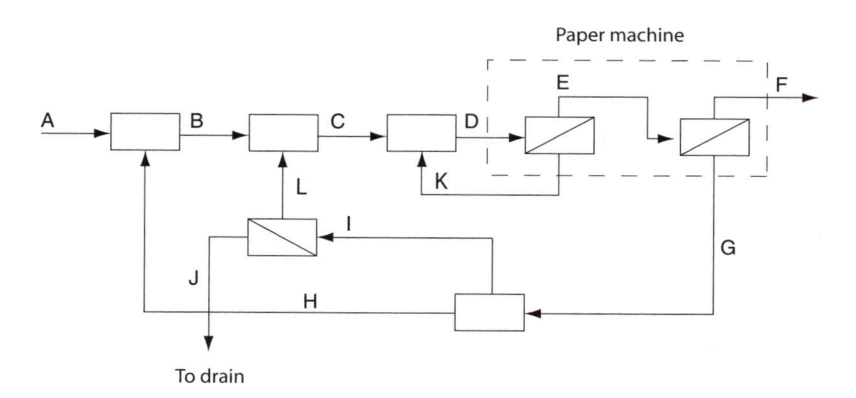

Figure 2.15: Mass flows around a paper machine

How much information do we need to solve the problem?

A quick analysis gives that we need 9 pieces of information: 2 for the feed, 2 for each of the 3 separators and 1 for the splitter (no information is needed for the 3 first mixers).

We get the same answer from the equation-counting method: There are 24 unknowns (2 in each of the 12 streams), 14 mass balance equations (2 for each for the 7 units), and for the splitter we know that $c_H = c_I$ (1 equation), that is, we have 15 equations. We then lack 24-15=9 pieces of information.

Exercise 2.17 * **Paper machine with fiber recycle (continued).**

(a) Formulate all the mass balances.

(b) Find all the stream amounts [kg/s] and fiber fractions [kg/kg] when we have the following 9 specifications: $A = 100, c_A = 0.1, F = 49, c_F = 0.2, H = 200, c_H = 0.015, L = 20, K = 5000, c_K = 0.002$. *What is the "wire retention"* $R_W = c_F F / c_D D$?

Example 2.15 Methanol process with recycle and purge. *Methanol is formed in a gas phase reactor from a feed with CO_2 and hydrogen according to the reaction*

$$CO_2 + 3H_2 \rightarrow CH_3OH + H_2O$$

The "fresh" feed (stream 0) consists of a stoichiometric mixture of CO_2 and hydrogen plus 5 mol% of some inert component (I). The product stream from the reactor is cooled so that all of the methanol and water is condensed and can be removed as liquid (stream 3) in the separator. The remaining gas is recycled to the reactor with the exception of a small purge stream which is necessary in order to avoid accumulation of inerts. The amount of purge is adjusted so that the combined feed to the reactor contains 20 mol% inerts. The conversion in the reactor (for each pass) is 60%. Calculate the required amount of fresh feed when the desired production rate of pure methanol is 2500 t/d (tons per day).

Analysis of the problem. *The flow sheet is shown in Figure 2.16. Do we have enough information? Let us do a quick analysis using Table 2.3:*

For the feed (stream 0) the composition is known (the ratio CO to H_2 is $1:3$ *since the feed is stoichiometric, and there are 5% inerts). The feed amount is not given, but instead the production rate of methanol is given (so we have enough information so far). For the mixing point, we need no more information (as mass balances give everything). In the reactor, 1 (independent) reaction takes place, so we need 1 piece of information and the conversion is indeed specified. In the separator, we must be able to determine the split of each component,*

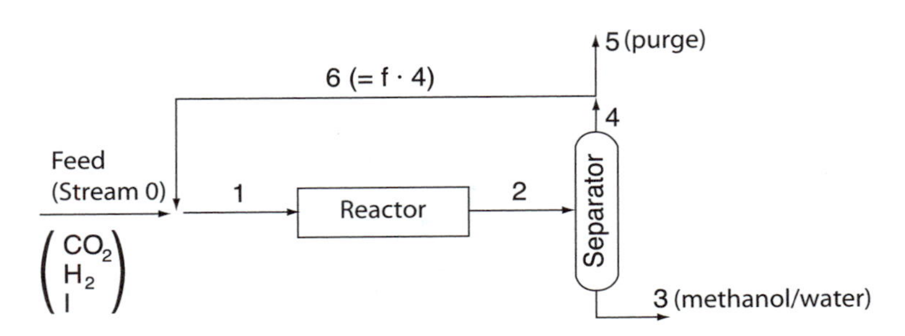

Figure 2.16: Flow sheet for a methanol process

and this is OK since it is given that the split is "perfect." For the purge, which is stream split, we need information about the split fraction. This is not specified but instead it is specified that the reactor feed contains 20% inerts. We can therefore, based on a quick analysis, conclude that we have enough information to solve the problem.

In principle, all that is needed is to formulate mass balance for the 5 components (CO_2, H_2, CH_3OH, H_2 and inert) for the 4 blocks (mixer, reactor, separator and purge), add the specified information and solve the problem. But we have many equations, 20 from the mass balances alone (although some of them "disappear" since ammonia and water are not recycled), and solving them is not a trivial matter unless a computer is used. Here, we want to use hand calculations, and the "trick" is to find a "smart" basis, that is, find a point in the process where we have enough information to start the calculations. Since one can relatively easily determine the composition of the reactor feed (stream 1), the easiest way is probably to start here, that is, we choose as a **basis** $n_{1,tot} = 100$ *[mol] (we could as well used units [mol/s]).*

Solution. *Since there is only one reaction and the feed is stoichiometric, and since CO_2 and H_2 split in the same way in the separator, the ratio between CO_2 and H_2 will remain stoichiometric in all streams, that is, in all streams we have*

$$n_{H_2} = 3n_{CO_2}$$

Since stream 1 contains 20% inert we therefore find for **stream 1** *(note that chosen $n_{1,\text{tot}} = 100$ mol):*

$$n_{CO_2,1} = 20, n_{H_2,1} = 60, n_{I,1} = 20 \quad [mol]$$

It is given that the conversion in the reactor is $X_A = 0.6$ (where A is CO_2 or H_2). More precisely, the definition of conversion is

$$X_A = \frac{\text{mol A reacted}}{\text{mol A supplied}} = \frac{n_{A,2} - n_{A,1}}{n_{A,1}}$$

or

$$n_{A,2} = n_{A,1}(1 - X_A)$$

Applying this for CO_2 and H_2 gives for **stream 2**

$$n_{CO_2,2} = 20 - 20 \cdot 0.6 = 8 \ mol$$

$$n_{H_2,2} = 60 - 60 \cdot 0.6 = 24 \ mol$$

Correspondingly, reactor mass balances for CH_3OH and H_2O give

$$n_{CH_3OH,2} = n_{H_2O,2} = 20 \cdot 0.6 = 12 \ mol$$

The mass balance for inert gives

$$n_{I,2} = n_{I,1} = 20 \ mol$$

*Water and methanol are condensed and removed as a liquid product (***stream 3***) and the remaining light components constitute* **stream 4**. *Mass balances for the separator give:*

$$n_{CO_2,4} = 8, n_{H_2,4} = 24, n_{I,4} = 20, n_{tot,4} = 52 \quad [mol]$$

A fraction f of stream 4 is recycled to give **stream 6**. *Mass balances for the stream split give*

$$n_{CO_2,6} = 8f, \ n_{H_2,6} = 24f, \ n_{I,6} = 20f, \ n_{tot,6} = 52f$$

We can now set up three mass balances around the mixing point (since we here have three components: CO_2, H_2 and I). However, we have already "used up" one of these balances since we fixed the stoichiometric ratio (H_2 to CO_2) in stream 1. We therefore only have two independent mass balances for the mixing point, which we choose to set up as total balance and the inert balance

$$\text{Total balance}: \quad n_0 + 52f = 100$$

$$\text{Inert balance}: \quad 0.05 n_0 + 20f = 20$$

*We then have 2 equations with 2 unknowns. Solution gives the recycle fraction $f = 0.862$ and a fresh feed amount $n_0 = 55.17 \ mol$. (***Comment***. If the reactor feed was not stoichiometric, we would have needed to introduce a parameter for the composition in stream 1 and would have ended up with three balance equations with three unknowns.)*

The mass balances with 100 mol in stream 1 as a basis are summarized in the table (all numbers in mol):

	Feed	*Stream* 1	*Stream* 2	*Stream* 3	*Stream* 4	*Stream* 5	*Stream* 6
CO_2	13.10	20	8	0	8	1.10	6.90
H_2	39.31	60	24	0	24	3.31	20.69
CH_3OH	0	0	12	12	0	0	0
H_2O	0	0	12	12	0	0	0
I	2.76	20	20	0	20	2.76	17.24
Sum	55.17	100	76	24	52	7.17	44.83

In order to get the actual amount of fresh feed, we need to scale all the numbers. The molar mass of methanol is $32 \cdot 10^{-3}$ kg/mol and we then find that 2500 t/d methanol corresponds to 904 mol/s, that is, the scale factor is $904/12 = 75.33$ (mol/s)/mol. The actual amount of fresh feed then is $55.17 mol \cdot 75.33 s^{-1} = 4156 \ mol/s$, and all the numbers in the table need to be multiplied by 75.33.

Example 2.16 Reactor with recycle (and a small warning). *The first Norwegian edition of this book presented as an introductory example an exercise which turned out to be under-specified, that is, it lacked information. As we will see there are 11 equations, but only 10 of them turn out to be independent, so we have $n_{\text{DOF}} = n_{\text{unknown}} - n_{\text{indep. equat.}} = 11 - 10 = 1$. This is not always easy to discover (no students reported anything suspicious to me!), so this example serves as a small warning.*

Flawed problem statement. *"Consider the continuous process in Figure 2.12 (page 62) where the amount of (unconverted) A in the product is 2 mol/s. A steady state situation is assumed. Formulate the equations and show that the solution given in the flow sheet in Figure 2.10 is correct."*

Solution to flawed problem statement. *The mass balances for components A and B give for the mixing point,*

$$n_{A,2} = n_{A,1} + n_{A,4} \qquad \text{[mol A/s]}$$
$$n_{B,2} = n_{B,1} + n_{B,4} \qquad \text{[mol B/s]}$$

for the reactor,

$$n_{A,3} = n_{A,2} - \xi \qquad \text{[mol A/s]}$$
$$n_{B,3} = n_{B,2} + 2\xi \qquad \text{[mol B/s]}$$

(where ξ is the extent of reaction and the factor 2 comes from the stochiometry, $A \rightarrow 2B$), and for the separator,

$$n_{A,4} + n_{A,5} = n_{A,3} \qquad \text{[mol A/s]}$$
$$n_{B,4} + n_{B,5} = n_{B,3} \qquad \text{[mol B/s]}$$

This gives 6 equations with 11 unknowns (amount of A and B in the 5 streams plus ξ). But we have 5 other pieces of information; 2 specifications for the feed and 1 for the product,

$$n_{A,1} = 101 \text{ mol A/s}, \quad n_{B,1} = 0, \quad n_{A,5} = 2 \text{ mol A/s}$$

together with 2 specifications for the separator,

$$\frac{n_{A,5}}{n_{A,5} + n_{B,5}} = 0.01, \quad \frac{n_{A,4}}{n_{A,4} + n_{B,4}} = 0.90$$

This gives 11 unknowns and 11 equations, that is, the problem seems to be solvable. Indeed, we find that the solution given in Figure 2.10 satisfies the equations (with $\xi = 99$).

However, as already indicated, there is a "small" problem here: the above equation system actually has an infinite number of solutions. This can be seen by trying to solve the equations in detail, rather than just checking that the solution given in Figure 2.10 satisfies the equations. After some substitution of variables, we end up with an equation of the kind $n_{A,4} = n_{A,4}$ or "$0 = 0$" (which of course is correct, but it has an infinite number of solutions). It turns out that we can freely choose the amount of the recycle stream (n_4) and always find a solution. For example, one possible solution is to have zero recycle ($n_{A,4} = n_{B_4} = 0$), where stream 2 is equal to stream 1 (the feed) and stream 3 is equal to stream 5 (the product).

If you know some linear algebra, you will see that the 6 mass balances and the 5 specifications give a set of linear equations which can be written in the form

$$Ax = b$$

where the vector x contains the 11 unknowns and A is a 11×11 matrix:

```
x = [nA1, nB1, nA2, nB2, nA3, nB3, nA4, nB4, nA5, nB5, xi]

A = [  -1    0    1    0    0    0   -1    0    0     0    0 ;
        0    1    0    1    0    0    0   -1    0     0    0 ;
        0    0   -1    0    1    0    0    0    0     0    1 ;
        0    0    0   -1    0    1    0    0    0     0   -2 ;
        0    0    0    0   -1    0    1    0    1     0    0 ;
        0    0    0    0    0   -1    0    1    0     1    0 ;
        1    0    0    0    0    0    0    0    0     0    0 ;
        0    1    0    0    0    0    0    0    0     0    0 ;
        0    0    0    0    0    0    0    0 0.99 -0.01    0 ;
        0    0    0    0    0    0  0.1 -0.9    0     0    0 ;
        0    0    0    0    0    0    0    0    1     0    0 ]

b = [0 0 0 0 0 0 101 0 0 0 2]'
```

Let us check that the solution given in the Figure 2.10 (page 55) satisfies these equations:

```
x0 =  [101   0   1001 100  902  298  900  100   2   198  99]'
A*x0 - b
% MATLAB answers: ans = 0-vector (so it's OK)
```

However, it is not the only solution. In general, as linear equation system has the solution $x = A^{-1}b$*, but in this case it turns out that the matrix is A is singular (and MATLAB issues a warning about this), that is, the 11 equations are not independent. This is confirmed by computing the rank of matrix A, which is 10 (and not 11 as it should to be full rank):*

```
% MATLAB:
x = inv(A)*b
% MATLAB answers:   Warning: Matrix is singular to working precision.
rank(A)
% MATLAB answers: ans = 10
```

Why is the problem statement flawed? *It is not immediately clear what the problem is, because all the information seems to be independent, but it is not. The problem is the following: We have specified* $n_{A,5} = 2$ *mol A/s. Since the feed is 101 mol A/s, it then follows that 99 mol A/s is converted, and the stoichiometry of the reaction* $A \rightarrow 2B$ *tells that 198 mol B/s is formed. Since there is only one exit stream, this B must end up in the product (stream 5). The fraction A in the product is therefore* $2/(2+198) = 0.01$*. In the problem statement, this is also given as a seemingly independent piece of information ("product with 1% A"), but it is not. If we for example change the specification for the amount A in the product from 2 mol/s to* $n_{A,5} = 3$ *mol/s, then we end up with two contradicting pieces of information, and the problem has no solution.*

In order to obtain a solvable problem we need to replace one of the specifications "product with 1% A" and "A in the product is 2 mol/s" by another piece of information. After a little thinking, we realize that we need some information that (indirectly) determines the amount of recycle (stream 4), for example, by specifying one of the following

- *Amount of stream 4 (Exercise 2.13).*
- *Conversion of A for each pass through the reactor (Exercise 2.14).*
- *Equilibrium constant for the reactor (Exercise 2.15).*

Exercise 2.18* *The equilibrium reactor with recycle in Example 2.11 (page 56) can be described by a set of linear equations* $Ax = b$*. Find A and b, and use MATLAB to find the solution* $x = A^{-1}b$*, and check that this is consistent with what we have previously found.*

2.11 Simulation versus design

Let us finally make a remark about the difference between simulation (analysis) and design in problem solving.

1. **Analysis / Simulation**. In *analysis*, the process and its equipment is given (so the equations are given) and we want to analyze the relationship between the variables; usually the stream data. All examples we have considered so far all come under the heading of analysis. A special case of analysis is *simulation* where the task is to compute the outflows (of a unit or process) given data about the inflows.
2. **Design**. In *design* (process design), the feed streams (inflows) and outlet streams (outflows) are given and we want to find a process where this can be achieved in a profitable way. In design, we need to make both structural and parametric decisions:

(a) First we need to determine the *structure* of the flow sheet (which units do we need and how they should be connected), that is, at this point the equations are not known. Systematic methods for determining the structure are often called *process synthesis.*

(b) With a given flow sheet (structure), the equations are known and we need to determine the parameters, that is, the dimensions of the equipment, for example the volume of a tank or the area of a heat exchanger. This is often called *(equipment) design.*

The final equations are the same in the cases of simulation and design – the difference is which variables are unknown; in simulation some of the flows are unknown and in design some of the equipment parameters are unknown.

2.12 Summary

When using the balance principle one needs to:

1. **Define the control volume (the system's boundary).** This is not as trivial as one might think, and choosing the "right" control volume can often simplify the further calculations.

2. **Define the period of time to be considered.** This is usually not very difficult:

   - For a **batch process**, it is often the period of time from filling the reactant (at initial time t_0) to draining of the product (at the end (*final*) time), that is, $\Delta t = t_f - t_0$.
   - For a **continuous process**, we usually set up balances at a given time t (and the balances are per unit of time). For the special (and very common) case of a steady-state process, the variables are constant and do not change with time.

3. **Formulate balance for quantity B.** Often it is not obvious which quantity(ies) to use. In general, for a given control volume, one can set up the following balances:

   - 1 *total* mass balance
   - $N_c - 1$ component balances (where N_c is the number of components)
     Note: The sum of all N_c component balances is the total mass balance
   - 1 energy balance (needed if we want to find the temperature or heat transfer or work)
   - 1 momentum balance or mechanical energy balance (needed if we want to find the relationship between flows and pressure)

3

Mass balances with reaction

In this chapter, we formulate steady state component balances for systems that undergo chemical reactions. We need information about the extent of reaction for each independent chemical reaction and we discuss alternative ways of specifying this. If you are not particularly interested in chemical reactions, then this chapter may be skipped.

3.1 Introduction

In order to describe a chemical reactor we need, in addition to the mass and energy balances, to know how much is converted in each independent chemical reaction. The information can be given in the form of:

1. Conversion or extent of reaction (overall description)
2. Assumption of chemical equilibrium (thermodynamic description)
3. Kinetic data and type of reactors (detailed description; see Chapter 10)

In this chapter we concentrate on the overall description, but at the end we consider an example where we assume chemical equilibrium and use the equilibrium constant.

3.2 The component balance

First, a reminder of the general balance equation (2.1):

$$\text{Change of inventory} = \text{In} - \text{Out} + \text{Generated} - \text{Lost}$$

Here, we use molar balances and combine the terms "Generated - Lost" in the term "(Net) generated by chemical reaction." Note that this term can be negative if the component is removed in the reaction. The mass balance for an arbitrary component A is then (over a period of time or per unit of time):

$$\boxed{\text{Change of A} = \text{In A} - \text{Out A} + \text{Net generated of A by reaction}}$$

Let

$$n_A = \text{inventory of component A} \quad [\text{molA}]$$
$$G_A = \text{net amount of A generated by reaction} \quad [\text{mol A; mol A/s}]$$

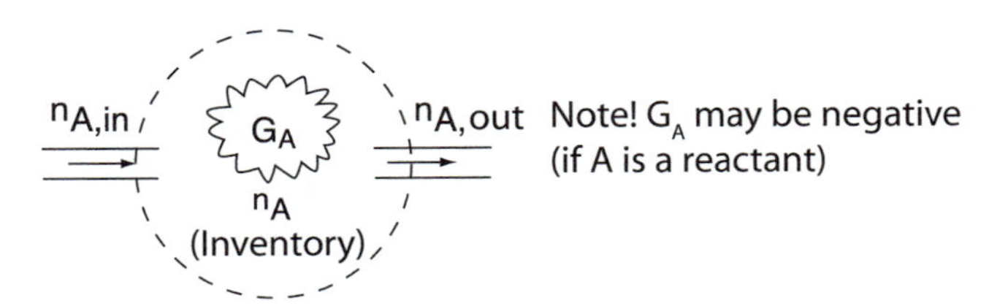

Figure 3.1: Component balance with chemical reaction

The general mass balance on molar basis is then

1. Over a time period Δt (for batch process);

$$\boxed{\Delta n_A = n_{A,\text{in}} - n_{A,\text{out}} + G_A} \quad \text{[mol A]} \tag{3.1}$$

2. At time t (rate for continuous process)

$$\boxed{\frac{dn_A}{dt} = n_{A,\text{in}} - n_{A,\text{out}} + G_A} \quad \text{[mol A/s]} \tag{3.2}$$

where we have omitted the dots in the last equation to simplify notation.

3.3 Steady-state component balance

We assume in the remainder of this chapter that there is no accumulation (that is, $\Delta n_A = 0$ or $dn_A/dt = 0$), such that the mass (molar) balances (3.1) and (3.2) for an arbitrary component A can be written

$$n_{A,\text{out}} = n_{A,\text{in}} + G_A \quad \text{[mol A; mol A/s]} \tag{3.3}$$

The standard notation in reaction engineering is to use the symbol $n_{A,0}$ for the inflow ($n_{A,\text{in}}$) and n_A for the outflow ($n_{A,\text{out}}$), and we adopt this here. The steady-state component balance (3.3) then becomes

$$\boxed{n_A = n_{A,0} + G_A} \quad \text{[mol A; mol A/s]} \tag{3.4}$$

where in this case

$n_A =$ amount of A in outstream (reactor product) [mol A; mol A/s]

$n_{A,0} =$ amount of A in instream (reactor feed) [mol A; mol A/s]

We can use the steady-state balance equation (3.4) for two cases – a batch reactor and a continuous reactor.

1. **Batch reactor.** At the start, we charge an amount $n_{A,0}$ [mol A] (in addition there can be other components). We then let the reactions proceed, and remove the reactor product n_A [mol A] at the end. We assume that the reactor is empty both before charging and removing so that there is no accumulation over the time period. The mass balance for component A over the reactor from beginning to end is then as given in (3.4) with units [mol A].

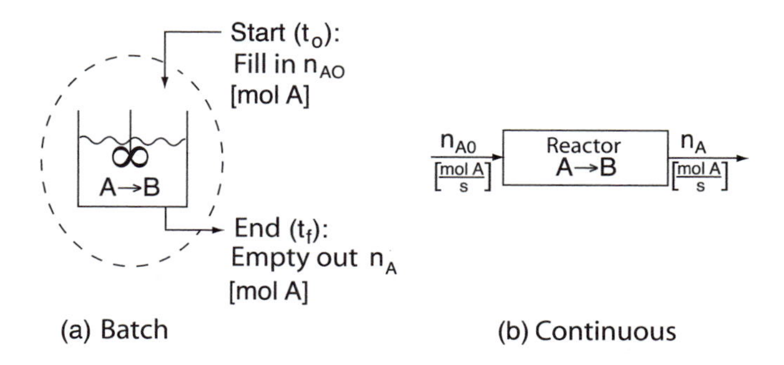

Figure 3.2: Cases where the balance (3.4) applies

2. **Continuous reactor without accumulation (steady-state process).** We have a continuous feed $n_{A,0}$ [mol A/s] (we omit the dot notation). In the reactor, chemical reactions occur such that the product stream is n_A [mol A/s]. We have no accumulation, that is, at any given time the amount of A in the reactor is constant. The mass balance for component A over the reactor (at some arbitrary time) is then as given in (3.4) [mol A/s].

3.4 Conversion and extent of reaction

The term G_A is included to account for the amount generated by chemical reaction. Here, we look at an overall description of the reactions. Two alternative ways of obtaining G_A is to specify

- Conversion X (for a component); this is commonly used as a specification.
- Extent of reaction ξ (for a reaction); this is recommended as an internal variable for calculations.

In order to define X and ξ, we first need to define the stoichiometric coefficient.

Stoichiometric coefficient ν (Greek letter nu) – results from writing the reaction equation in the form

$$0 = \sum_i \nu_i A_i$$

For example, for the reaction $2H_20 = 2H_2 + O_2$ is written as $0 = -2H_20 + 2H_2 + O_2$ and we get $\nu_{H_2O} = -2$, $\nu_{H_2} = 2$ and $\nu_{O_2} = 1$. Note that the stoichiometric coefficient is negative for reactants.

Extent of reaction ξ_j (Greek letter xi) – an extensive quantity that tells how far a given reaction j has proceeded:

$$\boxed{\xi_j = \frac{\text{mol component (A) generated in reaction } j}{\text{stoichiometric coefficient for component (A) in reaction } j}} \tag{3.5}$$

Note that this definition results in the same value for the extent of reactions, irrespective of which components in the reaction we consider. In the case of a single reaction, we find for an arbitrary component A

$$\xi = \frac{n_A - n_{A,0}}{\nu_A} \tag{3.6}$$

or

$$n_A = n_{A,0} + \underbrace{\nu_A \xi}_{G_A} \quad [\text{mol A; mol A/s}]$$

For multiple reactions, each with an extent of reaction ξ_j [mol], the amount of product is:

$$\boxed{n_A = n_{A,0} + \underbrace{\sum_j \nu_{A,j}\xi_j}_{G_A}} \quad [\text{mol A; mol A/s}] \tag{3.7}$$

where $\nu_{A,j}$ is the stoichiometric coefficient for component A in reaction j.

Note. For calculations, you are recommended to use the mass balance as given in (3.7), with the extents of reaction ξ_j as "internal variables."

Example 3.1 *A reactor is supplied with 4 mol of A and 2.5 mol of B. 3 mol of A is converted to the desired product D and 0.2 mol of A is converted to the undesired byproduct U, according to the following reactions:*

$$\text{reaction 1}: \quad 2A + B \rightarrow D$$
$$\text{reaction 2}: \quad A \rightarrow U$$

The extent of reaction for the two reactions is

$$\xi_1 = \frac{-3 \text{ mol}}{-2} = 1.5 \; mol, \quad \xi_2 = \frac{-0.2 \text{ mol}}{-1} = 0.2 \; mol$$

The mass balance (3.7) then gives the amount of component A in the product:

$$n_A = n_{A,0} + \underbrace{(-2)\xi_1 + (-1)\xi_2}_{G_A} = 4 \; mol - 3 \; mol - 0.2 \; mol = 0.8 \; mol$$

Similarly, the mass balance (3.7) gives for the other components:

$$n_B = n_{B0} + (-1)\xi_1 = 2.5 \; mol - 1.5 \; mol = 1.0 \; mol$$
$$n_D = n_{D0} + (+1)\xi_1 = 0 \; mol + 1.5 \; mol = 1.5 \; mol$$
$$n_U = n_{U0} + (+1)\xi_2 = 0 \; mol + 0.2 \; mol = 0.2 \; mol$$

The conversion is the fraction of a given compound that reacts (reactant). For reactant A, we define

$$\boxed{X_A = \frac{\text{mol reactant A consumed by reactions}}{\text{mol reactant A supplied to the system}}} = \frac{n_{A,0} - n_A}{n_{A,0}} \tag{3.8}$$

which gives the component mass balance:

$$n_A = n_{A,0} \underbrace{- n_{A,0} X_A}_{G_A} = n_{A,0}(1 - X_A) \quad [\text{mol A; mol A/s}] \tag{3.9}$$

Note that the conversion X_A is dimensionless and is always between 0 and 1.

A comparison of (3.9) and (3.7) gives

$$G_A = \sum_j \nu_{Aj}\xi_j = -n_{A,0}X_A \quad \text{[mol A]; [mol A/s]} \tag{3.10}$$

which provides the relationship between extent of reaction and conversion. For example, in the case of a single reaction we get

$$\xi = \frac{n_{A,0}X_A}{(-\nu_A)} \tag{3.11}$$

The conversion will generally differ for various reactants, and it is normally defined for the *limiting* reactant:

Limiting reactant (or key reactant) is the reactant that limits the maximum value of the extent of reaction of a given (desired) reaction. If $n_{i,0}$ is the amount of feed of component i, then total conversion of this compound corresponds to an extent of reaction $\frac{n_{i,0}}{|\nu_i|}$. The maximum extent of reaction is the minimum among these, that is,

$$\xi_{\max} = \min_{i=\text{reactant}} \{\frac{n_{i,0}}{|\nu_i|}\} = \frac{n_{LR,0}}{|\nu_{LR}|} \tag{3.12}$$

and the limiting reactant (LR) is the component i = LR that achieves the minimum value.

The conversion (for a given component A) is in (3.8) defined as the total conversion of A in *all reactions*, but sometimes we refer to the conversion $X_{A,j}$ for a component A in a *given reaction* j. From (3.11), this can be defined as $X_{A,j} = -\nu_{A,j}\xi_j/n_{A,0}$.

Example 3.1, continued. *The conversion of reactant A (in the two reactions) is*

$$X_A = \frac{3\ mol + 0.2\ mol}{4\ mol} = 0.8$$

The conversion of reactant A in reaction 1 (where A is a limiting reactant) is $X_{A,1} = \frac{3\ mol}{4\ mol} =$ 0.75. *The conversion of reactant B (which only takes part in reaction 1) is* $X_B = 1.5\ mol/$ $2.5\ mol = 0.60$.

Example 3.2 Reactor for production of acrylonitrile. *Acrylonitrile is produced from propylene, ammonia and oxygen:*

$$C_3H_6 + NH_3 + \frac{3}{2}O_2 \rightarrow C_3H_3N + 3H_2O$$

The feed contains 10 mol% propylene, 12% ammonia and 78% air. The conversion for the limiting reactant is 30%. Calculate the extent of reaction and product composition.

Solution. *The stoichiometric coefficients for the reaction are*

$$\nu_{C_3H_6} = -1,\ \nu_{NH_3} = -1,\ \nu_{O_2} = -1.5,\ \nu_{C_3H_3N} = 1,\ \nu_{H_2O} = 3$$

As a basis, we choose $n_0 = 100\ mol$ *feed. We assume that the composition of air is 79 mol%* N_2 *and 21%* O_2. *The feed to the reactor is then [mol]*

$$\begin{aligned} n_{C_3H_6,0} &= 10 \\ n_{NH_3,0} &= 12 \\ n_{O_2,0} &= 0.21 \cdot 78 = 16.4 \\ n_{N_2,0} &= 61.6 \end{aligned}$$

In order to find the limiting reactant, divide the feed amount by the stoichiometric coefficient for each component:

$$\frac{n_{C_3H_6,0}}{|\nu_{C_3H_6}|} = 10\ mol, \quad \frac{n_{NH_3,0}}{|\nu_{NH_3}|} = 12\ mol, \quad \frac{n_{O_2,0}}{|\nu_{O_2}|} = 10.93\ mol$$

Thus, propylene (C_3H_6) is the limiting reactant and the maximum obtainable extent of reaction is 10 mol. Since 30% of the limiting reactant is converted, this means that the extent of reaction for the reaction is

$$\xi = 0.3 \cdot 10\ mol = 3\ mol$$

When the extent of reaction is known, it is simple to obtain the product distribution from the mass balance (3.7) ("Out = In + Generated" [mol] for each component):

$$\begin{aligned}
n_{C_3H_6} &= n_{C_3H_6,0} - \xi = 10 - 3 = 7.0\ mol \\
n_{NH_3} &= n_{NH_3,0} - \xi = 12 - 3 = 9.0\ mol \\
n_{O_2} &= n_{O_2,0} - 1.5\xi = 16.4 - 1.5 \cdot 3 = 11.9\ mol \\
n_{C_3H_3N} &= n_{C_3H_3N,0} + \xi = 0 + 3 = 3.0\ mol \\
n_{H_2O} &= n_{H_2O,0} + 3\xi = 0 + 3 \cdot 3 = 9.0\ mol \\
n_{N_2} &= n_{N_2,0} + 0 = 61.6\ mol
\end{aligned}$$

Check: The conversion for propylene (limiting reactant) is

$$X_{C_3H_6} = \frac{n_{C_3H_6,0} - n_{C_3H_6}}{n_{C_3H_6,0}} = \frac{10\ mol - 7\ mol}{10\ mol} = 0.3$$

which is consistent with the given information.

3.5 Selectivity and yield

For cases with side (by) reactions, the selectivity or yield is often specified. Let us assume that reactant A (usually the limiting) is converted to a desired product in reaction 1 and to an undesired byproduct in reaction 2:

$$\begin{aligned}
-\nu_{A,1}\ A + \cdots &\rightarrow \text{Desired product} + \cdots; \quad \xi_1\ \text{[mol reacted]} \\
-\nu_{A,2}\ A + \cdots &\rightarrow \text{Undesired byproduct} + \cdots; \quad \xi_2\ \text{[mol reacted]}
\end{aligned}$$

The selectivity ϕ is the fraction of reactant (usually the limiting) that is converted to the desired product.

$$\boxed{\phi = \frac{\text{mol reactant converted to desired product}}{\text{mol reactant converted in total}}} = \frac{\nu_{A,1}\xi_1}{\nu_{A,1}\xi_1 + \nu_{A,2}\xi_2} \tag{3.13}$$

ϕ has a value between 0 and 1, and $\phi = 1$ if there is no side reaction.

The yield is the amount of (desired) product generated as a fraction of the maximum theoretically obtainable (usually for the limiting reactant),

$$\boxed{Y = \frac{\text{mol product generated}}{\text{mol product generated if all of the reactant generated product}}} \tag{3.14}$$

With a little thinking, we find that this is equal to the fraction of supplied reactant that forms the desired product, that is,

$$Y = \frac{\text{mol reactant converted to desired product}}{\text{mol reactant supplied}} = \frac{|\nu_{A,1}|\ \xi_1}{n_{A,0}} \tag{3.15}$$

Y has value between 0 and 1. Unless stated otherwise, the yield is usually defined for the limiting reactant.

The relation $Y = \phi X$ applies between yield, selectivity and conversion when one considers the limiting reactant:

$$\underbrace{\frac{\text{converted to desired}}{\text{supplied}}}_{Y} = \underbrace{\frac{\text{converted to desired}}{\text{total converted}}}_{\phi} \cdot \underbrace{\frac{\text{total converted}}{\text{supplied}}}_{X} \tag{3.16}$$

This is derived more exactly by combining (3.10), (3.13) and (3.15). We note that a high yield Y requires both a high conversion X and a high selectivity ϕ. If there are no side reactions, that is, $\phi = 1$, the yield of desired product is equal to the conversion of reactant.

Example 3.3 Dehydrogenation of ethane. *The following reactions take place in a continuous steady-state reactor*

$$C_2H_6 \rightarrow C_2H_4 + H_2 \quad [\xi_1 \text{ mol reacted}]$$

$$C_2H_6 + H_2 \rightarrow 2CH_4 \quad [\xi_2 \text{ mol reacted}]$$

where the last reaction is undesired. The feed contains 85 mol% ethane (C_2H_6) and the rest inerts (I). The conversion of ethane is 50.1% and the yield of ethylene (C_2H_4) is 47.1%. Calculate the composition of the reactor product, the selectivity and the yield.

Solution. *As a basis, we choose 100 mol feed. We use the extent of reaction for the two reactions to express the mass balances (3.4) for the components [mol]:*

$$\begin{aligned} n_{C_2H_6} &= 85 - \xi_1 - \xi_2 \\ n_{C_2H_4} &= \xi_1 \\ n_{H_2} &= \xi_1 - \xi_2 \\ n_{CH_4} &= 2\xi_2 \\ n_I &= 15 \end{aligned} \tag{3.17}$$

where we want to determine ξ_1 and ξ_2. The conversion of ethane is 0.501, that is, from (3.8)

$$X_{C_2H_6} = \frac{n_{C_2H_6,0} - n_{C_2H_6}}{n_{C_2H_6,0}} = \frac{85 - n_{C_2H_6}}{85} = 0.501$$

and we find $n_{C_2H_6} = 42.4$ mol. The yield of ethylene is 0.471, i.e., from (3.15)

$$Y = \frac{n_{C_2H_4} - n_{C_2H_4,0}}{n_{C_2H_6,0}} = \frac{n_{C_2H_4} - 0}{85} = 0.471 \tag{3.18}$$

we find $n_{C_2H_4} = 40.0$ mol, that is,

$$\xi_1 = 40.0 \ [mol]$$

The mass balance for ethane (3.17) then gives

$$\xi_2 = 85 - 40.0 - 42.4 = 2.6\ [mol]$$

and we can determine all the product amounts from the mass balances. We find [mol]

$$n_{C_2H_4} = 40,\ n_{H_2} = 37.4,\ n_{CH_4} = 5.2,\ n_I = 15,\ n_{tot} = 140$$

The composition of the reactor product is

$$30.3\% C_2H_6,\ 28.5\% C_2H_4,\ 26.7\% H_2,\ 3.7\% CH_4,\ 10.7\% I$$

The selectivity for ethylene from ethane is from (3.13)

$$\phi = \frac{\xi_1}{\xi_1 + \xi_2} = \frac{40\ mol}{42.6\ mol} = 0.939$$

Check: From (3.16), the yield of ethylene is $Y = \phi X = 0.939 \cdot 0.501 = 0.471$ which is consistent with the specification.

Example 3.1 (page 80), continued. *The selectivity for conversion of A to product D is, from (3.13),* $\phi = 3\ mol/3.2\ mol = 0.9375$. *From (3.15), the* yield *of D from A is* $Y = 3\ mol/4\ mol = 0.75$. *The conversion of reactant A in the two reactions is* $X_A = \frac{3\ mol + 0.2\ mol}{4\ mol} = 0.8$, *and as expected from (3.16) we have* $Y = \phi X_A$.

Exercise 3.1* *A reactor is supplied with 1 mol methanol (CH_3OH) and 0.7 mol O_2. 0.75 mol methanol is converted to the desired product (formaldehyde),*

$$CH_3OH + \frac{1}{2}O_2 = HCHO + H_2O$$

while 0.2 mol methanol is burnt in an undesired side reaction,

$$CH_3OH + \frac{3}{2}O_2 = CO_2 + 2H_2O$$

(a) Determine the limiting reactant and its conversion. (b) Calculate the extent of reaction for the reactions and the product distribution. (c) Determine the selectivity for methanol to formaldehyde and the yield. (d) Finally, check that $Y = \phi \cdot X$.

Comments.

1. Ideally, we want to maximize both the yield Y and the selectivity ϕ. However, these are often conflicting objectives because a high selectivity ϕ is often obtained with a low conversion X, whereas $Y = \phi X$ (3.16) says that a high yield is obtained with a high conversion. Usually, it is more important for industrial production to have a high selectivity rather than a high conversion. This is because unconverted reactant can be recycled, while "wrongly converted" reactant is often a loss. The reaction conditions are therefore often chosen such that we have high selectivity and low conversion (and thereby low yield) in the reactor. Note that this is for each pass (*per pass*) through the reactor, and with recycle the *overall yield* and conversion for the process are usually much higher.
2. Chemists (that is, those who work on a lab-scale) usually include separation losses etc. when calculating the yield, but this is not included in our definition since we refer to "mol product generated" and not "mol product."
3. **Warning.** There are many other terms and conflicting definitions for selectivity and yield, so be careful to check the definition when reading other literature. For example, some authors (e.g., Levenspiel) call "our" selectivity ϕ the *fractional yield* while yet others (e.g., Scott-Fogler) call it the *reaction yield.* The "selectivity" is then instead defined as $S =$ (mole reactant to desired product)/ (mole reactant to byproducts) or $S' =$ (mole desired product)/(mole byproduct). S and S' give different numerical values unless it takes the same amount of reactant to form 1 mol of desired product as to form 1 mol of byproduct. The quantities ϕ ("our selectivity") and S are closely related since $S = \frac{\phi}{1-\phi}$, but "our" selectivity ϕ has the advantage that its value is between 0 and 1, while S and S' are between 0 and ∞.

3.6 Reaction and recycle

In most processes with reaction, there is some recycle of un-converted reactant to the reactor. This reduces losses and increases the overall yield and conversion. For hand calculations, one must often combine the mass balances for several units (for example reactor, separator, purge, mixer) in order to solve the problem; see for example, Example 2.11 (page 56) and Example 2.15 (page 71) for details. Here are some additional exercises:

Exercise 3.2* Reactor with recycle. *Propane is de-hydrogenated to propylene in a catalytic reaction*

$$C_3H_8 \rightarrow C_3H_6 + H_2$$

The overall conversion of propane is 95% (for the overall process). The reactor product is separated into two streams: a "light" product with H_2, C_3H_6 and 0.555% of the C_3H_8 in reactor outlet, and a "heavy" recycle with the remaining C_3H_8 and 5% of the propylene in the reactor product. The recycle is fed back to the reactor.

(a) Draw a flow sheet and perform a quick analytic check if the problem is solvable, for example using Table 2.3.

(b) Calculate the composition of the product.

(c) What is the recycle ratio (amount recycled/amount fresh feed)?

(d) Find the conversion in the reactor.

Exercise 3.3 Production of bioproteins.

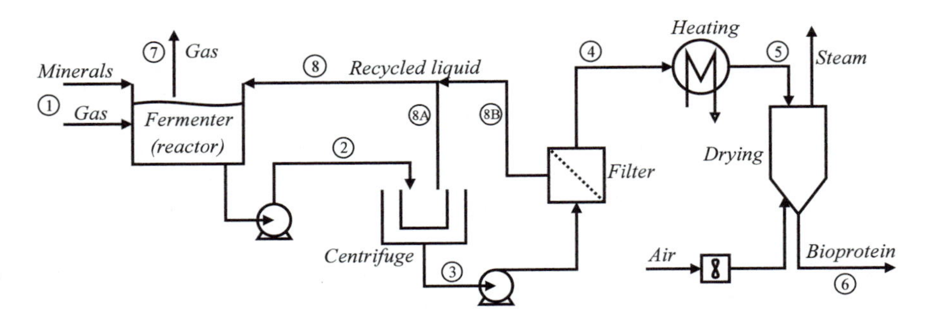

Figure 3.3: Plant for production of bioproteins from natural gas

Bioproteins can be produced from natural gas by fermentation. The following stream data for a 50000 ton bioprotein per year plant are found in a design report:

Stream	Total kg/h	Biomass weight%	Biomass $\frac{kmol}{h}$	CO_2 $\frac{kmol}{h}$	H_2O $\frac{kmol}{h}$	O_2 $\frac{kmol}{h}$	CH_4 $\frac{kmol}{h}$	NH_3 $\frac{kmol}{h}$
1 (feed)		0%	0	0		844	566	51.6
2—		2%	258	0		0	0	0
3		15%	258	0		0	0	0
4 = 5		20%	258	0		0	0	0
6 (product)	6347	100%	258	0	0	0	0	0
7 (waste gas)		0%	0		0			
8	286650	0%	0	0	15925	0	0	0

The overall reaction is given as

$$3O_2(g) + 2CH_4(g) + 0.2NH_3(g) \rightarrow CH_{1.8}O_{0.5}N_{0.2}(s) + CO_2(g) + 3.4H_2O(l)$$

where $CH_{1.8}O_{0.5}N_{0.2}$ is the bioprotein (biomass). The feed also contains some salts and nutrients which are not included.

Problem: *Fill in the table and correct any errors in the given mass balances and reaction stoichiometry (try to find the most likely errors).*

It is recommended that you start by answering the following questions: (i) Is the reaction stoichiometry correct, and if not, how can it be corrected? (i) What is the limiting reactant and what is its conversion? (iii) What is the extent of reaction? (iv) How many production hours are assumed per year?

3.7 Atomic balances

The balances given above are **component balances** (molecular balances), for example for ethylene or methane. One can also formulate atomic balances, for example balances on the atoms H and C. Note that the number of atoms is conserved, so there is no need for a term for "generated by reaction." However, the atomic balances do not give us any additional information, since we have indirectly "used up" the atomic balances when balancing the chemical reactions to obtain the stoichiometric coefficients; see Appendix A.2 (page 329.)

Example 3.2 (page 81), continued. *In order to illustrate the use of atomic balances, consider the atomic balance for H in the reactor for production of acrylonitrile. The following reaction takes place*

$$C_3H_6 + NH_3 + \frac{3}{2}O_2 \rightarrow C_3H_3N + 3H_2O \tag{3.19}$$

Four of the components contain hydrogen and the steady-state atomic balance for H gives ("Out H = In H"):

$$6n_{C_3H_6} + 3n_{NH_3} + 3n_{C_3H_3N} + 2n_{H_2O} = 6n_{C_3H_6,0} + 3n_{NH_3,0} + 3n_{C_3H_3N,0} + 2n_{H_2O,0}$$

and if we insert the numbers we previously obtained, we find that they satisfy this equation

$$\underbrace{6 \cdot 7 + 3 \cdot 9 + 3 \cdot 3 + 2 \cdot 9}_{96} = \underbrace{6 \cdot 10 + 3 \cdot 12 + 3 \cdot 0 + 2 \cdot 0}_{96}$$

As expected, the atomic balance does not give any new information compared to the component balances.

Thus, when using our "standard method" based on component balances, we don't need to use atomic balances. Nevertheless, there are cases where it may be simpler to use atomic balances. This is shown in the next example.

Example 3.4 Use of atomic balances. *For a process, the net (overall) feed and product streams are given in Table 3.1. We want to fill in the three missing numbers in the table. We solve the exercise using 1) component balances and 2) atomic balances, to confirm that the results are the same.*

Table 3.1: Stream data for Example 3.4

	Feed [mol]	Product [mol]
CH_4	110	0
CH_3OH	0	32
H_2O	200	n_{H2O}
O_2	$n_{O2,0}$	0
H_2	0	2
CO	0	1
CO_2	0	n_{CO2}

1. Solution with "standard method" (component balances and extent of reaction.) *As discussed in Section 3.8 (page 88), we can formulate 4 independent reactions for the 7 components given in the table. We choose to consider the following reactions:*

$$\begin{aligned} CH_4 + H_2O &= CO + 3H_2; \xi_1 \text{ [mol]} \\ CO + 2H_2 &= CH_3OH; \xi_2 \text{ [mol]} \\ CO + H_2O &= CO_2 + H_2; \xi_3 \text{ [mol]} \\ CO + \frac{1}{2}O_2 &= CO_2; \xi_4 \text{ [mol]} \end{aligned}$$

The component balances $n_A = n_{A,0} + \sum_j \nu_{A,j}\xi_j$ for the seven components give when inserting numbers from Table 3.1:

$$\begin{aligned}
\text{methane}: &\quad 0 = 110 - \xi_1 \\
\text{methanol}: &\quad 32 = 0 + \xi_2 \\
\text{water}: &\quad n_{H2O} = 200 - \xi_1 - \xi_3 \\
O_2: &\quad 0 = n_{O2,0} - 0.5\xi_4 \\
H_2: &\quad 2 = 0 + 3\xi_1 - 2\xi_2 + \xi_3 \\
CO: &\quad 1 = 0 + \xi_1 - \xi_2 - \xi_3 - \xi_4 \\
CO_2: &\quad n_{CO2} = 0 + \xi_3 + \xi_4
\end{aligned}$$

This gives 7 equations in 7 unknowns. First, we determine the extent of reactions:

- *The methane balance gives $\xi_1 = 110$ mol.*
- *The methanol balance gives $\xi_2 = 32$ mol.*
- *The H_2 balance gives $\xi_3 = 2 - 330 + 64 = -264$ mol.*
- *The CO balance gives $\xi_4 = -1 + 110 - 32 + 264 = 341$ mol.*

It is then simple to find the three missing amounts from the three remaining balances:

- $n_{H2O} = 200 - 110 + 264 = 354$ *mol.*
- $n_{O2,0} = 0.5 \cdot 341 = 170.5$ *mol.*
- $n_{CO2} = -264 + 341 = 77$ *mol.*

2. Solution using atomic balances. *We can set up three independent atomic balances for C, H and O. The atomic C-balance ("C in = C out") gives*

$$n_{CH4,0} + n_{CH3OH,0} + n_{CO,0} + n_{CO2,0} = n_{CH4} + n_{CH3OH} + n_{CO} + n_{CO2}$$

Inserted numbers from Table 3.1:

$$110 = 32 + 1 + n_{CO2}$$

and we find $n_{CO2} = 77$ [mol]. Similarly, the H-balance gives:

$$4 \cdot 110 + 2 \cdot 200 = 4 \cdot 32 + 2n_{H2O} + 2 \cdot 2$$

and we find $n_{H2O} = 354$. Finally, the O-balance gives:

$$200 + 2 \cdot n_{O2,0} = 32 + 354 + 1 + 2 \cdot 77 \quad \Rightarrow \quad n_{O2,0} = 170.5 \text{ [mol]}$$

As expected, the two methods are consistent, but we see that the direct use of the atomic balances is actually much simpler in this particular case.

3.8 Independent reactions and matrix formulation

This subsection is very interesting, especially if you like to write things compactly using matrices and vectors, but it can be skipped as it is not strictly needed for reading the rest of the book.

It is important to know the number of independent reactions because this tells us how many independent pieces of information we need to specify for the reactor at steady state. If we for example have 2 independent reactions then the conversion in a reactor is completely described by 2 extent of reactions (ξ_1 and ξ_2), by 2 conversions, or by specifying two product amounts. Let

N_c - no. of components that participate in reactions (excluding inerts)
N_a - no. of *independent* atomic balances for these N_c components
N_r - no. of *independent* chemical reactions for these N_c components

We then have (see Appendix A.2, page 329):

$$N_r = N_c - N_a \tag{3.20}$$

Example 3.5 *In Example 3.3 (page 83) with $N_c = 4$ components (C_2H_6, C_2H_4, H_2, CH_4) there are $N_a = 2$ independent atomic balances (for C and H), that is, there are $N_r = 4-2 = 2$ independent chemical reactions (as expected).*

Example 3.6 *We consider a methanol reactor with 7 components: CO, CO_2, H_2, CH_3OH, H_2O, CH_4 and N_2. However, the latter two (CH_4 and N_2) are chemically inert in this case, so there are only $N_c = 5$ components that participate in the chemical reactions. For these 5 components we can set up $N_a = 3$ independent atomic balances (for C, H and O). Thus, we can formulate $N_r = N_c - N_a = 5 - 3 = 2$ independent reactions.*

Example 3.7 *Note that the atomic balances can depend on each other, such that N_a is less than the number of atoms. For the components CH_4, CH_3OH and O_2 ($N_c = 3$) we have 3 atoms (C, H and O), but only $N_a = 2$ independent atomic balances (because C and H "follow each other" in the ratio 1:4 in CH_4 and CH_3OH, and thus may be viewed as a "combined atom" CH_4). We therefore have only $N_r = N_c - N_a = 1$ independent chemical reaction, which is $CH_4 + \frac{1}{2}O_2 = CH_3OH$.*

Now we get into the fun matrix part, which you may have to skip if you are not familiar with matrices.

Atom matrix A. In some cases, it may be difficult to find the number N_a of independent atomic balances. In such cases, one can obtain the atom (species) matrix A, which is simply a "table" of the number of atoms in each of the N_c components (species, molecules).

> A - matrix of chemical composition (atoms) of each component (1 column for each component; 1 row for each atom)

(Note that A depends on the order we choose to list the components, and the order we list the atoms). We then have that the number of independent atomic balances is $N_a = \text{rank}(A)$. The concept of rank of a matrix is known from mathematics.

Example 3.5 continued. *We want to check that there are indeed $N_a = 2$ independent atom balances. For the components*

$$C_2H_6,\ C_2H_4,\ H_2, CH_4$$

and atoms C and H, the atom matrix is

$$A = \begin{bmatrix} 2 & 2 & 0 & 1 \\ 6 & 4 & 2 & 4 \end{bmatrix}$$

As expected, we find $N_a = rank(A) = 2$ (this can be checked in MATLAB with the command `>>rank(A)`*).*

Example 3.8 *In Example 3.2 (page 81) only one reaction was specified for the following $N_c = 5$ components:*

$$C_3H_6,\ NH_3,\ O_2,\ C_3H_3N,\ H_2O$$

Can this be correct? The atom matrix for the five components and the four atoms H, C, N and O is:

$$A = \begin{bmatrix} 3 & 0 & 0 & 3 & 0 \\ 6 & 3 & 0 & 3 & 2 \\ 0 & 1 & 0 & 1 & 0 \\ 0 & 0 & 2 & 0 & 1 \end{bmatrix}$$

Here, column 1 *gives the atomic composition of C_3H_6, column 2 that of NH_3, etc. The 4 rows in A are independent, which is confirmed by computing $N_a = rank(A) = 4$, so the 4 atom balances are independent. This implies that we have only $N_r = N_c - N_a = 5 - 4 = 1$* independent *chemical reaction. Thus, the specified reaction (3.19) is the only one possible (try yourself if you doubt this).*

Stoichiometric matrix N. It is not always obvious whether a proposed reaction set contains independent reactions. To check this, we can compute the rank of the stoichiometric matrix N, where N is simply a "table" of the stoichiometric coefficients for the proposed reactions.

> N - matrix of stoichiometric coefficients for components (1 column for each component; 1 row for each reaction)

The number of independent reactions in the proposed set is $N_r = \text{rank}(N)$.

Example 3.5 further continued. *For the components C_2H_6, C_2H_4, H_2 and CH_4, we formulate two reactions*

$$C_2H_6 \rightarrow C_2H_4 + H_2$$

$$C_2H_6 + H_2 \rightarrow 2CH_4$$

The stoichiometric matrix becomes

$$N = \begin{bmatrix} -1 & 1 & 1 & 0 \\ -1 & 0 & -1 & 2 \end{bmatrix}$$

We find $\text{rank}(N) = 2$ *so the two reactions are independent. If we extend the reaction set with the reaction*

$$C_2H_4 + 2H_2 \rightarrow 2CH_4$$

we get the extended stoichiometric matrix

$$N_e = \begin{bmatrix} -1 & 1 & 1 & 0 & \\ -1 & 0 & 0 & 2 & \\ 0 & -1 & -2 & 0 & 2 \end{bmatrix}$$

We find $\text{rank}(N_e) = 2$, *so only 2 of the proposed 3 reactions are independent.*

Component mass balances in matrix form. With many components and reactions, it is convenient to write the steady-state component mass balances $n_A = n_{A,0} + G_A$ (3.4) ("Out A = In A + Generated A") in matrix form

$$\underline{n} = \underline{n}_0 + \underbrace{N^T \underline{\xi}}_{\underline{G}} \tag{3.21}$$

where

$$\underline{n} = \begin{bmatrix} n_A \\ n_B \\ \vdots \end{bmatrix}$$ - vector of component amounts in reactor product [mol; mol/s]

$\underline{n_0}$ - vector of component amounts in reactor feed [mol; mol/s]

$\underline{\xi}$ - vector of extents of reaction [mol; mol/s]

$\underline{G} = N^T \underline{\xi}$ - vector of component amounts generated in reactions [mol; mol/s]

Atom balances and consistency between A and N. If $\underline{n}$ is the vector of component amounts, then the vector of atom amounts is

$$\underline{a} = A\underline{n} \quad \text{[mol atoms; mol/s atoms]}$$

Now, multiply the component mass balance equation (3.21) on both sides by the matrix A to get $A\underline{n} = A\underline{n}_0 + AN^T\underline{\xi}$, or $\underline{a} = \underline{a}_0 + AN^T\underline{\xi}$ [mol atoms]. Now, since atoms are conserved quantities, we must have $\underline{a} = \underline{a}_0$ (same amount of atoms in product and feed). It then follows that $AN^T\underline{\xi} = 0$, and since this must hold for any $\underline{\xi}$, we must always have that $AN^T = 0$ or equivalently

$$NA^T = 0 \tag{3.22}$$

This consistency relationship between the atom matrix A and stoichiometric matrix N can be useful when checking for errors (for example in MATLAB; we must always have that `N*A'` gives a matrix with only 0's).

Example 3.5 even further continued. *For the reactions,*

$$C_2H_6 \rightarrow C_2H_4 + H_2$$

$$C_2H_6 + H_2 \rightarrow 2CH_4$$

we find that

$$NA^T = \begin{bmatrix} -1 & 1 & 1 & 0 \\ -1 & 0 & -1 & 2 \end{bmatrix} \begin{bmatrix} 2 & 6 \\ 2 & 4 \\ 0 & 2 \\ 1 & 4 \end{bmatrix} = \begin{bmatrix} -2+2+0+0 & -6-4+2+0 \\ -2+0+0+2 & -6+0-2+8 \end{bmatrix} = \begin{bmatrix} 0 & 0 \\ 0 & 0 \end{bmatrix}$$

and we have as expected $NA^T = 0$.

Exercise 3.4* *Propose a reaction set, that is, formulate two independent reactions for Example 3.6 (page 89. Find N, A, and compute* rank(N), rank(A) *and NA^T.*

Exercise 3.5 *In each of the following cases you should find the number of independent reactions and propose a reaction set. None of the components are chemically inert.*

(a) H_2, H_2O, NO and NO_2.*
(b) H_2, H_2O, O_2, NO, NO_2, and N_2.
(c) H_2, H_2O, O_2, NO, NO_2, HNO_3, NH_3 and N_2.*
(d) NO, NO_2 , N_2O, N_2, O_2, NH_3 and H_2O.

3.9 Reaction with chemical equilibrium

Here, we consider an example where the reaction is in equilibrium. The example demonstrates the usefulness of introducing the extent of reaction as an internal variable for the calculations. A nice additional example is the ammonia synthesis equilibrium reactor on page 176.

Example 3.9 Mass balance for methanol reactor with equilibrium. *We consider a methanol reactor where the reactor feed has the following composition (in mol%):*

$$\begin{aligned} CO &: 3.8\% \\ CO_2 &: 2.0\% \\ H_2 &: 76.4\% \\ CH_3OH &: 0.3\% \\ H_2O &: 0.2\% \\ CH_4 &: 16.8\% \\ N_2 &: 0.5\% \end{aligned}$$

Here, the latter two components are chemically inert in our reaction reactor. The reactor operates at 100 bar and the exit temperature is 270 ^oC. Methanol can be generated in the following two independent reactions

$$\begin{aligned} CO + 2H_2 &= CH_3OH && [\xi_1 \text{ mol reacted}] \\ CO_2 + 3H_2 &= CH_3OH + H_2O && [\xi_2 \text{ mol reacted}] \end{aligned}$$

We want to calculate the reactor product composition when it is given that the product contains 5.0% methanol and we assume the "shift" reaction

$$CO + H_2O = CO_2 + H_2$$

is in equilibrium at 270 ^oC with equilibrium constant $K = 30$.

Solution. *Let us first point out that the shift reaction ($CO + H_2O = CO_2 + H_2$) is equal to the difference of the two formation reactions for methanol, such that there are only two independent reactions (see also Example 3.6, page 89). Thus, if we introduced an extent of reaction ξ_3 also for the shift reaction, we would get an unnecessary extra variable. This would not in itself constitute a problem, but the resulting equation set would be undetermined. In order to get a unique solution, we would then have to "randomly" assign a value to one of the variables ξ_1, ξ_2 or ξ_3. In our case, where we do not use the shift reaction, this corresponds to setting $\xi_3 = 0$.*

Let us quickly analyze the solvability of the problem. We have given all information about the feed and have in addition two other pieces of information; and since there are two independent reactions we should, according to Table 2.3 (page 70), have enough information.

We choose as a basis n_0=100 [mol] (the feed). The mass balances for the components over the reactor give, with the extents of reactions as "internal" variables, the following product:

$$\begin{aligned} n_{CO} &= 3.8 - \xi_1 \\ n_{CO_2} &= 2.0 - \xi_2 \\ n_{H_2} &= 76.4 - 2\xi_1 - 3\xi_2 \\ n_{CH_3OH} &= 0.3 + \xi_1 + \xi_2 \\ n_{H_2O} &= 0.2 + \xi_2 \\ n_{CH_4} &= 16.8 \\ n_{N_2} &= 0.5 \\ n_{tot} &= 100 - 2\xi_1 - 2\xi_2 \end{aligned}$$

From the given information we further have that

$$\frac{n_{CH_3OH}}{n_{tot}} = 0.05$$

and since the shift reaction is in equilibrium, we get with the assumption of ideal gas (see page 386)

$$K = \frac{p_{CO_2}/p^{\ominus} \cdot p_{H_2}/p^{\ominus}}{p_{CO}/p^{\ominus} \cdot p_{H_2O}/p^{\ominus}} = 30$$

where $p^{\ominus}$=1 *bar and the partial pressure for component* i *is defined as* $p_i = n_i p/n_{tot}$, *where* p *is the (total) pressure. The last equation gives*

$$\frac{n_{CO_2} \cdot n_{H_2}}{n_{CO} \cdot n_{H_2O}} = 30$$

where the pressure p *dropped out because we have the same number of moles on both sides of the shift reaction. Now, inserting from the mass balances gives two equations in two unknowns (the extent of reactions* ξ_1 *and* ξ_2*):*

$$\frac{0.3 + \xi_1 + \xi_2}{100 - 2\xi_1 - 2\xi_2} = 0.05 \tag{3.23}$$

$$\frac{(2.0 - \xi_2)(76.4 - 2\xi_1 - 3\xi_2)}{(3.8 - \xi_1)(0.2 + \xi_2)} = 30 \tag{3.24}$$

These two equations are difficult to solve analytically, so we solve them numerically (this is a nice exercise in numerical mathematics). The following solution method can be used – although, I am afraid that the numerical mathematics people will not be too impressed if they see it ☺.

1. *Guess* ξ_1 *and calculate* ξ_2 *from (3.23)*
2. *From these calculate the left hand side (LHS) of (3.24) and check if it is close to* $K = 30$.

We find that:

$$\xi_1 = 1 \Rightarrow \xi_2 = 3.27 \Rightarrow LHS = -8.5$$
$$\xi_1 = 2 \Rightarrow \xi_2 = 2.27 \Rightarrow LHS = -4.0$$
$$\xi_1 = 3 \Rightarrow \xi_2 = 1.27 \Rightarrow LHS = 41.1$$
$$\xi_1 = 2.9 \Rightarrow \xi_2 = 1.37 \Rightarrow LHS = 29.5$$

which is close enough, that is, we have

$$\xi_1 = 2.9; \quad \xi_2 = 1.37 \quad [mol]$$

The product stream is then [mol]

$$\begin{aligned}
n_{CO} &= 3.8 - \xi_1 = 3.8 - 2.9 = 0.9 \\
n_{CO_2} &= 2.0 - \xi_2 = 2.0 - 1.37 = 0.63 \\
n_{H_2} &= 76.4 - 2\xi_1 - 3\xi_2 = 76.4 - 5.8 - 4.11 = 66.49 \\
n_{CH_3OH} &= 0.3 + \xi_1 + \xi_2 = 0.3 + 2.9 + 1.37 = 4.57 \\
n_{H_2O} &= 0.2 + \xi_2 = 0.2 + 1.37 = 1.57 \\
n_{CH_4} &= 16.8 \\
n_{N_2} &= 0.5 \\
n_{tot} &= 100 - 2\xi_1 - 2\xi_2 = 100 - 5.8 - 2.74 = 91.46
\end{aligned}$$

Exercise 3.6* **Equilibrium for synthesis gas reactor.** *The first step in the production of methanol and ammonia is to produce "synthesis gas" consisting of H_2, CO and CO_2. This is also a common process for producing hydrogen. Consider a case where the synthesis reactor operates at 21 bar and the feed is a mixture of water vapor and natural gas in the ratio 2.5:1. The natural gas consists of CH_4 (95.5 mol%), C_2H_6 (3.0%), C_3H_8 (0.5%), C_4H_{10} (0.4%) and N_2 (0.6%). The three "heavier" hydrocarbons ($n \geq 2$) are assumed to react completely (100% conversion) according to the reactions*

$$C_nH_{2n+2} + nH_2O \rightarrow nCO + (2n+1)H_2$$

In addition, we have the following equilibrium reactions:

$$CH_4 + H_2O = CO + 3H_2; \quad K_1 = 710$$

$$CO + H_2O = CO_2 + H_2; \quad K_2 = 0.81$$

where the latter is the shift reaction. Determine the composition of the product (synthesis gas). (Comment: The temperature at the reactor exit is 880 °C. The first reaction is actually not quite in equilibrium and to partly correct this, we have used the thermodynamic equilibrium constant at 864 °C).

(You get two equations with two unknowns that must be solved numerically, for example with MATLAB.)

3.10 Summary

Conversion (for a component) and extent of reaction (for a reaction) are two alternative ways of providing an overall description of chemical reactions. The conversion has the advantage that it is dimensionless, and for this reason it is often used when specifying the reactions. However, the extent of reaction is more practical for calculations. For this reason, it is recommended to use the extent of reaction as an "internal variable," especially when there are several reactions.

The values for the extent of reaction or conversion can be calculated from a more detailed description of the reaction process, such as equilibrium constants (thermodynamics) or from kinetic data. In the latter case, we also need to know the type and quantity of the reactor. This is dealt with in Chapter 10.

Exercise 3.7 *Define the following quantities:*

- *Stoichiometric coefficient, ν*
- *Number of independent reactions*
- *Limiting reactant*
- *Conversion, X (used in order to give information)*
- *Extent of reaction, ξ [mol; mol/s] (used during calculations)*
- *Selectivity, ϕ*
- *Yield, Y*

Note that the quantities X, ϕ and Y are dimensionless and take on values between 0 and 1. Show that $Y = \phi X$.

4

The energy balance

The energy balance (which is a generalization of the 1st law of thermodynamics) is needed in order to calculate temperature, heat transfer or work. In this chapter, we derive the energy balance for open flow systems, and show that the enthalpy H provides a practical way of combining a stream's internal energy U and flow work.

Before you start reading this chapter, you should make sure you are well acquainted with the contents of Appendix A which contains basic topics from physical chemistry and thermodynamics. In particular, check out the section on thermochemistry (page 357).

4.1 The general energy balance (open system)

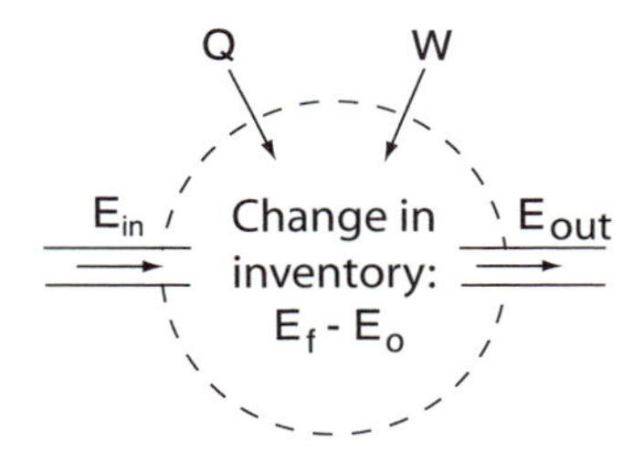

Figure 4.1: General energy balance

Energy (E) is a *conserved quantity*, and from (2.7), the **general energy balance** over a time period t_0 (initial state) to t_f (end state) is for an *open system*:

$$\text{Change in inventory of energy} = \text{Net supplied energy}$$

$$\boxed{\underbrace{E_f - E_0}_{\Delta E} = E_{\text{in}} - E_{\text{out}} + Q + W} \quad [J] \tag{4.1}$$

Here

- $E_0 = E(t_0)$: the system's (total) energy in its initial state
- $E_f = E(t_f)$: the system's (total) energy in its final state
- $\Delta E = E_f - E_0$: change in system energy over the time period
- E_{in} and E_{out}: energy "indirectly" supplied and removed by mass streams (bulk transport)

- Q: supplied heat from the surroundings (through wall)
- W: supplied work (from the surroundings).

In rate form, the energy balance at time t becomes (see proof of (2.8)):

$$\frac{dE}{dt} = \dot{E}_{\text{in}} - \dot{E}_{\text{out}} + \dot{Q} + \dot{W} \quad [J/s] \tag{4.2}$$

In the following, we will (as usual) omit the dots.

4.2 Energy forms

In this chapter, we consider the following contributions to the energy E (there are many more, which are usually not important for us):

(Total) Energy = internal energy + potential energy + kinetic energy

or

$$E = U + E_P + E_K$$

where E is (total) energy, U is internal energy, E_P is potential energy and E_K is kinetic energy. By kinetic energy (E_K), we mean motion on a macro scale, and not the unordered temperature-dependent molecular motions (for example translation, rotation and vibration) which are included in the internal energy U.

In addition, one can have other energy forms, such as electric energy, surface energy and rotational energy, see (A.21) page 345, but these are not considered in this chapter.

The internal energy (U) represents the energy of the molecules and includes most of the energy forms of interest for us, such as chemical energy, thermal energy, vaporization energy, etc. For ideal gases (and for most liquids and solids) the internal energy is only a function of temperature and composition, that is, it is independent of pressure (see page 353).

Potential (E_P) and kinetic energy (E_K) are usually neglected in "our" energy balances because their changes are small. This is illustrated by the following example.

Example 4.1 Waterfall. *To illustrate that changes in potential and kinetic energy can often be neglected when compared to typical changes in internal energy, we will calculate the increase in temperature when water falls down 100 m. We consider a mass m=1 kg.*

1. *At the top of the waterfall, water has a potential energy,*

$$E_{P,1} = mgh_1 \approx 1\ kg \cdot 10\ m/s^2 \cdot 100\ m = 1000 \text{ J}$$

   At the top, the velocity v_1 is small, and the kinetic energy is approximately zero ($E_{K,1} \approx 0$).

2. *During its fall, the potential energy is converted to kinetic energy*

$$E_{K,2} = m\frac{v_2^2}{2} \quad [\text{J}]$$

   More precisely, the energy balance (which we will return to soon) tells that $E_2 = E_1$. When we neglect friction losses, this gives $E_{K,2} = E_{P,1}$ or $\frac{mv_2^2}{2} = mgh_1$, and we find that the water velocity v_2 at the bottom of the waterfall is

$$v_2 = \sqrt{2gh_1} = \sqrt{2 \cdot 10m/s^2 \cdot 100m} = 45 \text{ m/s}$$

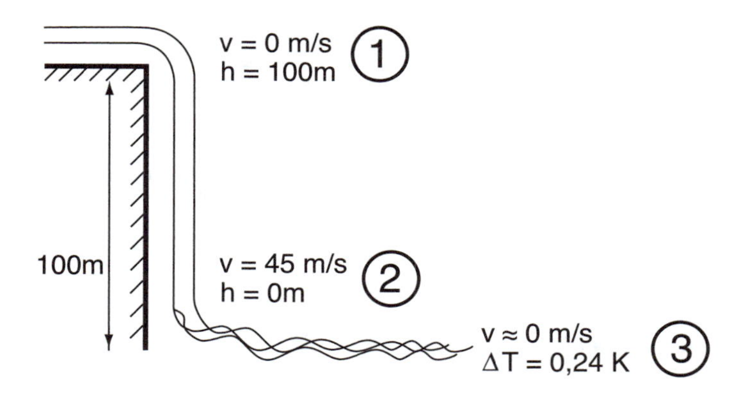

Figure 4.2: Waterfall: potential energy (1) → kinetic energy (2) → Internal (thermal) energy (3)

3. At the sudden "stop" at the bottom of the waterfall, the kinetic energy is converted to internal energy . More precisely, the energy balance gives $E_3 = E_2$, or $U_3 = U_2 + E_{K,2}$ because we assume $E_{K,3} \approx 0$. The change in the internal energy is $\Delta U = U_3 - U_2 = mC_V \Delta T$ where $C_V \approx C_p = 4180$ J/kg K for water. The energy balance then gives

$$mC_p\Delta T = \frac{mv_2^2}{2} \quad \Rightarrow \quad \Delta T = \frac{v_2^2/2}{C_p} = \frac{gh_1}{C_p} = \frac{10m/s^2 \cdot 100m}{4180J/kg\ K} = 0.24K$$

that is, the increase in temperature after the fall from 100 m is only $\Delta T = T_3 - T_2 = T_3 - T_1 = 0.24$ K.

Note that 0.24 K is a small number when compared to typical temperature changes in our processes. This is confirmed by computing the supplied specific energy (per kg) in the waterfall which is $E_{P,1}/m = gh_1 = 10\ m/s^2 \cdot 100\ m = 1000\ m^2/s^2 = 1$ kJ/kg. This is a small number compared to, for example, the heat of vaporization for water, which is 2450 kJ/kg.[1]

We summarize the waterfall example with the following conclusions:

- In process plants, height differences are typically less than 60 m (which is less than the waterfall of 100 m considered in the example), and **changes in potential energy can usually be neglected** compared to typical changes in internal (thermal) energy.
- In process plants, flow velocities are typically about 1 m/s for liquids and up to 20 m/s for gases at 1 bar. This is much less than the velocity of 45 m/s at the bottom of the waterfall in the example, and we therefore conclude that **changes in kinetic energy can usually be neglected** compared to typical changes in internal (thermal) energy.

In the following we will, unless otherwise stated, **neglect changes in kinetic and potential energy**, that is, we assume that the energy is equal to the internal energy,

$$E = U$$

[1] For comparison, we can calculate the energy released energy if a mass Δm disappears in a nuclear reaction following Einstein's famous formula: $\Delta E/\Delta m = c^2 = (3 \cdot 10^8 m/s)^2 = 9 \cdot 10^{16}$ J/kg $= 9 \cdot 10^{13}$ kJ/kg. This is about 10^{14} times larger than the potential energy in our waterfall (1000 J/kg). This explains why everyone in the 1950's, when the first nuclear power plants were build, assumed that nuclear energy would be the world's dominant future energy source.

With this assumption, the general energy balance (4.1) becomes

$$\boxed{\underbrace{U_f - U_0}_{\Delta U} = U_{\rm in} - U_{\rm out} + Q + W} \quad [J] \tag{4.3}$$

(but notice that we can always generalize this by replacing U by E).

4.3 Work forms

Work is organized transfer of energy, for example, when the system is moved under the influence of an external force. There are many forms of work (see page 344):

- **Volume change work** $W_{\Delta V}$ (system pV work) is the work associated with changes in the system's volume.
- **Flow work** $W_{\rm flow}$ (stream pV work) is the work associated with the volume displacements of streams that enter and exit the system.
- **Shaft work** W_s is the mechanical work supplied (or extracted if it negative) using movable machinery (pump, compressor, turbine) associated with changes in pressure.
- **Electrochemical work** $W_{\rm el}$ is the supplied work (or extracted if it negative) when the system (e.g., operating as a battery or fuel cell) is connected to an external electric circuit.
- and **other work** $W_{\rm other}$, for example surface work when the surface area changes or electromagnetic work.

The total work supplied to the system from the surroundings is

$$W = W_{\rm flow} + \underbrace{W_{\Delta V} + W_s + W_{\rm el} + W_{\rm other}}_{W_n} \tag{4.4}$$

where we have chosen to distinguish between the "useless" flow work $W_{\rm flow}$ associated with bringing the flows in and out, and the remaining "useful" (non-flow) work W_n. Included in W_n is work for the system's volume changes ($W_{\Delta V}$), mechanical shaft work (W_s), electrochemical work ($W_{\rm el}$), etc.

Sign of work W. Everyone agrees that Q is the heat supplied to the system from the surroundings. However, for work W, there are two sign conventions in common use. In this book, we use the newer IUPAC convention which says that W is *supplied* work. Thus, positive work means that work is supplied to the system (for example, with a compressor), while negative work means that the system performs work on the surroundings (for example, with a turbine or by increasing the system's volume or by taking out electrical work). [2]

---

[2] The **IUPAC convention**, which we use in this book, is that W is positive when work is *supplied* to the system This is consistent, because all transfers are defined as positive when supplied to the system. The **older convention**, still commonly used by mechanical engineers, is that the W is positive when the system *performs* work. This convention probably dates back to the introduction of steam engines about 300 years ago. In that era, engineers were particularly interested in heat engines for pumping water or driving machinery. The desired output was work W and the required input was heat Q. It made sense to describe both as positive quantities. The main practical implication of using the older sign convention is that the term $Q + W$ in the energy balance instead becomes $Q - W$.

pV**-work.** Let us take a closer look at the two terms $W_{\Delta V}$ and W_{flow} related to pressure-volume (pV) work:

1. **Volume change work** $W_{\Delta V}$ is the work related to changes in the system's volume. We have (see (A.26) for derivation):

$$W_{\Delta V} = -\int_{V_0}^{V_f} p_{\text{ex}} dV \quad [\text{J, J/s}] \tag{4.5}$$

   where p_{ex} is the pressure of the surroundings, V_0 is the volume in the initial state and V_f is the volume in the final state. Obviously, a volume increase requires that the system performs work on the surroundings, so the negative sign is due to the sign convention that W is supplied work. For the case of reversible volume change work, we have $p_{\text{ex}} = p$. For steady-state processes, the volume is constant so $W_{\Delta V} = 0$.
2. **The flow work** W_{flow} is the "useless" work that a stream performs as it enters (in) or exits (out) the system. We have (see a more detailed derivation below):

$$W_{\text{flow}} = W_{\text{flow,in}} - W_{\text{flow,out}} = p_{\text{in}} V_{\text{in}} - p_{\text{out}} V_{\text{out}} \quad [\text{J, J/s}] \tag{4.6}$$

   Here $W_{\text{flow,in}} = p_{\text{in}} V_{\text{in}}$ is the work that the inlet stream(s) with pressure p_{in} and volume V_{in} supplies to the system when it is "pushed into" (or forces itself into) the system. Correspondingly, the term $W_{\text{flow,out}} = p_{\text{out}} V_{\text{out}}$ is the work that the outlet stream(s) with pressure p_{out} and volume V_{out} performs on the surroundings as it is "pushed out of" the system.

The total energy supplied to the system from a stream is the sum of the contributions from the stream's internal energy and its flow work, and we recognize this sum as the stream's "enthalpy":

$$H_{\text{in}} = U_{\text{in}} + W_{\text{flow,in}} = U_{\text{in}} + p_{\text{in}} V_{\text{in}} \tag{4.7}$$

$$H_{\text{out}} = U_{\text{out}} + W_{\text{flow,out}} = U_{\text{out}} + p_{\text{out}} V_{\text{out}} \tag{4.8}$$

that is,

A stream's enthalpy is the sum of its internal energy and flow work

You may have wondered why for chemical reactions one usually specifies the enthalpy H and not the internal energy U. Now you know one reason: it is more practical to use enthalpy because H also includes the associated flow work for an open system.

Derivation of flow work and enthalpy, (4.6)-(4.8). Let's see what happens when a stream supplies a small mass dm_{in} [kg] to the system. First of all, the small mass carries an internal energy given by

$$dU_{\text{in}} = u_{\text{in}} \cdot dm_{\text{in}} \quad [J]$$

(where u_{in} [J/kg] is specific internal energy). In addition, the small mass has a volume dV_{in} [m^3], and when entering the small mass performs a work on the system because it must displace some mass that is already in the system. From Newton's 2nd law, work is force times distance, that is, we have (see Figure 4.3):

$$dW_{\text{flow in}} = F \cdot dl = p_{\text{in}} A \cdot \frac{dV_{\text{in}}}{A} = p_{\text{in}} dV_{\text{in}}$$

We introduce specific quantities for volume (v [m^3/kg]) and enthalpy (h [J/kg]). The total energy (internal energy + flow work) supplied with the small mass is then

$$dU_{\text{in}} + dW_{\text{flow in}} = dU_{\text{in}} + p_{\text{in}} dV_{\text{in}} = \underbrace{(u_{\text{in}} + p_{\text{in}} v_{\text{in}})}_{h_{\text{in}}[\text{J/kg}]} dm_{\text{in}} = dH_{\text{in}}$$

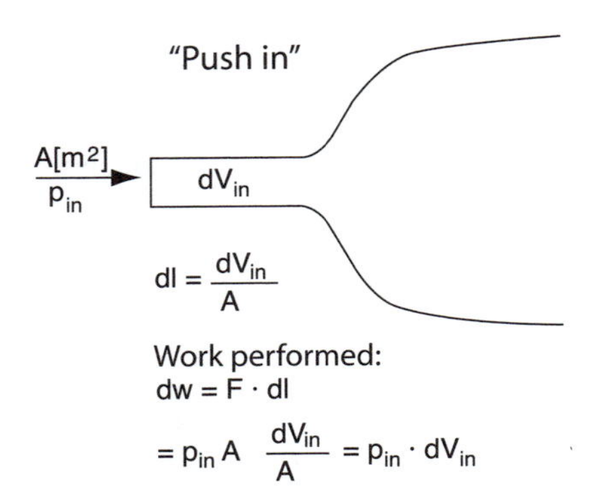

Figure 4.3: Work that a small mass supplies to the system as it enters the system

which is equal to the enthalpy of the small mass. We also see from this that flow work for the inflow(s) with volume $V_{\rm in}$ is equal to $W_{\rm flow,in} = p_{\rm in}V_{\rm in}$ [J]. A corresponding derivation can be made for the outlet stream(s), but with opposite sign. □

4.4 Alternative formulations of the energy balance

Here, we summarize the general energy balance and consider some different forms of it. The **general energy balance** for an open system is given in (4.1):

$$E_f - E_0 = E_{\rm in} - E_{\rm out} + Q + W \quad [J]$$

For cases where *changes in internal energy dominate* (that is, we neglect changes in the system's kinetic and potential energy, etc.), we can set $E \approx U$, and the energy balance is as given in (4.3):

$$U_f - U_0 = U_{\rm in} - U_{\rm out} + Q + W \quad [J]$$

There are many forms of work, but first of all let us divide the work into the "useless" flow work and the remaining "useful" non-flow work, that is,

$$W = W_{\rm flow} + W_n \tag{4.9}$$

where from (4.6) $W_{\rm flow} = p_{\rm in}V_{\rm in} + p_{\rm out}V_{\rm out}$. By introducing *the streams' enthalpy* defined by

$$H_{\rm in} = U_{\rm in} + p_{\rm in}V_{\rm in}; \quad H_{\rm out} = U_{\rm out} + p_{\rm out}V_{\rm out}$$

the general energy balance (4.3) for an open system can then be written in enthalpy form (see Figure 4.4):

$$\boxed{\underbrace{U_f - U_0}_{\Delta U} = H_{\rm in} - H_{\rm out} + Q + \underbrace{W_s + W_{\Delta V} + W_{\rm el} + W_{\rm other}}_{W_n}} \tag{4.10}$$

Here

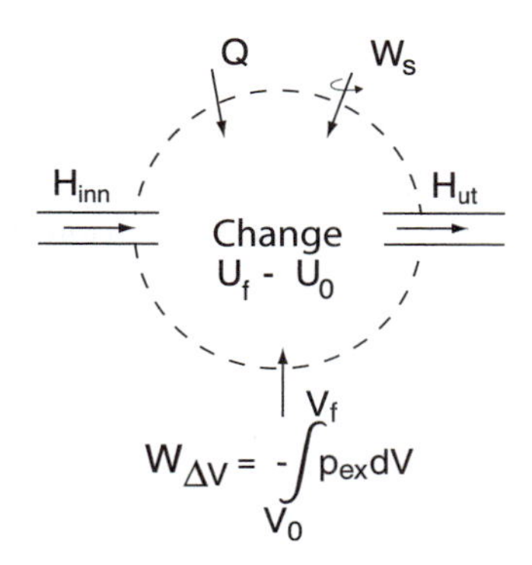

Figure 4.4: General energy balance in enthalpy form (kinetic and potential energy not included)

- $\Delta U \triangleq U_f - U_0$ is the change in the *system's* internal energy (over time period Δt)
- $H_{\rm in}$ and $H_{\rm out}$ are the enthalpies (sum of internal energy and flow work) that follow the mass flows
- Q is the supplied heat from the surroundings (through the wall)
- $W_{\Delta V} = -\int_{V_0}^{V_f} p_{\rm ex} dV$ is the work for changes in the system's volume

Special case I: Closed system

For a closed system, there is no mass exchange, so $H_{\rm in} = 0$, $H_{\rm out} = 0$ and $W_{\rm flow} = 0$ and the general energy balance (4.10) becomes:[3]

$$\boxed{\underbrace{U_f - U_0}_{\Delta U} = Q + W} \quad [J] \tag{4.11}$$

which is usually called the **1st law of thermodynamics**. Here, we have used the fact that $W = W_n$ since $W_{\rm flow} = 0$. For a closed system, $W_s = 0$ (since shaft work gives a volume change, which is already included in the term $dW_{\Delta V}$), and it follows that

$$W = W_n = W_{\Delta V} + W_{\rm el} + W_{\rm other} = -\int_{V_0}^{V_f} p_{\rm ex} dV + W_{\rm el} + W_{\rm other} \tag{4.12}$$

For the special case of a reversible process, $p_{\rm ex} = p$. The energy balance for a closed system is discussed in more detail on page 346.

A further special case is an **isolated system**, which is a closed system without heat and work exchange, that is, with $Q = 0$ and $W = 0$. The energy balance for an isolated system is that the internal energy is constant, i.e., $U_f = U_0$.

[3] Unless otherwise stated, ΔX indicates changes in the quantity X for the *system*. For example, $\Delta U = U_f - U_0$ is change in the system's internal energy. Sometimes this is indicated more explicitly by writing $\Delta X_{\rm system}$.

Special case II: No accumulation (steady-state process)

Another very important special case (the most important in this book) is a process where the system's variables (including energy) do not change over the time we observe the system. That is, we have no accumulation and $\Delta U_{\text{system}} = U_f - U_0 = 0$. Furthermore, $V_f = V_0$, so the volume change work is zero, that is, $W_{\Delta V} = -\int_{V_0}^{V_f} p_{\text{ex}} dV = 0$. The general energy balance (4.10) becomes

$$\boxed{\underbrace{H_{\text{out}} - H_{\text{in}}}_{\Delta H_{\text{flows}}} = Q + W_s} \quad [\text{J; J/s}] \tag{4.13}$$

where H_{in} and H_{out} are the enthalpies of the streams, Q is supplied heat and W_s is supplied shaft work. Note that this energy balance (4.13) applies both to a *steady state flow process* (then with unit [J/s]) and to a *batch process without accumulation* (then with unit [J]).

What happened to the W_{el} in (4.13)? I have simply been lazy and omitted it, that is, W_s in (4.13) should strictly speaking include the supplied electrochemical work W_{el} plus "other" work W_{other}. (W_{el} is important for many processes, such as fuel cells; but we can usually safely set $W_{\text{other}} = 0$ for steady-state processes). This laziness comes on top of already omitting kinetic and potential energy (E_K and E_P). In summary, from now on in this book, when you encounter a "general" energy balance, like in (4.13), use the following "**energy balance reading rule**" (unless the terms are already included):

- Shaft work W_s really means $W_s + W_{\text{el}}$ + other work forms.
- Internal energy U really means $E = U + E_K + E_P$ + other energy forms.
- Enthalpy H really means $H + E_K + E_P$ + other energy forms.

Another important special case of (4.13) is a **steady state adiabatic process without shaft (or electrochemical) work**. Here $Q = 0$ and $W_s = 0$ and the energy balance (4.13) is simply (see Figure 4.7)

$$\boxed{H_{\text{out}} = H_{\text{in}}} \quad \text{or} \quad \boxed{\underbrace{H_{\text{out}} - H_{\text{in}}}_{\Delta H_{\text{flows}}} = 0} \tag{4.14}$$

That is, the enthalpy of the outflow(s) is equal to the enthalpy of the inflow(s). The simple "enthalpy balance"[4] (4.14) is very important for practical calculations, and applies for example to

- mixing process (often with pressure change)
- flow through valve (always with pressure change)
- adiabatic process with phase change
- adiabatic chemical reactor

[4] We see that the energy balance (4.13) takes the form of a "enthalpy balance," but this is actually a misleading term since enthalpy is not a conserved quantity. For this reason, the expression "enthalpy balance" should be avoided or one should at least use quotation marks.

Comments on the general energy balance (4.10).

1. The general energy balance (4.10), and its corresponding dynamic version in (11.11), covers most of "our" cases.
2. The energy balance can be in units [J], [J/s], [J/mol] or [J/kg] depending on what we have chosen as a basis for mass.
3. Δ-variables are often used and the general energy balance (4.10) then becomes

$$\underbrace{\Delta U_{\text{system}}}_{U_f - U_0} + \underbrace{\Delta H_{\text{flows}}}_{H_{\text{out}} - H_{\text{in}}} = Q + W_s \tag{4.15}$$

4. We can also write the energy balances in **differential form** (this is common in thermodynamics). With our standard assumption of internal energy being the dominant energy form ($dE = dU$), the differential version of the general energy balance (4.3) is

$$dU_{\text{system}} + dU_{\text{flows}} = dQ + dW \tag{4.16}$$

Here, as above, we divide the work into "non-flow" and (useless) "flow" work

$$dW = dW_n + d(pV)_{\text{flows}}$$

By introducing the enthalpy for streams, we then get the differential version of (4.15):

$$dU_{\text{system}} + dH_{\text{flows}} = dQ + dW_n \tag{4.17}$$

where $dW_n = dW_s + dW_{dV} + dW_{\text{el}} + dW_{\text{other}}$ and $dW_{dV} = -p_{\text{ex}} dV$.

- For a **closed system**, we have $dH_{\text{flows}} = 0$ and $dW_s = 0$ (since shaft work is included in the term dW_{dV}), and with $dW_{\text{other}} = 0$, the energy balance becomes $dU_{\text{system}} = dQ + dW_{dV} + dW_{\text{el}} = dQ - p_{\text{ex}} dV + dW_{\text{el}}$.
- For a **steady state flow process**, we have $dU_{\text{system}} = 0$ and $dV = 0$, and the energy balance with $dW_{\text{other}} = 0$ becomes $dH_{\text{flows}} = dQ + dW_s + dW_{\text{el}}$.

5. **Warning.** When using Δ-variables, we have distinguished between the change in energy within the system limit (ΔU_{system}) and the change in energy for streams that enter and leave the system (ΔU_{flows}). However, this distinction is usually not made, which can easily lead to confusion. For example, the energy balance (4.13) for a steady-state flow process (open system) is often written in the form $\Delta H = Q + W_s$, which for cases without shaft work (for example for a heat exchanger or for a chemical reactor with heat exchange) gives

$$\Delta H = Q \quad \text{(steady} - \text{state flow process with no work)} \tag{4.18}$$

where it is understood that $\Delta H = \Delta H_{\text{flows}} = H_{\text{out}} - H_{\text{in}}$. Let us now consider a *closed system* where the energy balance (4.11) gives $\Delta U = Q - \int_{V_0}^{V_f} p_{\text{ex}} dV$ (where $\Delta U = U_f - U_0$). Assuming that volume changes occur reversibly such that $p_{\text{ex}} = p$, and assuming constant pressure p, the energy balance then becomes $\Delta U = Q^{\text{rev}} - p\Delta V$ (where $\Delta V = V_f - V_0$), which with the introduction of enthalpy $\Delta H = \Delta U + \Delta(pV) = \Delta U + p\Delta V$ gives

$$\Delta H = Q^{\text{rev}} \quad \text{(closed system with constant pressure)} \tag{4.19}$$

where it is understood that $\Delta H = \Delta H_{\text{system}} = H_f - H_0$. I recall that the similarity between the two energy balances (4.18) and (4.19) confused me when I was a student – so be attentive! The similarity has led many – including professors that teach physical chemistry – to erroneously believe that use of enthalpy in the energy balance assumes constant pressure. But this is not the case for a flow process (open system), where enthalpy enters the energy balance (4.18) as the sum of the stream's internal energy and flow work, and has nothing to do with constant pressure (end of Warning).

For most of the calculations in this book, we consider the special cases of steady-state process (equation 4.13) or closed system (equation 4.11), but let us first consider an example where we must use the energy balance in its general form (4.10).

Example 4.2 Filling a tank with gas.

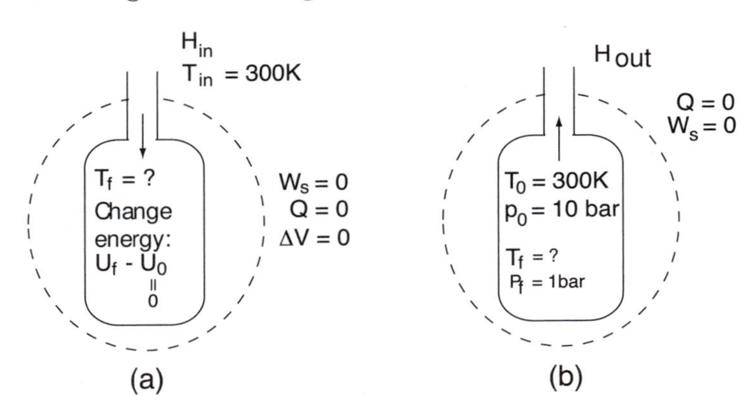

Figure 4.5: (a) Filling a tank with gas. (b) Emptying a tank with gas.

An empty (evacuated) tank is being filled with air at temperature $T_{\rm in} = 300$ *K as shown in Figure 4.5a. Find the final temperature* T_f *in the tank after the filling is complete. We assume ideal gas, adiabatic process* ($Q = 0$) *and* $\gamma = C_p/C_V = 1.4$ *constant. Although it may seem strange that no data are given for the external pressure or volume of the tank, it will become apparent that this limited amount of information is enough to solve the exercise. As a basis, let us assume that* 1 *mol gas is being filled.*

Solution. *The volume of the tank is constant* ($V_f = V_0$) *and there is no shaft work* ($W_s = 0$) *or other non-flow work. Furthermore,* $U_0 = 0$ *since initially there is no mass in the tank, and there is no outlet stream so* $H_{\rm out} = 0$. *The general energy balance (4.10) then becomes*

$$U_f = H_{\rm in} \tag{4.20}$$

that is, the internal energy in the tank is equal to the enthalpy of what has been filled in. We express (as usual) the internal energy by the enthalpy,

$$U_f = H_f - p_f V_f = H_f - RT_f \quad [J/mol]$$

where the last equality holds since $pV = RT$ *for 1 mol of ideal gas. The energy balance (4.20) then gives*

$$H_f - H_{\rm in} = RT_f$$

For an ideal gas, enthalpy is only a function of temperature (that is, independent of pressure) and with the assumption of constant heat capacity we have $H_f - H_{\rm in} = C_p(T_f - T_{\rm in})$ *(see also (A.49)), which is the enthalpy change when 1 mol of gas is taken from the in-state to the state* f. *The energy balance for 1 mol then becomes*

$$C_p(T_f - T_{\rm in}) = RT_f$$

which solved with respect to T_f *and inserted* $C_p - C_V = R$ *[J/mol K] gives*

$$T_f = \frac{C_p}{C_V} T_{\rm in} = \gamma T_{\rm in} \tag{4.21}$$

For air $\gamma = C_p/C_V = 1.4$, and we find $T_f = 1.4 \cdot 300 = 420$ K, that is, the temperature rises by 120 K to 420 K. The temperature increase is due to the flow work performed by the inflowing gas.

We assumed as basis 1 mol gas, but the answer is the same no matter what amount of air we fill in. This implies that the temperature in the tank immediately goes up to 420 K and stays there during the filling. This assumes that we initially have vacuum, neglect heat transfer to the tank's walls ($Q = 0$), and have constant heat capacity and ideal gas.

Example 4.3 Emptying a tank with gas. *A tank (container) contains a gas at 10 bar and 300 K (state 0), Figure 4.5b. The tank is emptied such that the final pressure becomes 1 bar. Calculate the final temperature in the tank (state f) when adiabatic process is assumed and $\gamma = C_p/C_V = 1.4$.*

Solution. *This resembles the previous example (except that the process is reversed and the temperature drops), but it is actually not quite as simple to solve (try yourself!). It turns out that we need to use the* differential version *of the energy balance and integrate in order to solve the exercise; and we find that the temperature drops gradually from 300 K and down to 155 K, according to (C.2), $T_f/T_0 = (p_f/p_0)^{\frac{R}{C_P}}$, where p_f is the dropping pressure in the tank. To read the rest of the story see page 393.*

It is interesting to summarize our findings for (a) filling and (b) emptying a tank with ideal gas, see Figure 4.5. (a) During filling, we find that the temperature "jumps" up to $T_f = \frac{C_p}{C_V} T_{\text{in}}$ and stays at this value during the rest of the filling. (b) During emptying, the temperature T_f drops gradually as the pressure drops.

4.5 Calculation of enthalpy

4.5.1 Standard method: Absolute enthalpy

In order to use the energy balance, (4.10), (4.11) and (4.13), we need to calculate the enthalpy H. Since enthalpy is a state function, it is in principle easily obtained from the stream data; the enthalpy of each stream (the "absolute" enthalpy) is the sum of the contributions from the components in the streams:

$$H_{\text{in}} = \sum_{\text{components } i} n_{i,\text{in}} \cdot H_m(i, \text{in}) \tag{4.22}$$

$$H_{\text{out}} = \sum_{\text{components } i} n_{i,\text{out}} \cdot H_m(i, \text{out}) \tag{4.23}$$

where $H_m(i)$ [J/mol] is the (partial) molar enthalpy for component i in the mixture (alternatively, we can use mass basis). Calculation of enthalpy is discussed in more detail in Appendix A.13 (page 354). In general, $H_m(i)$ is a function of the composition, temperature and pressure, but we can often introduce short-cuts:

- For liquids and solids, and for gases at moderate pressure, the dependency on pressure p can be neglected.
- For near-ideal mixtures (for example, ideal gas, solids and liquid mixtures of similar compounds), the heat of mixing can be neglected (that is, $H_m(i)$ is independent of composition) and we have $H_m(i) = H_m^*(i)$ where $H_m^*(i)$ is the enthalpy of pure component i.

Enthalpy must always be given relative to a reference state (where $H = 0$), and the reference state must be chosen so that all streams can be "formed" from it. It is always "safe" to use the elements at $T_{\text{ref}} = 298.15$ K and $p_{\text{ref}} = p^{\ominus} = 1$ bar as the reference state. For ideal gas at temperature T, we then have (see page 355 for more details on temperature dependency):

$$H_m(i,T) = H_m^*(i,T) = \Delta_f H_m^{\ominus}(i,298) + \int_{298}^{T} C_{p,m}(i,T)dT \quad [\text{J/mol}] \tag{4.24}$$

where $\Delta_f H_m^{\ominus}(i,298)$ is the heat of formation for pure component i as ideal gas at 298.15 K and 1 bar. As mentioned on page 365, there are many other possible choices for the reference state. In particular, for systems with no chemical reaction, it is cumbersome to go all the way back to the elements. It is nevertheless recommended that one normally chooses the elements at 298 K as reference state.

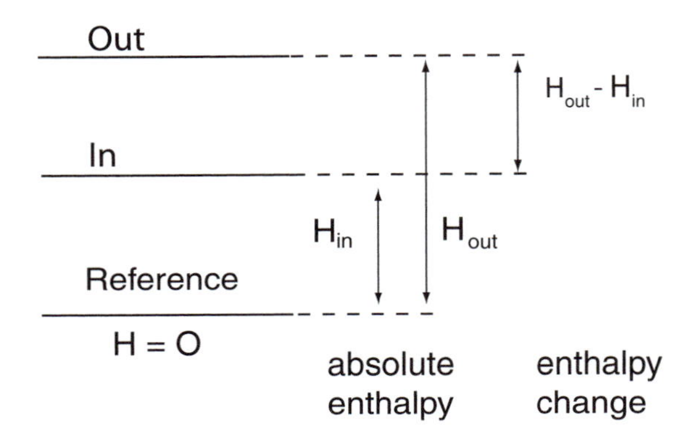

Figure 4.6: Alternative methods for evaluation of enthalpy.

4.5.2 Alternative method: Direct evaluation of enthalpy change (the "subprocess" method)

Instead of the standard approach with absolute enthalpies, there exists an alternative method ("method 2"), where one avoids introducing a reference state. The basis is that in many cases, for example in the steady-state energy balance (4.13), we do not need to evaluate absolute enthalpies, H_{in} and H_{out}, but rather their change, $H_{\text{in}} - H_{\text{out}}$ (see Figure 4.6). Since enthalpy is a state function, the change $H_{\text{out}} - H_{\text{in}}$ can be found as the sum of the enthalpy changes for a series of *imaginary individual subprocesses* that bring us from the inlet to the outlet streams,

$$H_{\text{out}} - H_{\text{in}} = \sum_i \Delta_i H$$

Here, $\Delta_i H$ is the enthalpy change for (the imaginary) subprocess i. Because we evaluate differences, there is no need to explicitly introduce reference states. This alternative method obviously gives the same result and is particularly well suited for hand calculations. (Actually, with some thinking, one realizes that the alternative method corresponds to choosing the combined inlet streams as the reference state, so it is all the same!)

In the examples below, we will use both methods for evaluation of enthalpy:

Method 1. Absolute enthalpy (this is the recommended standard method).

Method 2. Enthalpy change for subprocesses (this is often simpler for hand calculations and yields more physical insight).

4.6 Energy balance for mixing processes

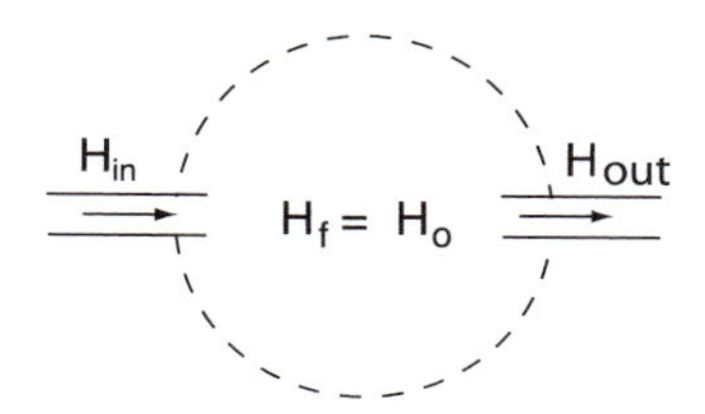

Figure 4.7: The steady state energy balance is $H_{out} = H_{in}$ [J/s] for an adiabatic process ($Q = 0$) without shaft work ($W_s = 0$).

In this section, we consider steady-state mixing processes (see Figure 4.7 where H_{in} may consist of several streams). For a mixing process, we have $Q = 0$ (adiabatic) and $W_s = 0$ (no shaft work) and the energy balance simply becomes (see (4.14)),

$$H_{out} = H_{in} \quad [\text{J; J/s; J/kg, J/mol, etc.}] \tag{4.25}$$

We next consider some examples. Make sure that you understand these!

Example 4.4 Mixing of cold and hot water. *Water with temperature T_1 (stream 1) is mixed with water with temperature T_2 (stream 2) to produce water with temperature T_3 (outlet stream). Formulate the energy balance for the process when constant heat capacity is assumed.*
Solution. *The mass balance gives (see Figure 4.8)*

$$m_3 = m_1 + m_2 \quad [kg]$$

and the energy balance $H_{out} = H_{in}$ gives

$$H_3 = H_1 + H_2$$

Method 1 (absolute enthalpies). *No reaction takes place and we choose to use water (pure component) as liquid at T_{ref} as the reference state. With constant heat capacity C_p [J/kg K], the (absolute) enthalpies for the three streams then become*

$$H_1 = m_1 C_p (T_1 - T_{ref})$$

$$H_2 = m_2 C_p (T_2 - T_{ref})$$

$$H_3 = m_3 C_p (T_3 - T_{ref})$$

and the energy balance gives

$$\underbrace{m_3 C_p(T_3 - T_{\text{ref}})}_{H_3} = \underbrace{m_1 C_p(T_1 - T_{\text{ref}})}_{H_1} + \underbrace{m_2 C_p(T_2 - T_{\text{ref}})}_{H_2} \quad [\text{J}] \tag{4.26}$$

Multiplying the mass balance $m_3 = m_1 + m_2$ by $C_p T_{\text{ref}}$ gives

$$m_3 C_p T_{\text{ref}} = m_1 C_p T_{\text{ref}} + m_2 C_p T_{\text{ref}}$$

and subtracting this from the energy balance (4.26) gives

$$m_3 C_p T_3 = m_1 C_p T_1 + m_2 C_p T_2$$

By introducing the mass balance $m_3 = m_1 + m_2$, this becomes

$$m_1 C_p(T_3 - T_1) + m_2 C_p(T_3 - T_2) = 0 \tag{4.27}$$

We note that the reference temperature T_{ref} drops out. This could alternatively have been found by using the "trick" of directly setting $T_{\text{ref}} = 0$. In general, the reference state always drops out for simple heating processes without phase transition and reaction.

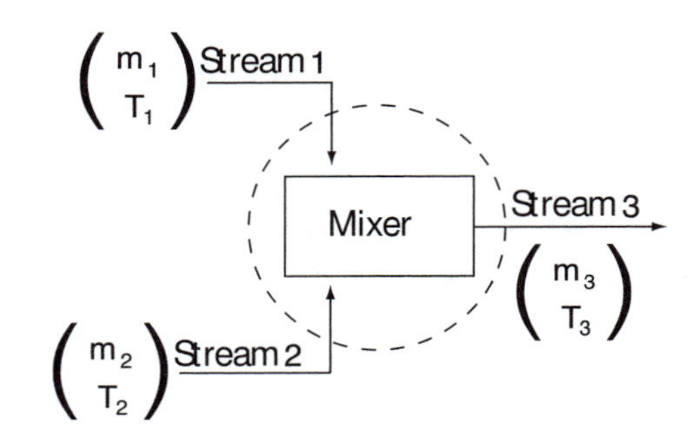

Figure 4.8: Flow sheet for mixing of cold and hot water

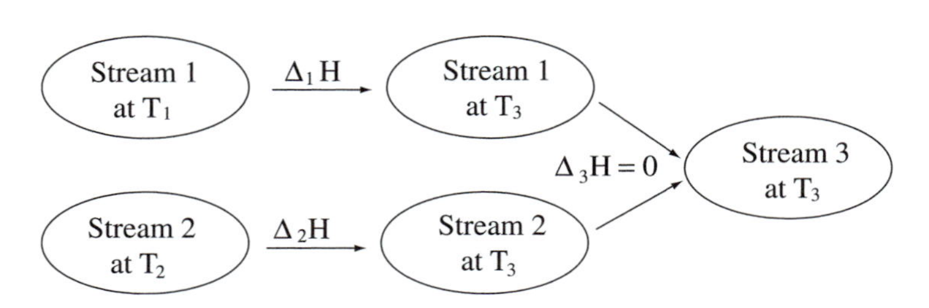

Figure 4.9: State diagram for mixing of cold and hot water: enthalpy change from inlet (streams 1 and 2) to outlet (stream 3) streams are obtained by adding imaginary subprocesses

Check with method 2: Enthalpy change for subprocesses. *As an alternative, we directly evaluate the enthalpy change $H_{\text{out}} - H_{\text{in}}$ (going from inlet streams 1 and 2 to outlet stream 3) by adding the following imaginary subprocesses as shown in Figure 4.9):*

1. *"Heat" (it could actually be cooling) stream 1 from T_1 to T_3: $\Delta_1 H = m_1 C_p(T_3 - T_1)$.*

2. "Heat" stream 2 from T_2 *to* T_3: $\Delta_2 H = m_2 C_p(T_3 - T_2)$.

3. Mix the two streams at temperature T_3*: There is no heat of mixing (since both are water at the same temperature), that is,* $\Delta_3 H = 0$.

Since enthalpy is a state function, we get $H_{\text{out}} - H_{\text{in}} = \Delta_1 H + \Delta_2 H + \Delta_3 H$*, and the energy balance* $H_{\text{out}} - H_{\text{in}} = 0$ *gives*

$$m_1 C_p(T_3 - T_1) + m_2 C_p(T_3 - T_2) = 0$$

which (of course) is identical to what we found in (4.27). Note that the alternative method is much simpler in this case.

Comment. Flow sheet and state diagram. At this point, it is useful to note the difference between the **flow sheet** in Figure 4.8 and the **state diagram** in Figure 4.9.

- In the *flow sheet*, which represents the actual process, the process units (where the changes take place) are blocks, while the streams (with their given states) are arrows (lines). The flow sheet is the most important in process engineering.
- In the *state diagram*, which represents imaginary changes from inlet to outlet streams, it is opposite: the states are circles and the changes are arrows.

Example 4.5 Mixing of two ideal gas streams with different compositions.[5] *10 mol/s of a gas stream at 300 K and 6 bar with heat capacity* $C_{p,1} = 30$ *kJ/mol K (stream 1) and 20 mol/s of another gas stream at 500 K and 5 bar with heat capacity* $C_{p,2} = 65$ *kJ/mol K (stream 2) are mixed to produce a combined stream ("out") at 4 bar. What is the temperature of the mixture when ideal gas and constant heat capacities are assumed?*

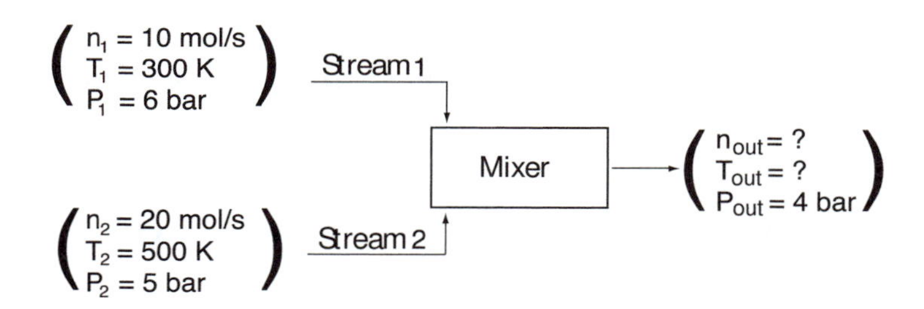

Figure 4.10: Flow sheet for mixing of two gas streams

Solution. *Let us first note that the pressures are not important, because enthalpy is independent of pressure for ideal gas (but we note that the outlet stream has a lower pressure than the inlet, which is reasonable). The mass balance gives (see Figure 4.10):*

$$n_{\text{out}} = n_1 + n_2 = 10 + 20 = 30 \quad [mol/s]$$

The energy balance (4.25) gives

$$H_{\text{out}} = H_1 + H_2 \quad [J/s]$$

Absolute enthalpies (method 1). *With the pure components at* T_{ref} *as the reference state and with the assumption of constant heat capacity, the absolute enthalpies of the streams are*

$$H_{\text{out}} = n_{\text{out}} C_{p,\text{out}}(T_{\text{out}} - T_{\text{ref}})$$

[5] We get the same result in this example if we consider ideal mixing of two liquid streams.

$$H_1 = n_1 C_{p,1}(T_1 - T_{\text{ref}}); \quad H_2 = n_2 C_{p,2}(T_2 - T_{\text{ref}})$$

where C_p [J/mol K] is the molar heat capacity. For an ideal mixture, we have that the heat capacity $C_{p,\text{out}}$ of the mixture is equal to the molar average value (see (A.51) on page 356), that is

$$C_{p,\text{out}} = \frac{n_1 C_{p,1} + n_2 C_{p,2}}{n_{\text{out}}} \quad [J/mol\ K]$$

Combining these equations gives

$$(n_1 C_{p,1} + n_2 C_{p,2}) T_{\text{out}} = n_1 C_{p,1} T_1 + n_2 C_{p,2} T_2$$

As expected, T_{ref} drops out, and we find that (sorry for being lazy here and not inserting units, but it works since I am sticking to SI units; except that energy is in kJ rather than in J):

$$T_{\text{out}} = \frac{n_1 C_{p,1} T_1 + n_2 C_{p,2} T_2}{n_1 C_{p,1} + n_2 C_{p,2}} = \frac{10 \cdot 30 \cdot 300 + 20 \cdot 65 \cdot 500}{10 \cdot 30 + 20 \cdot 65} = 462.5\ K$$

Check with method 2: Enthalpy change for individual subprocesses. *In order to form the outlet stream from the inlet streams, we consider the following idealized subprocesses:*

1. *Heat stream 1 from T_1 to T_{out}: $\Delta_1 H = n_1 C_{p,1}(T_{\text{out}} - T_1)$.*
2. *"Heat" stream 2 from T_2 to T_{out}: $\Delta_2 H = n_2 C_{p,2}(T_{\text{out}} - T_2)$ (this is actually cooling since $T_{\text{out}} < T_2$).*
3. *Mix the two streams at temperature T_{out}: For an ideal gas, the heat of mixing is zero and $\Delta_3 H = 0$.*

We then have $H_{\text{out}} - H_{\text{in}} = \Delta_1 H + \Delta_2 H + \Delta_3 H$ and the energy balance $H_{\text{out}} - H_{\text{in}} = 0$ gives

$$\underbrace{n_1 C_{p,1}(T_{\text{out}} - T_1)}_{\Delta_1 H > 0} + \underbrace{n_2 C_{p,2}(T_{\text{out}} - T_2)}_{\Delta_2 H < 0} = 0$$

One way to interpret this equation is that the heat $\Delta_2 H$ released when stream 2 is cooled to the mixture's temperature is balanced with the heat $\Delta_1 H$ needed to heat stream 1. Solving the equation gives $T_{\text{out}} = 462.5$ K, which agrees with our standard absolute enthalpy method.

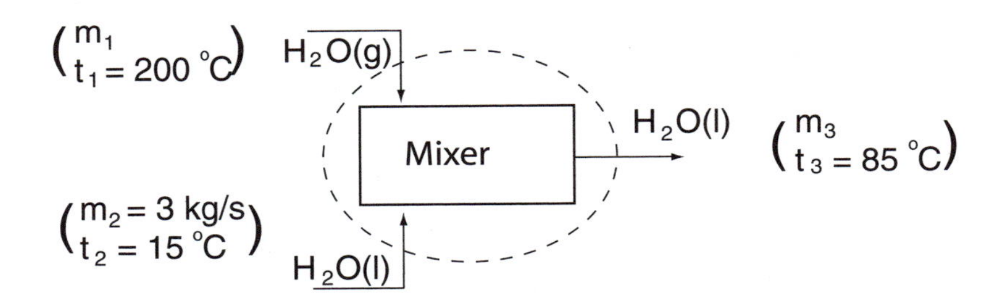

Figure 4.11: Flow sheet for heating water by mixing with steam

Example 4.6 Heating by mixing water with steam. *Steam (= water vapor) at 200 °C (stream 1) is mixed with 3 kg/s of cold water (liquid) at 15 °C (stream 2) to produce hot water at 85 °C (stream 3). How much steam (m_1 [kg/s]) is needed? (Data: The heat of vaporization for water at 100 °C is 40.6 kJ/mol, which is equal to 2257 kJ/kg. The heat capacities for water are assumed to be constant and equal to 4.18 kJ/kg K (liquid) and 1.87 kJ/kg K (vapor)).*

Solution. *The flow sheet is shown in Figure 4.11. We use "method 2" and directly evaluate the enthalpy change by adding the following imaginary subprocesses:*

1. *Stream 1 (steam) is cooled from* $t_1 = 200^oC$ *to* $t_b = 100^oC$: $\Delta_1 H = m_1 C_p(g)(t_b - t_1)$.
2. *Stream 1 is condensed at 100* ^oC: $\Delta_2 H = m_1(-\Delta_{\rm vap}H) = m_1(-2257)$ *[kJ] (the negative sign is because condensation is the opposite process of evaporation).*
3. *Stream 1 (which now is liquid) is cooled from* $t_b = 100^oC$ *to the final mixture temperature* $t_3 = 85^oC$: $\Delta_3 H = m_1 C_p(l)(t_3 - t_b)$.
4. *Stream 2 is heated from* $t_2 = 15^oC$ *to the mixture temperature* $t_3 = 85^oC$: $\Delta_4 H = m_2 C_p(l)(t_3 - t_2)$.
5. *The two streams (which are now both water at 85* ^oC*) are mixed to form the outlet stream. There is no heat of mixing, so* $\Delta_5 H = 0$.

Thus, we have $H_{out} - H_{in} = \sum_i \Delta_i H = \Delta_1 H + \Delta_2 H + \Delta_3 H + \Delta_4 H + \Delta_5 H$ *and since the energy balance for a mixing process is* $H_{out} - H_{in} = 0$*, we get (sorry again for being lazy and not inserting units, as I actually strongly recommend that you do):*

$$m_1 \cdot 1.87 \cdot (-100) + m_1(-2257) + m_1 \cdot 4.18 \cdot (-15) + 3 \cdot 4.18 \cdot 70 = 0 \quad [kJ/s]$$

and we find $m_1 = 0.350$ *kg/s.*

We note from the examples above that the alternative method with evaluation of enthalpy changes for subprocesses ("method 2") is often simpler when performing calculations by hand. However, in general (and particularly when using a computer and for dynamic processes) it is recommended to use absolute enthalpies for the streams ("method 1"), see (4.22)–(4.23).

Exercise 4.1 *Solve Example 4.6 by evaluating the absolute enthalpies ("method 1"). (a) Choose the reference state as liquid at* $t_{\rm ref} = 15^oC$*. (b) Choose the reference state as liquid at* $t_{\rm ref} = 85^oC$ *(the calculations in the latter case are almost identical to the ones given in Example 4.6).*

4.6.1 Exercises

Below are some relatively simple exercises, where it is necessary to combine the mass and energy balances.

Exercise 4.2* Mixing of hot saltwater. *2.5 kg/s of hot saltwater with 1 weight% salt and a temperature of 37* ^oC *is to be produced by mixing the following streams: Stream 1 is pure water at 8* ^oC*. Stream 2 contains 10% salt at 80* ^oC*. Stream 3 contains 1% salt at 50* ^oC*. How much is needed of each stream? State clearly all assumptions. (b) How much is needed if stream 3 contains 2% salt?*

Exercise 4.3 Process for producing oil-water mixture (emulsion). *1 kg/s of a water solution (emulsion) with 30 weight% oil and temperature 20* ^oC *is produced in a two-step process. In step 1, streams 1 and 2 are mixed and, in step 2, the product is cooled to the desired temperature. Stream 1 is pure water at 15* ^oC *and stream 2 is pure oil at 90* ^oC *(for simplicity, use physical data for water for all streams). Calculate the amount of streams 1 and 2 and also the cooling need.*

Exercise 4.4* A different process for producing oil-water mixture. *1 kg/s of a product (stream 4), which is a water solution with 30 weight% oil and temperature 20* ^oC*, is produced in a two-step process. In step 1, streams 1 and 2 are mixed and, in step 2, pure water (stream 3) is removed by decanting the water phase (since oil and water naturally form two phases, unless we make an emulsion by strongly mixing the two phases). Stream 1 is pure* *water at 15 °C and stream 2 is pure oil at 90 °C. Calculate the amount of the four streams. (This looks a lot like exercise 4.3 but note that you here are mixing directly to the desired temperature.)*

Exercise 4.5* *Repeat Exercise 4.4 but assume that the decanted stream (stream 3) contains 2 weight% oil.*

Exercise 4.6 Yet another process for oil-water mixture. *1 kg/s of a water solution with 30 weight% oil and a temperature 20 °C is produced in a two-step process. In step 1, streams 1 and 2 are mixed and, in step 2, pure water (stream 3) is removed by evaporation (the vapor is removed under vacuum). Stream 1 is pure water at 15 °C and stream 2 is pure oil at 90 °C. Calculate the amounts of the four streams. (This is very similar to exercise 4.3 but the cooling is here by evaporation. You can use physical data for pure water; see exercise A.5, for calculating the heat of vaporization of water at 20 °C.)*

4.6.2 Heat of mixing and temperature change during mixing

It is recommended that you read Appendix A.14 on thermochemistry before you continue.

In the above examples, we have neglected the heat of mixing and this is often a reasonable assumption if the components do not interact strongly (bond) with each other. However, there are important exceptions. For example, when mixing water with strong acids or bases, the heat of mixing (solution heat, dilution heat) can be considerable.

The **heat of mixing** $\Delta_{\text{mix}}H$ *is defined as the heat that must be supplied in order to keep a constant temperature when mixing streams with different compositions (but with the same temperatures).* A negative heat of mixing, $\Delta_{\text{mix}}H < 0$, means that the mixing generates heat, so cooling ($Q < 0$) is required to keep a constant temperature. The heat of mixing effect is caused by changes in the bonds between the molecules, and can be viewed as a "weak version" of the heat of reaction. Heat of mixing is further discussed in Appendix A.14.2 (page 359); some data are given on page 417.

Adiabatic temperature change. Now, assume $Q = 0$, and consider the following problem: What is the mixture's ("outlet stream's") temperature T_{mix} when we mix two (inlet) streams with the same temperature T_0? From the energy balance, enthalpy is constant for a mixing process, that is,

$$H_{\text{out}} - H_{\text{in}} = 0 \tag{4.28}$$

However, this does not mean that the temperature is constant. We evaluate the enthalpy change $H_{\text{out}} - H_{\text{in}} = \Delta_1 H + \Delta_2 H$ as the sum of two imaginary subprocesses:

1. Mixing at constant temperature (T_0): $\Delta_1 H = \Delta_{\text{mix}}H(T_0)$ = heat of mixing.
2. Take ("heat") the mixture from T_0 to T_{mix}: $\Delta_2 H = C_p(T_{\text{mix}} - T_0)$ (here we have assumed that C_p for the mixture is constant in this temperature region).

From (4.28), $\Delta_1 H + \Delta_2 H = 0$, and we derive that the adiabatic temperature change for the mixing process is:

$$\Delta T = T_{\text{mix}} - T_0 = \frac{-\Delta_{\text{mix}}H(T_0)}{C_p} \tag{4.29}$$

Example 4.7 Mixing of salt with water. *Calculate the temperature rise when we mix 100 g $CaCl_2$ ("road salt") and 0.9 kg water such that we end up with 1 kg of solution.*

Solution. *From the data on page 417, the heat of mixing for a dilute solution is -75 kJ/mol salt (a negative number means that heat is "released" during mixing) which, with a molecular weight 111 g/mol, is equal to -0.68 kJ/g salt, that is, for our case with 100 g salt per kg solution we have $\Delta_{\text{mix}}H = -68$ kJ/kg solution. The heat capacity for the solution is assumed to be the same as for water, that is, $C_p = 4.18$ kJ/kg K, and we assume that water and $CaCl_2$-salt are at the same temperature before the mixing. The temperature rise is then from (4.29)*

$$\Delta T = \frac{-\Delta_{\text{mix}}H}{C_p} = \frac{68\ kJ/kg}{4.18\ kJ/kg\ K} = 16\ K$$

Comment: On molar basis, we have 1.6% salt and 98.4% water so this is indeed a dilute solution.

Example 4.8 "Add acid to water, not water to acid." *Mixing acid and water can be dangerous! The heat of mixing is negative, $\Delta_{\text{mix}}H < 0$, which means that heat is released. Mixing may result in a large and sudden temperature increase that gives flash boiling, and you may splash acid all over yourself. To reduce the danger, you should* **add acid to water** *(and not the other way around), We want to check this well-known lab rule using the heat of mixing data given on page 417. More precisely, we want to calculate the adiabatic temperature rise when we (a) mix 1 g pure sulfuric acid (H_2SO_4) with 1 kg water, and (b) mix 1 g water with 1 kg pure sulfuric acid.*

Solution. *(a) "Acid to water": We have 1 g ($= \frac{1\ g}{98.1\ g/mol} = 0.0102$ mol) H_2SO_4 and 1 kg ($= \frac{1000\ g}{18\ g/mol} = 55.5$ mol) water, that is, the mol ratio between water and acid is $n = 55.5mol/0.0102mol = 5444$. From the data on page 417, the heat of mixing is -96 kJ/mol H_2SO_4, that is, $\Delta_{\text{mix}}H = -96 \cdot 0.0102 = -0.98$ kJ/ kg solution, and with a heat capacity $C_p = 4.18$ kJ/kg K (water) the temperature rise is $\Delta T = 0.98/4.18 = 0.23$ K.*

(b) "Water to acid": We have 1g (0.0555 mol) water and 1 kg (10.2 mol) H_2SO_4, that is, the mol ratio is $n = 0.0555/10.2 = 0.0054$. The data on page 417 give that the heat of mixing is -0.226 kJ/mol H_2SO_4, that is, $\Delta_{\text{mix}}H = -0.226 \cdot 10.2 = -2.30$ kJ/ kg solution, and, with a heat capacity $C_p = 1.42$ kJ/kg K (H_2SO_4), the adiabatic temperature rise is $\Delta T = 2.30/1.42 = 1.62$ K, that is, about 7 times higher (than adding the same amount of acid to water).

Conclusion: *The lab rule is confirmed.*

Of course, in the previous exercise with mixing of acid and water, the *final* temperature change is the same if we mix a given amount of acid and water, irrespective of how we do the mixing. However, if we start by adding water to acid, we get a large *initial* temperature increase, but then the temperature will drop as we further dilute it with water. This large temperature overshoot does not occur if we add acid to water. In the next exercise, you should find the final temperature change.

Exercise 4.7* *We want to find the (final) temperature change when we mix 0.5 kg pure sulfuric acid (H_2SO_4) with 1 kg of pure water to make a 33 weight% sulfuric acid solution. The feeds are at 20 °C. (a) The heat capacity for 33 weight% sulfuric acid is $c_P = 3$ kJ/kg K. Use the data on page 417 to show that $\Delta_{\text{mix}}H(20^oC) = -217$ kJ/kg solution. (b) Find the adiabatic temperature rise. (c) Determine the cooling need to keep the temperature at 20 °C in a process where we mix 0.5 kg/s pure sulfuric acid (H_2SO_4) with 1 kg/s of pure water.*

4.7 Valve: Isenthalpic pressure relief

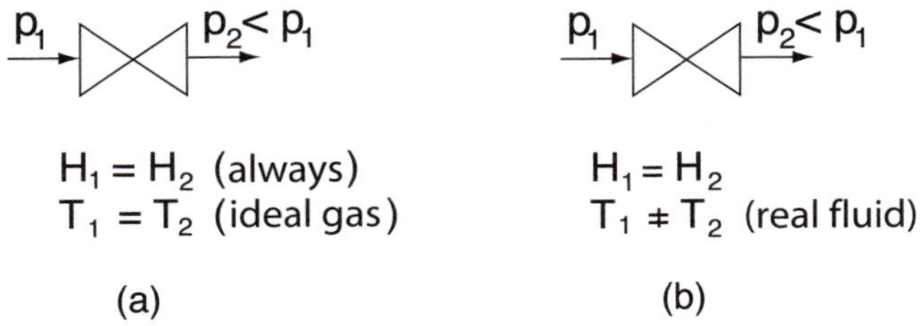

Figure 4.12: Pressure relief of (a) ideal gas and (b) real fluid

A very important industrial process is pressure relief (pressure reduction) through a valve (choke), see Figure 4.12. The process is adiabatic ($Q = 0$) and no shaft work ($W_s = 0$) is performed. We neglect changes in kinetic and potential energy (neglecting potential energy changes is safe for a valve, but for gases there is an increase in kinetic energy because the density drops over the valve). We then have for a steady-state process (without accumulation) $H_{\text{out}} = H_{\text{in}}$. In a valve, there is only one instream and one outstream, and to show this more clearly, we replace "in" by "1" and, correspondingly, "out" by "2." The energy balance then becomes

$$\boxed{H_2 = H_1} \tag{4.30}$$

that is, we have an *isenthalpic process*, where the stream's enthalpy is constant. This is a very important result and is used a lot! Note that the pressure changes.

Example 4.9 Expansion (pressure relief) of ideal gas in valve. *A gas stream at 10 bar and 303 K is relieved to 2 bar through a valve. What is the gas temperature after the valve if we assume ideal gas?*

The energy balance gives $H_2 = H_1$, and since $n_2 = n_1$ [mol/s] (mass balance), we get that the molar enthalpy is constant. For an ideal gas with constant composition, the (molar) enthalpy is only a function of temperature and we conclude that

$$\text{ideal gas}: \quad T_2 = T_1$$

that is, the temperature is constant (!) at 303 K. This assumes that we have ideal gas and can neglect changes in kinetic energy; see page 126 for an example of a leaking valve where we cannot neglect changes in kinetic energy.

For a **real gas**, the process is also isenthalpic, $H_2 = H_1$, but here we normally get some cooling over the valve ($T_2 < T_1$) because it usually takes some energy to pull the gas molecules apart – this is the so-called Joule-Thompson effect. Note the words "normally" and "usually" because there are also "unusual" conditions where the temperature may increase ($T_2 > T_1$) during expansion of gas in a valve (this is at very high pressures where it may take energy to push the molecules closer together; see the discussion in the next section).

When performing such calculations for real fluids, it is practical to use thermodynamic diagrams as discussed in the next section. The example with expansion of a gas is continued for a real gas (ammonia) in Example 4.10.

4.8 Real fluids: Thermodynamic state diagrams

Thermodynamic (state) diagrams are very useful for performing calculations for pure components. The diagrams also offer provide insight and it is recommended that you spend some time studying such diagrams. Three thermodynamic state diagrams are found in the Appendix:

- pH diagram for methane (page 418)
- pH diagram for ammonia (page 419)
- HS diagram for water (page 420)

Enlarged versions of these diagrams plus diagrams for additional components are found at the book's home page.

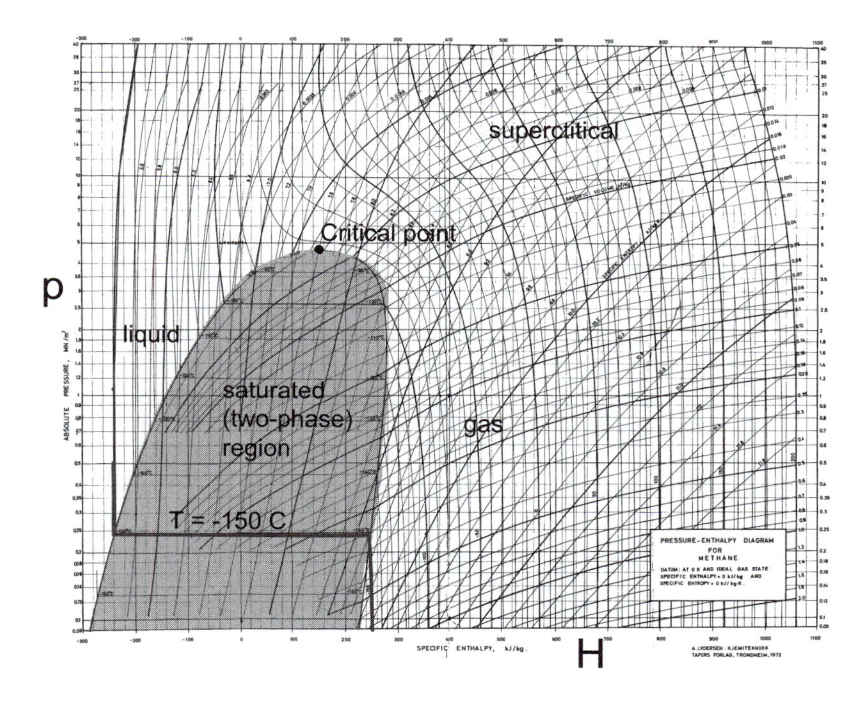

Figure 4.13: Pressure-enthalpy (pH) diagram for methane (see also page 418)

4.8.1 Pressure enthalpy (pH) diagram

For a given amount (say, 1 kg) of a pure component, it is sufficient to specify two independent variables to uniquely determine its state (see page 332). In the pH **(pressure-enthalpy) diagram** in Figure 4.13, the two independent variables are specific enthalpy h [kJ/kg] and pressure p [MN/m^2] (the latter on log scale). On the diagram are shown lines for constant temperature [^{o}C] (isotherms), constant specific entropy [kJ/kg K] and constant specific volume [m^3/kg]. We see that the isotherms are roughly vertical, except in the two-phase region where they are horizontal. In the

pH-diagram, we identify four regions; liquid at the left, two-phase in the lower middle, gas at the lower right and supercritical at the upper right. Actually, the liquid and gas regions are not as clearly defined as most people think, because if we go via the supercritical region, then it is possible to go from liquid to gas (and reverse) without observing any phase transition or discontinuity in properties.

How do we then define gas and liquid? Let us start with the two-phase "**saturated**" vapor-liquid region (gas in the saturated region is referred to as **vapor**) which has a clear boundary and is therefore well defined. **Liquid** is normally defined as the region where $T < T_c$ (critical temperature) and $p > p_{sat}(T)$. **Gas** is the region where $p < p_c$ (critical pressure) and $T > T_{sat}(p)$. The remaining region where $T > T_c$ and $p > p_c$ is the **supercritical** region. However, you should not be too concerned about the exact definitions because there is no *fundamental* difference between liquid and gas, at least not if the gas is at sufficiently high pressure so that we do not have ideal gas.

Exercise 4.8 *Use these definitions to identify more exactly the four regions (liquid, gas, saturated, supercritical) on the pH-diagram for methane.*

pH-diagram for methane. Let us take a closer look at the pH diagram in Figure 4.13 (for larger size, see page 418). The enthalpy and entropy are given with ideal gas at 0 K as the reference state, but this is irrelevant for use of the diagram. *Let us follow the isotherm at* -150^oC (which is lower than the critical temperature of -83^oC): We begin at 400 bar (=40 MN/m^2) (in the top left corner of the diagram). Here the enthalpy is -195 kJ/kg. The critical pressure for methane is 45 bar, so we are well above the critical pressure. We lower the pressure and follow the isotherm. At 10 bar the enthalpy has fallen to -240 kJ/kg, that is, we need to remove enthalpy (cool) to keep a constant temperature as pressure decreases (!); this is the opposite of what we "normally" expect for a non-ideal gas. The reason is that we are in the "unusual" high-pressure part of the diagram where the repulsive forces between the molecules dominate – the result is that the isotherms tilt to the right – and the **Joule-Thompson** temperature change through a valve is positive rather than negative. At lower pressures, still following the -150^oC isotherm, methane behaves more like one would expect for a liquid. First of all, the lines for constant volume are roughly vertical, that is, the volume is independent of pressure. Furthermore, the isotherms are roughly vertical, that is, enthalpy does not depend on pressure. We enter the two-phase region a little below 2.5 bar. Supply of heat (enthalpy) will here result in liquid evaporating under constant pressure and temperature. The isotherm is therefore horizontal in the two-phase region. Finally, all the liquid has evaporated and we are at the right of the two-phase region where we have gas. At this point the enthalpy has increased to about 246 kJ/kg. The isotherms are again almost vertical, which shows that the gas is close to ideal.

Let us now follow the isotherm at $0^o\,C$ (which is well above the critical temperature of -83^oC): Let us start at 400 bar where the enthalpy is 320 kJ/kg. When we lower the pressure at constant temperature, we remain now in the "gas" region the whole time, and since the isotherms tilt to the left, we need to supply heat (enthalpy) in order to keep a constant temperature. Thus, we are here in the more "normal" region where the attractive forces between the molecules dominate and we need to use energy to pull them apart – thus, the **Joule-Thompson effect** is negative, as we normally expect (see also Example 4.10). At 100 bar the enthalpy is 450 kJ/kg, and at 20 bar it is 560 kJ/kg. At lower pressures the isotherms becomes more vertical as the gas behaves more ideally. At 1 bar the enthalpy is about 570 kJ/kg.

4.8.2 Enthalpy-entropy (HS) diagram

Here, the two independent variables are specific enthalpy h [kJ/kg] and specific entropy s [kJ/kg K], and on the diagram are shown lines for constant pressure and temperature. The HS-diagram for water in the Appendix (page 420) shows the gas and supercritical regions together with parts of the two-phase region.

4.8.3 Cooling by expansion in a choke valve

There are two ways of generating cold by expanding a fluid to lower pressure in a choke valve (see Figure 4.12):

1. **Joule-Thompson effect.** If we expand a *real gas* in a valve, there is normally a negative temperature change ($\Delta T = T_2 - T_1 < 0$) because of the Joule-Thompson effect; see the expansion of ammonia gas in Example 4.10. Note that the term Joule-Thompson effect refers to a *gas*, or more generally a fluid that does not undergo a phase change. The Joule-Thompson effect is zero for an *ideal gas*.
2. **Flashing effect.** If we expand a *liquid* over a valve, then at a sufficiently low pressure p_2 we get evaporation where gas is formed. Since evaporation requires energy, this "flashing" will give in a negative temperature change (cooling), but this is *not* referred to as a Joule-Thompson effect. Normally, the cooling effect of flashing is significantly larger than the Joule-Thompson effect. For this reason, refrigerators and air condition systems are usually based on cycles where the main cooling is generated by flashing (sometimes referred to as "auto-refrigeration"); for more details see Example 8.6 (page 206).

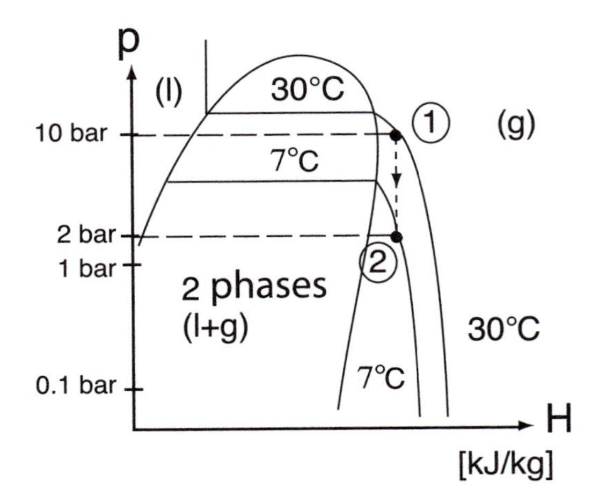

Figure 4.14: Joule-Thompson effect: Cooling by expanding ammonia gas from 10 bar (point 1) to 2 bar (point 2) over a valve ($H_1 = H_2$)

Example 4.10 Expansion of real gas (ammonia) over a valve. *Again, consider the expansion (pressure relief) from 10 bar to 2 bar in Example 4.9 (page 114), but now with a real gas (ammonia) instead of an ideal gas. The energy balance gives*

$$H_2 = H_1$$

and from the pressure-enthalpy diagram for ammonia (see Figure 4.14; more detailed on page 419), we read off the enthalpy at $T_1 = 303K$ *(30 °C) and 10 bar as* $H_1 = 1400$ *kJ/kg (this is compared to the reference state which here is saturated liquid at 0 °C, where for some unknown reason* $H = 100$ *kJ/kg has been chosen; but the absolute value is irrelevant for the calculations). The enthalpy is constant during the process, and we read from the diagram that the temperature at 2 bar drops to about* $t_2 = 8\,^oC$*, that is, we have a temperature drop (***Joule-Thompson effect***)*

$$\Delta T = T_2 - T_1 = -22 \text{ K}$$

We note that ammonia does not behave as an ideal gas in this region.

Flashing. *If we start with ammonia as saturated liquid at 10 bar (from the diagram this corresponds to a temperature of 24° C) and "flash" it down to 2 bar in the two-phase region, then the temperature drops to about −20° C, corresponding to a temperature change* $\Delta T = -44$ *K. Thus, flashing gives twice the temperature drop of a Joule-Thompson valve in this case.*

Exercise 4.9* *A stream of 10 kg/s of methane gas at 200 bar and 0 °C is throttled (choked) down to 1 bar. (a) What is the temperature after the choking? (b) How much heat must be supplied to get the temperature back to 0 °C?*

4.9 Energy balance with chemical reaction

We discussed the mass balance for systems with chemical reaction in Chapter 3. As shown in (3.7), for a *steady state process*, the component mass balance for an arbitrary component A, is for the case with reactions

$$n_{A,\text{out}} = n_{A,\text{in}} + \sum_{\text{reactions } j} \nu_{A,j}\xi_j \quad [\text{mol A; mol A/s}]$$

where ξ_j [mol; mol/s] is the extent of reaction for reaction j. Because the mass of a component is not a conserved quantity, we note that we need an "extra term" $G_A = \sum_j \nu_{A,j}\xi_j$ [mol A; mol A/s] to account for the chemical reactions.

On the other hand, energy is a conserved quantity, and chemical energy is included in the internal energy U (and is thereby included in the enthalpy H). It is therefore not necessary to include an "extra term" in the energy balance for cases with chemical reaction. Thus, the energy balance (4.13) for a *steady-state process* applies also to the case with reactions,

$$H_{\text{out}} = H_{\text{in}} + Q + W_s \quad [\text{J/s}] \tag{4.31}$$

It may seem a bit strange that there is no "extra term" for the heat of reaction, but this is because the enthalpy H (which normally has the elements as the reference) includes also the chemical energy. We will now illustrate the use of the energy balance, and its combination with the mass balance, for some cases with chemical reaction.

Example 4.11 Combustion of methane. *In an adiabatic combustion chamber at 1 bar, a continuous and complete combustion of methane (natural gas) takes place according to the reaction*

$$CH_4 + 2O_2 = CO_2 + 2H_2O(g)$$

The feed is methane and air at 25 °C and excess air is used such that there is 2.5 times more oxygen in the feed than what is needed stoichiometrically. The air is assumed to be 79% N_2 and 21% O_2. The remaining data can be taken from the table on page 416. **Task**: *Calculate the temperature T in the combustion chamber.*

Solution. *We choose as a basis 1 mol of CH_4 feed. From the stoichiometry, we then have that the feed (inflow) contains 5 mol O_2 and $\frac{0.79}{0.21} \cdot 5$ mol = 18.81 mol N_2. The extent of reaction is $\xi = 1$ mol, and from the mass balance ($n_A = n_{A,0} + \nu_A \xi$) the product stream becomes as shown in Table 4.11.*

Table 4.1: Stream data for combustion of methane

	Feed stream In [mol]	Product stream Out [mol]	$C_p^\ominus(298)$ [J/K mol]	$\Delta_f H^\ominus(298)$ [kJ/mol]
CH_4 (g)	1	0	35.31	−74.81
O_2 (g)	5	3	29.36	0
CO_2 (g)	0	1	37.11	−393.51
H_2O (g)	0	2	33.58	−241.82
N_2 (g)	18.81	18.81	29.13	0
T [K]	298	1382		
H [kJ]	−74.81	−74.81		

The (absolute) enthalpy of a stream is from (4.22) given as

$$H = \sum_{\text{components } i} n_i \cdot H_m(i, T) \tag{4.32}$$

where n_i is the number of moles of component i in the stream. From (4.24), the molar enthalpy of each component as ideal gas when we assume constant heat capacity is

$$H_m(i, T) = \Delta_f H_m^\ominus(i, 298) + C_{p,m}(T - 298.15) \quad \text{[J/mol]}$$

Here we have chosen the elements at 298 K as reference. Let us now put in numbers from Table 4.11. Since the inflow is at 298.15 K, there is no C_p-contribution here and thus

$$H_{\text{in}} = \sum_i n_{i,\text{in}} H_m(i, T_{\text{in}}) = 1 \cdot (-74.81) + 5 \cdot 0 + 18.81 \cdot 0 = -74.81 \text{ kJ}$$

or $H_{\text{in}} = -74810$ J. The enthalpy of the outlet stream is (everything is calculated in [J]):

$$H_{\text{out}} = \sum_i n_{i,\text{out}} H_m(i, T_{\text{out}}) = \underbrace{3 \cdot 0 + 1 \cdot (-393510) + 2 \cdot (-241820)}_{\text{heat of formation}}$$

$$+ \underbrace{(3 \cdot 29.36 + 1 \cdot 37.11 + 2 \cdot 33.58 + 18.81 \cdot 29.13)\ (T_{\text{out}} - 298.15)}_{C_p-\text{contribution}}$$

or $H_{\text{out}} = -877150 + 740.285 \cdot (T_{\text{out}}[K] - 298.15)$ [J]. The energy balance $H_{\text{out}} = H_{\text{in}}$ then gives that the (outlet) reactor temperature is

$$T_{\text{out}} = 298.15 + \frac{877150 - 74810}{740.285} = 298.15 + 1083.8 = 1382.0 \text{ K } (1109^oC)$$

Adiabatic temperature change for chemical reaction. Let us generalize the previous example. In an adiabatic reactor, the temperature will change due to the contribution from the heat of reaction. When constant heat capacity is assumed, the energy balance gives that the adiabatic temperature change is

$$\Delta_r T = T_r - T_0 = \frac{\sum_j \Delta_r H_j^{\ominus}(T_0)\xi_j}{C_p} \tag{4.33}$$

where T_r is the adiabatic reaction temperature, $\Delta_r H_j^{\ominus}(T_0)$ [J/mol] is the standard heat of reaction for reaction j at the feed temperature T_0, ξ_j [mol] is the extent of reaction for reaction j, $C_p = \sum_i n_i C_p^{\ominus}(i)$ [J/K] is the (total) heat capacity of the product and n_i [mol] is the composition of the product.

In Example 4.11, we have that $T_0 = 298.15$ *K and*

$$\Delta_r H^{\ominus}(298K) = \Delta_f H^{\ominus}(CO_2) + 2\cdot\Delta_f H^{\ominus}(H_2O) - \Delta_f H^{\ominus}(CH_4) - 2\Delta_f H^{\ominus}(O_2)$$

$$= (-393.51) + 2(-241.82) - (-74.81) - 2\cdot 0 = -802.34 \text{ kJ/mol} \tag{4.34}$$

and with ξ*=1 mol and* $C_p = 740.3$ *J/K, the adiabatic temperature change is* $\Delta_r T = \frac{802340}{740.3}K =$ 1084 *K (which, as expected, is the same as found above).*

Exercise 4.10 *Derive (4.33), for example, by adding the two imaginary subprocesses for (1) reaction at* T_0 *and (2) heating from* $T_{in} = T_0$ *to the reaction temperature* $T_{out} = T_0 + \Delta_r T$ *("method 2"; see also Figure 4.16).*

Example 4.12 Energy balance for methanol reactor. *This is a continuation of Example 3.9 on page 92. Methanol is formed from CO and* CO_2 *by two exothermic reactions,*

$$CO + 2H_2 = CH_3OH; \quad \Delta_r H^{\ominus}(298) = -90.1 \text{ kJ/mol}$$

$$CO_2 + 3H_2 = CH_3OH + H_2O; \quad \Delta_r H^{\ominus}(298) = -49.0 \text{ kJ/mol}$$

We assume ideal gas and thus neglect pressure's influence on enthalpy. Data for the composition of the total feed (stream 1+2) and the product (stream 3), together with thermodynamic data, are given in Table 4.2.

Table 4.2: Stream data for methanol reactor

Component	Stream 1+2 [mol]	Stream 3 [mol]	$C_p^{\ominus}(298)$ [J/K mol]	$\Delta_f H^{\ominus}(298)$ [kJ/mol]
CO(g)	3.8	0.9	29.14	−110.53
CO_2(g)	2.0	0.63	37.11	−393.51
H_2(g)	76.4	66.49	28.82	0
CH_3OH(g)	0.3	4.57	43.89	−200.66
H_2O(g)	0.2	1.57	33.58	−241.82
CH_4(g)	16.8	16.8	35.31	−74.81
N_2(g)	0.5	0.5	29.13	0
Total	100	91.46		

The reactions take place in a quench reactor where cold feed gas is injected along the reactor, see Figure 4.15. The feed to the reactor consists of a fraction f *at 60* ^oC *(stream 2) and a*

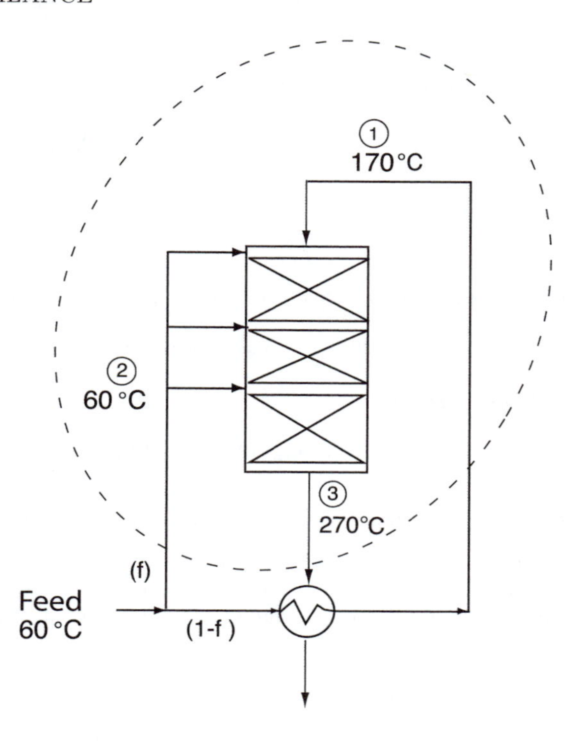

Figure 4.15: Methanol reactor with pre-heating of feed and quenching along the reactor

fraction $(1-f)$ *that is pre-heated to 170 °C (stream 1). The task is to find the value of* f *when it is given that the outlet temperature from the reactor is 270 °C (stream 3).*

Solution. *With 100 mol of total feed (given basis), we have*

$$n_1 = 100(1-f); \; n_2 = 100f; \; n_3 = 91.46 \quad [mol] \tag{4.35}$$

and the task is to find f. *The energy balance for an adiabatic reactor is* $H_{\text{out}} = H_{\text{in}}$, *that is,*

$$H_3 = H_1 + H_2 \quad [J]$$

$$n_3 H_{m,3} = n_1 H_{m,1} + n_2 H_{m,2} \quad [J] \tag{4.36}$$

where $H_{m,j}$ *is the molar enthalpy for stream* j. *Everything takes place in the gas phase so the "absolute" enthalpy for each stream* j *with the elements as the reference state is*

$$H_{m,j} = \Delta_{\text{f}} H_j^{\ominus}(298) + \int_{298.15K}^{T_j} C_{p,m,j}(T)dT \quad [\text{J/mol}] \tag{4.37}$$

where for our case with constant heat capacity

$$H_{m,j} = \Delta_{\text{f}} H_j^{\ominus}(298) + C_{p,m,j}(T_j - 298.15K)$$

Here, $H_{m,j}$ *[J/mol] is the enthalpy change to form 1 mol of* stream j *from the elements at 298.15 K and then heat the stream to* T_j. *Note that here we have introduced the enthalpy of formation* $\Delta_{\text{f}} H_j^{\ominus}(298)$ *[J/mol] for* stream j *and the heat capacity* $C_{p,m,j}$ *[J/mol K] for* stream j. *For ideal gas, these are equal to the molar average of the sum of the contributions for the components, see (A.43). For the feed streams (streams 1 and 2), we then get*

$$\Delta_f H_1^{\ominus}(298) = \frac{1}{n_{\rm tot}} \sum_i n_i \Delta_f H^{\ominus}(i) = \sum x_{i,1} \Delta_f H^{\ominus}(i)$$

$$= 0.038 \cdot (-110.53) + 0.02 \cdot (-393.51) + 0.764 \cdot 0 + 0.003 \cdot (-200.66)$$
$$+0.002 \cdot (-241.82) + 0.168 \cdot (-74.81) + 0.005 \cdot 0 = -25.72 \quad \text{kJ/mol}$$

Correspondingly, the molar enthalpy of formation for the product stream (stream 3) is:

$$\Delta_f H_3^{\ominus}(298) = -31.72 \quad \text{kJ/mol}$$

The molar heat capacity for streams 1 and 2 are

$$C_{p,m,1}(298) = C_{p,m,2}(298) = \sum_i x_i C_p^{\ominus}(i, 298) = 0.038 \cdot 29.14 + 0.02 \cdot 37.11 + 0.764 \cdot 28.82$$

$$+0.003 \cdot 43.89 + 0.002 \cdot 33.58 + 0.168 \cdot 35.31 + 0.005 \cdot 29.13 = 30.14 \ J/K \ mol$$

Correspondingly, for stream 3, $C_{p,m,3}(298) = 30.91$ *J/mol K. This is the heat capacity at 298 K, but we assume that the heat capacity is constant independent of the temperature.*

This gives the following "absolute" molar enthalpies for the three streams with the elements at 298.15 K as reference:

$$H_{m,1} = -25.72 + 30.14 \cdot 10^{-3} \cdot 145 = -25.72 + 4.37 = -21.35 \quad \text{kJ/mol}$$

$$H_{m,2} = -25.72 + 30.14 \cdot 10^{-3} \cdot 35 = -25.72 + 1.05 = -24.67 \quad \text{kJ/mol}$$

$$H_{m,3} = -31.72 + 30.91 \cdot 10^{-3} \cdot 245 = -31.72 + 7.57 = -24.15 \quad \text{kJ/mol}$$

Together with (4.35), this gives when inserted into the energy balance (4.36), $f = (22.09 - 21.35)/(24.67 - 21.35) = 0.22$, *that is, 22% of the feed must be supplied as cold quench gas.*

Check: Enthalpy change for subprocesses ("method 2"). *Here, we solve the problem by direct evaluation of the enthalpy changes. We know the heat of reaction at 298 K and to go from the inlet streams to the outlet stream, we imagine the following subprocesses (see Figure 4.16):*

1. *Produce the reactor feed at temperature* T_0 *(here we choose* $T_0 = 298.15$ *K):*
   (a) Take stream 1 from $T_1 = 443K$ *(170 °C) to 298.15 K (cooling):* $\Delta_{1,a}H = n_1 C_{p,m,1}(T_1 - 298.15) = 100(1-f) \cdot 30.14 \cdot (298.15 - 443) J = -437(1-f)\ kJ.$
   (b) Take stream 2 from $T_2 = 333K$ *(60 °C) to 298.15 K (cooling):* $\Delta_{1,b}H = n_2 C_{p,m,2}(T_2 - 298.15) = 100f \cdot \cdot 30.14 \cdot (298.15 - 333) J = -105f\ kJ.$
   (c) Mix the two streams at 298.15 K (there is no heat of mixing since the compositions are identical), $\Delta_{1,c}H = 0$.
2. *React at* T_0 *such that we form the outlet stream at* $T_0 = 298K$. *There are two reactions and the enthalpy change for this subprocess is:*

   $$\Delta_2 H = \xi_1 \Delta_r H_1^{\ominus}(298) + \xi_2 \Delta_r H_2^{\ominus}(298)$$

   From Example 3.9 (page 92) (and also from the data in the table), the extent of reactions are $\xi_1 = 2.9$ *mol and* $\xi_2 = 1.37$ *mol. The two heat of reactions can be obtained from the data for the heat of formation* $\Delta_f H^{\ominus}(i, 298)$ *given in Table 4.2. For the first reaction, we get*

   $$\Delta_r H_1^{\ominus} = \sum_i \nu_i \Delta_f H^{\ominus}(i) = (-1) \cdot -110.53 + (-2) \cdot 0 + (+1) \cdot (-200.66) = -90.13 \ kJ/mol$$

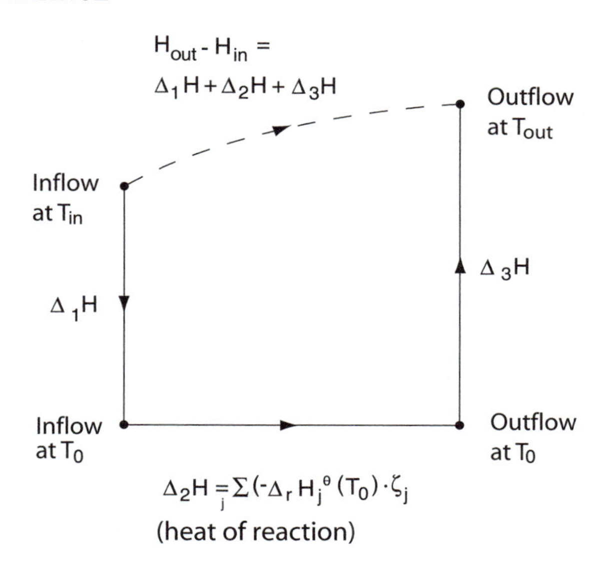

Figure 4.16: Energy balance for reactor with heat of reaction as subprocess

(that is, the reaction is exothermic). And, correspondingly, for the second reaction

$$\Delta_r H_2^\ominus = (-1) \cdot -393.51 + (-3) \cdot 0 + (+1) \cdot (-200.66) + (+1) \cdot (-241.82) = -48.97 \ kJ/mol$$

(also exothermic). This gives "the heat released by reaction" $\Delta_3 H = 2.9 mol \cdot (-90.13) kJ/mol + 1.37 mol \cdot (-48.97) kJ/mol = -328.47 \ kJ.$

3. *Heat stream 3 from* $T_0 = 298.15 \ K$ *to* $T_{\text{out}} = T_3 = 543K$ *(270 °C):* $\Delta_3 H = n_3 C_{p,m,3}(T_3 - 298.15) = 91.46 \cdot 30.91 \cdot (543 - 298.15) J = 692.6 \ kJ.$

We thus find that

$$H_{\text{out}} - H_{\text{in}} = \Delta_{1a} H + \Delta_{1b} H + \Delta_2 H + \Delta_3 H$$

and the energy balance $H_{\text{out}} - H_{\text{in}} = 0$ *gives*

$$\underbrace{-437(1-f) - 105f}_{\Delta_1 H} \quad \underbrace{-328.47}_{\Delta_2 H} \quad \underbrace{+692.6}_{\Delta_3 H} = 0 \quad [\text{kJ}]$$

Since the reactor operates adiabatically, we can state this with words: "The cooling needed for the subprocesses 1 and 2 (negative sign) must balance the heating needed for subprocess 3 (positive sign)." We solve the equation and find, as before, $f = 0.22$.

Comment. *The outlet stream (the reactor product) is at 270 °C and we could alternatively have chosen* $T_0 = T_{\text{out}} = 270$ *°C (see Figure 4.16). With data for heat of reaction at 270 °C (which we can easily find, see (A.59) and (A.60)), we can then compute* $H_{\text{out}} - H_{\text{in}}$ *by cosidering the following subprocesses:*

1. *(a) Take stream 1 from 170 °C to 270 °C (heating needed).*
   (b) Take stream 2 from 60 °C to 270 °C (heating needed).
2. *React at 270 °C (cooling needed since the reactions are exothermic).*

Note that the contribution $\Delta_3 H$ *in Figure 4.16 is zero since* $T_{\text{out}} = T_0$.

The solution is obviously the same, that is, $f = 0.22$ *(try it yourself!).*

A third alternative is to react at the feed temperature, that is, to choose $T_0 = T_{\text{in}}$*, but in this case this is impractical since the feed streams have different temperatures.*

Because enthalpy is a state function, we note that there are many ways to get the same answer.

Exercise 4.11 Styrene process.

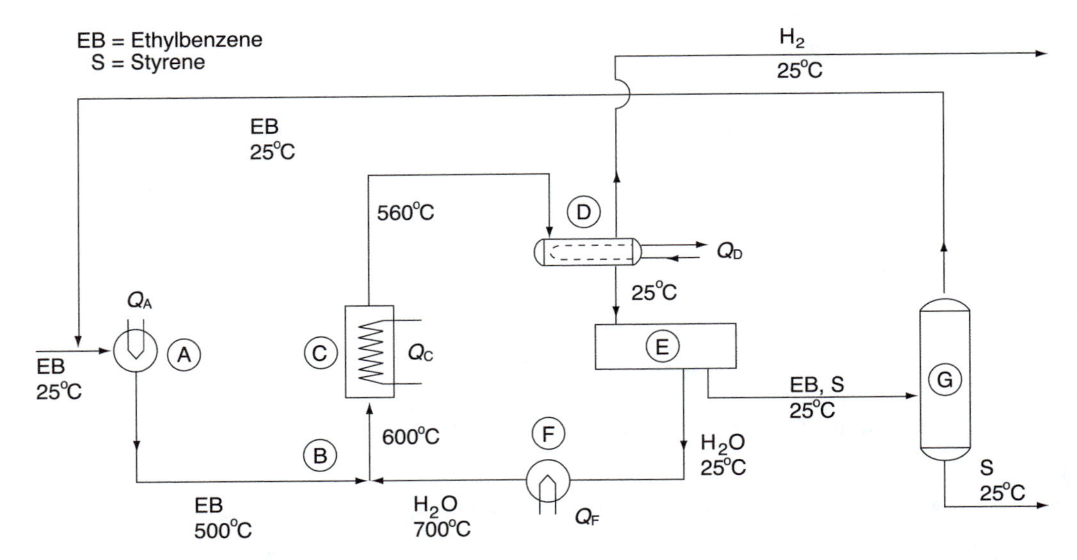

Figure 4.17: Simplified flow sheet for the styrene process

Ethyl benzene reacts to styrene in a catalytic gas phase reaction

$$C_8H_{10}(g) = C_8H_8(g) + H_2 \ ; \quad \Delta_r H^{\ominus}(600^oC) = +124.5 \text{ kJ/mol}$$

In Figure 4.17, a simplified version of the process' flow sheet is shown. The process takes place at 1 atm. Feed and recycled ethyl benzene in liquid phase at 25 °C are mixed and heated from 25 °C to 500 °C in an evaporator (A). Ethyl benzene, now in the gas phase, is mixed with steam (water vapor) at 700 °C, and fed to the reactor at 600 °C (B) (the steam is added to avoid undesired side-reactions and remove carbon from the catalyst). The reaction product is cooled to 25 °C (D), such that ethyl benzene, styrene and water are condensed, while H_2 is removed as a gas product. Water and hydrocarbons are insoluble and are separated in a liquid separator (E). The water is evaporated (F) and recycled to the reactor. The hydrocarbon stream is separated in a distillation column (G) into almost pure ethyl benzene (which is recycled to the evaporator) and pure styrene (product). The amount of styrene product should be 1 kg/s.

Data:
Ethyl benzene (EB): $C_p(l) = 182$ J/mol K; $\Delta_{\text{vap}}H = 36.0$ kJ/mol (at $t_b = 136^o C$); $C_p(g)$ [J/mol K] $= 118 + 0.30 \cdot t$ [°C].
Styrene (S): $C_p(l) = 209$ J/mol K; $\Delta_{\text{vap}}H = 37.1$ kJ/mol (at $t_b = 145$ °C); $C_p(g)$ [J/mol K] $= 115 + 0.27 \cdot t$ [°C].
Water (H_2O): $C_p(l) = 75.3$ J/mol K; $\Delta_{\text{vap}}H = 40.7$ kJ/mol (at $t_b = 100$ °C); $C_p(g)$ [J/mol K] $= 33.4 + 0.01 \cdot t$ [°C].
Hydrogen: $C_p(g) = 29.1$ J/mol K .

Task: *(a) Calculate the amount of feed and recycled ethyl benzene (all in mol/s).*
(b) Calculate the amount of water that circulates in the reactor system (in mol/s).

(c) Calculate the amount of supplied or removed heat (energy) in the evaporators (A and F) and in the reactor (C) (in J/s).

(d) Discuss the possibility of reducing the energy consumption.

(e) A feed stream to replace loss of water in the products is not shown in the flow sheet but should be included. How large must this make up feed be if the vapor pressure of water at 25 ^oC is 0.03 atm such that the hydrogen product contains 3 mol% of water?

4.10 Energy balance with kinetic and potential energy

So far in this chapter, we have neglected changes in kinetic and potential energy. With these contributions included, the general energy balance (4.10) becomes

$$\begin{aligned}(U + E_K + E_P)_f - (U + E_K + E_P)_0 &= (H + E_K + E_P)_{\text{in}} - (H + E_K + E_P)_{\text{out}} \\ &+ Q + \underbrace{W_s + W_{\Delta V} + W_{\text{el}} + W_{\text{other}}}_{W_n} \; [J] \quad (4.38)\end{aligned}$$

For a process stream with mass m [kg], the potential energy is

$$E_P = mgz \quad [J]$$

where z [m] is the height relative to a chosen reference and g [m/s^2] is the acceleration of gravity. The kinetic energy is

$$E_K = m\alpha\frac{v^2}{2} \quad [\text{J}]$$

where the average (mean) velocity v defined as

$$v[\text{m/s}] = \frac{\dot{V}\ [\text{m}^3/\text{s}]}{A\ [\text{m}^2]}$$

where $\dot{V}$ [m^3/s] is the volumetric flow and A [m^2] is the cross section (for example, the cross section of a pipe). The factor α corrects for the velocity not being the same all over the cross section. For turbulent flow (the most common case), the velocity profile is almost flat and $\alpha \approx 1$, while for laminar pipe flow the velocity is higher towards the middle of the pipe and we have $\alpha = 2$ (this is discussed in more detail in fluid mechanics).

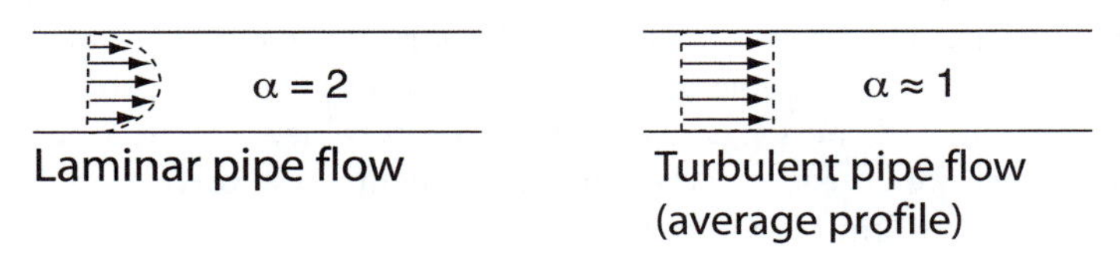

Figure 4.18: Velocity profiles for flow in pipe

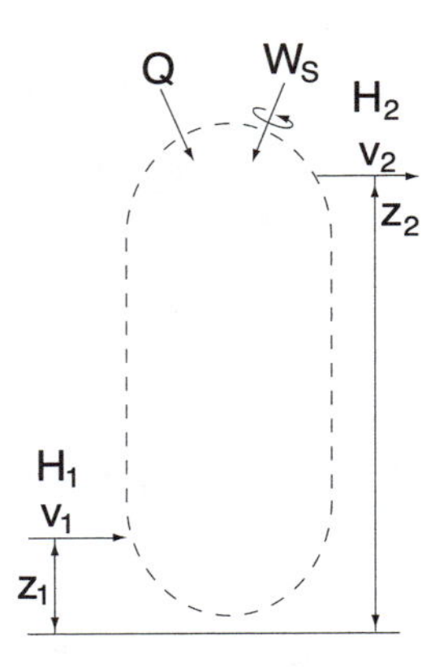

Figure 4.19: Steady-state energy balance with kinetic and potential energy

Steady-state process

Here, we again consider a steady state continuous process such that the left side of (4.38) is zero. We also assume $W_{el} = 0$ and $W_{other} = 0$. The steady-state energy balance with potential and kinetic energy included for the case with only one inlet stream and one outlet stream (see Figure 4.19) then becomes

$$\boxed{H_2 + m\alpha_2\frac{v_2^2}{2} + mgz_2 = H_1 + m\alpha_1\frac{v_1^2}{2} + mgz_1 + Q + W_s} \quad [J; J/kg; J/s] \qquad (4.39)$$

Here, we have replaced "out" by 2 and "in" by "1" in order to simplify the notation and clearly state that we only have one inlet stream and one outlet stream. It is trivial to generalize to the case with several streams.

As before, there are many choices of basis, for example mass basis [J/kg] (where $m = 1$ kg) or per unit of time [J/s] (where $m = \dot{m}$ [kg/s] is the mass flow).

The velocities can be found from the mass balance. The mass flow [kg/s] at the inlet is $\dot{m}_1 = \rho_1\dot{V}_1 = \rho_1 v_1 A_1$ and at the outlet $\dot{m}_2 = \rho_2\dot{V}_2 = \rho_2 v_2 A_2$, where v_1 and v_2 [m/s] are the average velocities. The mass balance $\dot{m}_1 = \dot{m}_2$ [kg/s] then gives the "continuity equation"

$$\rho_1 v_1 A_1 = \rho_2 v_2 A_2 \qquad (4.40)$$

For an **incompressible fluid** (most liquids), we have that $\rho_1 = \rho_2$, and we find that

$$v_1 A_1 = v_2 A_2 \quad [m^3/s] \qquad (4.41)$$

and we see that the velocity is uniquely determined by the cross section area.

Compared to changes in internal energy, the changes in potential and kinetic energy are usually negligible. This has already been illustrated using the waterfall example on page 96. However, if the velocity becomes very large (over 100 m/s), then the contribution from kinetic energy can become considerable. This is illustrated in the next example, where we consider icing that may occur on a valve that leaks because enthalpy is converted to kinetic energy.

Example 4.13 Leaking valve. *A leaking valve is often detected because ice is formed – also for cases where the gas temperature inside the pipe is well above 0 °C. This is because* *the gas velocity at the leakage point can become very large, and the energy needed for this is obtained by lowering the temperature.*

Consider an example where the temperature is 300 K and the pressure is 3 bar inside the pipe. We consider four different cases for the velocity of the gas leaking out: 100 m/s, 200 m/s, 300 m/s and equal to the speed of sound.

Here, the **speed of sound** c_s *for an ideal gas is equal to*

$$c_s = \sqrt{\gamma \frac{R}{M} T} = \sqrt{c_p(\gamma - 1)T} \quad [m/s] \tag{4.42}$$

where c_p *[J/kg K] is specific heat capacity,* $\gamma = c_p/c_V$ *and* M *is the molar mass [kg/mol]. We use numbers for air as the ideal gas and assume constant* $c_p = 1$ *kJ/kg,K and* $\gamma = 1.4$.

Exercise: *Calculate the temperature change* ΔT *for the leaking gas for the four cases.*

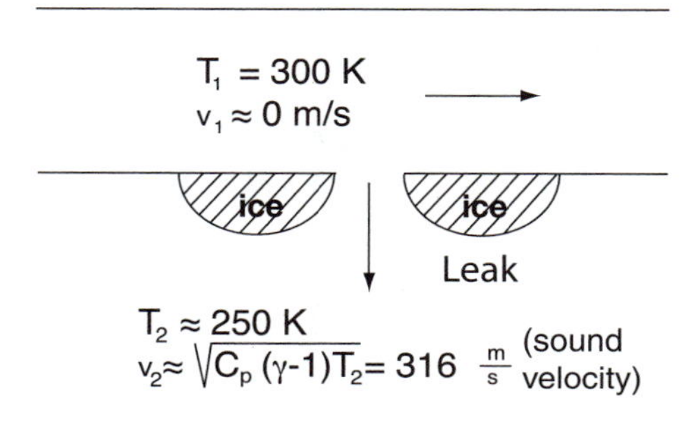

Figure 4.20: Leaking valve

Solution. *The energy balance (4.39) on mass basis becomes*

$$\boxed{h_2 + \alpha_2 \frac{v_2^2}{2} + g z_2 = h_1 + \alpha_1 \frac{v_1^2}{2} + g z_1 + w_s + q} \quad [J/kg]$$

Here, we have $w_s = 0$ *(no work) and* $q = 0$ *(adiabatic process) and we neglect potential energy* $g(z_2 - z_1) \approx 0$*). The energy balance gives*

$$h_2 + \alpha_2 \frac{v_2^2}{2} = h_1 + \alpha_1 \frac{v_1^2}{2} \quad [J/kg]$$

In our case, we can set $v_1 \approx 0$ *(inside the pipe) and we assume that we have turbulent flow at the leakage point such that* $\alpha_2 \approx 1$. *The energy balance then gives*

$$\text{Ideal and real gas}: \quad h_2 - h_1 = -\frac{v_2^2}{2}$$

Assuming ideal gas and constant heat capacity, we have further that $h_2 - h_1 = c_p(T_2 - T_1)$ *and we find*

$$\text{Ideal gas}: \quad \Delta T = T_2 - T_1 = -\frac{v_2^2}{2c_p} \tag{4.43}$$

For $v_2 = 100$ *m/s, we then get*

$$\Delta T = -\frac{100^2}{2 \cdot 1000} = -5 \ K$$

and the temperature for the other cases are given in the table:

v_2 $[m/s]$	$\Delta T = T_2 - T_1$ $[K]$
100	−5
200	−20
300	−45
316.2	−50

For the last case, when the velocity is the speed of sound, $v_2 = c_s = \sqrt{c_p(\gamma-1)T_2}$, we get from (4.43),

$$T_2 - T_1 = -\frac{v_2^2}{2c_p} = -\frac{(\gamma-1)T_2}{2} \tag{4.44}$$

which solved in respect to T_2 gives

$$T_2 = \frac{T_1}{1+\frac{\gamma-1}{2}} = \frac{T_1}{1.2} = 250\ K \quad \Rightarrow \quad T_2 - T_1 = -50K$$

Comment. *The velocity at the smallest cross section can never exceed the speed of sound. At* **sonic conditions** *($v_2 = c_s$ speed of sound), we find by combining (4.44) and (A.40) (which applies frictionless (reversible) flow), that the pressure ratio p_2/p_1 is equal to the* **critical pressure ratio** *Ψ, where*

$$\Psi = \left(\frac{2}{\gamma+1}\right)^{\frac{\gamma}{\gamma-1}} \tag{4.45}$$

If the pressure ratio exceeds the critical pressure ratio,

$$(p_2/p_1) \geq \Psi \tag{4.46}$$

we will have sonic conditions ($v_2 = c_s$) at the smallest cross section of the contraction (nozzle), provided we have a smooth nozzle with no friction. In reality, the speed v_2 will be lower. For air, $\gamma = 1.4$ and we get $\Psi = 0.53$. This means that if the surrounding pressure is p_0 (for instance, 1 bar in our case) and p_1 exceeds p_0/Ψ (that is, in our case if p_1 exceeds 1 bar/0.53 = 1.88 bar; which it does since $p_1 = 3$ bar), then for frictionless flow the pressure at the smallest cross section (at the leakage point) is equal to $p_2 = \Psi p_1$ (that is, $p_2 = 0.53 \cdot 3$ bar = 1.59 bar in our case) and here the velocity will be equal to the speed of sound (or lower).

Concluding our example, we see that it is likely that the velocity at the leakage point is close to the speed of sound (which is a little more than 300 m/s), and the resulting temperature drop of about 50K makes icing very likely to occur.

4.11 Summary of energy balance

The general energy balance over the time period from t_0 to t_f is

$$\underbrace{U_f - U_0}_{\Delta U = \Delta U_{\text{system}}} + \underbrace{H_{\text{out}} - H_{\text{in}}}_{\Delta H_{\text{flows}}} = Q + W_s + W_{\Delta V} + W_{\text{el}} + W_{\text{other}}$$

where $H_{\text{in}} = U_{\text{in}} + p_{\text{in}}V_{\text{in}}$ is the sum of the inlet flow's internal energy and flow work, and $H_{\text{out}} = U_{\text{out}} + p_{\text{out}}V_{\text{out}}$ is the sum of the outlet stream's internal energy and flow work. The above energy balance is for the most common case where the energy only includes internal energy (in the molecules). If other energy forms are important, such as kinetic or potential energy, then it is always possible to add these terms to U and H, see (4.39).

5

Heat exchange

So far, we have dealt generally with mass and energy balances. In this chapter, we show how to calculate the heat transfer (Q) that enters into the energy balances. This involves the very important unit operation of heat exchange.

5.1 Introduction

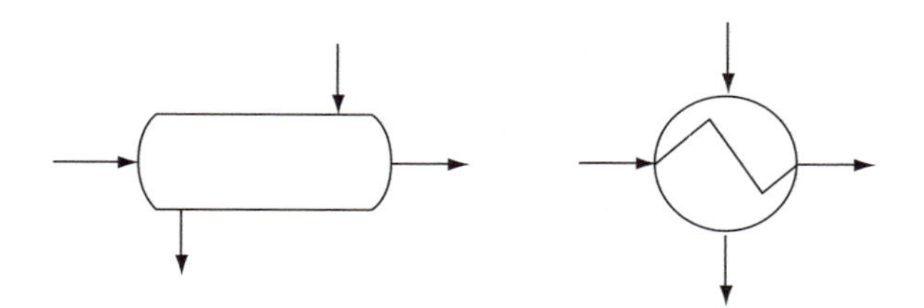

Figure 5.1: Flow sheet symbols for heat exchanger

In industrial processes, the heat transfer between streams is usually carried out in a **heat exchanger** where heat is transferred from the hot side (stream) to the cold side (stream) through the wall of the heat exchanger.

Some examples of heat exchange:

- heat, evaporate or melt a stream
- cool, condense or freeze a stream
- recover energy by matching energy excess in a hot stream with energy demand and a cold stream

Figure 5.2 shows the commonly used **shell and tube** countercurrent flow heat exchanger. Heat exchangers are also known from everyday life:

- Radiator in a car or a house where heat is transferred from hot water to air (liquid/gas heat exchanger).
- Cooling ribs in a refrigerator where heat is transferred from the hot refrigerant to the surrounding air (liquid/gas heat exchanger).
- Heat exchanger in building (cold climate) where heat is transferred from the hot outgoing to the cold incoming air (gas/gas heat exchanger).

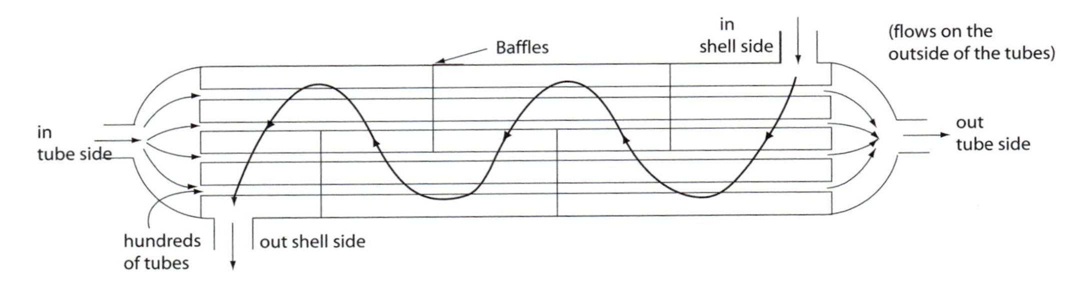

Figure 5.2: Shell and tube heat exchanger with countercurrent flow

Humans mainly use evaporation (sweating) for cooling (see page 36), but this is not a pure heat exchange process because there is also an exchange of mass.

In heat exchangers, a large contact area is desired to get good heat transfer. This applies in particular to gas/gas heat exchangers because the heat transfer coefficient to gas is usually small.

There are two main principles for contacting streams in a heat exchanger: **cocurrent flow** or **countercurrent flow**. Generally countercurrent flow is the most effective. A practical example of countercurrent flow is the foot of a duck, see Figure 5.3. In addition to heat transfer, the countercurrent flow principle is also used for mass

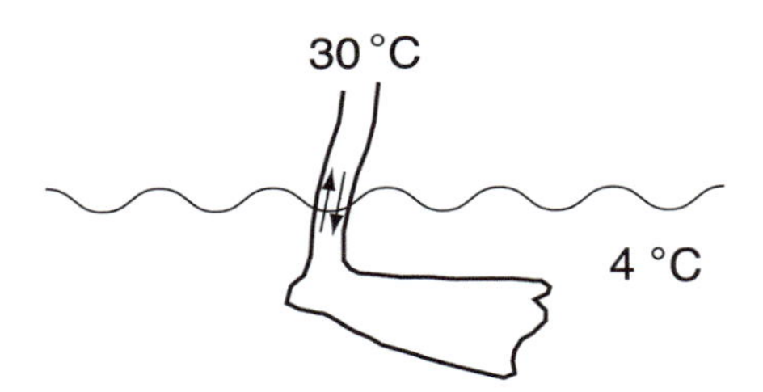

Figure 5.3: The countercurrent flow principle: When the duck is in cold water, the blood at the end of the foot is cold, but on its way back to the bird's body, the cold blood is heated by the countercurrent flow of hot blood.

transfer applications, for example, in distillation, absorption, extraction and membrane processes (see Example 2.4 and the following exercises). An example of a membrane process is the transfer of oxygen to the blood in the lungs.

The principle for co- and countercurrent heat exchange is shown in Figure 5.4 (the same principle applies to mass transfer, but with temperature replaced by concentration). In a **countercurrent** flow heat exchanger, the temperatures for the streams approach each other at the exit. There is a good utilization of the whole heat exchanger, in particular if the heat capacity flow rates, $(mc_p)_c$ and $(mc_p)_h$ [J/s K], are about the same for both streams, such that the temperature profiles are similar. Note that the exit temperature of the cold stream may be higher than the exit temperature of the hot stream, which of course is impossible with cocurrent flow.

Despite the fact that countercurrent flow is the most effective, it is quite common, due to economic and mechanical reasons, to use heat exchangers with a mix of counter and cocurrent flow, as for the **one shell pass and two tube passes** configuration

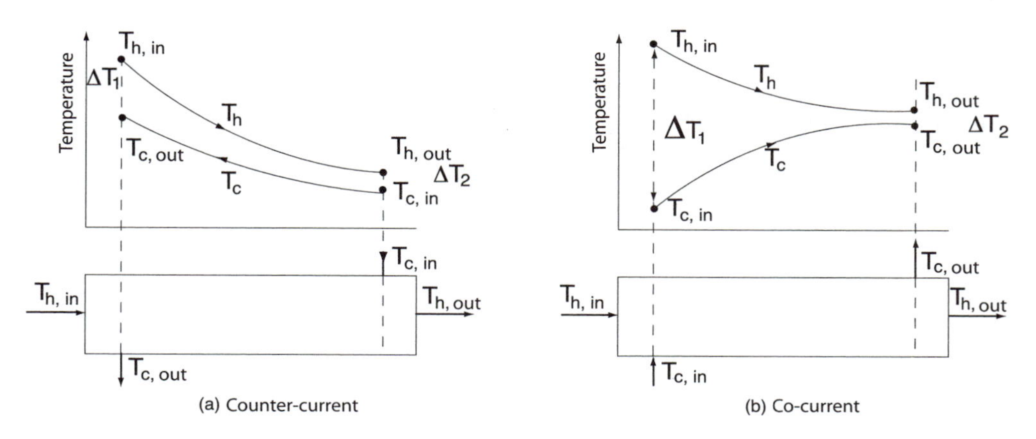

Figure 5.4: (a) Countercurrent flow and (b) cocurrent flow heat exchanger

discussed later; see Figure 5.7.

Sign of Q in this chapter (a remark on notation). Normally, Q is defined as the heat supplied to the system. However, in a heat exchanger, the system may be both the hot side and the cold side. Therefore, to avoid confusion, in this chapter, Q ([J/s=W]) is defined as the positive amount of heat transferred from the hot (h) to the cold (c) side. This means that the supplied heat on the cold side is Q, and the supplied heat on the hot side is $-Q$. Also note that in this chapter we omit the dot notation for rates, that is, we use Q and m in the meaning of $\dot{Q}$ and $\dot{m}$.

5.2 Calculation (design) of heat exchangers

Consider Figure 5.5 where a hot stream m_h [kg/s] is cooled from $T_{h,\text{in}}$ to $T_{h,\text{out}}$, by heat exchange with a cold stream m_c [kg/s], which is heated from $T_{c,\text{in}}$ to $T_{c,\text{out}}$.

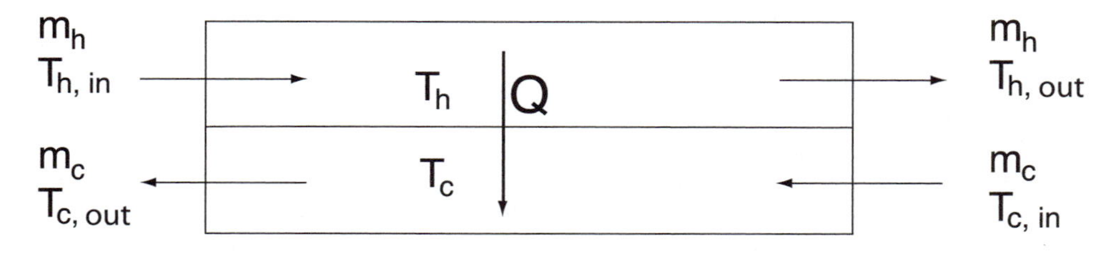

Figure 5.5: Schematic diagram of heat exchanger

Heat transfer

In the case of constant temperatures on the hot and cold sides, the heat transfer from the hot to the cold side is given by

$$Q = UA\Delta T \tag{5.1}$$

where

- $\Delta T = T_h - T_c$ – temperature difference between hot and cold sides.
- A [m^2] – area of the heat transfer surface.
- U [W/m^2, K] – overall heat transfer coefficient.

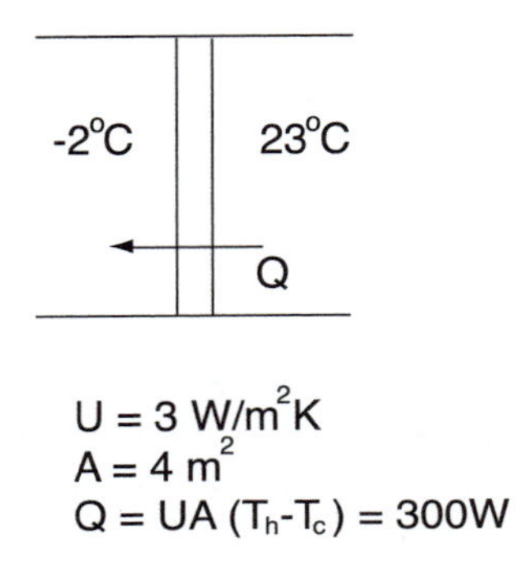

Figure 5.6: Heat loss through window

Example 5.1 Heat loss through window. *Indoor air with a constant temperature of 23° C exchanges heat through a window with outdoor air with constant temperature -2° C. The heat transfer coefficient is $U = 3$ W/m^{2}K and the area A of the "heat exchanger" (window) is 4 m^2. Calculate the heat transferred Q (the heat loss).*

Solution. *Since the temperature on both sides is constant, the heat transfer can be calculated from (5.1). We find*

$$Q = UA(T_h - T_c) = 3W/m^2K \cdot 4m^2 \cdot 25K = 300W$$

In the general case, the temperature difference $\Delta T = T_h - T_c$ varies with the position through the heat exchanger, see Figure 5.4, and we need to replace ΔT in (5.1) with the mean temperature difference $\overline{\Delta T}$. If the temperature difference at one end of the heat exchanger is ΔT_1 and at the other ΔT_2, then our first guess is to set $\overline{\Delta T}$ equal to the arithmetic mean, $(\Delta T_1 + \Delta T_2)/2$. However, the situation is not quite as favorable in reality because the temperatures "flatten out" in the region where ΔT is small. For ideal countercurrent flow (and also for ideal cocurrent flow) with *constant heat capacities*, the correct mean value is the **logarithmic mean temperature** difference defined by (for derivation see page 394):

$$\boxed{\Delta T_{\text{lm}} = \frac{\Delta T_1 - \Delta T_2}{\ln(\Delta T_1/\Delta T_2)}} \tag{5.2}$$

This is always lower than the arithmetic mean. The temperatures differences ΔT_1 and ΔT_2 between the hot and cold streams at the two ends (entrance and exit) (1 and 2) of the heat exchanger are given by:

- **Countercurrent flow**, see Figure 5.4(a),

$$\Delta T_1 = T_{h,\text{in}} - T_{c,\text{out}}, \quad \Delta T_2 = T_{h,\text{out}} - T_{c,\text{in}} \tag{5.3}$$

- **Cocurrent flow**, see Figure 5.4(b),

$$\Delta T_1 = T_{h,\text{in}} - T_{c,\text{in}}, \quad \Delta T_2 = T_{h,\text{out}} - T_{c,\text{out}} \tag{5.4}$$

In order to adjust ΔT_{lm} for deviation from ideal countercurrent flow (or ideal cocurrent flow), we introduce the factor $F \leq 1$, and we then generally have for cases with constant heat capacities

$$\boxed{Q = UA\Delta T_{\text{lm}}F} \tag{5.5}$$

For ideal countercurrent and ideal cocurrent flow, $F = 1$. For the common "mixed" countercurrent flow case with a heat exchanger with a "U-turn" on the tube side (**one shell pass and two tube passes**), you can use Figure 5.7 to find F as a function of the "efficiency" P and heat capacity flow ratio Z.

Energy balances

The heat transfer is calculated using (5.5). In order to find all temperatures, we need, in addition, the energy balances for the hot and cold sides. Assuming steady state and no heat loss to the surroundings, the energy balance (4.13) for the **cold side** is

$$(H_{\text{out}} - H_{\text{in}})_c = Q$$

$$m_c\,(h_{c,\text{out}} - h_{c,\text{in}}) = Q \tag{5.6}$$

where h_c [J/kg] is specific enthalpy on cold side and $Q > 0$ is the heat supplied (from the hot side).

1. For cases where we have, on the cold side, evaporation at constant temperature T_c (for example, boiling of water at 100 ^{o}C), we get $h_{c,\text{out}} - h_{c,\text{in}} = \Delta_{\text{vap}}h(T_c)$ [J/kg] = heat of vaporization for cold fluid.
2. For cases where we have a fluid with constant heat capacity $c_{p,c}$ [J/kg K], we get $h_{c,\text{out}} - h_{c,\text{in}} = c_{p,c}(T_{c,\text{out}} - T_{c,\text{in}})$ and the energy balance for the cold side becomes

$$\boxed{Q = m_c c_{p,c}(T_{c,\text{out}} - T_{c,\text{in}})} \tag{5.7}$$

The energy balance for the **hot side** (*hot*) is correspondingly

$$(H_{\text{out}} - H_{\text{in}})_h = -Q$$

$$m_h\,(h_{h,\text{out}} - h_{h,\text{in}}) = -Q \tag{5.8}$$

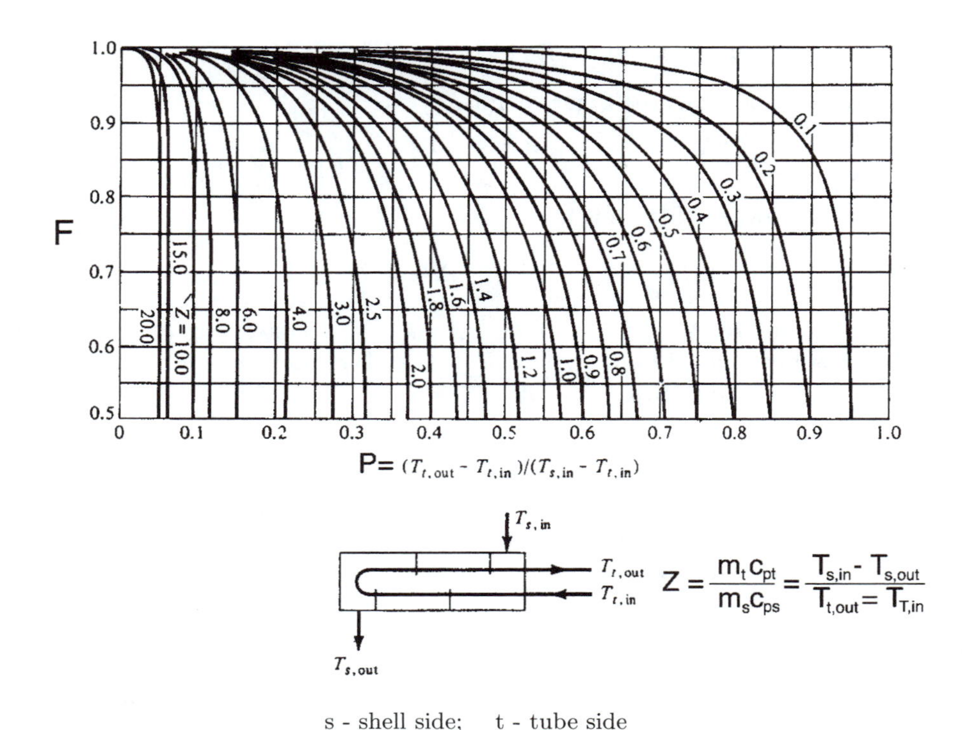

Figure 5.7: Correction factor F for countercurrent heat exchanger with "U-turn" on the tube side (*one shell pass and two tube passes*). $F = 1$ for ideal countercurrent flow.

1. For cases with, on the hot side, condensation at constant temperature T_h (for example, condensation of water vapor at 120 ^{o}C and 2 bar), we have $h_{h,\text{out}} - h_{h,\text{in}} = -\Delta_{\text{vap}} h(T_h)$ [J/kg] = the condensation heat for the fluid on hot side.
2. For cases where we have a fluid with constant heat capacity $c_{p,h}$ [J/kg K] on the hot side, we get

$$\boxed{Q = m_h c_{p,h}(T_{h,\text{in}} - T_{h,\text{out}})} \tag{5.9}$$

In summary, the necessary equations for a heat exchanger with no heat loss and constant heat capacities are given by (5.2), (5.5), (5.7) and (5.9). These equations are well suited for **design** where the objective is to find A. However, for the "**simulation**" case where A is given, it is recommended to use the rearranged equations in Section 5.3.

Example 5.2 Area of countercurrent flow heat exchanger *(see Figure 5.5). A hot stream with mass flow m_h = 1 kg/s, heat capacity $c_{p,h}$ = 4 kJ/kg K and temperature $T_{h,\text{in}} = 360$ K is to be cooled to $T_{h,\text{out}} = 330$ K by heat exchange with a cold stream with mass flow m_c = 2 kg/s, heat capacity $c_{p,c}$ = 3 kJ/kg K and temperature $T_{c,\text{in}} = 320$ K. The overall* *heat transfer coefficient U is 500 W/m². Calculate the area of the heat exchanger when ideal countercurrent flow is assumed.*

Solution. *First, we need to use the energy balances to find Q and the outlet temperature on the cold side. The energy balance (5.9) on the hot side gives*

$$Q = m_h c_{p,h}(T_{h,\mathrm{in}} - T_{h,\mathrm{out}}) = 1kg/s \cdot 4kJ/kg\ K \cdot (360K - 330K) = 120kW$$

The energy balance (5.7) on the cold side gives

$$Q = m_c c_{p,c}(T_{c,\mathrm{out}} - T_{c,\mathrm{in}}) \quad \Rightarrow \quad T_{c,\mathrm{out}} - T_{c,\mathrm{in}} = \frac{Q}{m_c c_{p,c}} = \frac{120}{2 \cdot 3} = 20\ K$$

that is, $T_{c,\mathrm{out}} = 340\ K$*. The temperature differences between hot and cold sides at the two ends of the heat exchanger are then according to (5.3) (countercurrent flow):*

$$\Delta T_1 = T_{h,\mathrm{in}} - T_{c,\mathrm{out}} = 360K - 340K = 20K$$

$$\Delta T_2 = T_{h,\mathrm{out}} - T_{c,\mathrm{in}} = 330K - 320K = 10K$$

and the logarithmic mean temperature difference is

$$\Delta T_{\mathrm{lm}} = \frac{\Delta T_1 - \Delta T_2}{\ln(\Delta T_1/\Delta T_2)} = 14.4K$$

(which we note, as expected, is slightly lower than the arithmetic mean value of 15K). From (5.5), the required area of the heat exchanger, assuming ideal countercurrent flow (F = 1), is:

$$A = \frac{Q}{U\Delta T_{\mathrm{lm}}} = \frac{120 \cdot 10^3 W}{500(W/m^2\ K) \cdot 14.4K} = 16.6m^2$$

Heat exchangers: Some comments and typical numbers

1. The **overall heat transfer coefficient** U for industrial heat exchangers is typically in the range from 5 W/m² K and up to 4000 W/m² K; see also Table 5.1. The lowest values are for gas/gas heat exchangers at low pressure.
2. In English literature one often finds U given in Btu/(hr)(sqft)(°F) and to *convert* to W/m² K, you need to multiply by 5.6784. For example, the design book by Seider et al. (*Process design principles*, Wiley, 1999) states that typical U-values are 200 Btu/hr ft² F for reboilers, and 90 Btu/hr ft² F for air coolers, which in SI units is 1136 W/m² K and 511 W/m², respectively.
3. To compare, it is possible to have U less than 1 W/m² K for a high-quality double or triple glass window with only natural convection (little wind).
4. For heat loss from the human body, U can typically vary from 2 W/m² K (well dressed and gentle breeze) to 20 W/m² K (swimming in water); see page 35 for details.
5. The area of industrial heat exchangers is typically from 1 m² to 3000 m² (they are usually divided into several smaller units if they become too large). However, spiral wound heat exchangers, of the kind used in LNG processes, can have areas up to 50000 m². Typical heat transfer rates (Q) in industrial heat exchangers vary from 10 kW to 100000 kW = 100 MW.
6. The expression for ΔT_{lm} in (5.2) on page 394 is derived with the assumption of constant heat capacities for the streams. The expression also applies if there is constant temperature on one of the sides, such as during boiling and condensation (which is equivalent to an infinite heat capacity).

Table 5.1: Typical overall heat transfer coefficients for heat exchangers in the gas industry. Source: *SI Engineering Data Book*, Gas Processors Suppliers Association, Tulsa, USA, 1980

Use		U [W/m^2 K]
water	air (1 bar)	110-140
water	gas (1.7 bar)	200-225
water	gas (3 bar)	225-285
water	gas (6 bar)	340-400
water	gas (8 bar)	450-570
water	oil	740-850
water	condensating propane	710-765
water	condensating naphtha	400-450
water	water	1000-1140
oil	oil	450-570
propane (liq)	propane (liq)	625-740
gas (1 bar)	gas (1 bar)	5-40
gas (1.7 bar)	gas (4.4 bar)	280-400
gas (8 bar)	gas (8 bar)	340-450
gas (8 bar)	propane (liq)	340-450
boiling liquid	oil	510-680
boiling liquid	condensating water vapor	800-900

7. For $1/1.4 < \Delta T_1/\Delta T_2 < 1.4$ (that is, for cases where the temperature difference between the two sides is fairly constant), the error is less than 1% if the logarithmic temperature difference $\Delta T_{\rm lm}$ is replaced by the arithmetic mean

$$\Delta T_m = \frac{\Delta T_1 + \Delta T_2}{2}$$

The arithmetic mean is better than ΔT_m for numerical calculations. For example, ΔT_m has the disadvantage that it becomes 0/0 for $\Delta T_1 = \Delta T_2$, although the correct value in this case clearly is $\Delta T_m = \Delta T_1 = \Delta T_2$.
8. We can interpret $1/U$ as the overall "resistance" to the heat transfer. The overall resistance can be calculated by adding the "resistances" on the cold side ($1/h_c$), the hot side ($1/h_h$) and through the wall (d/k), that is,

$$\frac{1}{U} = \frac{1}{h_c} + \frac{1}{h_h} + \frac{d}{k} \quad [\mathrm{m^2\ K/W}] \tag{5.10}$$

where h_c [W/m^2 K] is the heat transfer coefficient on cold side, h_h on hot side, k [W/m K] is thermal conductivity and d [m] is the wall thickness. The values for h_c and h_h can be found from published correlations. (Note. Do not confuse this h with specific enthalpy.) Note that h has to do with the heat transfer between a flow and a wall, while U is for heat transfer from one flow to another flow. h is called the **heat transfer coefficient** and U the **overall heat transfer coefficient**.
9. The most common kind of heat exchanger in the process industry is the shell and tube heat exchanger, see Figure 5.2. The heat transfer is usually best on the inside of the tubes, and we therefore prefer to have the gas on the tube side and the

liquid on the shell side ("the outside of the tubes") in a gas/liquid heat exchanger. Another issue to consider is that the heat loss to the surroundings is reduced by having the cold stream on the shell side.

10. The temperature difference $\Delta T = T_h - T_c$ varies along the heat exchanger, for example as shown in Figure 5.4 for countercurrent flow and cocurrent flow. If ΔT is small, a large heat exchanger area is required. A common **rule of thumb** for design, in order to avoid large heat exchangers, is to require $\Delta T > \Delta T_{\min}$ throughout the exchanger, with $\Delta T_{\min} = 10$ K being a typical value.

Example 5.3 Cocurrent flow. *Repeat Example 5.2 for the case with cocurrent flow. What is the area of the heat exchanger?*

The answer is that this is impossible with a cocurrent flow heat exchanger. The energy balance calculations are the same as with cocurrent flow, and we find that the required temperature difference at the outlet end is $\Delta T_2 = T_{h,\text{out}} - T_{c,\text{out}} = 330K - 340K = -10K$, *which is negative and thereby impossible. Another way of showing that this is impossible is to assume infinite area. Then the exit streams will have the same temperature, that is,* $T_{h,\text{out}} = T_{c,\text{out}} = T$, *and the energy balance becomes*

$$Q = m_c c_{p,c}(T - T_{c,\text{in}}) = m_h c_{p,h}(T_{h,\text{in}} - T)$$

which gives

$$T = \frac{m_h c_{p,h} T_{h,\text{in}} + m_c c_{p,c} T_{c,\text{in}}}{m_h c_{p,h} = m_c c_{p,c}} = \frac{4 \cdot 360 + 6 \cdot 320}{4 + 6} = 336\ K$$

while we wanted to cool to $T_{h,\text{out}} = 330\ K$. *In comparison, if we for this example use a countercurrent flow heat exchanger with infinite area, then we achieve* $T_{h,\text{out}} = T_{c,\text{in}} = 320K$.

Example 5.4 Heat exchanger with two tube pass. *Repeat Example 5.2 for the case with a "U-turn" on the tube side (two tube pass) as shown in Figure 5.7, and with the hot stream on the tube side. What is the area of the heat exchanger?*

Here, we have a mix of co- and countercurrent flows, so in the light of the previous example with cocurrent flow, we expect that the common "U turn" heat exchanger is not well suited in this case. To use the diagram in Figure 5.7, we need the "efficiency" P and the mC_p ratio Z (note that we have the hot stream on the tube side, that is, we have $h = t$ (tube) and $c = s$ (shell)):

$$P = \frac{T_{t,\text{out}} - T_{t,\text{in}}}{T_{s,\text{in}} - T_{t,\text{in}}} = \frac{330 - 360}{320 - 360} = 0.75; \quad Z = \frac{m_t c_{p,t}}{m_s c_{p,s}} = \frac{1 \cdot 4}{2 \cdot 3} = 0.67$$

These values of P and Z are outside the diagram, but by extrapolating we can guess that F is lower than 0.2. That is, we need a heat exchanger that has an area at least 5 times larger than for ideal countercurrent flow. We can therefore conclude that a countercurrent flow heat exchanger should be used for this specific application.

Example 5.5 Area of heat exchanger for cooling synthesis gas. *Figure 5.8 shows a countercurrent flow heat exchanger where 8500 kmol/h (100000 kg/h) of hot synthesis gas with temperature 870° C is cooled to 360° C by evaporation of water on the other side at constant temperature 308° C (the pressure on the water side is 99 bar, that is, we generate what is often called high pressure steam). The overall heat transfer coefficient is 300 W/m^2 K. The mean heat capacity on the gas side is* $C_p = 36$ *J/mol K and the heat of vaporization of water at 99 bar is 1350 kJ/kg. Find the required area for the exchanger and calculate the amount of steam produced [t/h].*

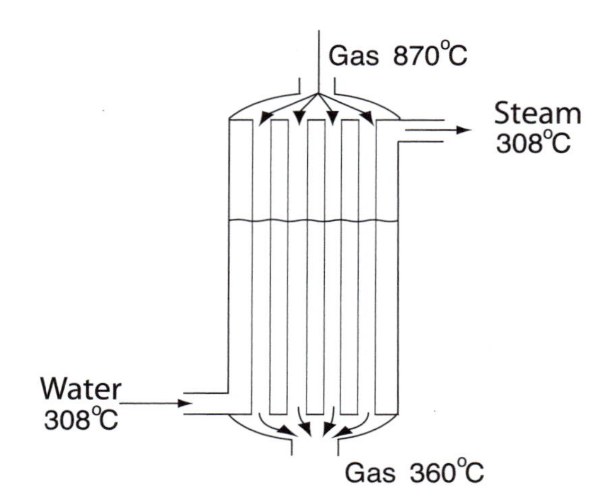

Figure 5.8: Heat exchanger for cooling down synthesis gas

Solution. *The temperature differences at the two heat exchanger ends are*

$$\Delta T_1 = 870 - 308 = 562\ K; \quad \Delta T_2 = 360 - 308 = 52\ K$$

which gives the logarithmic mean temperature difference

$$\Delta T_{\rm lm} = \frac{\Delta T_1 - \Delta T_2}{\ln(\Delta T_1/\Delta T_2)} = 214 \text{ K}$$

(this is significantly less than the arithmetic mean of 307 K because the temperature on the hot side "flattens" towards the exit of the heat exchanger). The energy balance (5.9) on the hot side gives

$$Q = n_h C_{p,h}(T_{h,\rm in} - T_{h,\rm out}) = (8500/3.6) \cdot 36 \cdot (870 - 360) = 43.4 \cdot \underbrace{10^6 \text{J/s}}_{\text{MW}}$$

(here 3.6 is the conversion factor from kmol/h to mol/s) and the area of the heat exchanger becomes

$$A = \frac{Q}{U\Delta T_{\rm lm}} = \frac{43.4 \cdot 10^6}{300 \cdot 214} = 676\ m^2$$

The energy balance for the water side where we have evaporation is

$$H_{out} - H_{in} = m \cdot \Delta_{\rm vap} H = Q \quad \Rightarrow \quad m = \frac{Q}{\Delta_{\rm vap} H} = \frac{43.4 \cdot 10^6}{1350 \cdot 10^3} = 32.1 \text{ kg/s}$$

that is, the amount of high pressure steam is m = 32.1 kg/s = 115.7 t/h. Note that in this case with constant temperature on one side there is no difference between co- and countercurrent flow, at least not from a thermodynamic point of view.

Next, some exercises.

Exercise 5.1* *A gas stream of 700 mol/s and C_p = 40 J/mol K is cooled from 400 °C to 210 °C in a countercurrent heat exchanger by heating liquid water from 100 °C to 350 °C.*

(a) Formulate the total energy balance [J/s] and calculate the amount of water, when the specific heat capacity for water is 4.18 kJ/kg K.

(b) Calculate UA for the heat exchanger assuming ideal countercurrent flow.

(c) The water flow increases such that the exit temperature of the gas is reduced from 210 ^oC to 180 ^oC. Calculate the exit temperature for water and flow of water (you can assume that the inlet temperatures are unchanged, the gas flow is unchanged and UA is constant).

Exercise 5.2* Comparison of cocurrent flow and countercurrent flows. *A hot liquid stream with temperature 37.5 ^oC is cooled to 20 ^oC in a heat exchanger where the heat transferred is 73.3 kW. Cooling water with specific heat capacity 4.18 kJ/kg K is used and its temperature goes from 12°C (in) to 17.5^{o}C (out). The overall heat transfer coefficient is $U = 200 W/m^2 K$. Calculate (a) the amount of cooling water, and the area of the heat exchanger for the cases with (b) countercurrent flow, (c) cocurrent flow and (d) two tube pass with the hot stream on the tube side.*

5.3 Simulation of heat exchangers

In the above examples the temperature were given and we calculated the area of the heat exchangers – this is called **design**. But during operation, the equipment (heat exchanger) is given and we want to calculate what happens (find the exit temperature) – this is called **simulation**. The same equations can be used – the difference is which quantity is unknown. Let us consider an example.

Example 5.6 Calculation of the exit temperature for a given heat exchanger (simulation). *A cold stream (5 kg/s, c_p = 1.5 kJ/kg K, 20 ^oC) is to be heat exchanged with a hot stream (3 kg/s, c_p = 1.2 kJ/kg K, 70 ^oC) in a countercurrent flow heat exchanger with U=150 W/m^2 K and A = 90 m^2.* **Problem**: *Calculate the exit temperatures and heat transferred.*

Solution. *Constant heat capacities are assumed. As shown above, a countercurrent flow heat exchanger is described by the following equations (which are easy to remember!)*

$$m_c c_{p,c}(T_{c,\text{out}} - T_{c,\text{in}}) = Q \tag{5.11}$$

$$m_h c_{p,h}(T_{h,\text{out}} - T_{h,\text{in}}) = -Q \tag{5.12}$$

$$Q = UA\Delta T_{\text{lm}} \tag{5.13}$$

$$\Delta T_{\text{lm}} = \frac{\Delta T_1 - \Delta T_2}{\ln(\Delta T_1/\Delta T_2)} \tag{5.14}$$

where ΔT_1 and ΔT_2 are the temperature differences between the hot and cold sides at the two ends of the heat exchanger. The following data are given:

$$m_c = 5kg/s; \; c_{p,c} = 1500J/kg\ K; \; t_{c,\text{in}} = 20^oC$$

$$m_h = 3kg/s; \; c_{p,h} = 1200J/kg\ K; \; t_{h,\text{in}} = 70^oC$$

This gives four equations with four unknowns ($Q, T_{c,\text{out}}, T_{h,\text{out}}, \Delta T_{\text{lm}}$), so it should be possible to find a solution. However, the equations are difficult to solve analytically. It is possible to reformulate the equations such that they are more suited for simulation (see Section 5.3), but let us first illustrate how (5.11)-(5.14) can be solved by iteration ("trial and error"):

1. *Guess a value of ΔT_{lm}.*
2. *Calculate Q from (5.13).*
3. *Calculate $T_{c,\text{out}}$ from (5.11).*

4. Calculate $T_{h,\text{out}}$ from (5.12).
5. Calculate ΔT_{lm} from (5.14) and compare with the guess.
6. If the solution is not converged, go to step 1 and make a new guess (note that the "direct substitution" method, where we go to step 2 with the new ΔT_{lm}, does not work in this case).

The following MATLAB program can be used:

```
% Data:
mh=3; cph=1200; thi=70;
mc=5; cpc=1500; tci=20;
U = 150; A=90;
%  1, Guess logarithmic mean temp
dtlm0 = 5;
% 2. Amount of transferred heat
Q = U*A*dtlm0
% 3,4 . From the energy balances on cold and hot side
tcu = tci + Q/(mc*cpc)
thu = thi - Q/(mh*cph)
% 5. Calculate the logarithmic mean temperature difference (countercurrent
% flow)
dt1 = thi - tcu
dt2 = thu - tci
dtlm = (dt1 - dt2)/ log(dt1/dt2)
```

We start by guessing ΔT_{lm}= 5 K. This gives a new calculated value of 35.9 K. Some more values:

Guessed ΔT_{lm}	Calculated ΔT_{lm}
5	35.9
10	20.8
15	unphysical
12	13.5
12.5	11.21
12.2	12.61
12.3	12.17

We consider this to be good enough, that is, $\Delta T_{\text{lm}} = 12.3$ K. We find that

$$Q = 166.1\ kW;\ t_{h,\text{out}} = 23.9^oC;\ t_{c,\text{out}} = 42.1^oC$$

Transformed equations for simulation (ϵ-NTU method)

In the above simulation example, we solved the heat exchanger equations (5.11)–(5.14) by iteration, but this is not necessary. Using some algebra, we can derive the following equivalent simple expressions which are suitable for *simulation of a given heat exchanger*:[1]

$$T_{h,\text{out}} = (1-\epsilon_h)T_{h,\text{in}} + \epsilon_h T_{c,\text{in}} \quad (5.15)$$

$$T_{c,\text{out}} = \epsilon_c T_{h,\text{in}} + (1-\epsilon_c)T_{c,\text{in}} \quad (5.16)$$

where for *ideal countercurrent flow* the "efficiencies" ϵ_h and ϵ_c can be calculated from the following formulas (which you are not likely to be able to remember, as opposed

[1] For more about the ϵ-NTU method and applying it to other heat exchangers, see for example: A. F. Mills, *Heat and Mass Transfer*, Irwin (1995).

to (5.11)–(5.14)!):

$$N_{tu} = \frac{UA}{m_c c_{p,c}} \quad (5.17)$$

$$C = \frac{m_c c_{p,c}}{m_h c_{p,h}} \quad (5.18)$$

$$\epsilon_c = \frac{1 - \exp(-N_{tu}(C-1))}{C - \exp(-N_{tu}(C-1))} \quad (5.19)$$

$$\epsilon_h = \epsilon_c \cdot C \quad (5.20)$$

For *ideal cocurrent flow*, (5.19) is replaced with

$$\epsilon_c = \frac{1 - \exp(-N_{tu}(C+1))}{C+1} \quad (5.21)$$

These equations are easy to program, for example in MATLAB or in a spreadsheet. All of the above variables are dimensionless. N_{tu} is the **number of heat transfer units (NTUs)** and C is the ratio of the heat flow capacities for hot and cold sides. A larger value for N_{tu} gives ϵ_c and ϵ_h closer to 1.

Note from (5.15)–(5.16) that it is very simple to find the exit temperature when we have first determined the efficiencies ϵ_h and ϵ_c. The efficiencies are always between 0 and 1, which is reasonable since the exit temperatures must necessarily be bounded by the inlet temperatures.

Example 5.7 Simulation of heat exchanger. *Let us continue Example 5.6 (page 139). First, calculate the number of heat transfer units N_{tu}, the heat capacity ratio C and the efficiencies*

$$N_{tu} = \frac{UA}{m_c c_{p,c}} = \frac{150 \cdot 90}{5 \cdot 1500} = 1.8$$

$$C = \frac{m_c c_{p,c}}{m_h c_{p,h}} = \frac{5 \cdot 1500}{3 \cdot 1200} = 2.083$$

$$\epsilon_c = \frac{1 - \exp(-N_{tu}(C-1))}{C - \exp(-N_{tu}(C-1))} = \frac{1 - 0.142}{2.083 - 0.142} = 0.442$$

$$\epsilon_h = \epsilon_c C = 0.442 \cdot 2.083 = 0.920$$

Equations (5.15)–(5.16) then give the exit temperatures (here, we use lower case letter for temperature since they are in ^oC),

$$t_{h,\text{out}} = 0.080 t_{h,\text{in}} + 0.920 t_{c,\text{in}} = 0.080 \cdot 70^o + 0.920 \cdot 20^o = 23.98^o C$$

$$t_{c,\text{out}} = 0.442 t_{h,\text{in}} + 0.558 t_{c,\text{in}} = 0.442 \cdot 70^o + 0.558 \cdot 20^o = 42.09^o C$$

which is the same as we found in Example 5.6 by solving equations (5.11)–(5.14) by iteration.

Comments on the ϵ-NTU method:

1. If there is no heat exchange (that is, $UA = 0 \Rightarrow N_{tu} = 0$), then both efficiencies ϵ_c and ϵ_h are 0, and as expected we have that $T_{h,\text{out}} = T_{h,\text{in}}$ and $T_{c,\text{out}} = T_{c,\text{in}}$. For an infinitely large heat exchanger (that is, $UA \to \infty$), N_{tu} goes to infinity and one

of the two efficiencies goes to 1, that is, the temperatures will approach each other at one end of the heat exchanger (both efficiencies go to 1 at the same time only if the streams have identical heat capacity flow rates mc_p such that $C = 1$).

2. From (5.15)–(5.16), we note that there is a "linear" relationship between the inlet and outlet temperatures. This implies that a given (e.g., 1 degree) increase in an inlet temperature will always result in the same increase in the exit temperatures (which is less than 1 degree and given by the efficiencies).

3. For the special case of $C = 1$, we have parallel temperature profiles on the two sides and the formula for ϵ_c in (5.19) cannot be used. One approach is to change the value of C slightly, for example to $C = 0.9999$, but alternatively one can for $C = 1$ use $\epsilon_c = N_{tu}/(N_{tu} + 1)$. The special case with condensing vapor on hot side can be approximated by using a very large $m_h c_{p,h}$, and the special case with boiling liquid on cold side can be approximated by setting a very large $m_c c_{p,c}$.

4. It follows from $Q = m_h c_{p,h}(T_{h,\text{in}} - T_{h,\text{out}})$ that the amount of heat transferred is

$$Q = \epsilon_h m_h c_{p,h} \left(T_{h,\text{in}} - T_{c,\text{in}}\right) = \epsilon_c m_c c_{p,c} \left(T_{h,\text{in}} - T_{c,\text{in}}\right) \tag{5.22}$$

5. We have chosen to introduce *two* efficiencies (ϵ_h and ϵ_c), but it is common to use a single efficiency $\epsilon \triangleq \max\{\epsilon_h, \epsilon_c)$ which is the largest of the two. We then have that

$$Q = \epsilon C_{\min}(T_{h,\text{in}} - T_{c,\text{in}}) \tag{5.23}$$

where $C_{\min} \triangleq \min\{m_c c_{p,c}, m_h c_{p,h}\}$ and we have that the efficiency $\epsilon \to 1$ for an infinitely large heat exchanger.

Exercise 5.3 *In a cocurrent flow heat exchanger, the inlet temperatures are 90 ^oC (hot side) and 25 ^oC (cold side). At the exit, the temperatures are 55 ^oC (hot side) and 53.5 ^oC (cold side). (a) What are the exit temperatures if you "switch" and instead use countercurrent flow (with the same heat exchanger)? (b) How much more heat is transferred (in %)? (Hint: It might be worthwhile to choose a basis, for example $Q = 100[W]$, for the case with countercurrent flow, and start by using the "design formulas" to calculate UA and the heat capacity flow rates (mC_p) on hot and cold side.*

6

Compression and expansion

To increase the pressure of a stream (compression), we normally need to supply work. If we reduce the pressure of a stream (expansion), we can potentially extract work. Here, we consider the calculation of this "shaft work" W_s for a steady-state continuous process.

6.1 Introduction

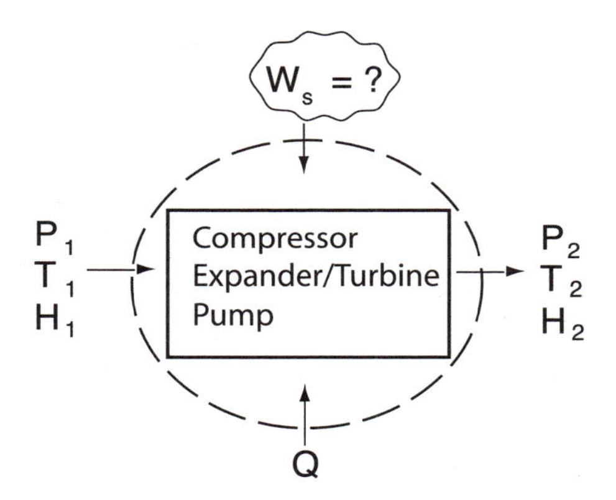

Figure 6.1: Steady-state energy balance for process with shaft work

The energy balance for a steady-state process where a stream is taken from state 1 (in) to state 2 (out) can, as shown in (4.13), be written

$$H_2 - H_1 = Q + W_s \tag{6.1}$$

where we have neglected changes in kinetic and potential energy, and have neglected electrochemical work. In order to compress a gas or pump a liquid (increase the pressure from p_1 to $p_2 > p_1$), we need to supply mechanical shaft work ($W_s > 0$). On the other hand, we can, by expanding a stream to a lower pressure ($p_2 < p_1$) in a turbine, remove mechanical shaft work ($W_s < 0$).

We want to derive simple expressions for calculating W_s. This is done by first obtaining the ideal (reversible) work W_s^{rev} that would have been obtained in a reversible (frictionless; lossless) machine, and then introducing an efficiency η to find the actual work. The efficiency is typically around 75%.

Comment: In order to calculate the efficiency and obtain a relationship between pressure rise and rotation speed for a given compressor, we would need a quite detailed description of the insides of these machines. This is outside the scope of this book, which deals with mainly mass and energy balances, and for us it is enough to use thermodynamic considerations.

6.2 Compression (increase of pressure)

A compressor is a machine (unit) that increases pressure of a *gas* stream by *supplying* mechanical work ($W_s > 0$). The flow sheet symbol for a compressor is two lines where the distance becomes narrower. This illustrates that the fluid takes less space when it is compressed, see Figure 6.2(a). Two simple examples of a compressor are a bicycle pump and a kitchen fan.

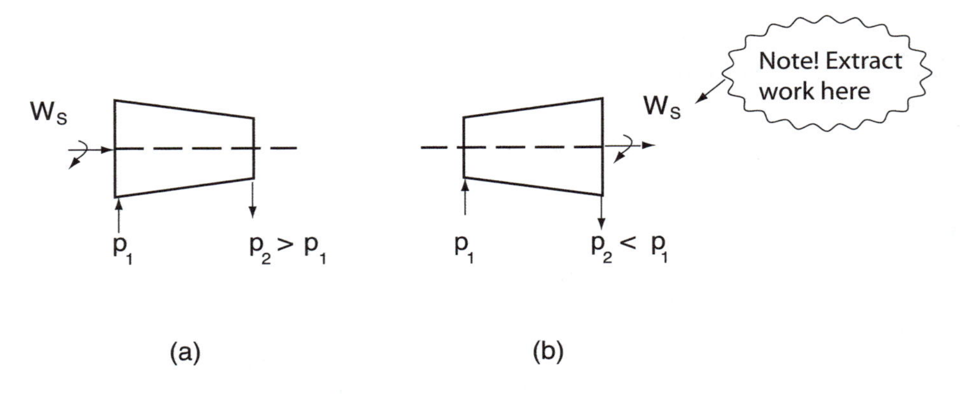

Figure 6.2: (a) Compressor (b) Turbine

If the reversible compression work for an ideal compressor is W_s^{rev}, then the actual compression work is

$$W_s = \frac{W_s^{\text{rev}}}{\eta} \tag{6.2}$$

where $\eta \leq 1$ is the (thermodynamic) efficiency of the compressor. In an actual compressor, part of the work, $W_s - W_s^{\text{rev}}$ (which is a positive number), is lost as friction heat.

Equation (6.2) also applies to a **pump** which is a unit that increases the pressure of a *liquid.* The work is usually much smaller here because a liquid has a much smaller volume than a gas.

6.3 Expansion in turbine

A turbine is a unit that removes mechanical work ($W_s < 0$) by expanding a fluid from a high to a low pressure. Note that we use the same term, turbine, for both gas and liquid. The flow sheet symbol for a turbine expands outwards to show that the gas takes more space when expanded, see Figure 6.2(b).

In the ideal reversible case, a turbine is the opposite of a compressor. That is, by reversing the gas stream, it is possible to remove the same amount of mechanical work that has to be supplied during compression. Therefore, we use the same formulas for calculating the reversible work $W_s^{\rm rev}$ in the two cases.

However, in practice, we have friction here too, and if the ideal **expansion work** is $W_s^{\rm rev}$, then the actual work that can be extracted is given by

$$W_s = \eta \cdot W_s^{\rm rev} \tag{6.3}$$

where $\eta \leq 1$ is the efficiency of the turbine. (Note that we here multiply by η, whereas we divide by η for a compressor or pump). In an actual turbine, part of the theoretical work, $W_s - W_s^{\rm rev}$ (which is a positive number since $W_s^{\rm rev}$ is more negative), is lost as friction heat.

6.4 Reversible shaft work

Here, we derive the ideal (reversible) shaft work $W_s^{\rm rev}$ for pressure changes for a continuous process (open system). The shaft work is supplied or removed using a cyclic machine, such as a piston compressor or a rotating compressor.

Consider a fluid (gas or liquid) with pressure p_1 that is compressed or expanded to a pressure p_2 in a steady state open process. The ideal (reversible) shaft work that must be supplied is (see derivation below)

$$\boxed{W_s^{\rm rev} = \int_{p_1}^{p_2} V dp} \quad [\text{J/s; J; J/mol; J/kg}] \tag{6.4}$$

where V [m^3/s; m^3; m^3/mol; m^3/kg] is the volume of the fluid that is compressed or expanded. In general, the volume V will change through the machine as a function of the actual T and p. This dependency is particularly important for gases; for an ideal we use pV^c = const where c is the polytrope exponent ($c = 1$ for isothermal process and $c = \gamma = C_p/C_V$ for adiabatic process) – see details below. For liquids we can usually assume that we have an incompressible fluid where V is constant, that is, independent of both T and p.

Comment: The shaft work $W_s^{\rm rev}$ in (6.4) is for a *pressure change*. It must not be confused with the work $W_{\Delta V}$ for a *system volume change*, which for a reversible process is $W_{\Delta V}^{\rm rev} = -\int pdV$ (see A.27), where V represents the volume of the system (machine) and not the volume of the stream. For a steady state process, the volume of the system is constant, that is, $W_{\Delta V} = 0$.

Derivation of (6.4)

To derive (6.4), consider an idealized process where n mol of a fluid with pressure p_1 is compressed to pressure p_2 using an ideal reversible piston compressor ("ideal bicycle pump"), see Figure 6.3. Note that the piston compressor (the system) returns to its original state after each cycle, so the system itself has no volume changes since the time period we consider is a whole cycle. Even though this derivation is for compression, we may reverse the process so the same expression holds for expansion.

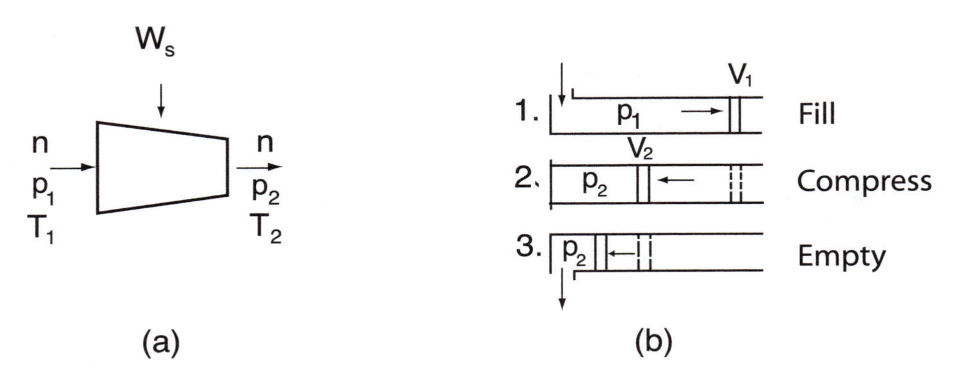

Figure 6.3: Reversible shaft work: (a) Flow sheet compressor. (b) Cycle for piston compressor with three subprocesses

The cycle that we are considering consists of the following three reversible steps (subprocesses):

1. The cylinder (e.g., bicycle pump) is filled at constant pressure p_1 (the process is adiabatic and the volume increases from 0 to V_1, where V_1 is the volume of the fluid in state 1).
2. The gas is compressed from p_1 to p_2 by pushing the piston such that the volume decreases from V_1 to V_2 (this process can be adiabatic or possibly involve cooling during the compression; this is not specified here – it will follow indirectly from the relationship between p and V used later).
3. The gas in the cylinder (bicycle pump) is emptied at constant pressure p_2 (the process is adiabatic and the volume decreases from V_2 to 0). (The cycle is now completed and we are back to start.)

Each subprocess is reversible, so that the outer (surrounding) pressure applied by the piston equals the system's pressure ($p_{\rm ex} = p$). The supplied work from the piston in the compressor in each subprocess is then given by (A.27) (see also Figure A.7, page 346):

$$W = -\int pdV \quad [J]$$

where p is the system's pressure (inside the cylinder). The supplied shaft work in each of the three steps is then:

1. Filling of the cylinder (work is performed)

$$W_1 = -\int_0^{V_1} p_1 dV = -p_1 V_1$$

2. Compression (work is supplied)

$$W_2 = -\int_{V_1}^{V_2} pdV = \int_{V_2}^{V_1} pdV$$

(Note that $V_1 > V_2$ during compression.)

3. Emptying (work is supplied)

$$W_3 = -\int_{V_2}^{0} p_2 dV = p_2 V_2$$

W_1 and W_3 are the flow works for pushing the stream in and out, whereas W_2 is the "actual" compression work.

The total supplied shaft work (which is a "piston work" in our idealized process) is the sum of these three works

$$W_s^{\rm rev} = W_1 + W_2 + W_3 = -p_1 V_1 + \int_{V_2}^{V_1} pdV + p_2 V_2 = \int_{p_1}^{p_2} Vdp \tag{6.5}$$

The last equality follows from considering the areas in Figure 6.4(a). Alternatively, we can integrate by parts, which gives $\int_1^2 d(pV) = p_2 V_2 - p_1 V_1 = \int_1^2 pdV + \int_1^2 Vdp$. □

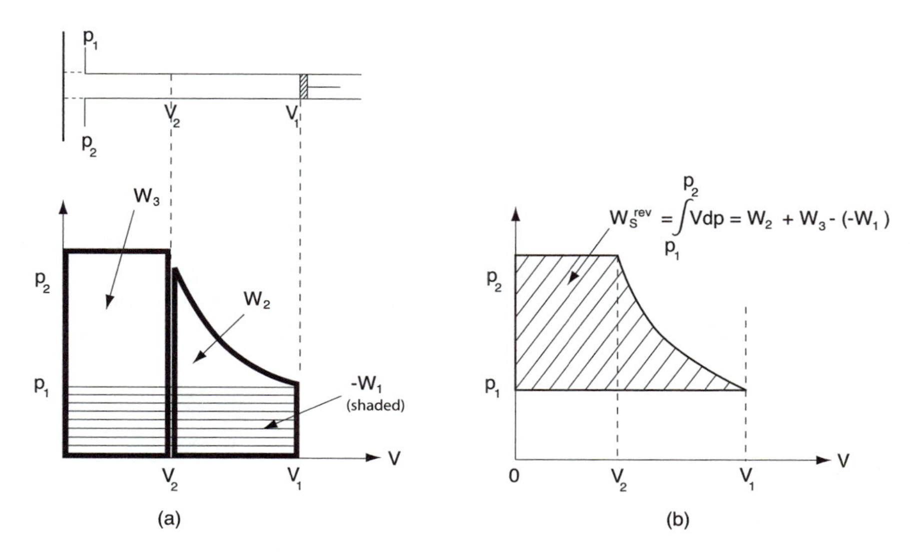

Figure 6.4: Derivation of reversible shaft work: (a) Work for each subprocess. (b) Work for the whole cycle

Comments on (6.4)

- Although the derivation is given for a piston compressor, W_s^{rev} in (6.4) also gives the shaft work for reversible pressure change for other types of equipment, for instance a radial compressor.
- The derivation also applies to expansion (reverse the derivation).
- The derivation of W_s^{rev} gives the shaft work for a *steady state continuous process* (open system) where the inflow is in state 1 and the outflow is in state 2. For a change from state 1 to state 2 in a *closed system*, the work for the change in volume is given by W_2 in process step 2.
- The compression can occur adiabatically ($Q = 0$) or possibly we can have cooling ($Q < 0$) or heating ($Q > 0$) – this follows indirectly from the relationship between p and V used when calculating W_s^{rev} in (6.4); see more details below.
- The energy balance for the process is $H_2 = H_1 + W_s + Q$. Note that the "internal" (useless) flow work for pushing the stream in and out of the machine (the compressor) is included in the enthalpies H_1 and H_2.
- If you know some thermodynamics, then (6.4) can alternatively be derived by considering the change of entropy for a reversible process: Start with the energy balance (6.1) for a continuous process without accumulation. This can be written in differential form as $dH = dQ + dW_s$. For a reversible process, we have that $dQ^{\text{rev}} = TdS$ which gives $dW_s^{\text{rev}} = dH - TdS$. Here, enthalpy is a state function, and from (B.68) we have $dH = TdS + Vdp$, which inserted gives $dW_s^{\text{rev}} = Vdp$.

6.5 Reversible shaft work for ideal gas

From (6.4), the ideal work for pressure change in 1 mole of fluid is

$$W_s^{\text{rev}} = \int_{p_1}^{p_2} V dp \quad [\text{J/mol}] \tag{6.6}$$

Here, V is the molar volume [m^3/mol]. In order to calculate W_s^{rev}, we must integrate Vdp, that is, we need to know the relation between V and p during the compression. Expression (6.6) applies to any fluid (including real gas), but we here consider an ideal gas where $pV = RT$ and consider three cases for how the compression or expansion occurs:

Isothermal: pV=constant

Adiabatic (isentropic): pV^{γ}=constant, where $\gamma = C_p/C_V$.

Polytropic: pV^c=constant, where c is the polytropic coefficient, $1 \leq c \leq \gamma$, obtained by cooling for compression or heating for expansion.

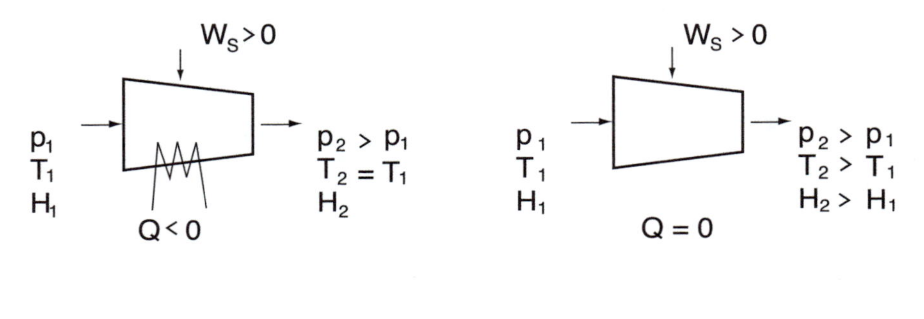

Figure 6.5: (a) Isothermal and (b) adiabatic compression. (For ideal gas (a) $H_2 = H_1$ and thereby $Q = -W_s$)

6.5.1 Isothermal process

In this case, we have $T_1 = T_2 = T_0$. For compression this is obtained by cooling and for expansion by heating, see Figure 6.5(a). For n mol of ideal gas with constant temperature, we have that

$$pV = nRT_0 = \text{constant}$$

which gives $\int V dp = nRT_0 \int \frac{dp}{p} = nRT_0 \ln p$ and we have from (6.6) that

$$\boxed{W_s^{\text{rev}}/n = RT_0 \ln\left(\frac{p_2}{p_1}\right)} \quad [J/mol] \tag{6.7}$$

Furthermore, for an ideal gas, the enthalpy is only a function of the temperature, that is, $H_1 = H_2$ and the energy balance (6.1) gives that $Q = -W_s$, that is, the heat Q is equal to performed work $(-W_s)$.

6.5.2 Adiabatic process (isentropic)

We consider an adiabatic reversible state change of an ideal gas and assume constant heat capacity. For this process, $pV^{\gamma}=$ constant or equivalently, see (A.40),

$$\frac{T_2}{T_1} = \left(\frac{p_2}{p_1}\right)^{\frac{\gamma-1}{\gamma}} \tag{6.8}$$

where $\gamma = C_p/C_V$. Alternatively, the exponent can be written

$$\frac{\gamma-1}{\gamma} = \frac{C_{p,m} - C_{V,m}}{C_{p,m}} = \frac{R}{C_{p,m}}$$

where $C_{p,m}$ [J/K mol] is the molar heat capacity. To find the shaft work, calculate T_2 from (6.8) and insert this into the energy balance. Since the heat capacity C_p [J/K kg; J/K mol] is assumed constant, the energy balance (6.1) gives

$$W_s^{\text{rev}} = H_2 - H_1 = mC_p(T_2 - T_1) \quad [J; J/s] \tag{6.9}$$

Depending on the units for C_p, the mass m can be on mass basis [kg; kg/s] or on mole basis [mol, mol/s] (in the latter case, the symbol n is often used instead of m). Using (6.8), we then have the adiabatic reversible work for an ideal gas with constant heat capacity

$$\boxed{W_s^{\text{rev}} = mC_pT_1\left[\left(\frac{p_2}{p_1}\right)^{\frac{\gamma-1}{\gamma}} - 1\right]} \quad [J; J/s] \tag{6.10}$$

Exercise 6.1 * Alternative derivation. *Integrate $W_s^{\text{rev}} = \int_{p_1}^{p_2} V dp$ in (6.6) using $pV^{\gamma} = p_1V_1^{\gamma} =$ constant, and show that you get (6.10).*

6.5.3 Polytropic process

There is also a third case, namely a **polytropic process** where there is some cooling (compression) or some heating (expansion) but not enough to keep the temperature constant. The polytropic process is calculated using (6.10), but with the polytrope exponent c instead of γ. We have that $1 \leq c \leq \gamma$ where $c = 1$ applies to an isothermal process, and $c = \gamma$ applies to an adiabatic process.

6.6 Actual work and examples

To find the actual work W_s for a compressor, pump or turbine, use the following procedure

1. Calculate the reversible work $W_s^{\rm rev}$ (sometimes denoted W_s'). The reversible outlet temperature is also often denoted with $'$, for instance, T_2'.
2. Calculate the actual work from the efficiency η:

$$\text{Compressor}: \quad W_s = W_s^{\rm rev}/\eta$$

$$\text{Turbine}: \quad W_s = W_s^{\rm rev} \cdot \eta$$

3. Calculate the actual outlet temperature T_2 using the energy balance

$$W_s = mC_p(T_2 - T_1)$$

We are often a bit sloppy with the sign for W_s, but note that W_s represents supplied work and should be negative for a turbine.

Example 6.1 Compression of ideal gas (important example). *$n = 1700$ mol/s of an ideal gas with $\gamma = C_p/C_V = 1.38$ at $p_1 = 20$ bar and 50 °C is compressed to $p_2 = 100$ bar. Calculate the ideal (reversible) compression work for (see figures 6.6- 6.9):*

1. *One-step adiabatic compression from 20 to 100 bar.*
2. *Two-step adiabatic compression with intermediate pressure 45 bar without intermediate cooling. Here the compression first occurs from 20 bar to 45 bar and then from 45 bar to 100 bar.*
3. *Two-step adiabatic compression with intermediate pressure 45 bar and intermediate cooling down to 50 °C. Here the compression first occurs from 20 bar to 45 bar, then the gas is cooled to 50 °C, before it is compressed from 45 bar to 100 bar.*
4. *Isothermal compression at 50 °C.*
5. *Finally, find for case 1 the actual exit temperature when the adiabatic efficiency is $\eta = 0.72$.*

Solution. *It is given that $\gamma = C_p/C_V = 1.38$. From this, it follows that $\frac{\gamma-1}{\gamma} = 0.275$ and $C_p = R\frac{\gamma}{\gamma-1} = 30.2$ J/mol K.*

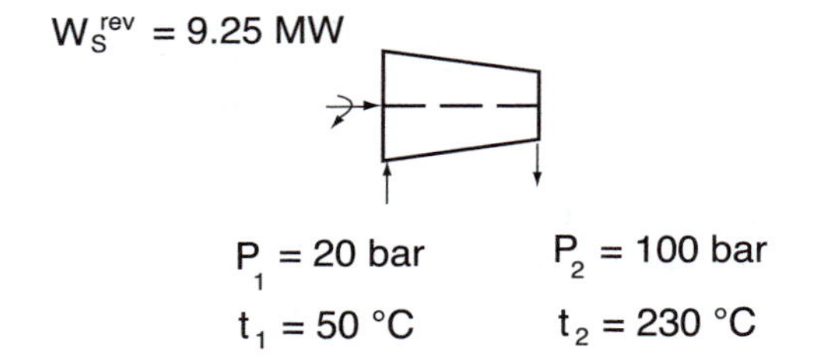

Figure 6.6: Adiabatic compression

1. **One-step adiabatic compression from 20 to 100 bar**. *The "ideal" temperature T_2 after a reversible compression is from (6.8),*

$$\frac{T_2}{T_1} = \left(\frac{p_2}{p_1}\right)^{\frac{\gamma-1}{\gamma}} = \left(\frac{100}{20}\right)^{0.275} = 1.558 \tag{6.11}$$

that is, $T_2 = 323 \cdot 1.558 = 503.1$ K (230 °C). The ideal (reversible) compression work is found from the energy balance (6.1)

$$W_s^{\rm rev} = H_2 - H_1 = nC_p(T_2 - T_1) =$$

$$1700\ mol/s \cdot 30.2\ J/K\ mol(503\ K - 323\ K) = 9.25 \cdot 10^6\ J/s = 9.25\ MW$$

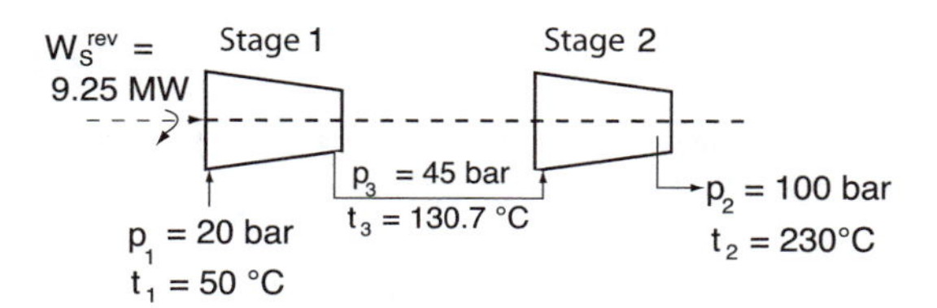

Figure 6.7: Two-step adiabatic compression without intermediate cooling

2. **Two-step adiabatic compression with intermediate pressure 45 bar without intermediate cooling.** *Here, the gas is first compressed from* $p_1 = 20$ *bar to* $p_3 = 45$ *bar and then from* $p_3 = 45$ *bar to* $p_2 = 100$ *bar. The answer is of course the same as in case 1, since for the first step*

$$\frac{T_3}{T_1} = \left(\frac{p_3}{p_1}\right)^{\frac{\gamma-1}{\gamma}}$$

and for the second step

$$\frac{T_2}{T_3} = \left(\frac{p_2}{p_3}\right)^{\frac{\gamma-1}{\gamma}}$$

which, when multiplied, gives the same final temperature as in (6.11), and from the energy balance the work must be the same.

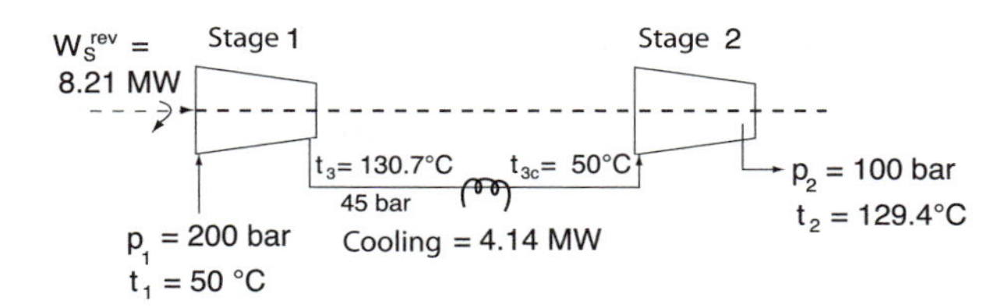

Figure 6.8: Two-step adiabatic compression without intermediate cooling

3. **Two-step adiabatic compression with intermediate pressure** $p_3 = 45$ **bar and intermediate cooling to** $t_{3c} = 50^o$**C**. *The ideal temperature after the first step is*

$$\frac{T_3}{T_1} = \left(\frac{p_3}{p_1}\right)^{\frac{\gamma-1}{\gamma}} = \left(\frac{45}{20}\right)^{0.275} = 1.2498$$

which gives $T_3 = 1.2498 \cdot 323.15 = 403.9K$ *(that is, 130.7* ^oC*). The energy balance gives*

$$W_{s1}^{\text{rev}} = nC_p(T_3 - T_1) = 1700 \cdot 30.2 \cdot 80.7 = 4.14 \text{ MW}$$

The gas is then cooled from $T_3 = 403.9K$ *to* $T_{3c} = 323$ *K (50*^{o} *C). The temperature after the second compression step is then*

$$\frac{T_2}{T_{3c}} = \left(\frac{p_2}{p_3}\right)^{\frac{\gamma-1}{\gamma}} = \left(\frac{100}{45}\right)^{0.275} = 1.2456$$

which gives $T_2 = 402.5$ *K (129.4 °C) and the energy balance gives*

$$W_{s2}^{\text{rev}} = nC_p(T_2 - T_{3c}) = 1700 mol/s \cdot 30.2 J/K\ mol \cdot 79.4 K = 4.07\text{MW}$$

The total work with intermediate cooling is 4.14 MW + 4.07 MW = 8.21 MW which is 11% lower than the 9.25 MW for one-step adiabatic compression.

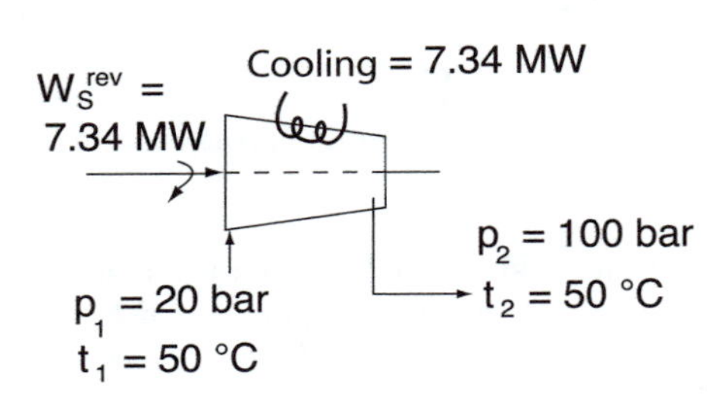

Figure 6.9: Isothermal compression

4. **Isothermal compression at 50 °C.** *The isothermal compression work is from (6.7)*

$$W_s^{\text{rev}} = nRT_1 \ln\left(\frac{p_2}{p_1}\right) = 1700 \cdot 8.31 \cdot 323 \cdot \ln 5 = 7.34\ MW$$

which is 21% lower than for the one-step adiabatic compression. Note that isothermal compression at 50 °C is equivalent to the limit of adiabatic compression with an infinite number of steps, each with intermediate cooling to 50 °C.

5. *Finally, let us find the actual exit temperature for* **one-step adiabatic compression**. *The efficiency is* $\eta = 0.72$, *that is, the actual compression work is*

$$W_s = \frac{W_s^{\text{rev}}}{\eta} = \frac{9.25}{0.72} = 12.85 MW$$

The actual exit temperature T_2 *is then from the energy balance* $W_s = nC_p(T_2 - T_1)$,

$$T_2 = 323 + \frac{12.85 \cdot 10^6}{1700 \cdot 30.2} = 323 + 250.2 = 573.2\ K$$

that is, $t_2 = 300^o C$, *while we found* $t_2' = 230^o C$ *for a reversible compression.*

In summary, we have for this example (see also Figure 6.10):
Adiabatic compression with 1 step: $W^{\text{rev}} = 9.25\ MW$.
Adiabatic compression with 2 steps: $W^{\text{rev}} = 9.25\ MW$.
Adiabatic compression with 2 steps with intermediate cooling: $W^{\text{rev}} = 8.21\ MW$.
Isothermal compression (same as infinite steps with imtermediate cooling): $W^{\text{rev}} = 7.34\ MW$.

We see from the above example that it is *optimal to cool as much as possible* during compression in order to reduce the compression work. This is also clear from the general formula $W_s^{\text{rev}} = \int V dp$ in (6.8), because cooling reduces the volume V. The most favorable is isothermal compression at a low temperature, but this requires continuous cooling, which is difficult in practice. Multi-step adiabatic compression with *intermediate cooling* between each step is therefore used in practice. The compression work savings by use of intermediate cooling is graphically illustrated in Figure 6.10

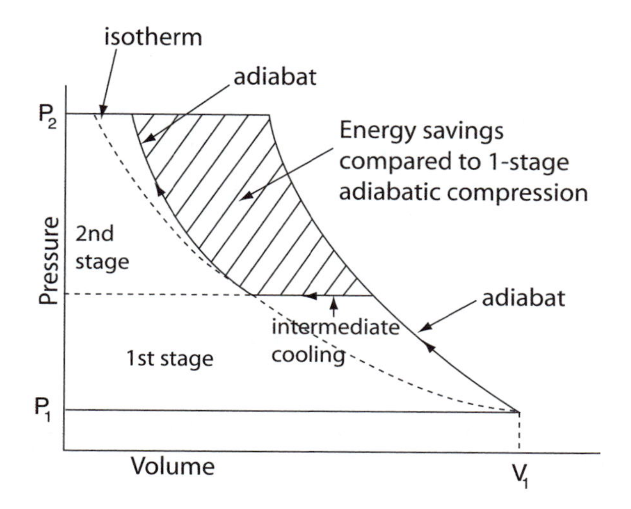

Figure 6.10: Energy gain for two-step adiabatic compression with intermediate cooling

(remember that $W_s^{\text{rev}} = \int V dp$). From the figure, we see again that with sufficiently many steps, adiabatic compression with intermediate cooling gives the same work as isothermal compression.

It can be shown that, for a given number of steps with intermediate cooling, it is optimal to use roughly the same compression ratio in each step. The compression ratio p_2/p_1 is typically about 2 per step.

Correspondingly, for a turbine, one should heat as much as possible in order to extract more work. In practice, expansion is performed in multiple steps with intermediate heating.

Exercise 6.2* *50 mol/s of an ideal gas with heat capacity $C_p = 30$ J/mol K is isothermally compressed (400 K) from 3 bar to 30 bar. The efficiency of the compressor is 0.7. Calculate the required cooling duty.*

Exercise 6.3 *A 6000 kmol/h gas stream with temperature 50 °C is compressed from 19 to 99 bar in two steps. The first step compresses the gas stream adiabatically to 45 bar. The gas is then cooled to 70 °C before being adiabatically compressed to 99 bar. Calculate the ideal (reversible) compression work in the two steps.*

Data: Assume ideal gas with constant heat capacity $C_p = 30.5$ J/mol,K.

Exercise 6.4* Compressor. *In order to produce vacuum and suck off the vapor in Exercise 4.6 (see page 112), a compressor that takes the gas from pressure p_3 to 1 atm is needed.*

Data: The vapor pressure of water at 20 °C is $p_3 = 2.337$ kPa, and the compressor has an efficiency of 0.6 (for both adiabatic and isothermal compression). The heat capacity of the gas is 33.6 J/mol K and ideal gas can be assumed. Note that 1 atm is equal to 101.33 kPa and that the amount of compressed gas is 0.0297 kg/s which corresponds to 1.65 mol/s.

Calculate the compression work for both (a) adiabatic and (b) isothermal compression.

6.7 Pump work

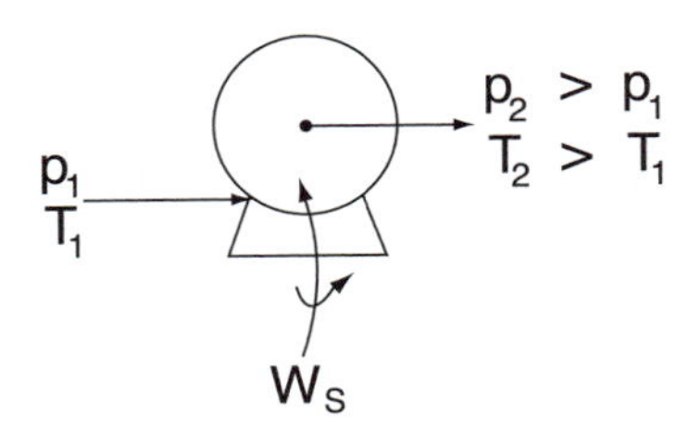

Figure 6.11: Pump

"Compression" (i.e., increase of pressure) of a liquid is usually called **pumping**. We often use mass basis for liquids and by introducing $V = m/\rho$ and $w_s^{\rm rev} = W_s^{\rm rev}/m$, (6.4) can be written as

$$w_s^{\rm rev} = \int_{p_1}^{p_2} \frac{1}{\rho} dp \quad [\text{J/kg}] \tag{6.12}$$

where ρ [kg/m^3] is the density and we use a lower case letter (w_s) to explicitly show that it is in mass basis. The density ρ [kg/m^3] of most liquids can be assumed constant, that is, independent of changes in pressure and temperature. From (6.12) the ideal (reversible) pump work assuming constant ρ (incompressible fluid) is

$$w_s^{\rm rev} = \frac{p_2 - p_1}{\rho} \quad [J/kg] \tag{6.13}$$

This work is per kg of pumped fluid. We can find the pump power [J/s] by multiplying with the mass flow $\dot{m}$ [kg/s]:

$$\boxed{\dot{W}_s^{\rm rev} = \frac{p_2 - p_1}{\rho} \dot{m} = (p_2 - p_1)\dot{V}} \quad [J/s] \tag{6.14}$$

where $\dot{V}$ [m^3/s] $= \dot{m}/\rho$ is the volumetric flow, which is constant throughout the process for an incompressible fluid. The actual pump work is

$$W_s = W_s^{\rm rev}/\eta \quad [\text{J, J/s}] \tag{6.15}$$

where η is the efficiency of the pump.

Remark. Note that $W_s^{\rm rev}$ in (6.14) gives the required work for increasing the pressure of the liquid. If the pump is also used to lift the liquid to a higher level (i.e., increase its potential energy) and/or to increase the velocity of the liquid (i.e., increase its kinetic energy), then also these terms, which we neglected in the energy balance, must also be included. (This is discussed in Chapter 9 on mechanical energy balance.

It may also be interesting to determine the temperature rise $T_2 - T_1$ for pumping, although it is usually small. If we assume adiabatic process ($Q = 0$) and constant heat capacity, then the energy balance (6.1) gives

$$\dot{m} c_p (T_2 - T_1) = W_s \quad [\text{J/s}] \tag{6.16}$$

Note that we can also run the pump in reverse such that we get a *turbine* where we extract work. Water turbines are used to produce electric power in hydroelectric power plants.

Example 6.2 Pumping. *1.5 kg/s water is pumped from 2 bar to 80 bar in a pump with 65% efficiency. Calculate the work and the temperature rise.*

Solution. *The density of water is 1000 kg/m^3 so the volumetric flow is $\dot{V} = 0.0015$ m^3/s. The pump work is then*

$$\dot{W}_s = \frac{\dot{V}(p_2 - p_1)}{\eta} = \frac{0.0015 \cdot (80 \cdot 10^5 - 2 \cdot 10^5)}{0.65} = 18000 \ J/s = 18 \ kW$$

The specific heat capacity for water is $c_p = 4180$ [J/kg,K], the mass flow is $\dot{m} = 1.5$ kg/s, and we find that the temperature rise is

$$T_2 - T_1 = \frac{\dot{W}_s}{\dot{m} c_p} = \frac{18000}{1.5 \cdot 4180} = 2.87 \ K$$

The use of (6.14) to calculate the heart's pump work is discussed in Exercise 1.27 on page 34.

6.8 Compression and expansion of real gases

The formulas given above are for ideal gases and for incompressible fluids (liquids). For real fluids, the reversible shaft work can in principle be calculated from (6.4),

$$W_s^{\text{rev}} = \int_{p_1}^{p_2} V dp$$

Example 6.3 Isothermal expansion of real gas. *We want to use (6.4) to calculate the reversible turbine work that may be obtained when $n = 6$ mol/s of gas expands isothermally at 30° C from 15 bar to 1 bar. The gas is assumed to follow the* **van der Waals equation of state**, *$p = nRT/V - an^2/V^2$, where $a = 0.68$ Pa m^6 mol^{-2}.*

Solution. *The reversible work is $W_s^{\text{rev}} = \int_{p_1}^{p_2} V dp$. It is easier to integrate over volume than over pressure, so we differentiate the equation of state to get $dp = [-nRT/V^2 + 2an^2/V^3]\, dV$. We then get with T constant, $W_s^{\text{rev}} = \int_{p_1}^{p_2} V dp = \int_{V_1}^{V_2} [-nRT/V + 2an^2/V^2]\, dV = -nRT \ln V_2/V_1 - 2an^2(1/V_2 - 1/V_1)$. The initial and final volumes are found from the equation of state which is a quadratic equation in V:*

$$p(V/n)^2 - RT(V/n) + a = 0$$

For the two values of p, we find the value of V for the gas phase is $V_1/n = 1.34 \cdot 10^{-3}$ m^3/mol and $V_2/n = 24.92 \cdot 10^{-3}$ m^3/mol, or $V_1 = 8.05 \cdot 10^{-3}$ m^3/s and $V_2 = 149.5 \cdot 10^{-3}$ m^3/s. The work is then (SI units):

$$W_s^{\text{rev}} = -6 \cdot 8.31 \cdot 303.15 \ln(149.5/8.05) - 2 \cdot 0.68 \cdot 6^2 \left(\frac{1}{149.5e-3} - \frac{1}{8.05e-3} \right) =$$
$$= -44160 J/s + 5754 J/s = -38406 J/s$$

In comparison, the work for an isothermal expansion of an ideal gas between the same pressures is, from (6.7), $W_s^{\text{rev}} = nRT \ln p_2/p_1 = 6 \cdot 8.31 \cdot 303.15 \ln(1/15) = -40932$ J/s.

In practice, the work for a real gas is rarely calculated in this way by integrating (6.4). Rather, from the exact relation $dW_s^{\rm rev} = Vdp = dH - TdS$ (see B.68), it follows that it is possible to find the work $W_s^{\rm rev}$ by first obtaining the changes in *enthalpy* (H) and *entropy* (S); see Section 7.2.4 page 169. (The concept of entropy is discussed in Chapter 7 and Appendix B, and if you are not familiar with entropy, it is recommended that you wait with reading the rest of this chapter.)

We consider two important cases: (1) Reversible isothermal process, and (2) Reversible adiabatic process.

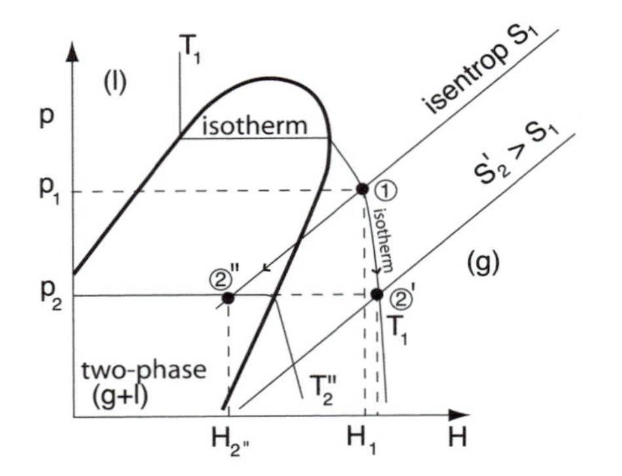

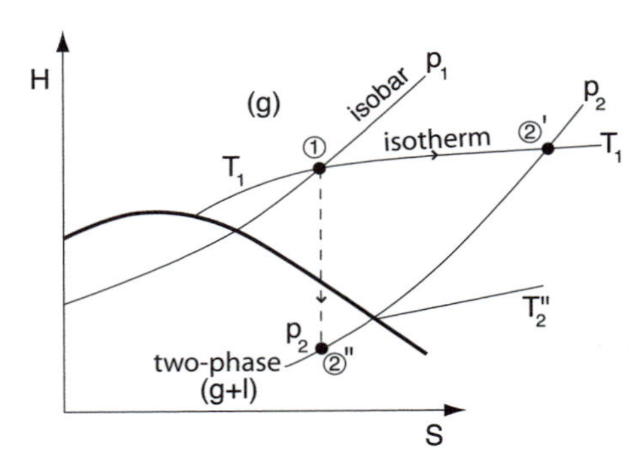

(a) Expansion in pH-diagram
②' : after isotherm
②'' : after isentrop (rev. adiabatic process)

(b) Same expansion in HS-diagram
②' : after isotherm
②'' : after isentrop (rev. adiabatic process)

Figure 6.12: Use of thermodynamic diagram to calculate reversible expansion work for real gas: (a) pH diagram. (b) HS diagram.

6.8.1 Reversible isothermal compression/expansion of real gas

We consider a steady-state process from state 1 ("inlet stream") to state 2 ("outlet stream"). The energy balance (6.1) gives

$$H_2 - H_1 = Q + W_s \quad [J]$$

For a **reversible isothermal process** at constant temperature, $T_1 = T_2 = T_0$, the entropy change, according to (7.4), is $\Delta S = S_2 - S_1 = Q^{\rm rev}/T_0$, that is, $Q^{\rm rev} = T_0(S_2 - S_1)$, which inserted in the energy balance gives

$$\text{Real gas}: \quad \boxed{W_s^{\rm rev} = (H_2 - H_1) - T_0(S_2 - S_1)} \tag{6.17}$$

Both the enthalpy H_i [J] and the entropy S_i [J/K] are state variables. For a real gas multicomponent mixture, we normally use an equation of state (page 342), together with data for ideal gas heat capacity, to compute the enthalpy and entropy changes. For pure components, it is practical and insightful to use thermodynamic diagrams (page 115), as shown in Figure 6.12. From a given value for p_1 and $T_1 = T_0$, we read off the values for H_1 and S_1. We then follow a line for constant temperature T_0 until we arrive at p_2, where we can read off H_2 and S_2 for the outlet stream (denoted 2' in the figure).

Example 6.4 Isothermal reversible compression of methane (ideal and real gas). *1700 mol/s methane gas is compressed isothermally at 50° C ($T_0 = 323K$) from $p_1 = 20$ bar (state 1) to $p_2 = 100$ bar (state 2). Calculate the reversible compression work for (a) ideal gas, (b) real gas (use the pH-diagram for methane, on page 418).*

Solution. *(a) For an ideal gas, we have from (6.7)*

$$W_s^{\text{rev}} = nRT_0 \ln(p_2/p_1) = -1700 \cdot 8.31 \cdot 323 \cdot \ln(100/20) \text{ J/s} = 7.34 \text{ MW}$$

Comment: (6.7) also follows from (6.17) since we, for an ideal gas with constant temperature, from (7.11) have that

$$H_2 - H_1 = 0; \quad S_2 - S_1 = -nR\ln(p_2/p_1)$$

(b) For a real gas, we read at 50 °C from the pH-diagram for methane (points 1 and 2' in Figure 6.12a):

$$h_2 - h_1 = 600 - 663 = -63\text{kJ/kg}, \quad s_2 - s_1 = 9.2 - 10.2 = -1.0\text{kJ/kg K}$$

The mass flow is $m = 1700$ *mol/s* $\cdot 16 \cdot 10^{-3}$ *kg/mol* $= 27.2$ *kg/s, and from (6.17) we find*

$$W_s^{\text{rev}} = m(h_2 - h_1) - mT_1(s_2 - s_1) = 27.2(-63 + 323 \cdot 1.0)kJ/s = 7.07MW$$

which is 4% lower than the value for ideal gas.

6.8.2 Reversible adiabatic (isentropic) compression/expansion of real gas

For a **reversible adiabatic process**, we have $Q^{\text{rev}} = 0$ and since the entropy change is $dS = dQ^{\text{rev}}/T$ (see (7.3) on page 163), it follows that $dS = 0$, i.e., the entropy is constant ("isentropic process") and $S_1 = S_2$. The shaft work is then from the energy balance

$$\text{Real gas}: \quad \boxed{W_s^{\text{rev}} = (H_2 - H_1)_S} \tag{6.18}$$

For a real gas multicomponent mixture, we normally use an equation of state (page 342), together with data for ideal gas heat capacity, to compute the enthalpy change at constant entropy. For a pure component, it is practical and insightful to use thermodynamic diagrams as shown in Figure 6.12: From a given value for p_1 and $T_1 = T_0$, we read off H_1. We then follow a line for constant entropy until we arrive at p_2, where we can read H_2 (and T_2) for the outlet stream (denoted 2" in Figure 6.12).

Example 6.5 Adiabatic reversible compression of methane (ideal gas and real gas). *We want to find the work for reversible adiabatic (isentropic) compression of 1700 mol/s methane from 50 °C ($T_1 = 323K$) and $p_1 = 20$ bar (state 1) to $p_2 = 100$ bar for (a) ideal gas and (b) real gas. In addition, find the actual work and the outlet temperature T_2 when the compressor has an efficiency of 85%.*

For the ideal gas case, $C_{P,m} = 35.3$ J/mol K is assumed constant. This is a continuation of Example 6.4 where we considered isothermal compression.

Solution. *(a) For an ideal gas with constant heat capacity, we have from (6.8) for an isentropic process*

$$T_2' = T_1 \left(\frac{p_2}{p_1}\right)^{\frac{R}{C_{p,m}}} = 323 \left(\frac{100}{20}\right)^{0.235} = 471.8K$$

and (6.9) gives

$$W_s^{\text{rev}} = (H_2' - H_1)_S = nC_{p,m}(T_2' - T_1) = 1700 \cdot 35.3 \cdot (471.8 - 323)J/s = 8.92MW$$

As expected, this is higher than for isothermal compression (7.34 MW) since cooling is favorable for compression. The actual work is $W_s = W_s^{\text{rev}}/\eta = 8.92$ *MW / 0.85 = 10.49 MW. From the energy balance, we then have that* $T_2 - T_1 = 10.49e6/(1700 \cdot 35.3) = 174.8$ *K, i.e., the actual exit temperature is* $T_2 = 497.8$ *K (224.8° C).*

(b) For a real gas, we read from the pH-diagram for methane (points 1 and 2" in Figure 6.12a):

$$(h_2' - h_1)_s = 967 - 663 = 304\text{kJ/kg}$$

The mass flow is $m = 27.2$ *kg/s and from (6.18) we have that*

$$W_s^{\text{rev}} = m(h_2' - h_1)_s = 27.2 \cdot 304 \text{ kJ/s} = 8.27 \text{ MW}$$

which is 7% lower than the value for ideal gas. The reversible temperature obtained from the diagram is $T_2' = 179^o C$ *(452 K). Now consider the actual work and temperature. The actual work is*

$$W_s = W_s^{\text{rev}}/\eta = 8.27\text{MW}/0.85 = 9.73\text{MW}$$

The actual specific enthalpy change is $(h_2 - h_1) = (h_2' - h_1)/\eta = 304/0.85 = 358$ *kJ/kg, which gives* $h_2 = 1021$ *kJ/kg. The pressure of the outlet stream is* $p_2 = 100$ *bar, and we can, from the pH diagram for methane, read off the actual temperature* $T_2 = 470$ *K (197° C). This is 28 K lower than the value 498 K found for ideal gas.*

Exercise 6.5* Adiabatic expansion of steam in turbine. *15 t/h of intermediate pressure steam at 30 bar and 450 °C (state 1) is expanded in a (adiabatic) condensing turbine (that is, with cooling at the exit of the turbine, where the vapor is condensed). With maximum cooling at the exit, we are able to reach 0.04 bar (which is the vapor pressure for water at 30 °C). (a) Calculate the work extracted in the turbine,* $W_s = H_2 - H_1$*, when it is given that the outstream at 0.04 bar (state 2) contains 5% liquid. (b) What is the (adiabatic) efficiency of the turbine (use the expansion from 1 to 2" in Figure 6.12b in order to calculate the reversible work)?* Data: *HS diagram for water.*

Exercise 6.6 Compression of methane. *1700 mol/s methane is compressed isothermally at 50° C from 20 bar to 100 bar. Find the reversible compression work for (a) ideal gas and (b) real gas (use pH-diagram for ammonia).*

Exercise 6.7 Expansion of steam in turbine. *15 t/h medium pressure steam at 30 bar and 450° C is expanded adiabatically in a condensing turbine (that is, there is cooling at the turbine outlet where the steam is condensed). With maximum cooling, the outlet turbine pressure is 0.04 bar (which is the saturation pressure for water at 30° C). Find the work that is extracted from the turbine when the outlet stream from the turbine contains 5% liquid væske. What is the turbine efficiency?* Data: *HS-diagram for steam.*

Exercise 6.8 Refrigerator (cooling cycle). *A home refrigerator uses* $C_2H_2F_4$ *(***R134a***) as the refrigerant (cooling medium). A cycle consists of the following four steps (see a similar cycle in Figure 8.7 on page 207):*

(a) Adiabatic compression: Saturated vapor at -10°C and 2 bar (state 1) is compressed to 10 bar (state 2 which is gas).

(b) Cooling ribs at the back of the refrigerator: Cooling/condensing at constant pressure down to a final temperature of 30°C (state 3 which is liquid).

(c) Valve: Expansion down to 2 bar (state 4 which contains about 20% gas).

(d) Take heat from the inside of the refrigerator: Evaporation at constant pressure (from state 4 to state 1).

Data: *pH-diagram for R134a (available at the book's home page). Assume that the compressor has a thermodynamic efficiency of 70%.*

- *Draw a flowsheet.*
- *Find the enthalpy change in each step (a, b, c, d) for each kg of refrigerant R134a that cycles. What is the temperature in state 2?*
- *What is the cooling duty (Q_C) when the power consumption in the compressor is $W = 300$ W, and what is the mass flow of refrigerant?*
- *Find the coefficient of performance $\mathrm{COP_C} = Q_C/W$.*
- *What is the theoretically maximum COP_C if the room temperature is 22°C and the temperature inside the refrigerator is 5°C. Why is this not 100%?*

Exercise 6.9 Compression of ammonia. *100 mol/s ammonia gas at 1 bar and 303 K (stream 1) is used to produce 100 mol/s ammonia gas at 10 bar and 303 K (stream 2). Calculate the reversible compression work for the following four processes assuming* ideal gas *and $C_p(g) = 37.4$ J/mol K:*

(a) Adiabatic compression with subsequent cooling to 303 K.

(b) Two-step adiabatic compression with pressure rise $p_{out}/p_{in} = 10^{1/2}$ in each step and intermediate cooling to 303 K.

(c) 12-step adiabatic compression with pressure rise $p_{out}/p_{in} = 10^{1/12}$ in each step and intermediate cooling to 303 K.

(d) Isothermal compression (at 303 K).

(e) **Real gas:** *Use the pressure-enthalpy diagram for ammonia to calculate the reversible work for (i) two-step adiabatic compression and (ii) isothermal compression.*

(f) Suggest an alternative process where cooling (condensation), pumping and heating (evaporation) is used instead of compression and cooling. What is the ideal pump work when the liquid density of ammonia is about 0.8 kg/l? What drawback does this process have?

7

Entropy and equilibrium

We need the state function entropy in order to proceed. Roughly speaking, entropy is a measure of the degree of disorder and the second law of thermodynamics states that the total entropy increases for all natural processes, and it reaches its maximum when we have equilibrium. With this as a starting point, we give in this chapter, among other things, the theoretical basis for use of equilibrium constants. Most of the basic material on entropy is collected in Appendix B, which should be read before you start this chapter.

7.1 The laws of thermodynamics

The second law of thermodynamics and the concept of entropy is discussed in more detail in Appendix B. In this chapter, we concentrate on process engineering applications, but let us start with a summary of the four laws of thermodynamics.

Zeroth law of thermodynamics *Two systems, both in equilibrium with a third system, must also be in equilibrium with each other.*

> **Example 7.1** *If we have a system with ice, liquid water and gas (all H_2O), where ice and liquid water are in equilibrium (the temperature must then be $0\ ^oC$), and in addition the water in the gas phase is in equilibrium with the liquid water (as given by the vapor pressure of liquid water), then the water in the gas phase (vapor pressure) is also in equilibrium with the ice (as given by the vapor pressure of ice).*

First law of thermodynamics This law is about the conservation of energy: *For a closed system, the change in the system's (internal) energy U is the sum of heat Q and work W, that is,*

$$\Delta U = Q + W \tag{7.1}$$

The first law of thermodynamics is a special case of the general energy balance, which has already been discussed in great detail in Chapter 4, see (4.1) and (4.10).

Second law of thermodynamics This law is about the inevitable decline in the "quality" of energy (from "high-quality" energy (exergy) that can be converted to useful work to "low-quality" thermal energy). There are many ways of stating the second law, but the following statement seems fairly obvious:

- *Heat cannot spontaneously flow from a material at lower temperature to a material at higher temperature*

Example. *Consider a home refrigerator. It would be fantastic if we could make a "self-cooling" refrigerator where we somehow could make heat flow from a low temperature at $0^o C$ (inside the refrigerator) to a high temperature at $25^o C$ (in the room). Unfortunately, this is impossible according to the second law of thermodynamics. To make it happen in practice, we need to supply some work* *(see Figure 8.6 on page 205).*

From the above fairly obvious statements, it is possible, by introducing the state function entropy S as a measure for the "degree of disorder," to derive the following alternative statement of the **second law of thermodynamics**:

- *For every real process, the total entropy of the system (S) and surroundings ($S_{\rm sur}$) must always increase, that is,*

$$\Delta S_{\rm total} = \Delta S + \Delta S_{\rm sur} \geq 0 \tag{7.2}$$

The total entropy change is zero only for the (imaginary) case of a reversible process. The second law of thermodynamics only applies to macroscopic processes, that is, when we consider the average behavior of a large number of particles (molecules).

If you think that the statement (7.2) of the second law, and the whole concept of entropy, is a bit fuzzy, then take a look at Appendix B. If this still does not help, then you may consult a book on classical thermodynamics that shows in detail how you can go from one of the more obvious statements (e.g., "heat cannot flow from low to high temperature") to the entropy version of the second law. I very much like the treatment on pages 26–38 in the excellent book of Denbigh.[1] Alternatively, you may dig into the subject of statistical thermodynamics, which provides another way of deriving the second law, and also provides a means to compute the entropy by calculating the probability of the system's state, based on a detailed description of the individual molecules.

Some other consequences of the second law are:

- A perpetual motion machine (*perpetuum mobile*) is impossible.
- The maximum obtainable "efficiency" for a process where heat is converted (partially) to work is $1 - T_C/T_H$ (the Carnot factor), where T_H is the temperature of the heat source and T_C is the temperature of the coolant. This result is derived in Chapter 8 that deals with "work from heat."

Example. *We have available an amount of heat $Q_H = 5$ GJ at 100 $^o C$, and we have cooling water at 5 $^o C$. From this, it is possible to extract a maximum work of*

$$|W| = Q_H(1 - \frac{T_C}{T_H}) = 5(1 - \frac{278}{373}) = 5 \cdot 0.255 = 1.27 \; GJ$$

Since energy is a conserved quantity (1st law), the remaining heat must be removed by cooling, $|Q_C| = |Q_H| - |W| = 5 - 1.273 = 3.73$ GJ.

[1] K. Denbigh, *The principles of chemical equilibrium*, Cambridge Press, 4th Ed, 1981.

Third law of thermodynamics At $T = 0$ K, all processes cease and the system entropy reaches a minimum. For a perfect crystal at O K, there is only one way to arrange the atoms so that the entropy ("degree of disorder") in this state can be set to zero.

The third law implies that it is meaningful to assign an absolute value for the entropy of each component (where $S = 0$ for the component as a perfect crystal at 0 K). Nevertheless, since we in this book are only interested in entropy changes, we normally do not use this; instead we (somewhat arbitrarily) set $S = 0$ for the elements in their standard state at 298.15 K and 1 bar.

7.2 Calculation of entropy

The entropy of a system can, as mentioned above, be theoretically calculated from statistical mechanics by considering the probability of the system's state on the microscopic level. From this, it is clear that the entropy is a state function.

How can we compute changes in the system's entropy by considering changes at the macroscopic level? Well, since we know that entropy is a state function, let us consider a reversible process. How can the entropy ("degree of disorder") change for such a system? It can not be caused by internal processes, since these are assumed to be reversible. Entropy changes must therefore be caused by interactions with the surroundings, which for a closed system involves transfer of work W and heat Q. Now, work is by definition "organized energy transfer" so this does not change the disorder (entropy). Thus, the only remaining source of change in disorder is the heat transfer Q, which is "disorganized" energy transfer and thus involves a transfer of disorder (entropy). Thus, we have for a reversible process that the only way to increase the system's entropy is by supplying heat Q. However, by how much does the entropy increase (quantitatively)? Intuitively, the increase in disorder (entropy) for a given Q is larger when the system temperature T is low. This intuition is correct, and it turns out that the entropy increase is given by Q/T. In summary, the change in a system's entropy can be computed as follows:

The system's entropy S is a state function. For a small change in the system state, the entropy change is

$$dS = dQ^{\text{rev}}/T \tag{7.3}$$

where dQ^{rev} is the heat supplied to the system in an (imaginary) reversible process, *and T is the system's temperature. For a large change between two states,*

$$\Delta S = \int \frac{dQ^{\text{rev}}}{T} \tag{7.4}$$

For a system with constant temperature, $\Delta S = \Delta Q^{\text{rev}}/T$.

By applying (7.4) to idealized reversible processes, we can find how the system's entropy depends on temperature, pressure and composition, $S(T, p, n_i)$. This is shown in the next section. However, let us first show how to calculate the entropy of the surroundings.

7.2.1 Entropy of the surroundings

The surroundings is also a system, so (7.4) gives $\Delta S_{\mathrm{sur}} = \int dQ_{\mathrm{sur}}^{\mathrm{rev}}/T_{\mathrm{sur}}$. When calculating the entropy change of the surroundings, we assume that all internal changes in the surroundings are reversible. This is to avoid that irreversibilities in the surroundings can somehow compensate for "impossible" irreversible processes inside the system (see also page 373). We then have $dQ_{\mathrm{sur}}^{\mathrm{rev}} = dQ_{\mathrm{sur}} = -dQ$ (because the heat supplied to the surroundings is minus the heat supplied to the system) and derive

$$\Delta S_{\mathrm{sur}} = \int \frac{-dQ}{T_{\mathrm{sur}}} \tag{7.5}$$

where T_{sur} is the temperature of the surroundings and dQ is the (actual) heat supplied to the system from the surroundings (note that it says dQ and not dQ^{rev}). If the temperature T_{sur} of the surroundings is constant, then

$$\Delta S_{\mathrm{sur}} = \frac{-Q}{T_{\mathrm{sur}}} \tag{7.6}$$

The second law of thermodynamics (7.2) can now be written in the following useful form:

$$\Delta S_{\mathrm{total}} = \Delta S + \int \frac{-dQ}{T_{\mathrm{sur}}} \geq 0 \tag{7.7}$$

Here, ΔS is a state function (independent of the process), whereas Q (dQ) depends on the actual (real) process. Note that we can have several kinds of surroundings, for example a cold reservoir with constant temperature $T_{\mathrm{sur}} = T_C$ and a hot reservoir with temperature $T_{\mathrm{sur}} = T_H$, and we may then need to add terms in (7.5) and (7.7).

For a completely reversible process we have $\Delta S_{\mathrm{total}} = 0$, which assumes that both the heat transfer *and* the processes within the system are reversible. We sometimes assume that only the heat transfer to the surroundings is reversible; this corresponds to assuming $T_{\mathrm{sur}} = T$ where T is the system's temperature.

The next section shows how to calculate ΔS for the system.

7.2.2 Calculation of system entropy

Entropy is a state function, and the entropy can therefore be calculated from (7.4) by considering idealized reversible processes. This is shown in more detail in Appendix B.4 (page 377); here we summarize these findings.

Entropy change for phase transition (given pressure). From (B.11)

$$\boxed{\Delta_{\mathrm{trs}} S = \frac{\Delta_{\mathrm{trs}} H}{T_{\mathrm{trs}}}} \tag{7.8}$$

where $\Delta_{\mathrm{trs}} H$ [J] is the enthalpy change for the phase transition, and T_{trs} is the temperature (which is indirectly determined by given pressure).

Example 7.2 *For water at 1 atm, we have* $T_{\mathrm{trs}} = T_b = 373.15$ *K (100 °C) for the phase transition between liquid and gas (evaporation), and we find* $\Delta_{\mathrm{vap}} S = \Delta_{\mathrm{vap}} H / T_b = (40.68 \cdot 10^3$ *J/mol)/(373.15 K) = 109.0 J/mol K. We note that the entropy increases for the phase transitions that require supply of heat.*

Entropy's dependency on temperature (given pressure). For a system (gas, liquid, solid) without phase transition, (B.12) gives

$$\boxed{S(T,p_0) - S(T_0,p_0) = \int_T^{T_0} C_p(T)\frac{dT}{T}} \quad [\text{J/K}] \tag{7.9}$$

where C_p [J/K] is the system's heat capacity. We note that entropy increases with temperature. If for simplicity we assume that C_p is independent of temperature, we get

$$S(T,p) - S(T_0,p) = C_p \ln \frac{T}{T_0} \quad [J(K] \tag{7.10}$$

Example 7.3 *Consider a gas with constant heat capacity C_p=30 J/mol K. The entropy increase when the gas is heated from $T_1 = 300$ to $T_2 = 400$ K is $\Delta S = C_p \ln \frac{T_2}{T_1} = 30 \ln \frac{400}{300} =$ 8.63 J/mol K.*

Entropy's dependency on pressure (given temperature). For liquids and solids, the pressure dependency is usually negligible. However, for gases it is important, and for **ideal gas** (B.14)) gives:

$$\boxed{S(T_0,p) - S(T_0,p_0) = -nR \ln \frac{p}{p_0}} \quad [\text{J/K}] \tag{7.11}$$

Example 7.4 *The molar entropy decrease for an isothermal compression of an ideal gas from 2 bar to 4 bar is: $\Delta S = -R \ln \frac{p_2}{p_1} = -R \ln \frac{4}{2} = -5.76$ J/mol K.*

Non-ideal conditions. For a real fluid, the entropy change can be found from thermodynamic state diagrams (this is most practical for pure components) or calculated from an equation of state.

Entropy's dependency on composition (given pressure and temperature). For an **ideal mixture**, the *entropy of the mixture, relative to that of the pure components* (in the same phase and same T and p), is from (B.27),

$$\boxed{\underbrace{S(T_0,p_0) - \sum_i n_i S^*_{m,i}(T_0,p_0)}_{\Delta_{\text{mix}} S} = -R\sum_i n_i \ln x_i = -nR\sum_i x_i \ln x_i} \quad [\text{J/K}] \tag{7.12}$$

Here $S^*_{m,i}$ [J/mol K] is the entropy of pure component i at the same pressure and temperature, n_i is the number of moles of component i in the mixture, $n = \sum_i n_i$ is the total number of moles and $x_i = n_i/n$ is the molar fraction. Since $\ln x_i$ is negative, we note (as expected) that the entropy increases when we mix the pure components.

Example 7.5 *We mix 0.2 mol O_2 and 0.8 mol N_2 (pure components) in order to produce 1 mol mixture. If ideal gas (and thereby ideal mixture) is assumed, the entropy of mixing is at constant pressure and temperature is $\Delta_{\text{mix}} S = -R\,(0.2 \ln 0.2 + 0.8 \ln 0.8) = -R\,(0.2 \cdot (-1.61) + 0.8(-0.2231)) = -R(-0.3219 - 0.1785) = 0.5004R = 4.16$ J/K mol.*

Mixing is obviously an irreversible process, so the entropy will always increase during mixing. If we, for example, mix two streams (1 and 2) to produce a product (mix), then from (7.12) the *entropy change for ideal mixing of two streams* is

$$\Delta S = S_{\text{mix}} - S_2 - S_1 = R\left[-\sum_i n_{i,\text{mix}} \ln x_{i,\text{mix}} + \sum_i n_{i,2} \ln x_{i,2} + \sum_i n_{i,1} \ln x_{i,1}\right] \tag{7.13}$$

Since $n_{i,\text{mix}} = n_{i,1} + n_{i,2}$ (mass balance for each component i), this can be written as

$$\Delta S = S_{\text{mix}} - S_2 - S_1 = R\left[\sum_i n_{i,1} \ln(x_{i,1}/x_{i,\text{mix}}) + \sum_i n_{i,2} \ln(x_{i,2}/x_{i,\text{mix}})\right] \quad \text{[J/K]} \tag{7.14}$$

where mix indicates the mixture (product) and 1 and 2 the different streams (which we naturally can have several). ΔS calculated from (7.14) is always positive – this is not quite obvious just by looking at the equation, but it must be this way since mixing is an irreversible process; see also Examples 7.7 and 7.8.

Summary: Entropy of ideal gas mixture. By adding the contributions (7.9), (7.11) and (7.12), we find that, for an ideal gas mixture, the entropy is

$$\boxed{S(T,p) = \underbrace{\sum_i n_i S^*_{m,i}(T_0,p_0)}_{S^*(T_0,p_0)} + \int_T^{T_0} C_p(T)\frac{dT}{T} - nR\ln\frac{p}{p_0} - nR\sum_i x_i \ln x_i} \quad [J/K] \tag{7.15}$$

where $S^*(T_0, p_0)$ is the entropy of the pure components at T_0 and p_0, and

$$C_p[\text{J/K}] = n_i C_{p,m}(i)$$

is the mixture's heat capacity. (7.15) can be used to calculate the entropy of an ideal gas stream with given T, p and composition (n_i).

On the other hand, note that the enthalpy of an ideal gas mixture is independent of pressure and composition:

$$H(T,p) = \underbrace{\sum_i n_i H^*_{m,i}(T_0,p_0)}_{H^*(T_0,p_0)} + \int_T^{T_0} C_p(T)T \quad [J] \tag{7.16}$$

The formulas (7.15) and (7.16) also apply to ideal liquid mixtures, except that the contribution from pressure on the entropy can be neglected (that is, we do not include the term $-nR\ln p/p_0$).

Value of $S^*(T_0, p_0)$. According to the third law of thermodynamics, the entropy S is equal to 0 for a perfect crystal at $T = 0$ K ("perfect order"), and this state is often chosen as the reference. Alternatively, the elements at $T_0 = 298.15$ K and $p_0 = 1$ bar are chosen as reference. This is practical, because this reference state is usually chosen for enthalpy, and we have

$$S^*(T_0, p_0) = \sum_i n_i \Delta_f S^{\ominus}_m(i, 298)$$

$$H^*(T_0, p_0) = \sum_i n_i \Delta_f H_m^\ominus(i, 298)$$

Note that the choice of reference state does not matter for the final answer because we are only interested in entropy and enthalpy changes (see also Example A.21, page 366).

Non-ideal conditions. For real fluid mixtures, we can replace $\ln x_i$ in the expression above by $\ln a_i$ where a_i is the activity. However, this is really just the definition of activity and does not bring us much further. For practical calculations, the entropy (and the activity) for real mixtures can be calculated, for example, from an equation of state with the use of thermodynamic relationships (which are straightforward, but are not covered in this book). These calculations are relatively involved and generally require the use of a computer.

7.2.3 Examples: Entropy change in irreversible process

We calculate the entropy change for some real (irreversible) processes and find, as expected, that the total entropy change is positive.

Example 7.6 Entropy change in heat exchanger. *Calculate the entropy change for the heat exchanger in Example 5.6 (page 139). Here,*

$$m_c = 5kg/s;\ c_{p,c} = 1500J/kg\ K;\ T_{c,\text{in}} = 293K; T_{c,\text{out}} = 315K$$

$$m_h = 3kg/s;\ c_{p,h} = 1200J/kg\ K;\ T_{h,\text{in}} = 343K; T_{h,\text{out}} = 297K$$

Note that we use absolute temperature [K] because this is always *required in thermodynamic calculations.*

Solution. *We assume that the heat loss for the heat exchanger to the surroundings is zero such that $\Delta S_{\text{sur}} = 0$. From (B.13), the entropy changes for the cold and hot sides of the process are*

$$\Delta S_c = m_c c_{p,c} \ln \frac{T_{c,ut}}{T_{c,inn}} = 5 \cdot 1500 \ln \frac{315}{293} W/K = 543.0\ W/K$$

$$\Delta S_h = m_h c_{p,h} \ln \frac{T_{h,ut}}{T_{h,inn}} = 3 \cdot 1200 \ln \frac{297}{343} W/K = -518.4\ W/K$$

Heat exchange is an irreversible process so, as expected from the second law of thermodynamics, the total entropy change is positive:

$$\Delta S_{\text{total}} = \Delta S + \Delta S_{\text{sur}} = \Delta S_c + \Delta S_h + 0 = 543.0\ J/K - 518.4\ W/K = 24.6\ W/K$$

Note that, as expected from the energy balance (4.13) with $Q = 0$ and $W_s = 0$, we have that

$$\Delta H = \Delta H_c + \Delta H_h = m_c c_{p,c}(T_{c,out} - T_{c,in}) + m_h c_{p,h}(T_{h,out} - T_{h,in}) =$$

$$5 \cdot 1500(315 - 293)J + 3 \cdot 1200(297 - 343)W = 165kW - 165kW = 0$$

We note, as expected, that the energy is constant (first law of thermodynamics), whereas the total entropy increases (second law).

Example 7.7 Entropy change for mixing two streams. *We mix 1 mol methane (stream 1) and 9.52 mol air (stream 2) to produce a product (stream "mix") that has a stoichiometric composition for combustion. The air is assumed to contain 21% O_2 and 79% N_2 and the* *combustion reaction is* $CH_4 + 2O_2 = CO_2 + 2H_2O$. *The amount and composition of the three streams are then*

Stream j	n_j	x_{CH_4}	x_{O_2}	x_{N_2}
1	1.0	1.0	0	0
2	9.52	0	0.21	0.79
mix	10.52	0.095	0.19	0.715

If we assume constant temperature and pressure and assume ideal mixture, the entropy change is from (7.13) given by

$$\begin{aligned}\Delta S &= 8.31 \cdot [-10.52(0.095 \ln 0.095 + 0.19 \ln 0.19 + 0.715 \ln 0.715) \\ &+ 9.52(0.21 \ln 0.21 + 0.79 \ln 0.79) + 1.0 \ln 1] = 68.10 - 40.66 - 0 = 27.44 \text{ J/K}\end{aligned}$$

Alternatively from (7.14):

$$\Delta S = 8.31 \cdot [1.0 \ln(1.0/0.095) + 2.0 \ln(0.21/0.19) + 7.52 \ln(0.79/0.715)] = 27.44 \text{J/K}$$

Example 7.8 Entropy change for mixing exhaust gas with infinite amount of air. *After the combustion in the previous example, we have 10.52 mol of exhaust gas (flue gas) with (mole fractions)*

$$x_{CO_2} = 0.095, x_{H_2O} = 0.19, x_{N_2} = 0.715$$

The corresponding amounts are:

$$n_{CO_2} = 1.0 \text{ mol}, n_{H_2O} = 2.0 \text{ mol}, n_{N_2} = 7.52 \text{ mol}$$

At the exit of the chimney, this exhaust gas is mixed with an infinite amount of air (the surroundings) with composition:

$$x_{CO_2} = 0.0004, x_{H_2O} = 0.01, x_{N_2} = 0.78, x_{O_2} = 0.21$$

(note that we have included the contents of CO_2 *and* H_2O *in the air – otherwise the entropy change would be infinite). The entropy change for the mixing exhaust gas with air is obtained from (7.14) by adding the entropy change for the three components in the stream*

$$\Delta S = -R\,(1.0 \ln(0.0004/0.095) + 2.0 \ln(0.01/0.19) + 7.52 \ln(0.78/0.715)) = 89.0 \text{J/K}$$

Note that there is no entropy change for the air because with an infinitely large amount, the composition of the air is constant.

Example 7.9 Distillation. *In a distillation column (see page 22) the opposite process of mixing takes place. The "mixed" feed stream is separated into a "light" and a "heavy" product. The entropy ("degree of disorder") for the products is less than that for the feed,*

$$\Delta S < 0$$

so this is apparently a process that violates the second law of thermodynamics (??!). However, in order to accomplish the separation, we need to supply heat $|Q_H|$ *in the reboiler and remove heat* $|Q_C|$ *in the condenser (we use absolute signs to show that these heats are positive quantities in the indicated "direction"). The two heats are roughly the same, that is,* $|Q_H| \approx |Q_C|$, *but the heat supply is at a higher temperature (at the boiling for the heavy product) than the heat removal (which is at the boiling point for the light product), that is,* $T_H > T_C$. *Thus, there is a net entropy increase in the surroundings,*

$$\Delta S_{\text{sur}} = \int \frac{-dQ}{T_{\text{sur}}} = \frac{-|Q_H|}{T_H} + \frac{|Q_C|}{T_C} \approx |Q_H| \left(-\frac{1}{T_H} + \frac{1}{T_C}\right) > 0 \tag{7.17}$$

We can calculate ΔS for the system using the formulas for mixing presented above. We will then find that the second law of thermodynamics is indeed satisfied,

$$\Delta S_{\text{total}} = \Delta S + \Delta S_{\text{sur}} > 0$$

For example, consider a distillation column separating a mixture of 40 mol-% methanol and 60 mol-% water into (almost) pure components; see also Example 8.20. The entropy change for the separation of 1 mol feed into the pure components is the negative of the entropy of mixing in (7.12),

$$\Delta S = R \sum x_i \ln x_i = R(0.4 \ln 0.4 + 0.6 \ln 0.6) = -0.67R = -5.57\text{J/mol K}$$

The boiling points for the pure components are 373 K (water) and 338 K (methanol), and in Example 8.20 (page 224) it is given that the heat supply is $Q_H = 31.8$ kJ/mol (for 1 mol liquid feed). Assuming that the heat transfer to the surroundings is reversible, we can set $T_H = 373K$ and $T_C = 338$ K and derive from (7.17)

$$\Delta S_{\text{sur}} = |Q_H| \left(-\frac{1}{T_H} + \frac{1}{T_C} \right) = 31.8E3 \left(-\frac{1}{373} + \frac{1}{338} \right) = 8.82 \text{ J/mol K}$$

Thus, we have

$$\Delta S_{\text{total}} = \Delta S + \Delta S_{\text{sur}} = -5.57 \text{ J/mol K} + 8.82 \text{ J/mol K} = 3.25 \text{ J/K mol}$$

which as expected is positive. The source for the irreversibility is the mixing of streams with different composition and temperature inside the distillation column. Distillation is discussed in more detail on page 224.

Exercise 7.1 Compression and expansion of ideal gas. *50 mol/s of an ideal gas at 300K and 1 bar (stream 1) is compressed to 10 bar (stream 2). The compression occurs adiabatically and the thermodynamic efficiency is 80%. Stream 2 is expanded in a valve down to 1 bar (stream 3) and is then cooled to 300 K (stream 4). Constant heat capacity $C_p = 30$ J/mol K is assumed.*

(This is seemingly an idiotic process, but maybe something occurs with stream 2, for example a chemical reaction, which we are not interested in here).

(a) Make a table that shows temperature, pressure, enthalpy [J/mol] and entropy [J/mol K] for the four streams. Choose 298.15 K and 1 bar as the reference state.

(b) Calculate the supplied heat and work [J] in the process.

(c) Calculate the total entropy increase [J/mol K] in each of the three process steps (compression, pressure relief, cooling), that is, calculate the entropy change in the process plus the surroundings (the surroundings are assumed to be at 1 bar and 280K).

7.2.4 Work W_s for reversible compression and expansion

With the state function entropy, we are now in position to derive expressions for the reversible shaft work for compression and expansion of real fluids. The energy balance (1st law) for a steady-state process gives $W_s = \Delta H - Q$. For a reversible process we get from (7.3) that $Q^{\text{rev}} = \int T dS$ (2nd law), and the reversible shaft work is

$$W_s^{\text{rev}} = \Delta H - \int T dS$$

For an **isothermal process** we then get

$$W_s^{\text{rev}} = \Delta H - T\Delta S \quad (T \text{ constant}) \tag{7.18}$$

For an **adiabatic process** we have $Q^{\text{rev}} = 0$ and the entropy is constant ($\Delta S = 0$) and we get

$$W_s^{\text{rev}} = \Delta H \quad (S \text{ constant}) \tag{7.19}$$

These expressions are very important for practical calculations, and we already used them in Section 6.8 (page 155). Now that you have learned about entropy, you should go back and revisit this material.

7.2.5 Work and entropy change for adiabatic process

Let us look at a steady-state adiabatic process in more detail. For an adiabatic process, we have that $Q = 0$. But this doesn't necessarily mean that the entropy change is 0; this is only the case if the process is, in addition, reversible. We illustrate this by considering an adiabatic pressure relief (expansion) from a pressure p_1 (state 1) to a lower pressure p_2 (state 2) for two alternative processes (see Figure 7.1):

(a) Reversible adiabatic expansion in a turbine

(b) Irreversible adiabatic expansion in a valve

We show that the first process is isentropic and the second is isenthalpic. *This section is mostly repetition, but make sure that you know and understand it!* First, a reminder about the energy balance (4.13) for a steady-state adiabatic process,

$$W_s = \Delta H = H_2 - H_1 \quad \text{(adiabatic process)} \tag{7.20}$$

where W_s is supplied shaft work, and we have assumed that the inflow is in state 1 and the outflow is in state 2.

(a) Reversible adiabatic expansion. We consider an expansion from pressure p_1 (state 1) to pressure p_2 (state 2). The process is adiabatic and reversible, that is, $Q^{\text{rev}} = 0$, and it follows from the definition of entropy in (7.4) that the entropy change is 0, that is,

$$\Delta S = S_2 - S_1 = \int \frac{dQ^{\text{rev}}}{T} = 0 \quad \text{(real and ideal gas)}$$

This means that the process is **isentropic**, and this can be used to find the temperature change and thereby the enthalpy change.

For an ideal gas with constant heat capacity, the relationship between temperature and pressure for an isentropic process is given by (6.8),

$$\frac{T_2}{T_1} = \left(\frac{p_2}{p_1}\right)^{R/C_p} \quad \text{(ideal gas)}$$

which with $p_2 < p_1$ gives $T_2 < T_1$ (the temperature drops). During a reversible adiabatic expansion both the temperature and pressure drop, and the changes

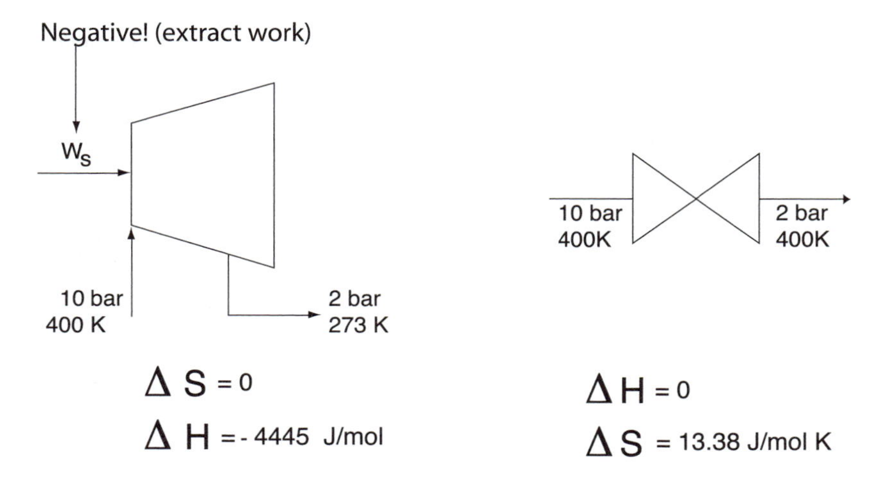

Figure 7.1: Adiabatic expansion: (a) Reversible in turbine (isentropic). (b) Irreversible in valve (isenthalpic).

are such that the entropy increase for the pressure drop is equal to (but with opposite sign) the entropy reduction for the temperature drop. For an ideal gas with constant heat capacity, it further holds that $H_2 - H_1 = mC_p(T_2 - T_1)$ and it follows that

$$W_s = \Delta H = C_p(T_2 - T_1) \quad \text{(ideal gas)}$$

which with $T_2 < T_1$ gives a negative number (work is extracted).

(b) Adiabatic pressure relief over valve (sometimes called a Joule-Thompson valve, if the objective is to lower the temperature for a real gas). In this case, no shaft work is extracted, so $W_s = 0$, and it follows from the energy balance (7.20) that

$$W_s = \Delta H = H_2 - H_1 = 0 \quad \text{(real and ideal gas)}$$

We have, in other words, that the enthalpy is constant – such a process is called **isenthalpic.** This can be used to determine the temperature change and the entropy change.

For an ideal gas, the enthalpy is only a function of the temperature and we therefore have that

$$T_2 = T_1 \quad \text{(ideal gas)}$$

that is, the temperature before and after the valve is the same. (For a *real gas*, the enthalpy also depends on pressure, and we end up with a temperature change; see Example 7.11 below).

But this process is not reversible (have you ever seen a gas that flows by itself from a low to a high pressure?), so for the valve the entropy increases. For example, for an ideal gas, the entropy change is from (7.11)

$$\Delta S = S(p_2) - S(p_1) = -R \ln \frac{p_2}{p_1} \quad [\text{J/mol K}] \quad \text{(ideal gas)}$$

which is positive since $p_1 > p_2$.

Example 7.10 Isentropic and isenthalpic pressure change for ideal gas. *Consider a continuous steady-state process where an ideal gas at 400 K and 10 bar is expanded to a pressure of 2 bar in an adiabatic process. Assume that the heat capacity is constant* $C_p = 35$ *[J/mol K] (which is the value of ammonia at 298 K). Calculate the change in entropy and enthalpy (per mol) for the following cases:*

(a) The gas is expanded in a turbine without loss (isentropic process).

(b) The pressure relief occurs over a valve (isenthalpic process).

Solution. *For the "inlet" state, we have* p_1*=10 bar and* $T_1 = 400K$*, and for the "outlet" state,* $p_2 = 2$ *bar and the temperature* T_2 *is to be found (it differs for processes (a) and (b)).*

(a) **Turbine**: *For a reversible adiabatic process, the entropy is constant,* $\Delta S = 0$*, and from (6.8) we have*

$$T_2 = T_1 \left(\frac{p_2}{p_1}\right)^{R/C_p} = 400 \left(\frac{2}{10}\right)^{8.31/35} = 273 \ K$$

The enthalpy change which is extracted as work is then

$$W_s = \Delta H = C_p(T_2 - T_1) = 35 \cdot (273 - 400) = -4445 \ J/mol$$

(remember the convention of W_s *being supplied work, that is, a negative value for* W_s *means that a work is extracted).*

(b) **Valve**: *For a pressure relief over a valve, no work is extracted (*$W_s = 0$*) and it follows from the energy balance that the enthalpy is constant,* $\Delta H = 0$*. Furthermore, for an ideal gas, enthalpy is only a function of temperature, therefore we find that the temperature is constant*

$$T_2 = T_1 = 400 \ K$$

The entropy change is from (B.14) given by

$$\Delta S = -R \ln \frac{p_2}{p_1} = -8.31 \ln \frac{2}{10} = 13.38 \ [J/molK]$$

that is, the entropy increases as expected.

The above calculations are for ideal gas, and in order to perform calculations for real gases, it is practical to use thermodynamic diagrams that express all of the state variables as functions of two independent variables, for example as a function of p and H or H and S. Using such diagrams, we can also handle cases where liquid is formed.

Example 7.11 Isentropic and isenthalpic pressure change for real gas. *Let us repeat the calculations from Example 7.10, but instead of assuming ideal gas we use the pressure-enthalpy diagram for* **ammonia** *on page 419. Before the expansion, the temperature is* $T_1 = 400K = 127\ ^oC$ *and the pressure is* $p_1 = 10$ *bar (point 1). After the expansion to* $p_2 = 2$ *bar, we can read off the values for the two cases (points 2a and 2b) as shown in Figure 7.2:*

(a) **Isentropic expansion in turbine.**

We follow the line for constant entropy down to pressure $p_2 = 2$ *bar and read off* $t_2 = 5^o C$ *(278 K) and* $H_2 - H_1 = 1398 - 1643 = -245$ *kJ/kg. This gives, with a molar mass of 17 g/mol,* $W_s = \Delta H = -4165$ *J/mol. The extracted (performed) work is 6% lower than for ideal gas, where we found* $W_s = -4445$ *J/mol. This is because, for real gas, we have to use some of the energy to pull the molecules apart.*

(b) **Isenthalpic expansion in valve.** *We follow the line for constant enthalpy down to pressure* $p_2 = 2$ *bar and read off* $t_2 = 119^o$ *C (392 K) and* $S_2 - S_1 = 6.87 - 6.11 = 0.76$ *kJ/kg K. This gives, with a molar mass 17 g/mol,* $\Delta S = 12.92$ *J/mol K. We find that we get some cooling during the expansion (from 127 °C to 119 °C). This is one of the reasons why the entropy increase (*$\Delta S = 12.92$ *J/mol K) is smaller than for ideal gas (*$\Delta S = 13.38$ *J/mol K).*

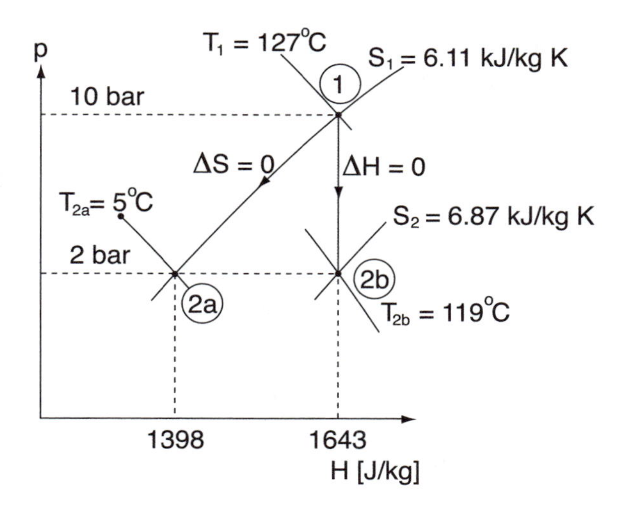

Figure 7.2: *pH* diagram for ammonia: (a) Isentropic and (b) Isenthalpic expansion

Exercise 7.2 *Consider a continuous steady-state process where the pressure is reduced from 3 bar to 2 bar using a valve.*
(a) Explain why the enthalpy is constant (which assumptions must be made?).
(b) Is the enthalpy constant if the velocity changes (before and after the valve) ?
(c) Is the process reversible? Is the entropy constant?

7.3 Equilibrium

For any real process, the second law of thermodynamics states that the total entropy must always increase, see (7.2). It then follows that, if we leave a system to itself, it will end up in an *equilibrium state* where the total entropy ("degree of disorder") has reached its maximum. This can be used to derive equilibrium conditions as shown in detail in Appendix B.6 (page 383), but before we get to these, let us mention Le Chatelier's principle.

7.3.1 Le Chatelier's principle

Le Chatelier's principle is a useful tool for understanding how systems in equilibrium respond to changes. The principle was stated in 1885 and says:

> *If a system in equilibrium is subject to a change (a "disturbance"), the equilibrium will shift in a direction that tends to counteract the initial change.*

For example, if we increase the temperature of a system in equilibrium, the equilibrium will shift such that heat is absorbed and the temperature is lowered. Similarly, if we increase the pressure of a system in equilibrium, the equilibrium will shift such that the pressure is lowered.

Le Chatelier's principle applies if we consider the dominant effect for a system in equilibrium; if several counteracting effects occur at the same time, then the principle may not to hold for the individual effects. Some examples of predictions that follow from Le Chatelier's principle are:

- For an exothermic reaction (where heat is generated), a higher reaction temperature shifts the equilibrium towards the reactants. This is because, according to Le Chatelier's principle, heat is "absorbed" by the reaction to counteract the increase in temperature.
- For an endothermic reaction, a higher reaction temperature shifts the equilibrium to the products.
- For a gas phase chemical reaction: When the pressure is increased, the equilibrium shifts to the side with the fewest number of moles. This is because, according to Le Chatelier, this counteracts the increase in pressure. For example, the reaction $N_2 + 3H_2 = 2NH_3$ is shifted toward the ammonia product when the pressure is increased.

Le Chatelier's principle can be justified by imagining the opposite: If, for example, a disturbance leads to an increased temperature, and the system responds by increasing the temperature even more, then this leads to a cascade of temperature increases that results in instability.

7.3.2 Chemical equilibrium

Le Chatelier's principle is useful for giving qualitative predictions of how the equilibrium is shifted, but it gives no quantitative information. Quantitative predictions about the equilibrium condition can, as mentioned, be derived by *maximizing the total entropy.* In Appendix B (see page 384) we prove the following:

> For systems at a **given pressure** p **and temperature** T, maximizing the total entropy (of the system and surroundings) is equivalent to **minimizing the system's Gibbs energy** $G \triangleq H - TS$.

Thus, at equilibrium, the Gibbs energy G (which is a state variable!) reaches a minimum where it no longer changes. In partuculat, if we have an equilibrium between two states (e.g., two phases or two sides of a chemical reaction), then the Gibbs energy for these two states must be the same, and we have (see (B.31)):

$$\Delta G = 0 \quad \text{(equilibrium at given } T \text{ and } p\text{)}$$

Applied to a chemical reaction at a given T and p, this gives the equilibrium condition $\Delta_r G = 0$. Here $\Delta_r G$ is the difference in Gibbs energy between the products and reactants. It is often practical for chemical reactions to introduce the equilibrium constant K, defined by

$$\ln K \triangleq -\frac{\Delta_\mathrm{r} G^\ominus(T)}{RT} \tag{7.21}$$

where $\Delta_\mathrm{r} G^\ominus(T)$ is standard change in Gibbs energy for the reaction at T and $p^\ominus = 1$ bar. Let us consider the general reaction

$$0 = \nu_A A + \nu_B B + \nu_C C + \nu_D D$$

where we assume that A and B are reactants (that is, ν_A and ν_B are negative) and C and D are products. For example, for the reaction

$$CO_2 + 3H_2 = CH_3OH + H_2O$$

we have that $\nu_A = -1, \nu_B = -3, \nu_C = 1$ and $\nu_D = 1$. By introducing the entropy and its dependency on composition, the equilibrium condition $\Delta_r G = 0$ can be written as (see Appendix B.6, page 386):

$$K = Q \triangleq \prod_i a_i^{\nu_i} = \frac{a_C^{\nu_C} a_D^{\nu_D}}{a_A^{-\nu_A} a_B^{-\nu_B}} \quad \text{(given } T, p) \tag{7.22}$$

where a_i is the activity relative to the component's standard state that was used when calculating the equilibrium constant K. A very common standard state is ideal gas at $p^\ominus = 1$ bar.

Comment on ideal gas assumption. The equilibrium condition in the form (7.22) is often called the **law of mass action**, which was expressed empirically by Norwegian chemists Guldberg and Waage in 1864 based on arguments about reaction rates (see page 258), that is, independent of the thermodynamics. They expressed the law as follows:

> At a specific temperature, the rate of a chemical reaction is proportional to the product of the concentration of the reacting substances. During the reaction, the rate decreases because the concentration of the reacting substances decreases. The rate of the reverse reaction, on the other hand, increases, and eventually it reaches an equilibrium where both rates are the same size. There is now a chemical equilibrium.

Ideal gas and pressure dependency

For an *ideal gas*, the activity is $a_i = p_i/p^\ominus$ where $p^\ominus = 1$ bar, and the equilibrium condition (7.22) becomes

$$K = \frac{\left(\frac{p_C}{p^\ominus}\right)^{\nu_C} \left(\frac{p_D}{p^\ominus}\right)^{\nu_D}}{\left(\frac{p_A}{p^\ominus}\right)^{-\nu_A} \left(\frac{p_B}{p^\ominus}\right)^{-\nu_B}} \tag{7.23}$$

By introducing the mole fraction $x_i = p_i/p$, this can be written as

$$\frac{x_C^{\nu_C} x_D^{\nu_D}}{x_A^{-\nu_A} x_B^{-\nu_B}} = \underbrace{K \cdot \left(\frac{p^\ominus}{p}\right)^{\Delta_r \nu}}_{=K_x(T,p)} \tag{7.24}$$

where $\Delta_r \nu = \sum_i \nu_i = \nu_C + \nu_D - |\nu_A| - |\nu_B|$ is the mole number change during the reaction. From (7.24), we see that

- K_x increases (the reaction is shifted to the product) when the pressure p increases for a reaction with negative mole number change (for example, for the reaction $N_2 + 3H_2 = 2NH_3$ where $\Delta_r \nu = -2$).
- K_x decreases (the reaction is shifted to the reactant) when pressure p increases for a reaction with positive mole number change (for example, the reaction $CH_4 + H_2O = CO + 3H_2$ where $\Delta_r \nu = 2$).

This is consistent with what we can derive from Le Chatelier's principle, but (7.24) expresses it quantitatively.

The temperature dependency of the equilibrium constant

The equilibrium constant $K(T)$ is a function of temperature, and according to **van't Hoff's** exact equation (B.53), the following applies

$$\frac{d\ln K}{dT} = \frac{\Delta_r H^\ominus(T)}{RT^2} \tag{7.25}$$

From van't Hoff's equation, we see that

- K increases (the reaction is shifted to the product) when T increases for an endothermic reaction with $\Delta_r H^\ominus > 0$.
- K decreases (the reaction is shifted to the reactant) when T increases for an exothermic reaction with $\Delta_r H^\ominus < 0$.

This is well known and also follows from Le Chatelier's principle.

Example 7.12 Equilibrium for ammonia synthesis. *Consider the reaction*

$$N_2 + 3H_2 = 2NH_3$$

We assume ideal gas, constant heat capacity and use the following data:

i	$C_p^\ominus(i,298)$ $[J/K, mol]$	$\Delta_f H^\ominus(i,298)$ $[kJ/mol]$	$\Delta_f G^\ominus(i,298)$ $[kJ/mol]$
$H_2(g)$	28.82	0	0
$N_2(g)$	29.13	0	0
$NH_3(g)$	35.06	−46.11	−16.41

(a) Calculate the standard enthalpy, entropy, Gibbs energy and the equilibrium constant for the reaction at 298 K, 400K, 500K, 600K, 700K and 800K. (Hint: With the assumption of constant heat capacity, you can use (B.56) and (B.57)).

(b) Calculate the equilibrium composition for a stoichiometric mixture at 700 K and pressures of 100 bar, 200 bar and 300 bar.

(c) In part (a), we assumed constant heat capacity when calculating the equilibrium constant. Compare the value of the equilibrium constant at 700 K with what you find with the three other alternative assumptions (1, 2 and 4) given in Appendix B.6.2 (page 388).

Solution. *(a) The standard enthalpy of reaction at 298 K is*

$$\Delta_r H^\ominus(298) = \sum_i \nu_i \Delta_f H^\ominus(i,298) = 2\cdot(-46.11) - 1\cdot 0 - 3\cdot 0 = -92.22\ kJ/mol$$

Similarly, the standard Gibbs reaction energy at 298 K is

$$\Delta_r G^\ominus(298) = \sum_i \nu_i \Delta_f G^\ominus(i,298) = 2\cdot(-16.41) - 1\cdot 0 - 3\cdot 0 = -32.82\ kJ/mol$$

From this, we can calculate standard entropy of reaction at 298 K

$$\Delta_r S^\ominus(298) = \frac{\Delta_r H^\ominus(298) - \Delta_r G^\ominus(298)}{298.15\ [K]} = \frac{-92.22\cdot 10^3 - (-32.82)\cdot 10^3}{298.15} = -199.3\ J/molK$$

The change in standard heat capacity for the reaction at 298 K is

$$\Delta_r C_p^\ominus = \sum_i \nu_i C_p^\ominus(i,298) = 2\cdot 35.06 - 3\cdot 28.82 - 1\cdot 29.13 = -45.47\ J/mol\ K$$

With the assumption of constant heat capacity C_p, $\Delta_r H^{\ominus}(T)$ can be calculated from (B.56) and $\Delta_r S^{\ominus}(T)$ from (B.57), and we find for the reaction $N_2 + 3H_2 = 2NH_3$:

T	$\Delta_r H^{\ominus}(T)$	$\Delta_r S^{\ominus}(T)$	$\Delta_r G^{\ominus}(T)$	$K(T)$
$[K]$	$[J/mol]$	$[J/molK]$	$[J/mol]$	$[-]$
298	−92220	−199.3	−32820	$5.62 \cdot 10^5$
400	−96858	−212.7	−11772	34.52
500	−101400	−222.9	10025	0.0896
600	−105950	−231.2	32738	0.0014
700	−110500	−238.2	56213	$6.36 \cdot 10^{-5}$
800	−115050	−244.2	80339	$5.65 \cdot 10^{-6}$

(b) For the ammonia reaction, $\nu_{N_2} = -1$, $\nu_{H_2} = -3$ and $\nu_{NH_3} = 2$, and from (7.23) we have at chemical equilibrium for an ideal gas

$$K = \frac{\left(\frac{p_{NH_3}}{p^{\ominus}}\right)^2}{\left(\frac{p_{N_2}}{p^{\ominus}}\right)\left(\frac{p_{H_2}}{p^{\ominus}}\right)^3}$$

where $p^{\ominus} = 1$ bar, and the partial pressure is defined by

$$p_i = y_i\, p = \frac{n_i}{n_{tot}} p$$

Assume that we start with 1 mol N_2 and 3 mol H_2 (basis) and that the extent of reaction is ξ [mol]. Then, the material balance gives that

$$n_{N_2} = 1 - \xi; \quad n_{H_2} = 3 - 3\xi; \quad n_{NH_3} = 2\xi$$

$$n_{tot} = n_{N_2} + n_{H_2} + n_{NH_3} = 4 - 2\xi$$

Combining the above equations and data gives one equation with one unknown (ξ). We may solve the equation numerically by iterating on ξ, as shown in the MATLAB code below. At equilibrium, we then find at 700 K for the three pressures:

$p[bar]$	$\xi[mol]$	x_{N_2}	x_{H_2}	x_{NH_3}
100	0.299	0.206	0.618	0.176
200	0.429	0.182	0.545	0.273
300	0.507	0.165	0.496	0.339

We find, as expected, that the reaction to ammonia is favored by high pressure. The reaction is exothermic so the equilibrium conversion to ammonia is favored by low temperature, but industrial reactors nevertheless operate at high temperature to make the reaction go sufficiently fast (that is, such that we approach the equilibrium state).

We used MATLAB for the calculations:

```
In general:
The MATLAB command fsolve('file',x0) solves by iteration the equation f(x)=0 and returns x.
Here the file.m contains an algorithm for calculating z=f(x) (what we desire is z=0).

For our case: We write

> x = fsolve('nh3eq',0.1)

where x0=0.1 is the starting value for the extent of reaction.
MATLAB will then solve to find the extent of reaction x = 0.4295 (with our numbers).
We have in advance saved the following file nh3eq.m:
```

```
% This is file: nh3eq.m
% Calculates the value of the function z=f(x), where
%    x - extent of reaction
%    z = Q-K - deviation from equilibrium
function z=f(x)
p=200;              % the pressure is here 200 bar (can be changed)
nn2=1-x;
nh2=3-3*x;
nnh3=2*x;
ntot=nn2+nh2+nnh3;
pn2=p*nn2/ntot;
ph2=p*nh2/ntot;
pnh3=p*nnh3/ntot;
Q = (pnh3^2 / (pn2 * ph2^3));
K=6.36e-5;     % K is here at T= 700 K (can be changed)
z=Q-K;
```

Comment. *In this example, we assumed ideal gas, which may seem unreasonable at such high pressures. However, the temperature is also high, so the deviation from ideal gas is not as large as one may expect. For more exact calculations, it is recommended to replace the partial pressure p_i by the fugacity $f_i = \phi_i p_i$. The fugacity coefficient ϕ_i can be calculated from an equation of state such as the SRK equation.*

(c) Starting from $K_1 = 5.62 \cdot 10^5$ at $T_1 = 298.15$ K, we calculate the equilibrium constant at $T_2 = 700$ K using the four alternative methods given in Appendix B.6.2, page 388. Note that the exact method gives $K(700) = 8.25 \cdot 10^{-5}$ (see method 4 below).

1. *The very rough assumption of a constant equilibrium constant equal to $5.62 \cdot 10^5$ (298.15 K) is completely unacceptable in this case, because the heat of reaction is far from zero. It gives a value for K at 700 K which is almost a factor 10^{10} too high.*
2. *With the assumption of constant $\Delta_r H^\ominus$ (and constant $\Delta_r S^\ominus$), the integrated van't Hoff's equation gives (B.55):*

$$\ln K(700) = \ln(5.62 \cdot 10^5) - \frac{-92220}{8.31}\left(\frac{1}{700} - \frac{1}{298}\right) = 13.24 - 21.35 = -8.12$$

   that is, $K(700) = e^{-8.12} = 2.99 \cdot 10^{-4}$ (which is a factor 3.6 too high).
3. *As shown in subtask (b), the assumption of constant $\Delta_r C_p^\ominus$, independent of temperature, gives $K(700) = 6.36 \cdot 10^{-5}$ (which is a factor 1.3 too low).*
4. *Exact calculations using temperature-dependent C_p-data and (B.58) and (B.59) give $K(700) = 8.25 \cdot 10^{-5}$ (see MATLAB code below).*

In conclusion, the results computed assuming constant C_p are probably OK (within 30% for K) for most engineering calculations, but data for $C_p(T)$ should be used for more accurate calculations.

```
% MATLAB code for exact calculation of K(T) for reaction N2 + 3 H2 = 2 NH3
hf298_h2 = 0; gf298_h2 = 0; sf298_h2 = 0;
hf298_n2 = 0; gf298_n2 = 0; sf298_n2 = 0;
hf298_nh3 = -46110; gf298_nh3 = -16410; sf298_nh3 = (hf298_nh3-gf298_nh3)/298.15;
% Cp-data from Reid, Prausnitz & Pauling, 1987: cp(T) = cpa + cpb*T + cpc*T^2 + cpd*T^3
cpa_h2 = 2.714e1; cpb_h2 = 9.274e-3; cpc_h2 = -1.381e-5; cpd_h2 = 7.645e-9;
cpa_n2 = 3.115e1; cpb_n2 =-1.357e-2; cpc_n2 =  2.680e-5; cpd_n2 =-1.168e-8;
cpa_nh3= 2.731e1; cpb_nh3= 2.383e-2; cpc_nh3=  1.707e-5; cpd_nh3=-1.185e-8;
% Integrate to find H(T) and S(T) for each component
T=700; T0=298.15;
hfT_h2 = hf298_h2 + cpa_h2*(T-T0) + cpb_h2*(T^2-T0^2)/2 + cpc_h2*(T^3-T0^3)/3 + cpd_h2*(T^4-T0^4)/4;
hfT_n2 = hf298_n2 + cpa_n2*(T-T0) + cpb_n2*(T^2-T0^2)/2 + cpc_n2*(T^3-T0^3)/3 + cpd_n2*(T^4-T0^4)/4;
hfT_nh3= hf298_nh3+ cpa_nh3*(T-T0)+ cpb_nh3*(T^2-T0^2)/2+ cpc_nh3*(T^3-T0^3)/3+ cpd_nh3*(T^4-T0^4)/4;
sfT_h2 = sf298_h2 + cpa_h2*log(T/T0) + cpb_h2*(T-T0) + cpc_h2*(T^2-T0^2)/2 + cpd_h2*(T^3-T0^3)/3;
sfT_n2 = sf298_n2 + cpa_n2*log(T/T0) + cpb_n2*(T-T0) + cpc_n2*(T^2-T0^2)/2 + cpd_n2*(T^3-T0^3)/3;
sfT_nh3= sf298_nh3+ cpa_nh3*log(T/T0)+ cpb_nh3*(T-T0)+ cpc_nh3*(T^2-T0^2)/2+ cpd_nh3*(T^3-T0^3)/3;
```

```
% H, S, G and K for reaction N2 + 3 H2 = 2 NH3
dhrT = 2*hfT_nh3 - hfT_n2 - 3*hfT_h2
dsrT = 2*sfT_nh3 - sfT_n2 - 3*sfT_h2
dgrT = dhrT - T*dsrT
K = exp(-dgrT/(8.3145*T))
```

Exercise 7.3* NO_x **equilibrium.** *A gas mixture at 940 °C and 2.5 bar consists of 5%* O_2*, 11%* NO*, 16%* H_2O *and the rest* N_2*. The formation of* NO_2 *is neglected, and you need to check whether this is reasonable by calculating the ratio (maximum) between* NO_2 *and* NO *that one would get if the reaction*

$$NO + 0.5O_2 = NO_2$$

was in equilibrium at 940 °C. **Data.** *Assume constant heat capacity and use data for ideal gas from page 416.*

Exercise 7.4 *Consider the gas phase reaction*

$$4NH_3 + 5O_2 = 4NO + 6H_2O$$

(a) Calculate standard enthalpy, entropy, Gibbs energy and the equilibrium constant for the reaction at 298 K and 1200 K.
(b) Calculate the equilibrium composition at 1200 K and 8 bar when the feed consists of 10 mol-% ammonia, 18 mol-% oxygen and 72 mol-% nitrogen.
(c) What is the feed temperature if the reactor operates adiabatically?
Data. *Assume constant heat capacity and use data for ideal gas from page 416.*

7.4 Introduction to vapor/liquid equilibrium

Phase equilibrium, and in particular vapor/liquid-equilibrium (VLE), is important for many process engineering applications. The thermodynamic basis for phase equilibrium is the same as for chemical equilibrium, namely that the Gibbs energy G is minimized at a given T and p (see page 174).

7.4.1 General VLE condition for mixtures

Vapor/liquid-equilibrium (VLE) for mixtures is a large subject, and we will here state the general equilibrium condition, and then give some applications. The fact that the Gibbs energy G is minimized at a given temperature T and pressure p implies that a necessary equilibrium condition is that G must remain constant for any small perturbation, or mathematically $(dG)_{T,p} = 0$ (see page 385). Consider a small perturbation to the equilibrium state where a small amount dn_i of component i evaporates from the liquid phase (l) to the vapor/gas phase (g). The necessary equilibrium condition at a given T and p then gives

$$dG = (\bar{G}_{g,i} - \bar{G}_{l,i})dn_i = 0 \tag{7.26}$$

where $\bar{G}_i$ [J/mol i] is the partial Gibbs energy, also known as the **chemical potential**, $\mu_i \triangleq \bar{G}_i$. Since (7.26) must hold for any value of dn_i, we derive the equilibrium

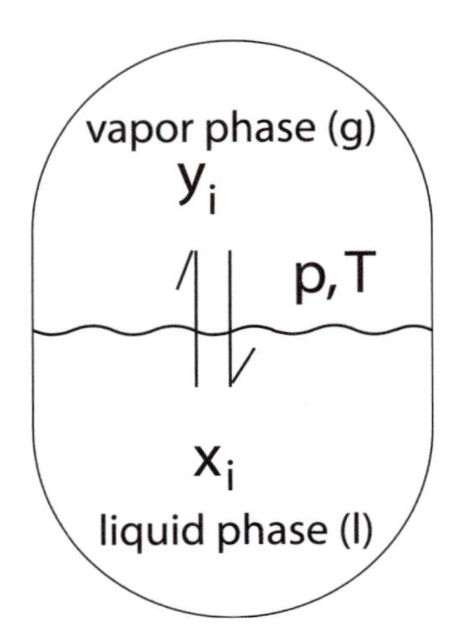

Figure 7.3: Vapor/liquid equilibrium (VLE)

condition $\bar{G}_{g,i} = \bar{G}_{l,i}$. That is, the VLE-condition is that the chemical potential for any component i is the same in both phases,

$$\mu_{g,i} = \mu_{l,i} \tag{7.27}$$

7.4.2 Vapor pressure of pure component

Let us first consider VLE for a pure component. The component vapor pressure $p^{\mathrm{sat}}(T)$ is the equilibrium (or **saturation**) pressure for the pure liquid at temperature T. As the temperature increases, the molecules in the liquid phase move faster and it becomes more likely that they achieve enough energy to escape into the vapor phase, so the vapor pressure increases with temperature. For example, the vapor pressure for water is 0.0061 bar at 0 ^{o}C, 0.03169 bar at 25 ^{o}C, 1.013 bar at 100 ^{o}C, 15.54 bar at 200 ^{o}C and $p_c = 220.9$ bar at $T_c = 374.1^o$C (critical point).

As the temperature and resulting vapor pressure increases, the molecules come closer together in the gas phase, and eventually we reach the **critical point** (at temperature T_c and pressure p_c), where there is *no difference between the liquid and gas phases.* For a pure component, the **critical temperature** T_c is the highest temperature where a gas can condense to a liquid, and the vapor pressure is therefore only defined up to T_c. The corresponding **critical pressure** p_c is typically around 50 bar, but it can vary a lot, e.g., from 2.3 bar (helium) to 1500 bar (mercury).

For a pure component, the *exact* **Clapeyron equation** provides a relationship between vapor pressure and temperature,

$$\frac{dp^{\mathrm{sat}}}{dT} = \frac{\Delta_{\mathrm{vap}} S}{\Delta_{\mathrm{vap}} V} = \frac{\Delta_{\mathrm{vap}} H}{T \Delta_{\mathrm{vap}} V} \tag{7.28}$$

Here $\Delta_{\mathrm{vap}} H = H_g - H_l$ [J/mole] is the heat of vaporization at temperature T and

$\Delta_{\text{vap}} V = V_g - V_l$ [m^3/mol] is the difference in molar volume between the phases. An equivalent expression applies for the vapor pressure over a pure solid.

Derivation of (7.28): From (7.27) the necessary equilibrium condition is $G_g = G_l$. Assume that there is a small change in T which results in a small change in p. From (B.66), the resulting changes in Gibbs energy are $dG_l = V_l dp - S_l dT$ and $dG_g = V_g dp - S_g dT$. Since the system is still in equilibrium after the change, we must have $dG_l = dG_g$ which gives $(V_g - V_l)dp - (S_g - S_l)dT$. The Clapeyron equation follows by noting that $\Delta_{\text{vap}} S = \Delta_{\text{vap}} H/T$, see (7.8).

In most cases, we have $V_g \gg V_l$, and for ideal gas we have $V_g = p/RT$ and from (7.28) we then derive, by using $\frac{1}{p}dp = d\ln p$, the *approximate* **Clausius-Clapeyron equation**,

$$\frac{d\ln p^{\text{sat}}(T)}{dT} = \frac{\Delta_{\text{vap}} H(T)}{RT^2} \tag{7.29}$$

which applies for a pure component at low pressure, typically less than 10 bar. If the heat of vaporization $\Delta_{\text{vap}} H$ is constant (independent of T; which indeed is somewhat unrealistic since it decreases with temperature and is 0 in the critical point), we derive from (7.29) the *integrated Clausius-Clapeyron equation*,

$$p^{\text{sat}}(T) = p^{\text{sat}}(T_0) \exp\left[-\frac{\Delta_{\text{vap}} H}{R}\left(\frac{1}{T} - \frac{1}{T_0}\right)\right] \tag{7.30}$$

which is sometimes used to compute the vapor pressure at temperature T given $p^{\text{sat}}(T_0)$ at temperature T_0. However, (7.30) is not sufficiently accurate for practical calculations, so instead empirical relationships are used. A popular one is the **Antoine equation**, [2]

$$\ln p^{\text{sat}}(T) = A - \frac{B}{T + C} \tag{7.31}$$

Note that (7.30) is in the form (7.31) with $A = \ln p^{\text{sat}}(T_0) + \Delta_{\text{vap}} H/RT_0$, $B = \Delta_{\text{vap}} H/R$ and $C = 0$. Antoine parameters for some selected components are given in Table 7.2 (page 190).

Example 7.13 *For water, we find in an older reference book the following Antoine constants: $A = 18.3036$, $B = 3816.44$ and $C = -46.13$. This is with pressure in [mm Hg] and temperature in [K] (note that these Antoine parameters are different from those given in Table 7.2). The vapor pressure at 100 °C is then*

$$p^{\text{sat}}(373.15\ K) = e^{18.3036 - \frac{3816.44}{(373.15-46.13)}} = 759.94 \text{ mmHg} = \frac{759.94}{750.1} \text{ bar} = 1.013 \text{ bar}$$

which agrees with the fact that the boiling temperature for water is 100 °C at 1 atm = 1.01325 bar.

Engineering rule for vapor pressure of water. The following simple formula, which is easy to remember, gives surprisingly good estimates of the vapor pressure for water for temperatures from 100^{0}C (the normal boiling point) and up to 374$^o C$ (the critical point):

$$p^{\text{sat}}_{H_2O}[\text{bar}] = \left(\frac{t[^o C]}{100}\right)^4 \tag{7.32}$$

[2] Numerical values for the three Antoine constants A, B and C are found in many reference books (for example, B.E. Poling, J.M. Prausnitz, J.P. O'Connell, *The properties of gases and liquids*, 5th Edition, McGraw-Hill, 2001.

This formula is very handy for engineers dealing with steam at various pressure levels. For example, from the formula we estimate $p^{\text{sat}} \approx 1$ bar at 100°C (the correct value is 1 atm = 1.013 bar) and $p^{\text{sat}} \approx 2^4 = 16$ bar at 200°C (the correct value is 15.53 bar).

Exercise 7.5* *Test the validity of the simple formula (7.32), by comparing it with the following experimental vapor pressure data for water:*

$t[^oC]$	0	25	50	75	100	120	150	200	250	300	374.14(t_c)
$p[bar]$	0.00611	0.03169	0.1235	0.3858	1.013	1.985	4.758	15.53	39.73	85.81	220.9(p_c)

Also test the validity of the two alternative sets of Antoine constants for water (given in Example 7.13 and Table 7.2).

Exercise 7.6* **Effect of barometric pressure on boiling point.** *Assume that the barometric (air) pressure may vary between 960 mbar (low pressure) and 1050 mbar (high pressure). What is the corresponding variation in boiling point for water?*

Comment. Note the similarity between Clausius-Clapeyron's equation (7.29) for the temperature dependency of vapor pressure,

$$\frac{d \ln p^{\text{sat}}(T)}{dT} = \frac{\Delta_{\text{vap}} H(T)}{RT^2}$$

and van't Hoff's equation (7.25) for the temperature dependency of the chemical equilibrium constant K,

$$\frac{d \ln K}{dT} = \frac{\Delta_{\text{r}} H^{\ominus}(T)}{RT^2}$$

This is of course not a coincidence, because we can view evaporation as a special case of an endothermic "chemical reaction."

7.4.3 VLE for ideal mixtures: Raoult's law

Here, we consider vapor/liquid equilibrium of mixtures; see Figure 7.3 (page 180). Let

x_i - mole fraction of component i in the liquid phase
y_i - mole fraction of component i in the vapor phase

The simplest case is an *ideal liquid mixture* and *ideal gas* where **Raoult's law** states that for any component i, the partial pressure $p_i = y_i p$ equals the vapor pressure of the pure component i multiplied by its mole fraction x_i in the liquid phase, that is,

$$\text{Raoult's law}: \qquad y_i p = x_i p_i^{\text{sat}}(T) \tag{7.33}$$

A simple molecular interpretation of Raoult's law is that in an ideal liquid mixture the fraction of i-molecules at the surface is x_i, so the partial pressure $p_i = y_i p$ is reduced from $p_i^{\text{sat}}(T)$ (pure component) to $x_i p_i^{\text{sat}}(T)$ (ideal mixture).

Thermodynamic derivation of Raoult's law. A thermodynamic derivation is useful because it may later be generalized to the non-ideal case. We start from the general VLE condition $\mu_{g,i} = \mu_{l,i}$ in (7.27), which says that the chemical potential (= partial Gibbs energy) for each component is the same in both phases at the given p and T. Now, Gibbs energy is a state function, and we can also imagine another route for taking component i from the liquid to the vapor phase, consisting of four steps (all at temperature T): (1) Take component i out of the liquid mixture. From (B.41) the change in chemical potential for this "unmixing" is $\Delta\mu_{i,1} = -RT \ln a_i$ where the activity is $a_i = \gamma_i x_i$. For an **ideal liquid mixture** the activity coefficient is 1, $\gamma_i = 1$. (2) Take the pure component as liquid from pressure p to the saturation pressure $p_i^{\rm sat}(T)$. Since the liquid volume is small this gives a very small change in chemical potential, known as the **Poynting factor**, which we here neglect, i.e., $\Delta\mu_{i,2} \approx 0$. (3) Evaporate the pure component at T and $p_i^{\rm sat}(T)$. Since we have equilibrium ($\Delta G = 0$) there is no change in the chemical potential, $\Delta\mu_{i,3} = 0$. (4) In the gas phase, go from pure component at pressure $p_i^{\rm sat}(T)$ to a mixture at p where the partial pressure is p_i. From (B.40), the change in chemical potential for an **ideal gas** is $\Delta\mu_{i,4} = RT\ln(p_i/p_i^{\rm sat}(T))$. Now, since the initial and final states are in equilibrium, the sum of the change in chemical potential for these four steps should be zero and we derive $-RT\ln x_i + RT\ln(p_i/p_i^{\rm sat}(T)) = 0$ and Raoult's law follows.

7.4.4 Relative volatility

The relative volatility α is a very useful quantity. For example, it is used for short-cut calculations for distillation columns.[3] For a mixture, the **relative volatility** α between the two components L (the "light" component) and H (the "heavy" component) is defined as

$$\alpha \triangleq \frac{y_L/x_L}{y_H/x_H} \tag{7.34}$$

For an **ideal mixture** where Raoult's law (7.33) applies, we then have

$$\alpha = \frac{y_L/x_L}{y_H/x_H} = \frac{p_L^{\rm sat}(T)}{p_H^{\rm sat}(T)} \tag{7.35}$$

that is, α equals the ratio between the pure component's vapor pressures. Furthermore, we see from (7.30) that if the heat of vaporization for the two components are similar, then α does not change much with the temperature.

The approximation of **constant relative volatility** (independent of composition and temperature) is often used in practical calculations, and is based on the following assumptions

- Ideal liquid mixture such that Raoult's law applies (α is then independent of composition)
- The components have similar heat of vaporization (α is then independent of temperature)

These assumptions generally hold well for separation of "similar" components. However, the assumption of constant α is poor for many non-ideal mixtures. For example, for a mixture that forms an azeotrope, like water and ethanol, we have $\alpha = 1$ at the azeotropic point, with $\alpha > 1$ on one side and $\alpha < 1$ on the other side of the azeotrope (that is, even the order of "heavy" (H) and "light" (L) depends on the liquid composition).

Relative volatility from boiling point data. For ideal mixtures that follow Raoult's law, the following approximate relationship between the relative volatility α and the boiling point difference $T_{bH} - T_{bL}$ for the components applies:

$$\alpha \approx \exp\left[\frac{\Delta_{\rm vap}H}{RT_b} \cdot \frac{T_{bH} - T_{bL}}{T_b}\right] \tag{7.36}$$

[3] For more on distillation see, for example, I.J. Halvorsen, S. Skogestad: "Distillation Theory," *Encyclopedia of Separation Science*, D. Wilson (Editor-in-chief), Academic Press, 2000 (available at S. Skogestad's homepage).

Here $T_b = \sqrt{T_{bH} \cdot T_{bL}}$ is the geometric mean boiling point, and $\Delta_{\rm vap}H$ is the average heat of vaporization for the two components at the average boiling point T_b. From Trouton's rule (see page 378), a typical value is $\frac{\Delta_{\rm vap}H}{RT_b} \approx \frac{85\text{J/mol K}}{8.31\text{J/mol K}} = 10.2$.

Derivation of (7.36). We assume that Raoult's law holds such that (7.35) holds. If we assume that the heat of vaporization is independent of temperature, then the integrated Clausius-Clapeyron equation (7.30) gives for component L if we choose $T = T_{bH}$ and $T_0 = T_{bL}$:

$$p_L^{\rm sat}(T_{bH}) = p_L^{\rm sat}(T_{bL}) \exp\left[-\frac{\Delta_{\rm vap}H_L}{R}\left(\frac{1}{T_{bH}} - \frac{1}{T_{bL}}\right)\right]$$

In practice, $\Delta_{\rm vap}H_L$ depends on temperature, so an average value for the temperature interval from T_{bL} to T_{bH} should be used. At the normal boiling points, $p_L^{\rm sat}(T_{bL}) = p_H^{\rm sat}(T_{bH}) = 1$ atm, and the relative volatility at $T = T_{bH}$ becomes

$$\alpha = \frac{p_L^{\rm sat}(T_{bH})}{p_H^{\rm sat}(T_{bH})} = \exp\left[-\frac{\Delta_{\rm vap}H_L}{R}\left(\frac{1}{T_{bH}} - \frac{1}{T_{bL}}\right)\right]$$

A similar expression for α at $T = T_{bL}$ is derived by considering component H, and combining the two yields (7.36). □

Example 7.14 *Let us use (7.36) to calculate an approximate value for relative volatility for the mixture methanol (L) - ethanol (H). We obtain the following data for the pure components*

$$\textit{Methanol}: \quad T_{bL} = 337.8\text{K};\ \Delta_{\rm vap}H_L(T_{bL}) = 35.2\ kJ/mol$$

$$\textit{Ethanol}: \quad T_{bH} = 351.5\text{K};\ \Delta_{\rm vap}H_B(T_{bH}) = 40.7\ kJ/mol$$

The geometric mean boiling point is $T_b = 344.6$ *K, the average heat of vaporization is* $\Delta_{\rm vap}H = (\Delta_{\rm vap}H_L(T_{bL} + \Delta_{\rm vap}H_B(T_{bH})/2 = 37.9$ *kJ/mol and we get* $\Delta_{\rm vap}H/RT_b = 13.25$ *(which is higher than the value of 10.2 according to Trouton's rule). The boiling point difference is 13.7 K, and assuming ideal mixture, (7.36) gives* $\alpha \approx \exp\frac{12.90 \cdot 13.7}{344.6} = 1.69$. *The experimental value is about 1.73.*

We emphasize that the simplified formula (7.36) is primarily intended to provide insight, and one should normally obtain experimental data for the vapor/liquid equilibrium or use a more exact model.[4]

7.4.5 Boiling point elevation and freezing point depression

Consider a mixture consisting mainly of a volatile component (the solvent A) with some dissolved non-volatile component (the solute B). For example, this could be a mixture of water (A) and sugar (B). Such a solution has a higher boiling point than the pure component (e.g., water), and we want to find the boiling point elevation ΔT_b. For a *dilute* ideal mixture (solution) with mole fraction x_B of the non-volatile component, we derive that the **boiling point elevation** is

$$\Delta T_b = T_b - T_b^* = \frac{RT_b^{*2} x_B}{\Delta_{\rm vap}H} \tag{7.37}$$

where T_b^* is the boiling point of the pure component A, and T_b is the boiling point of the mixture. If the solution is *not dilute* then x_B should be replaced by $\ln\frac{1}{1-x_B}$.

[4] A comprehensive reference work for experimental vapor/liquid equilibrium data for mixtures is: J. Gmehling and U. Onken, *Vapor-liquid equilibrium data collection*, Dechema Chemistry Data Series (1977–).

Proof of (7.37). For an ideal mixture (solution), Raoult's law (7.33) gives that the partial pressure of the solvent (A) is $p_A = (1-x_B)p_A^{\rm sat}(T)$ where p_A is equal to the total pressure p since the other component is non-volatile. At the boiling point of the mixture, the total pressure is $p_0 = 1$ atm and we get $p_0 = (1-x_B)p_A^{\rm sat}(T_b)$. Here, from the integrated Clausius-Clapeyron equation (7.30) we have for the solvent $p_A^{\rm sat}(T_b) = p_A^{\rm sat}(T_b^*)\exp\left[-\frac{\Delta_{\rm vap}H}{R}\left(\frac{1}{T_b}-\frac{1}{T_b^*}\right)\right]$. Here, $p_A^{\rm sat}(T_b^*) = p_0 = 1$ atm (since the vapor pressure of a pure component is 1 atm at the normal boiling point), and by combining and taking the log on both sides we derive

$$\ln\frac{1}{1-x_B} = \frac{\Delta_{\rm vap}H}{R}\frac{T_b - T_b^*}{T_b^* T_b}$$

(7.37) follows by assuming a dilute solution ($x_B \to 0$) where $\ln\frac{1}{1-x_B} \approx x_B$ and $T_b^* \approx T_b$. An alternative derivation is to start from the general equilibrium condition $\mu_{g,A} = \mu_{l,A}$ in (7.27). Here $\mu_{l,A} = \mu_{l,a}^* + RT\ln x_A$ for an ideal mixture and $\mu_{g,A} = \mu_{g,A}^*$ because B is non-volatile. Using $\mu_{g,A}^* - \mu_{l,a}^* = \Delta_{\rm vap}G$, etc. leads to the desired results; for details see a physical chemistry textbook.

The reason for the boiling point elevation is that the dissolved components (B) make it more favorable from an entropy point of view for the solvent to remain the liquid phase. The same argument (that the solvent likes to remain in the liquid phase) also applies for freezing, and it can be proved that for a dilute ideal mixture the **freezing (melting) point depression** is

$$\Delta T_m = T_m^* - T_m = \frac{RT_m^{*2} x_B}{\Delta_{\rm fus}H} \tag{7.38}$$

where T_m^* is the melting (freezing) point of the pure component, T_m the melting point of the mixture and $\Delta_{\rm fus}H$ is the heat of melting.

In both (7.37) and (7.38), x_B is the sum of the mole fractions of all dissolved components (non-volatile or non-freezing). If a component dissociates (e.g., into ions), then this must be taken into account (see the sea water example below).

Remark. Note that both the boiling point elevation (7.37) and the freezing point depression (7.38) depend only on the concentration (mole fraction x_B) of the dissolved component (solute), and not on what component we have. Another such property is the osmotic pressure over an ideal membrane (see page 382). These three properties are referred to as **colligative solution properties.** They can, for example, be used to determine the molar mass (M) of a molecule (see Exercise 7.7).

Example 7.15 Boiling point elevation and freezing point depression of seawater. *We first need to find the mole fraction x_B of dissolved components. We assume that the salinity of seawater is 3.3%, that is, 1 l seawater contains 33 g/l of salt (NaCl). Since the molar mass of NaCl is 58.4 g/mol, we have that 33 g/l corresponds to (33 g/l) / (58.4 g/mol) = 0.565 mol/l of NaCl. However, when dissolved in water, NaCl splits in two* **ions,** *Na^+ and Cl^-. Now, 1 l of water is 55.5 mol (= (1000 g) / (18 g/mol)). Thus, 1 l of seawater consists of approximately 0.565 mol/l Na^+, 0.565 mol/l Cl^- and 55.5 mol water, and the corresponding mole fractions are approximately 0.01 (Na^+), 0.01 (Cl^-) and 0.98 (H_2O). The total mole fraction of dissolved components in seawater is then* $x_B = x_{Na} + x_{Cl} = 0.01 + 0.01 = 0.02$.

Water has a boiling point of $T_b^* = 373.15$ K (100 °C) and the heat of vaporization at the boiling point is $\Delta_{\rm vap}H = 40.66$ kJ/mol. Thus, from (7.37) the boiling point elevation is

$$\Delta T_b = T_b - T_b^* = \frac{8.31 \cdot 373.15^2 \cdot 0.02}{40666} K = 0.57K$$

so the **boiling point of seawater** *is about 100.57 °C.*

Water has a freezing (melting) point of $T_m^* = 273.15$ *K (0 °C) and the heat of* **fusion (melting)** *at the freezing point is* $\Delta_{\rm fus}H = 6.01$ *kJ/mol. Thus, from (7.38) the freezing point depression is*

$$\Delta T_m = T_m^* - T_m = \frac{8.31 \cdot 273.15^2 \cdot 0.02}{6010} K = 2.06K$$

so the **freezing point of seawater** *is about* $-2.06^o C$.

Exercise 7.7 *Adding 7 g of an unknown solute to 100 g water gives a boiling point elevation of 0.34 °C. Estimate the molar mass of the unknown solute, and the corresponding freezing point depression.*

7.4.6 VLE for dilute mixtures: Henry's law

Raoult's law cannot be used for "supercritical" components ("gases"), where T is above the critical temperature T_c for the component. This is because $p^{\rm sat}(T)$ is only defined for $T \leq T_c$. However, also supercritical components have a solubility in liquids. For example CO_2 can be dissolved in water at 50°C even though the critical temperature for CO_2 is 31°C. "Fortunately," the concentration in the liquid phase of supercritical (and other "light") components is usually low. For sufficiently dilute mixtures (low concentrations), there is a generally linear relationship between a component's gas phase fugacity ("thermodynamic partial pressure") and its liquid concentration, even for nob-ideal mixtures. This gives Henry's law, which also applies to supercritical components,

$$\text{Henry's law}: \quad f_i^V = H_i(T) \cdot x_i \quad (x_i \to 0) \tag{7.39}$$

Here, Henry's constant H_i [bar] is a function of temperature only (at least at pressure below 50 bar; at very high pressures we need to include the "Poynting factor" for the pressure's influence on the liquid phase). If the pressure p is sufficiently low, we can assume ideal gas phase where $f_i^V = p_i = y_i p$ (the partial pressure), and Henry's law (7.39) becomes

$$y_i = \frac{H_i}{p} x_i \quad (x_i \to 0, \text{ low } p) \tag{7.40}$$

Henry's law on the form (7.40) is valid for dilute solutions ($x_i < 0.03$, typically) and low pressures ($p < 20$ bar, typically). For an ideal mixture (liquid phase), Henry's constant H_i equals the component's vapor pressure (compare (7.33) and (7.40)), and Henry's constant is therefore expected to increase with temperature. Thus, the solubility is expected to be lower at high temperature. However, there are exceptions to this rule, as seen below for the solubility of H_2 andN_2 in ammonia.

Water. Henry's constant for the solubility of some gases in water at 0°C and 25°C is given in Table 7.1. Note from the critical data on page 416 that most of these gases are supercritical at these temperatures. Thus, they cannot form a pure liquid phase, but they can dissolve in liquid water. In all cases, Henry's constant increases with temperature. For example, for the solubility of CO_2 in water, Henry's constant increases from 740 bar at 0°C, to 1670 bar at 25°C and to 3520 bar at 60 °C.

Ammonia. The following values for Henry's constant for the solubility of H_2 and N_2 in ammonia were obtained using the SRK equation of state with interaction parameters $k_{ij} = 0.226$ between ammonia and nitrogen and $k_{ij} = 0$ between ammonia

Table 7.1: Henry's constant for the solubility of some gases in water

component i	H_i [bar] (0^oC)	H_i [bar] (25^oC)
H_2	58200	71400
N_2	53600	84400
CO	35700	60000
O_2	25800	44800
CH_4	22700	41500
C_2H_4	5570	11700
CO_2	740	1670
Cl_2	–	635
H_2S	270	545

and hydrogen:

Component i	H_i [bar] (-25^oC)	H_i [bar] (25^oC)
H_2	48000	15200
N_2	26000	8900

We note that H_i for both components *decrease* by a factor of about 3 as the temperature is increased from -25^oC to 25^oC. We then have the unexpected result that the solubility of these gases in ammonia is higher at high temperature.

Example 7.16 *The partial pressure of CO_2 over a water solution at 25 ^oC is 3 bar. Task: (a) Calculate the concentration of CO_2 in the solution [mol/l]. (b) Find the volume of $CO_2(g)$ at 1 atm and 25 ^oC that is dissolved in 1 l solution.*

Solution. *(a) We assume ideal gas and dilute solution. From Henry's law, we have that $p_i = H_i x_i$, where $H_i = 1670$ bar (Table 7.1) and $p_i = 3$ bar. This gives $x_i = 3/1670 = 0.0018$ [mol CO_2/ mol] (which confirms that we have a dilute solution). In 1 l of solution the amount of water is (1 kg)/ ($18 \cdot 10^{-3}$ kg/mol) = 55.5 mol. That is, the concentration of CO_2 is $c_i = x_i \cdot 55.5$ mol/l = 0.10 mol/l.*

(b) The molar volume of an ideal gas at 1 atm and 25 ^oC is $V_m = RT/p = 8.31 \cdot 298.15/1.01325 \cdot 10^5 = 0.02445\ m^3/mol$ = 24.45 l/mol. In 1 l solution, there is 0.10 mol CO_2, and the corresponding volume of this as gas at 1 atm is then 2.45 l.

7.4.7 VLE for real (non-ideal) mixtures

In this section, we summarize the equations used for calculation of vapor/liquid equilibrium for non-ideal mixtures. It is intended to give an overview, and you need to consult other books for practical calculations. Three fundamentally different methods are

1. Based on K values
2. Based on activity coefficients (for non-ideal mixtures of sub-critical components at moderate pressures)
3. Based on the same equation of state for both phases (for moderately non-ideal mixtures at all pressures)

1. K-value

The K-value is defined for each component as the ratio

$$K_i = \frac{y_i}{x_i} \tag{7.41}$$

where x_i is the mole fraction in the liquid phase and y_i is the mole fraction in the gas phase at equilibrium. Generally, the "K value" is a function of temperature T, pressure p and composition (x_i and y_i). For ideal liquid mixtures and ideal gas, we have from (7.33) that $K_i = p_i^{\text{sat}}(T)/p$, that is, the K value is independent of composition. For dilute mixtures, even non-ideal, we have from Henry's law (7.40) that $K_i = H_i(T)$. More generally, the K-value can be calculated from one of the two methods given below.

2. Activity coefficient

This method provides a generalization of Raoult's law to **non-ideal mixtures** and to real gases. From the general VLE-condition $\mu_{g,i} = \mu_{l,i}$ we derive for mixtures of subcritical components: (the proof follows the derivation given for Raoult's law on page 182)

$$\underbrace{\phi_i^V \cdot y_i \cdot p}_{f_i^V} = \underbrace{\gamma_i \cdot x_i \cdot \phi_i^{\text{sat}} \cdot p_i^{\text{sat}}(T) \cdot \exp\left[\frac{1}{RT}\int_{p_i^{\text{sat}}}^{p} \bar{V}_i^L dp\right]}_{f_i^L} \tag{7.42}$$

where f_i^V is the fugacity in the vapor phase and f_i^L is the fugacity in the liquid phase. The fugacity coefficients $\phi_i^V(T, p, y_i)$ and $\phi_i^{\text{sat}}(T)$ are 1 for ideal gases, and for real gases their value are usually computed from an equation of state for the gas phase, e.g., SRK. The **activity coefficients** γ_i depend mainly on the liquid composition (x_i) and are usually computed from empirical equations, such as the *Wilson*, NRTL, UNIQUAC and UNIFAC equations, based on experimental interaction data for all binary combinations. The exception is the UNIFAC equation which only requires interaction data for the groups in the molecule. The last exponential term is the so-called Poynting factor for the pressure's influence on the liquid phase (see the derivation for Raoult's law on page 182). It is close to 1, except at high pressures above about 50 bar.

At moderate pressures (typically, less than 10 bar) we can assume ideal gas, $\phi_i^V = 1$ and $\phi_i^{\text{sat}} = 1$, and from (7.42) we derive a commonly used relation:

$$\text{Nonideal mixture at moderate pressures}: \quad y_i p = \gamma_i x_i p_i^{\text{sat}} \tag{7.43}$$

For low concentrations of supercritical components we can use Henry's law, $y_i p = H_i x_i$. For an ideal liquid mixture we have $\gamma_i = 1$ and we rederive from (7.43) Raoult's law: $y_i p = x_i p_i^{\text{sat}}$.

3. Same equation of state for both phases

For mixtures that do deviate too much from the ideal (for example, for hydrocarbon mixtures), we can use the same reference state (ideal gas) and the same equation of state for both phases (for example, the SRK equation), and the VLE-condition $\mu_{g,i} = \mu_{l,i}$ gives

$$\phi_i^V y_i = \phi_i^L x_i \tag{7.44}$$

where the fugacity coefficients ϕ_i^V and ϕ_i^L are determined from the equation of state. The K value is then $K_i = \phi_i^L/\phi_i^V$. Note that (7.44) can also be used for supercritical components.

7.5 Flash calculations

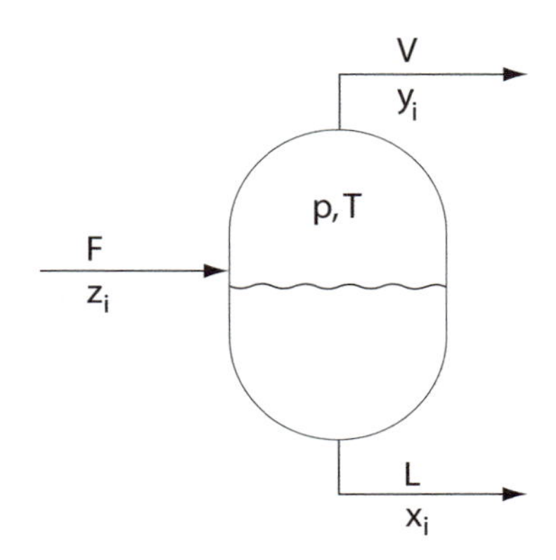

Figure 7.4: Flash tank

Flash calculations are used for processes with vapor/liquid-equilibrium (VLE). A typical process that requires flash calculations, is when a feed stream (F) is separated into a vapor (V) and liquid (L) product; see Figure 7.4.

In principle, flash calculations are straightforward and involve combining the VLE-equations with the component mass balances, and in some cases the energy balance. Some flash calculations are (with a comment on their typical numerical solution or usage):

1. Bubble point at given T (easy)
2. Bubble point at given p (need to iterate on T)
3. Dew point at given T (easy)
4. Dew point at given p (need to iterate on T)
5. Flash at given p and T (relatively easy)
6. Flash at given p and H ("standard" flash, e.g., for a flash tank after a valve)
7. Flash at given p and S (e.g., for condensing turbine)
8. Flash at given U and V (e.g., for dynamic simulation of an adiabatic flash drum)

The last three flashes are a bit more complicated as they require the use of the energy balance and relationships for computing H, S, etc. The use of flash calculations is best illustrated by some examples. Here, we assume that the VLE is given on K-value form, that is,

$$y_i = K_i x_i$$

Table 7.2: Data for flash examples and exercises: Antoine parameters for $p^{sat}(T)$, normal boiling temperature (T_b) and heat of vaporization $\Delta_{vap}H(T_b)$ for selected components. **Data**: Poling, Prausnitz and O'Connell, *The properties of gases and liquids*, 5th Ed., McGraw-Hill (2001).

```
% log10(psat[bar])=A-B/(T[K]+C)          Tb[K]          dvapHb [J/mol]
A1=3.97786;    B1=1064.840;  C1=-41.136;  Tb1=309.22;  dvapHb1=25790;  % pentane  C5H12
A2=4.00139;    B2=1170.875;  C2=-48.833;  Tb2=341.88;  dvapHb2=28850;  % hexane   C6H14
A3=3.93002;    B3=1182.774;  C3=-52.532;  Tb3=353.93;  dvapHb3=29970;  % cyclohex C6H12
A4=5.20277;    B4=1580.080;  C4=-33.650;  Tb4=337.69;  dvapHb4=35210;  % methanol CH3OH
A5=5.11564;    B5=1687.537;  C5=-42.98;   Tb5=373.15;  dvapHb5=40660;  % water    H2O
A6=4.48540;    B6= 926.132;  C6=-32.98;   Tb6=239.82;  dvapHb6=23350;  % ammonia  NH3
A7=3.92828;    B7= 803.997;  C7=-26.11;   Tb7=231.02;  dvapHb7=19040;  % propane  C3H8
A8=4.05075;    B8=1356.360;  C8=-63.515;  Tb8=398.82;  dvapHb8=34410;  % octane   C8H18
A9=4.12285;    B9=1639.270;  C9=-91.310;  Tb9=489.48;  dvapHb9=43400;  % dodecane C12H26
A10=3.98523; B10=1184.24;  C10=-55.578; Tb10=353.24; dvapHb11=30720;  % benzene  C6H6
A11=4.05043; B11=1327.62;  C11=-55.525; Tb11=383.79; dvapHb11=33180;  % toluene  C7H8
```

where y_i is the vapor phase mole fraction and x_i the liquid phase mole fraction for component i. In general, the "K-value" K_i depends on temperature T, pressure p and composition (both x_i and y_i). We mostly assume ideal mixtures, and use Raoult's law. In this case K_i depends on T and p only:

$$\text{Raoult's law}: \quad K_i = p_i^{sat}(T)/p$$

In the examples, we compute the vapor pressure $p^{sat}(T)$ using the Antoine parameters given in Table 7.2.

7.5.1 Bubble point calculations

Let us first consider bubble point calculations, In this case the liquid-phase composition x_i is given (it corresponds to the case where V is very small ($V \gtrsim 0$) and $x_i = z_i$ in Figure 7.4). *The bubble point of a liquid is the point where the liquid just starts to evaporate (boil)*, that is, when the first vapor bubble is formed. If the temperature is given, then we must lower the pressure until the first bubble is formed. If the pressure is given, then we must increase the temperature until the first bubble is formed. In both cases, this corresponds to adjusting T or p until the computed sum of vapor fractions is just 1, $\Sigma y_i = 1$ or

$$\Sigma_i K_i x_i = 1 \tag{7.45}$$

where x_i is given. For the ideal case where Raoult's law holds this gives

$$\Sigma_i \underbrace{x_i p_i^{sat}(T)}_{p_i} = p \tag{7.46}$$

Example 7.17 Bubble point at given temperature T. *A liquid mixture contains 50% pentane (1), 30% hexane (2) and 20% cyclohexane (3) (all in mol-%), i.e.,*

$$x_1 = 0.5; \quad x_2 = 0.3; \quad x_3 = 0.2$$

At $T = 400$ K, the pressure is gradually decreased. What is the bubble pressure and composition of the first vapor that is formed? Assume ideal liquid mixture and ideal gas (Raoult's law).

Solution. *The task is to find a p that satisfies (7.46). Since T is given, this is trivial; we can simply calculate p from (7.46). We start by computing the vapor pressures for the three components at $T = 400K$. Using the Antoine data in Table 7.2, we get:*

$$p_1^{\text{sat}}(400K) = 10.248 \text{ bar}$$

$$p_2^{\text{sat}}(400K) = 4.647 \text{ bar}$$
$$p_3^{\text{sat}}(400K) = 3.358 \text{ bar}$$

At the bubble point, the liquid phase composition is given, so the partial pressure of each component is

$$p_1 = x_1 p_1^{\text{sat}} = 5.124 \text{ bar}$$
$$p_2 = x_2 p_2^{\text{sat}} = 1.394 \text{ bar}$$
$$p_3 = x_3 p_3^{\text{sat}} = 0.672 \text{ bar}$$

Thus, from (7.46) the bubble pressure is

$$p = p_1 + p_2 + p_3 = 7.189 \text{ bar}$$

Finally, the vapor composition (composition of the first vapor bubble) is

$$y_1 = \frac{p_1}{p} = 0.713; \quad y_2 = \frac{p_2}{p} = 0.194; \quad y_3 = \frac{p_3}{p} = 0.093$$

For calculation details see the MATLAB code:

```
T=400; x1=0.5; x2=0.3; x3=0.2
psat1=10^(A1-B1/(T+C1)), psat2=10^(A2-B2/(T+C2)), psat3=10^(A3-B3/(T+C3))
p1=x1*psat1, p2=x2*psat2, p3=x3*psat3, p=p1+p2+p3
y1=p1/p, y2=p2/p, y3=p3/p
```

Example 7.18 Bubble point at given pressure p. *Consider the same liquid mixture with 50% pentane (1), 30% hexane (2) and 20% cyclohexane (3) (all in mol-%). A $p = 5$ bar, the temperature is gradually increased. What is the bubble temperature and composition of the first vapor that is formed?*

Solution. *In this case, p and x_i are given, and (7.46) provides an implicit equation for T which needs to be solved numerically, for example, by iteration. A straightforward approach is to use the method from the previous example, and iterate on T until the bubble pressure is 5 bar (for example, using the MATLAB code below). We find $T = 382.64$ K, and*

$$y_1 = \frac{p_1}{p} = 0.724; \quad y_2 = \frac{p_2}{p} = 0.187; \quad y_3 = \frac{p_3}{p} = 0.089$$

```
% MATLAB:
x1=0.5; x2=0.3; x3=0.2; p=5;
T=fzero(@(T) p-x1*10^(A1-B1/(T+C1))-x2*10^(A2-B2/(T+C2))-x3*10^(A3-B3/(T+C3)) , 400)
```

7.5.2 Dew point calculations

Let us next consider **dew point** calculations. In this case the vapor-phase composition y_i is given (it corresponds to the case where L is very small ($L \gtrsim 0$) and $y_i = z_i$ in Figure 7.4). The dew point of a vapor (gas) is the point where the vapor just begins to condense, that is, when the first liquid drop is formed. If the temperature is given, then we must increase the pressure until the first liquid is formed. If the pressure is given, then we must decrease the temperature until the first liquid is formed. In both cases, this corresponds to adjusting T or p until $\Sigma x_i = 1$ or

$$\Sigma_i y_i / K_i = 1 \tag{7.47}$$

where y_i is given. For an ideal mixture where Raoult's law holds this gives

$$\Sigma_i \frac{y_i}{p_i^{\text{sat}}(T)} = \frac{1}{p} \tag{7.48}$$

Example 7.19 Dew point at given temperature T. *A vapor mixture contains 50% pentane (1), 30% hexane (2) and 20% cyclohexane (3) (all in mol-%), i.e.,*

$$y_1 = 0.5; \quad y_2 = 0.3; \quad y_3 = 0.2$$

At T = 400 K, the pressure is gradually increased. What is the dew point pressure and the composition of the first liquid that is formed? Assume ideal liquid mixture and ideal gas (Raoult's law).

Solution. *The task is to find the value of p that satisfies (7.48). Since T is given, this is trivial; we can simply calculate $1/p$ from (7.48). With the data from Example 7.17 we get:*

$$\frac{1}{p} = \frac{0.5}{10.248} + \frac{0.3}{4.647} = \frac{0.2}{3.358} = 0.1729\text{bar}^{-1}$$

and we find $p = 5.78$ bar. The liquid phase composition is $x_i = y_i p / p_i^{\text{sat}}(T)$ and we find

$$x_1 = \frac{0.5 \cdot 5.78}{10.248} = 0.282, \quad x_2 = \frac{0.3 \cdot 5.78}{4.647} = 0.373, \quad x_3 = \frac{0.2 \cdot 5.78}{3.749} = 0.345$$

```
% MATLAB:
T=400; y1=0.5; y2=0.3; y3=0.2
psat1=10^(A1-B1/(T+C1)), psat2=10^(A2-B2/(T+C2)), psat3=10^(A3-B3/(T+C3))
p=1/(y1/psat1 + y2/psat2 + y3/psat3)
x1=y1*p/psat1, x2=y2*p/psat2, x3=y3*p/psat3
```

Example 7.20 Dew point at given pressure p. *Consider the same vapor mixture with 50% pentane (1), 30% hexane (2) and 20% cyclohexane (3). At $p = 5$ bar, the temperature is gradually decreased. What is the dew point temperature and the composition of the first liquid that is formed?*

Solution. *In this case, p and y_i are given, and (7.48) provides an implicit equation for T which needs to be solved numerically (e.g., using the MATLAB code below). We find $T = 393.30$ K, and from $x_i = y_i p / p_i^{\text{sat}}(T)$ we find*

$$x_1 = 0.278; \quad x_2 = 0.375; \quad x_3 = 0.347$$

```
% MATLAB:
y1=0.5; y2=0.3; y3=0.2; p=5;
T=fzero(@(T) 1/p-y1/10^(A1-B1/(T+C1))-y2/10^(A2-B2/(T+C2))-y3/10^(A3-B3/(T+C3)) , 400)
```

Example 7.21 Dew point with non-condensable components. *Calculate the temperature and composition of a liquid in equilibrium with a gas mixture containing 10% pentane (1), 10% hexane and 80% nitrogen (3) at 3 bar. Nitrogen is far above its critical point and may be considered non-condensable.*

Solution. *To find the dew-point we use* $\Sigma_i x_i = 1$. *However, nitrogen is assumed non-condensable so* $x_3 = 0$. *Thus, this component should not be included in (7.48), which becomes*

$$\frac{y_1}{p_1^{\rm sat}(T)} + \frac{y_2}{p_2^{\rm sat}(T)} = \frac{1}{p}$$

Solving this implicit equation in T *numerically (e.g., using the MATLAB code below) gives* $T = 314.82K$ *and from* $x_i = y_i p/p_i^{\rm sat}(T)$ *the liquid composition is*

$$x_1 = 0.245; \quad x_2 = 0.755; \quad x_3 = 0$$

7.5.3 Flash with liquid and vapor products

Next, consider a **flash** where a feed F (with composition z_i) is split into a vapor product V (with composition y_i) and a liquid product (with composition x_i); see Figure 7.4 on page 189. For each of the N_c components, we can write a material balance

$$Fz_i = Lx_i + Vy_i \tag{7.49}$$

In addition, the vapor and liquid is assumed to be in equilibrium,

$$y_i = K_i x_i$$

The K-values $K_i = K_i(T, P, x_i, y_i)$ are computed from the VLE model. In addition, we have the two relationships $\Sigma_i x_i = 1$ and $\Sigma_i y_i = 1$. With a given feed (F, z_i), we then have $3N_c+2$ equations in $3N_c+4$ unknowns $(x_i, y_i, K_i, L, V, T, p)$. Thus, we need two additional specifications, and with these the equation set should be solvable.

pT-flash

The simplest flash is usually to specify p and T (pT-flash), because K_i depends mainly on p and T. Let us show one common approach for solving the resulting equations, which has good numerical properties. Substituting $y_i = K_i x_i$ into the mass balance (7.49) gives $Fz_i = Lx_i + VK_i x_i$, and solving with respect to x_i gives $x_i = (Fz_i/(L + VK_i)$. Here, introduce $L = F - L$ (total mass balance) to derive

$$x_i = \frac{z_i}{1 + \frac{V}{F}(K_i - 1)}$$

Here, we cannot directly calculate x_i because the vapor split V/F is not known. To find V/F we may use the relationship $\Sigma_i x_i = 1$ or alternatively $\Sigma_i y_i = \Sigma_i K_i x_i = 1$. However, it has been found that the combination $\Sigma_i(y_i - x_i) = 0$ results in an equation with good numerical properties; this is the so-called **Rachford-Rice** flash equation[5]

$$\Sigma_i \frac{z_i(K_i - 1)}{1 + \frac{V}{F}(K_i - 1)} = 0 \tag{7.50}$$

which is a monotonic function in V/F and is thus easy to solve numerically. A physical solution must satisfy $0 \leq V/F \leq 1$. If we assume that Raoult's holds, then K_i depends

[5] Rachford, H.H. and Rice, J.D.: "Procedure for Use of Electrical Digital Computers in Calculating Flash Vaporization Hydrocarbon Equilibrium," *Journal of Petroleum Technology*, Sec. 1, p. 19, Oct. 1952.

on p and T only: $K_i = p_i^{\text{sat}}(T)/p$. Then, with T and p specified, we know K_i and the Rachford-Rice equation (7.50) can be solved for V/F. For non-ideal cases, K_i depends also on x_i and y_i, so one approach is add an outer iteration loop on K_i.

Example 7.22 ***pT*-flash.** *A feed F is split into a vapor product V and a liquid product L in a flash tank (see Figure 7.4 on page 189). The feed is 50% pentane, 30% hexane and 20% cyclohexane (all in mol-%). In the tank, $T = 390K$ and $p = 5$ bar. For example, we may have a heat exchanger that keeps constant temperature and a valve on the vapor product stream that keeps constant pressure. We want to find the product split and product compositions. Assume ideal liquid mixture and ideal gas (Raoult's law).*

Comment. *This is a quite close-boiling mixture and we have already found that at 5 bar the bubble point temperature is 382.64 K (Example 7.18) and the dew point temperature is 393.30 K (Example 7.20). The temperature in the flash tank must be between these temperatures for a two-phase solution to exist (which it does in our case since $T = 390$ K).*

Solution. *The feed mixture of pentane (1), hexane (2) and cyclohexane (3) is*

$$z_1 = 0.5; \quad z_2 = 0.3; \quad z_3 = 0.2$$

We have $K_i = p_i^{\text{sat}}(T)/p$ and at $T = 390K$ and $p = 5$ bar, we find with the Antoine parameters in Table 7.2:

$$K_1 = 1.685, \quad K_2 = 0.742, \quad K_3 = 0.532$$

Now, z_i and K_i are known, and the Rachford-Rice equation (7.50) is solved numerically to find the vapor split $V/F = 0.6915$. The resulting liquid and vapor compositions are (for details see the MATLAB code below):

$$x_1 = 0.3393, \quad x_2 = 0.3651, \quad x_3 = 0.2956$$

$$y_1 = 0.5717, \quad y_2 = 0.2709, \quad y_3 = 0.1574$$

```
% MATLAB:
z1=0.5; z2=0.3; z3=0.2; p=5; T=390;
psat1=10^(A1-B1/(T+C1)); psat2=10^(A2-B2/(T+C2)); psat3=10^(A3-B3/(T+C3));
K1=psat1/p; K2=psat2/p; K3=psat3/p; k1=1/(K1-1); k2=1/(K2-1); k3=1/(K3-1);
% Solve Rachford-Rice equation numerically to find a=V/F:
a=fzero(@(a) z1/(k1+a) + z2/(k2+a) + z3/(k3+a) , 0.5)
x1=z1/(1+a*(K1-1)), x2=z2/(1+a*(K2-1)), x3=z3/(1+a*(K3-1))
y1=K1*x1, y2=K2*x2, y3=K3*x3
```

Example 7.23 Condenser and flash drum for ammonia synthesis. *The exit gas from an ammonia reactor is at 250 bar and contains 61.5% H_2, 20.5% N_2 and 18% NH_3. The gas is cooled to 25°C (partly condensed), and is then separated in a flash drum into a recycled vapor stream V and a liquid product L containing most of the ammonia. We want to calculate the product compositions (L and V) from the flash drum.*

Data. *In spite of the high pressure, we assume for simplicity ideal gas. Use vapor pressure data for ammonia from Table 7.2 and Henry's law coefficients for N_2 and H_2 from page 187. For ammonia, we assume ideal liquid mixture, i.e., $\gamma_{NH3} = 1$ (which is reasonable since the liquid phase is almost pure ammonia).*

Solution. *The feed mixture of H_2 (1), N_2 (2) and NH_3 (3) is*

$$z_1 = 0.615, \quad z_2 = 0.205, \quad z_3 = 0.18$$

For ammonia, we have at $T = 298.15$ K and $p = 250$ bar (Raoult's law):

$$K_3 = \frac{p_3^{\text{sat}}(T)}{p} = \frac{9.83 \text{ bar}}{250 \text{ bar}} = 0.0393$$

For H_2 and N_2, we have from the given data for Henry's coefficient at 25^oC (298.15 K):

$$K_1 = \frac{H_1(T)}{p} = \frac{15200 \text{ bar}}{250 \text{ bar}} = 60.8$$

$$K_2 = \frac{H_2(T)}{p} = \frac{8900 \text{ bar}}{250 \text{ bar}} = 35.6$$

Now, z_i and K_i are known, and the Rachford-Rice equation (7.50) is solved numerically to find the vapor split $V/F = 0.8500$. The resulting liquid and vapor compositions of the products are

$$x_1 = 0.0119, \quad x_2 = 0.0067, \quad x_3 = 0.9814$$
$$y_1 = 0.7214, \quad y_2 = 0.2400, \quad y_3 = 0.0386$$

This agrees well with flow sheet data from a commercial ammonia plant.

Other flashes

For other flashes, like the pH-**flash** (which is relevant for an adiabatic flash tank), one must include also the energy balance. For example, for an adiabatic flash tank, the steady-state energy balance gives that the enthalpy H is constant. That is, $H_{\text{in}} = H_{\text{out}}$, and we get

$$\underbrace{Fh_F}_{H} = Vh_V + Lh_L \tag{7.51}$$

where h_V and h_L [kJ/mol; kJ/kg] depend primarily on T, but in general also on x_i, y_i and p. One solution approach is to use the pT-flash described above, and iterate on T in an outer loop until the requirement on H is satisfied. Another approach is to solve the equations simultaneously, as shown for the dynamic adiabatic flash of methanol and ethanol in Example 11.18 (page 317).

7.5.4 Flash exercises

Exercise 7.8* Bubble and dew point at given temperature. *A hydrocarbon mixture contains 10% propane, 80% hexane and 10% dodecane. (a) Find the bubble point pressure at 300 K. (b) Find the dew point pressure at 300 K.*

Exercise 7.9* Bubble and dew point at given pressure. *A hydrocarbon mixture contains 10 mol-% propane, 80% hexane and 10% dodecane. (a) Find the bubble point temperature at 1 bar. (b) Find the dew point temperature at 1 bar.*

Exercise 7.10 Bubble point at given pressure. *A liquid mixture contains 4 mol-% hexane and the rest is octane. What is the composition of the first vapor formed if the total pressure is 1 atm?*

Exercise 7.11* Flash at given p and T. *A feed to a flash tank is 100 mol/s and contains 10% propane, 80% hexane and 10% dodecane. Find the amount of vapor product and the compositions when $T = 350K$ and $p = 2bar$.*

Exercise 7.12 Flash calculation for binary mixture. *Calculate the amount of liquid that will remain at equilibrium when a mixture of 7 kg hexane and 3 kg toluene is vaporized at 95^oC and 1.5 bar.*

Data: Molecular weights are 86.17 and 92.13.

Exercise 7.13 * Bubble and dew point calculations. *(a) A gas mixture of 15 mol-% benzene, 5 mol-% toluene and the rest nitrogen is compressed isothermally at 100°C until condensation occurs. What will be the composition of the initial condensate?*

(b) Calculate the temperature and composition of a vapor in equilibrium with a liquid that is 25 mol-% benzene and 75 mol-% toluene at 1 atm. Is this a bubble point or a dew point?

(c) Calculate the temperature and composition of a liquid in equilibrium with a gas mixture containing 15 mol-% benzene, 25 mol-% toluene and the rest nitrogen (which may be considered non-condensable) at 1 atm. Is this a bubble point or a dew point?

Exercise 7.14 Condenser for exhaust gas. *The exhaust gas from a natural gas power plant is at 1 bar and contains 76% N_2 (1), 12% O_2 (2), 4% CO_2 (3) and 8% H_2O (4). The gas is cooled to 25°C (partly condensed), and is then separated in a flash drum into a gas product V and a liquid product L containing most of the water. Find the compositions of the product streams. Are we able to remove any significant amount of CO_2 in the water?*

Data: Use pure component vapor pressure data for water and Henry's law coefficients for the gas components (see page 187).

8

Work from heat

Can we convert heat into work? The answer is "yes," but with clear limitations which can be derived from the second law of thermodynamics. More precisely, if we have available heat at temperature T_H and cold at a lower temperature T_C, then the maximum fraction of heat that can be converted into work using a *heat engine* is $1 - T_C/T_H$ (the Carnot factor). The reverse processes of obtaining *cold from heat* (refrigeration) and obtaining *heat from cold* (heat pump), which require work, are also limited by the Carnot factor. In practice, the efficiency is lower because of irreversibility that results in "lost work" or exergy (availability) losses. A thermodynamic (exergy) analysis can help to identify the main losses in the process. It can also be used to evaluate the overall efficiency of a technology.

8.1 Thermodynamics

This chapter is based on thermodynamics (see Appendices A and B). Thermodynamics is an extremely important and useful tool for engineers, but it has it limitations:

> Thermodynamics can tell us whether a given process is theoretically possible, but in can not tell us how or if the process can be realized in practice.

The start of thermodynamics as a science was the industrial use of steam engines in England from around 1710 and onwards. The steam engine was able to convert heat (generated by combustion of coal) into useful work by utilizing pressure differences. For a long time, one did not know how much work one could theoretically obtain from 1 kg coal, and how the process should be operated to maximize the efficiency. In this chapter, we want to give the answer to these questions.

The main difficulty was a very limited understanding of heat and energy and the relation between the two. To illustrate this, it was only in 1759 that Joseph Black found that there is a difference between temperature and heat. And it took almost another 100 years, with the formulation of the first law of thermodynamics, before it was finally established that if there is no change in the system's state then the net supplied heat equals the performed work. The first and second laws of thermodynamics were formulated by Carnot in 1850. This shows that thermodynamics is not an easy topic, but it is very important and sets clear limitations for our lives.

8.2 Heat engine and the first law

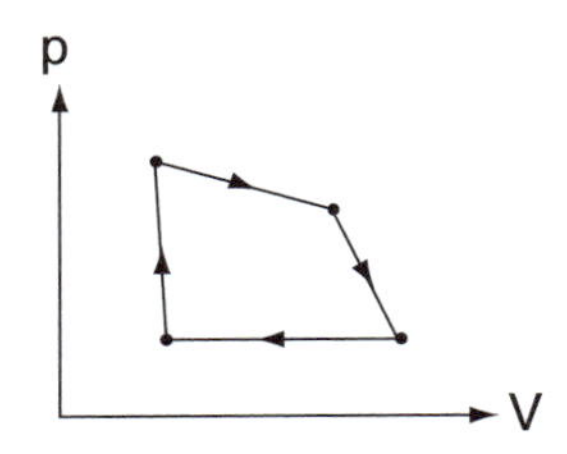

Figure 8.1: Cyclic process

The function of a **heat engine** is to convert heat into work. Examples of heat engines are steam engines, gas turbines, and gasoline and diesel engines (see Section 8.3 on page 203). A heat engine operates cyclically, that is, the engine (system) continuously passes through a **cyclic process** where it, after a completed cycle, returns to its original state, see Figure 8.1. If we consider the system over a complete cycle, then

$$\Delta U = 0, \;\; \Delta V = 0 \;, \Delta S = 0 \quad \text{etc.}$$

A cyclic process does not exchange mass with the surroundings, so it is a *closed system*, and from Chapter 4 the energy balance (first law of thermodynamics) is $\Delta U = Q + W$. If we consider changes over a complete cycle, then $\Delta U = 0$ and we derive

$$(-W) = Q \tag{8.1}$$

Note that with our sign convention for work, $(-W)$ *is performed work* by the system (the cyclic process) on the surroundings. Since $\Delta V = 0$ for a cyclic process, there is no work term related to changes in the system's volume, so all the work W is shaft work, that is, $W = W_s$.[1] From (8.1), it seems that for any cyclic process, the supplied heat $Q > 0$ can be extracted as useful work ($(-W) > 0$). This sounds too good to be true, and unfortunately things are not quite as simple. The important point is that $Q = Q_H + Q_C$ is the *net* supplied heat, and we will show that in order to convert high-temperature heat Q_H to work W, the second law of thermodynamics requires that some heat Q_C must be removed by cooling at a lower temperature, see Figure 8.2. From the first law of thermodynamics the performed work is $(-W) = Q_H + Q_C$, or equivalently with absolute values (so that there is no doubt about the signs), see Figure 8.2,

$$|W| = |Q_H| - |Q_C| \tag{8.2}$$

The net heat supplied to the cyclic process, $|Q_H| - |Q_C|$, is always less than the supplied heat $|Q_H|$ (we use absolute value to be sure there is no doubt about this). In the next section, we will show that the maximum performed work by the cycle equals $|Q_H|$ multiplied by the Carnot factor.

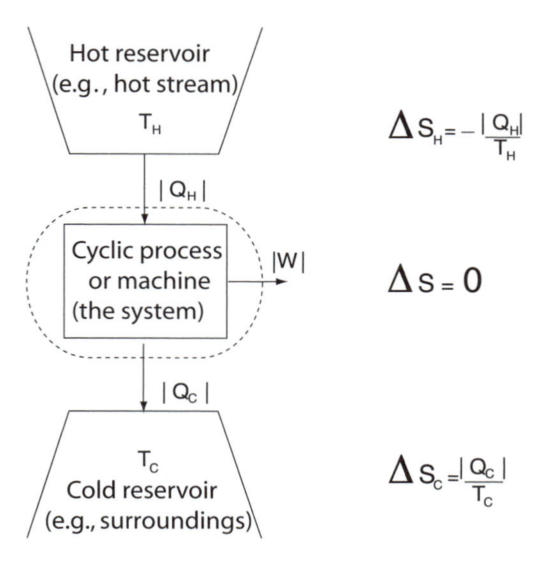

Figure 8.2: Heat engine that converts heat Q_H to work W

8.3 Heat engine and the second law

Let us consider the following problem:

> A given amount of heat Q_H is available at temperature T_H (H for *hot*) and we have access to cooling at temperature T_C (C for *cold*). How much of the heat can we convert to work, i.e., what is the maximum value of $|W|/|Q_H|$?

The problem is illustrated in Figure 8.2, where we use the absolute value so that there is no doubt about the signs for the various variables. We assume that we have a hot reservoir[2] (heating) at temperature T_H and a cold reservoir (cooling) at temperature T_C. We assume that the machine (the system) passes though a cyclic process. Let us start by calculating the entropy changes in the system (the machine) and in the two reservoirs for a reversible process (perfect machine). Since the machine is cyclic, the change in entropy in the system is zero, that is,

$$\Delta S = 0$$

The change in entropy in the cold and hot reservoirs (the surroundings) is, from (7.5), given by

$$\Delta S_H = -\frac{|Q_H|}{T_H} \quad \text{and} \quad \Delta S_C = \frac{|Q_C|}{T_C}$$

where T_H and T_C are the temperatures of the two reservoirs. We assume that the cyclic machine operates reversibly. This implies, among other things, that the heat

[1] In all of this chapter, we use W in the meaning "shaft work" W_s.

[2] A reservoir is a body with an infinite heat capacity or a fluid that is condensed/boiled such that the temperature is constant.

transfer is reversible so that the temperature of the system (machine) is T_H when heat is supplied to the system, and is T_C when heat is removed from the system. From the second law of thermodynamics (page 372), the total entropy change of the system and surroundings is zero for a reversible process:

$$\Delta S_{\text{total}} = \Delta S_H + \Delta S_C + \Delta S = 0$$

This gives for an ideal (reversible) machine:

$$\Delta S_{\text{total}} = -\frac{|Q_H|}{T_H} + \frac{|Q_C|}{T_C} = 0 \quad \Rightarrow \quad \boxed{\left(\frac{|Q_C|}{|Q_H|}\right)_{\text{rev}} = \frac{T_C}{T_H}} \tag{8.3}$$

Equation (8.3) says that for an ideal engine, the fraction of heat supplied at T_H that must be removed as cooling at T_C is the ratio between the two absolute temperatures. Furthermore, from the first law of thermodynamics (8.2),

$$|W| = |Q_H| - |Q_C|$$

and (8.3) gives for an ideal (reversible) cycle

$$\boxed{\left(\frac{|W|}{|Q_H|}\right)_{\text{rev}} = \frac{T_H - T_C}{T_H} = 1 - \frac{T_C}{T_H}} \tag{8.4}$$

which is always between 0 (for $T_H = T_C$) and 1 (for $T_H \gg T_L$). This is also known as the **Carnot factor** or Carnot "efficiency,"[3]

$$\eta_{\text{Carnot}} = 1 - \frac{T_C}{T_H} \tag{8.5}$$

To maximize the theoretical fraction of heat $|Q_H|$ that can be converted to work $|W|$, we want T_H to be as high as possible and than T_c as low as possible, corresponding to a Carnot factor close to 1. These results, which originate from the study of steam engines, are extremely fundamental and form the historical basis for thermodynamics as a subject.

Alternatively, if we have available a lot of heat and the cooling capacity Q_C is limited, then the following ratio is of interest,

$$\left(\frac{|W|}{|Q_C|}\right)_{\text{rev}} = \frac{T_H - T_C}{T_C} = \frac{T_H}{T_C} - 1 \tag{8.6}$$

which is always between 0 (for $T_H = T_C$) and infinity (for $T_H \gg T_C$). Again, we find that we can extract more work when there is a large temperature difference between the hot and cold reservoirs.

Note. It is emphasized that one must *always* use the *absolute* temperature T [K] when calculating the Carnot factor.

[3] I am not very happy about the commonly used term "Carnot efficiency," because an efficiency should – in my opinion – be 1 for an ideal (reversible) process. I therefore recommend using the term "Carnot factor;" more about this on page 211.

Example 8.1 Arctic versus tropical cooling water. *We have available a heat reservoir at 400 °C and a cold "tropical" reservoir (cooling) at 25 °C. How much heat can be extracted as work? What is the answer if we instead have "arctic" cooling at 5 °C?*

In the case with cooling at 25 °C (tropical conditions), the Carnot factor is

$$\eta = \frac{|W^{\text{rev}}|}{|Q_H|} = 1 - \frac{T_C}{T_H} = 1 - \frac{298}{673} = 0.557$$

that is, 55.7% of the heat can in theory be removed as work. For the case with cooling at 5 °C (arctic conditions), the Carnot factor is

$$\eta = \frac{|W^{\text{rev}}|}{|Q_H|} = 1 - \frac{T_C}{T_H} = 1 - \frac{278}{673} = 0.587$$

that is, 58.7% of the heat can in theory be removed as work. We can in other words extract about 3% more of the heat as work by using cold "arctic" cooling water instead of using cooling water at 25 °C.

The cold cooling water is a potential advantage in locating a thermal power plant in a cold climate. However, in practice the difference is not quite as large as in the above example. This is because, in warmer climates, colder cooling water (below 25 °C) can be obtained by using a cooling tower (where some water evaporates such that the water is cooled and the air is saturated). The dimensions of these cooling towers are often enormous (diameter and height up to 100 m) and they usually dominate the appearance of a nuclear or coal power plant. If someone sees one of these cooling towers with all the "smoke" coming out, they may think there is a serious pollution problem, but it is just water vapor.

Example 8.2 Log-mean Carnot factor. *The Carnot factor $1 - \frac{T_C}{T_H}$ applies to the case where T_C and T_H are constant. Let us consider two common cases, where either T_C or T_H vary linearly with the amount of heat Q transferred. In this case, the correct Carnot factor involves the logarithmic mean temperature difference.*

(a) T_H varies linearly with Q, T_C constant. *This is the case if we have a hot stream with constant heat capacity which is cooled from T_{H1} to T_{H2} and we have available cooling at constant temperature T_C. For a small supplied amount of heat $dQ_H = C_p dT_H$, it follows from (8.4) that the maximum work is $dW = \left(1 - \frac{T_C}{T_H}\right) dQ_H = \left(1 - \frac{T_C}{T_H}\right) C_p dT_H$. With constant heat capacity C_p, the maximum work that can be removed in the temperature interval from T_{H1} to T_{H2} is $|W| = C_p \int_{T_{H1}}^{T_{H2}} \left(1 - \frac{T_C}{T}\right) dT = C_p\left(T_{H2} - T_{H1} - T_C \ln T_{H_2}/T_{H1}\right)$. The heat supplied then is $Q_H = \int_{T_{H1}}^{T_{H2}} C_p dT = C_p(T_{H2} - T_{H1})$ and we find that the mean Carnot factor is*

$$\eta_{\text{Carnot}} = \frac{|W^{\text{rev}}|}{Q_H} = 1 - \frac{T_C}{T_{H,lm}}, \quad \text{where} \quad T_{H,lm} = \frac{T_{H2} - T_{H1}}{\ln(T_{H_2}/T_{H1})} \tag{8.7}$$

is the logarithmic mean temperature on the hot side.

(b) T_C varies linearly with transferred amount of heat, T_H constant. *This is the case if we have a cold stream with constant heat capacity which is heated from T_{C1} to T_{C2}. A similar derivation gives that the mean Carnot factor is $\eta = 1 - T_{C,lm}/T_H$, where $T_{C,lm} = (T_{C2} - T_{C1})/\ln(T_{C2}/T_{C1})$ is the logarithmic mean temperature on the cold side.*

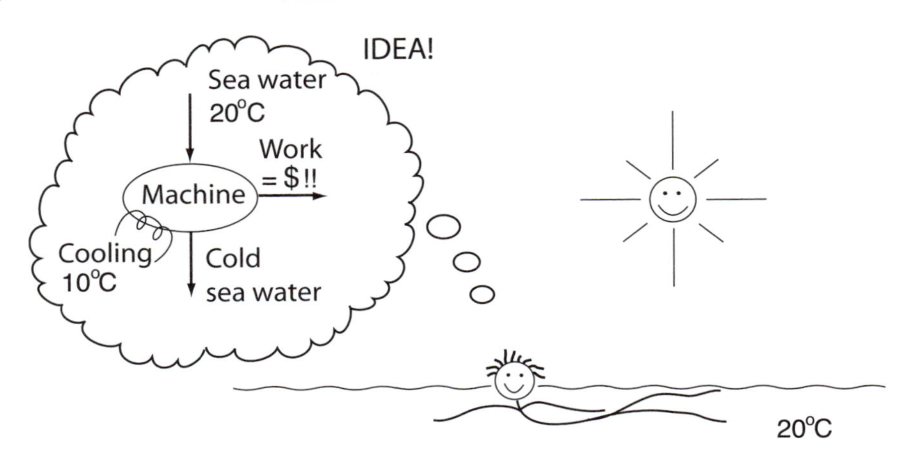

Figure 8.3: A summer idea

Example 8.3 A summer idea. *It is in the middle of the Norwegian "summer" and the temperature in the sea is 20 °C. You have available drinking water (maximum m = 2 kg/s) at 5 bar and 10 °C. You are not paying anything for the water and you come up with the ingenious idea (in your opinion) of using heat from the sea (your hot reservoir) to produce work by using the drinking water for cooling. What is the maximum work you can extract?*
Solution. *The amount of drinking water is the limiting resource here. Let us assume that the drinking water (cooling water) is heated by the seawater from 10 °C to 18 °C (this requires a rather large heat exchanger), that is,* $Q_C = mC_p\Delta T = 2{\cdot}4180{\cdot}8 = 66880$ *J/s = 66.9 kW. This looks quite promising, but how much work can actually be removed? In our case,* T_C *varies between 283 and 291 K, and since* C_p *is constant, the Carnot factor can, from Example 8.2, be found using the logarithmic mean temperature* $T_{C,lm} = (291-283)/\ln(291/283) = 286.98$ *K. The mean Carnot factor is then* $1 - 286.98/293 = 0.0205$ *(which is not very promising), and the work that can be removed is from (8.6)*

$$|W^{\text{rev}}| = \left(\frac{T_H}{T_{C,lm}} - 1\right)|Q_C| = \left(\frac{293}{286.98} - 1\right) 66880\ W = 1403\ W$$

that is, only 1.4 kW. Perhaps, we should instead try to use the pressure, i.e., we utilize the pressure difference $\Delta p = 5-1 = 4$ *bar in a water turbine. From (6.14), the maximum work is* $\frac{m}{\rho}\Delta p = \frac{2\ kg/s}{1000\ kg/m^3}\cdot 4\cdot 10^5\ N/m^2 = 800\ J/s$*, that is, 0.8 kW (which is even less). Conclusion: This is not worth it so we abandon the project and take a swim instead.*

From this example we see that heat engines are not very effective when the temperature difference is small. It illustrates the fact that *a given amount of heat is generally more valuable at high temperature than at low temperature.*

Exercise 8.1* *(a) An inventor claims to have produced a machine that takes heat at 200 °C and removes 50% of it as work. Is this possible? (Hint: Calculate the theoretical cooling temperature).*

(b) Is a reversible process in equilibrium? What is the entropy change for a reversible process?

Exercise 8.2 *500 mol/s of a product (P) is produced in a strongly exothermic reaction with* $\Delta_r H^\ominus = -200$ *kJ/mol P. Two alternative processes have been developed:*

1. A gas-phase reaction at 400 °C.
2. A liquid-phase reaction at 100 °C.

Which alternative would you prefer if it is desirable to utilize the heat to produce work in a heat engine? (Quantify your answer by calculating the maximum work that can be removed if we have cooling available at 15 °C).

Alternative heat engines

There are many heat engine processes that can be used to convert heat to work.

- One is the theoretical **Carnot cycle**, which is studied in the Appendix on page 373 using an ideal gas as the working fluid. This cycle is not used in practice.
- For practical purposes it is better to use a cycle that involves condensation and evaporation, for example, the **Rankine cycle**, which is discussed for a natural gas combustion power plant on page 227 using water (steam) as the working fluid. (The reverse of the Rankine cycle is the **Rankine refrigeration cycle** which is commonly used in refrigeration processes; see page 206 and Figure 8.7.)
- In an **internal combustion engine**, there is no working fluid. The most common is the **Otto cycle** used in ordinary gasoline engines. The combustion is initiated by a spark and is almost instantaneous, so we can assume that the heat is supplied at constant volume. For the Otto cycle, we find that the thermodynamic efficiency $\nu = |W|/|Q_H|$ is limited by the compression ratio, which should be as large as possible. However, pre-ignition of the air-gasoline mixture limits the compression ratio in most engines to about 10 or less.
- Another internal combustion engine is the **Diesel cycle**. Here, there is no spark so the combustion (heat supply) is slower and the system expands (the piston moves) during combustion. The result is that the Diesel cycle actually has a lower efficiency for a given compression ratio than the Otto cycle. However, the Diesel engine is more efficient in practice, because it can operate at much higher compression ratios, up to about 20. A good discussion of this and other cycles is given in thermodynamics textbooks, e.g., in the book by Smith and van Ness (see page vii).
- In the **combustion gas turbine** (**Brayton cycle**), the air and fuel are compressed before combustion and the product is expanded in a gas turbine to produce work. In an ideal Brayton cycle, the compression is isentropic, the combustion is at constant pressure and the expansion is isentropic back to the starting pressure. The efficiency is limited primarily by the combustion temperature, which should be as high as possible. A detailed example of a gas turbine in a natural gas power plant is given on page 226.
- In a **combined cycle power plant**, there is a Brayton cycle (combustion gas turbine) followed by a Rankine cycle that uses steam turbines to extract additional work. Combined cycle power plants are usually powered by natural gas, although other fuels, such as oil and coal, may also be used.

8.4 Reverse heat engine: Refrigeration and heat pump

Heat engines produce work. The *reverse* happens in refrigerators and heat pumps which produce:

- Cold from work (refrigeration)
- Heat from cold + work (heat pump)

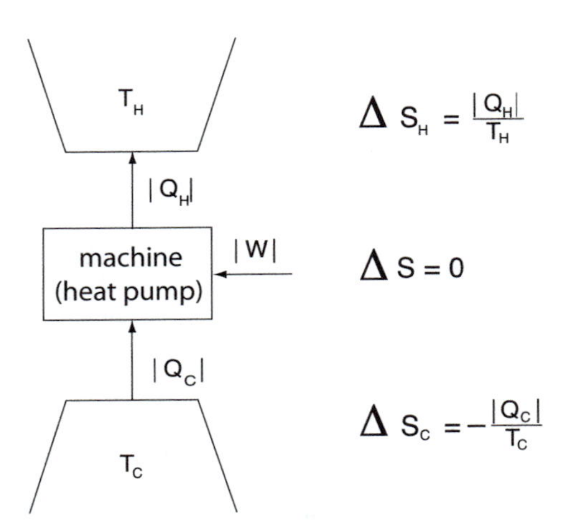

Figure 8.4: Heat pump that takes heat from low temperature (T_c) to high temperature (T_H)

In both cases they key is to use work to transfer ("pump") heat from low to high temperature (against the "natural" direction); see Figure 8.4, This is the opposite of a heat engine, and the expressions for an ideal (reversible) process are therefore the same as in (8.3) – (8.6), but the flows for heat and work go in the opposite direction.

From (8.6), we have, for an ideal refrigerator or heat pump, that

$$\left(\frac{|W|}{|Q_C|}\right)_{\text{rev}} = \frac{T_H}{T_C} - 1 \tag{8.8}$$

where everything now flows in the direction given in Figure 8.4. This is the minimum work that must be supplied in order to "pump" heat $|Q_C|$ from low temperature T_C to high temperature T_H. We want the supplied work $|W|$ to be as small as possible, so for refrigerators and heat pumps it is favorable to have *a small temperature difference between T_H and T_C*, which is the opposite of what we found for a heat engine.

From a thermodynamic point of view, a refrigerator and a heat pump are the same, but different terms are used depending on whether the primary objective is to remove heat Q_C at a low temperature T_C (cooling process) or supply heat Q_H at high temperature T_H (heating process; heat pump system). The (energy) efficiency of such systems are often reported in terms of the **coefficient of performance** (COP). The COP for a cooling (refrigeration) process and a heating (heat pump) process are, respectively,

$$\text{COP}_\text{H} = \frac{|Q_H|}{|W|} \leq \frac{1}{1 - \frac{T_C}{T_H}} \tag{8.9}$$

$$\text{COP}_\text{C} = \frac{|Q_C|}{|W|} \leq \frac{1}{\frac{T_H}{T_C} - 1} \tag{8.10}$$

The maximum values for a reversible process are given on the right and follow from (8.8). Notice that $\text{COP}_\text{H} = \text{COP}_\text{C} + 1$, and, in general, COP_H is always larger than 1, whereas COP_C is usually larger than 1. Also notice that COP_H is the inverse of the Carnot factor.

Example 8.4 Heat pump. *From a theoretical point of view, heat pumps are ideal for*

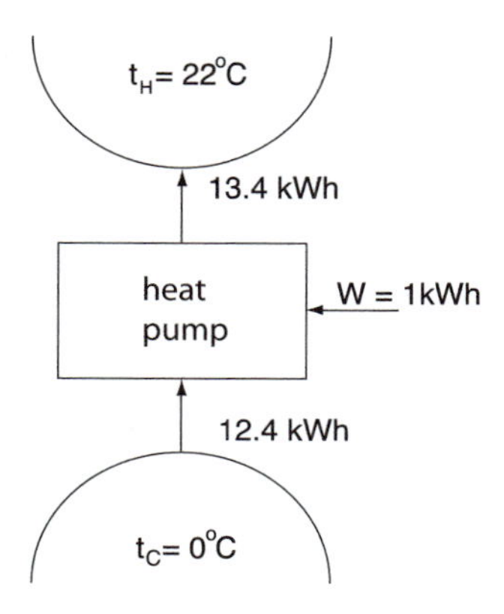

Figure 8.5: Ideal (reversible) heat pump for home heating

heating buildings where the energy source is electric energy. This is because electric energy can be converted almost 100% into mechanical work W using an electric motor. Let us consider a specific case where we have available a cold reservoir (for example, our garden) at temperature $t_C = 0\ ^oC$ and want to "pump" heat from the cold reservoir to the room temperature $t_H = 22\ ^oC$ (we use lower-case t's because we are not using absolute temperature). From (8.4), we have for the most favorable (reversible) conditions that the coefficient of performance is

$$\mathrm{COP_H} = \frac{|Q_H|}{|W|^{\mathrm{rev}}} \leq \frac{1}{1-\frac{T_C}{T_H}} = \frac{1}{1-\frac{273}{295}} = 13.4$$

This means that, if we supply 1 of kWh work (electric energy) for the heat pump, then we can in theory extract $Q_H = 13.4$ kWh as heat in the room. This looks very favorable, but before you rush and buy a heat pump, you should be aware the actual COP may only be about 30% of this. Also, if you consider using your garden as a cold reservoir, then you will need to bury a lot of piping if you want to avoid permafrost in your garden.

Example 8.5 Refrigerator. *A common refrigeration process is a home refrigerator that*

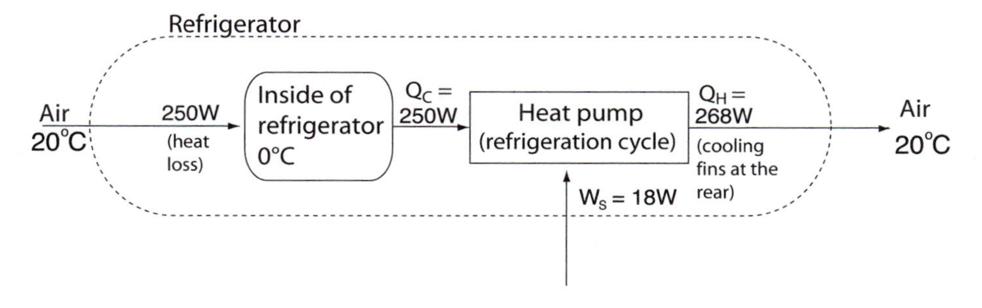

Figure 8.6: Ideal (reversible) home refrigerator

takes heat from a low temperature (inside the refrigerator) and transfers it to a higher temperature (room temperature). Consider a refrigerator with inside temperatures 0 ^oC and *room temperature 20 °C. The heat loss (actually, "cold loss") from this quite large refrigerator is 250 W. What is the minimum power required?*

Solution. *Our "system" is the heat pump (refrigeration cycle), for which* $Q_C = 250$ *W,* T_C*=273 K (0° C) and* $T_H = 293$ *K (20° C). In the ideal case, we then have from (8.6) that the work required for the refrigerator is* $|W^{\rm rev}| = |Q_C|(T_H/T_c - 1) = 250(293/273 - 1) = 250{\cdot}0.073 = 18.3W$*, corresponding to a coefficient of performance* $COP_C = |Q_C|/|W| = 13.7$*. This is very favorable, but the actual work will be much larger. This is because, in practice, the heat transfer is not reversible (i.e., in practice the refrigerator operates over a much wider temperature interval) and there are friction losses in the compressor; see Example 8.6 for details.*

Most refrigeration (cooling) processes use the vapor-compression cycle (**Rankine refrigeration cycle**) as discussed in the next example.

Example 8.6 Refrigerator with ammonia as refrigerant. *An old household refrigerator has a refrigeration cycle that operates with ammonia as the working fluid (refrigerant). The vapor-compression cycle (Rankine refrigeration cycle) consists of the following four subprocesses (see Figure 8.7 for a flow sheet and corresponding pressure enthalpy diagram):*

Subprocess a. *Compressor: Saturated ammonia vapor at -10 °C/3 bar (state 1) is compressed to 16 bar (state 2, which is gas).*

Subprocess b. *Condenser (heat exchanger or cooling at the outside back of the refrigerator): The superheated ammonia gas is cooled, condensed at a constant temperature of 40° C and finally subcooled to 30° C (stream 3).*

Subprocess c. *Choke valve: Expansion of ammonia liquid down to 3 bar (state 4, which contains about 15% gas)*

Subprocess d. *Evaporator (heat exchanger that takes heat from the inside of the refrigerator): Evaporation of the remaining liquid ammonia at constant pressure of 3 bar (from state 4 to state 1)*

Data: *pH-diagram for ammonia. Assume the compressor has an efficiency of 70%.*
Problem*:*

(i) *Calculate the enthalpy changes for each step (a,b,c,d) per kg ammonia that goes around in the cycle.*

(ii) *What is the cooling effect* (Q_C) *and the mass flow of ammonia, when the power consumption (work) in the compressor is* $W = 80$ *W ?*[4]

(iii) *What is the coefficient of performance* $\mathrm{COP_C} = Q_C/W$*?*

(iv) *What is the theoretical highest COP if the room temperature is 22 °C* ($T_H = 295K$) *and the temperature in the refrigerator is 5 °C* ($T_C = 278K$)*?*

Solution. *Enthalpies are found using the pH-diagram for ammonia on page 419 (see also Figure 8.7):*
State 1: *Saturated vapor at 3 bar/*$-10°$ *C:* $h_1 = 1350$ *kJ/kg*
State 2': *Follow the line for constant entropy from point 1 and to 16 bar:* $h_2' = 1590$ *kJ/kg (temperature* $t_2' = 115°C$*)*
State 2: *Efficiency is 70%, that is,* $h_2 = h_1 + (h_2' - h_1)/0.70 = 1350 + 240/0.7 = 1693$ *kJ/kg (temperature* $t_2 = 150° C$*)*
State 3: *Liquid at 30 °C and 16 bar:* $h_3 = 240$ *kJ/kg*

[4] I am sorry for the possible confusion, but note that W is the symbol for work, whereas W is the unit of Watt (1 W = 1 J/s).

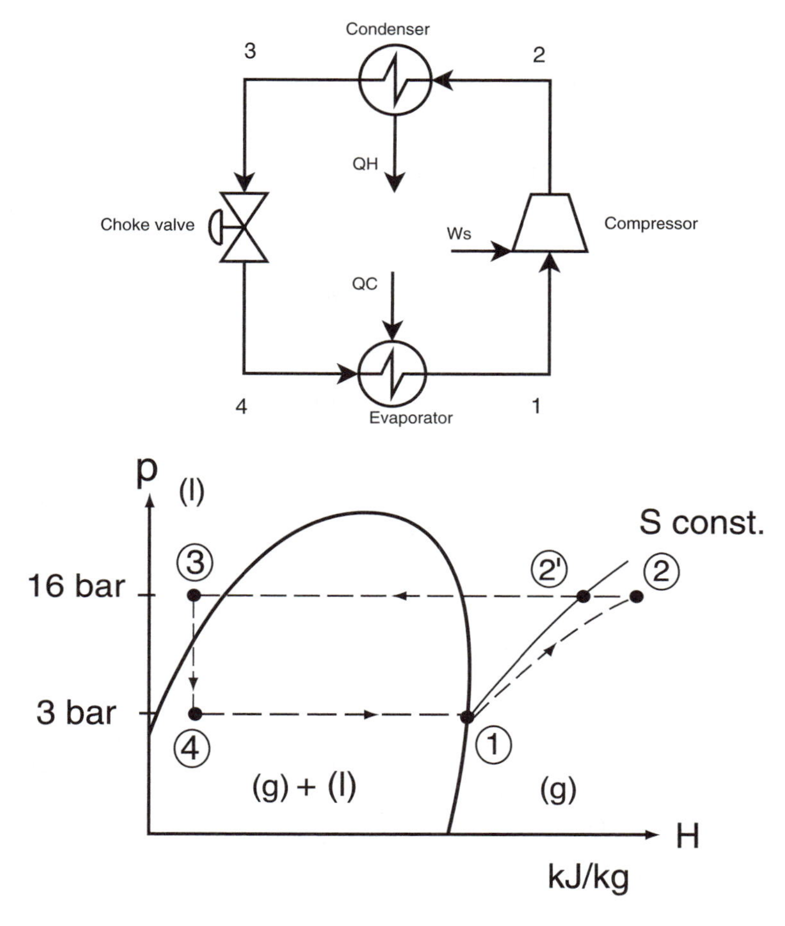

Figure 8.7: Refrigeration (cooling) using vapor-compression cycle with ammonia as the refrigerant

State 4: *Follow line for constant enthalpy down to saturated conditions at 3 bar:* $h_4 = h_3 = 240$ *kJ/kg*

Note that the temperature during condensation (at 16 bar between states 2 and 3) is $40\,^oC$*, while the temperature during the evaporation (at 3 bar between states 4 and 1) is* -10^oC.

(i) *Enthalpy change in each subprocess:*

Subprocess a: $w = \Delta h_a = 1693 - 1350 = 343$ *kJ/kg (power usage in compressor)*

Subprocess b: $\Delta h_b = 240 - 1693 = -1453$ *kJ/kg (heat transferred to the room)*

Subprocess c: $\Delta h_c = 0$ *(valve)*

Subprocess d: $q_C = \Delta h_d = 1350 - 240 = 1110$ *kJ/kg (heat that is removed from the refrigerator)*

(ii) *Actual* $\text{COP}_\text{C} = Q_C/W = q_c/w = 1110/343 = 3.23$

(iii) *With* $W = 80\ W$:

- *Cooling duty:* $Q_C = 3.23 \cdot 80\ W = 258\ W$.

- *Mass flow of ammonia:* $m = \frac{80 \text{ J/s}}{343 \cdot 10^3 \text{ J/kg}} = 0.233 \cdot 10^{-3}$ kg/s $= 0.84$ *kg/h*

(iv) *Theoretically best COP for an ideal refrigerator that operates between 5 °C and 22 °C is*

$$\text{COP}_{\max} = \frac{Q_C}{W}_{\text{rev}} = \frac{1}{\frac{T_H}{T_C} - 1} = \frac{T_C}{T_H - T_C} = \frac{278K}{17K} = 16.4$$

The reasons for the deviation between the actual COP of 3.23 and the theoretical COP of 16.4 are the following sources of irreversibility:

Subprocess a: *Efficiency of 70% in the compressor.*

Subprocess b: *Temperature difference between hot side (where temperature varies from 150 °C to 30 °C) and cold side (22 °C).*

Subprocess c: *Valve is irreversible. Could instead use turbine to extract work.*

Subprocess d: *Temperature difference between hot side (5 °C) and cold side (−10° C).*

A more detailed thermodynamic analysis of the lost work for this ammonia refrigeration cycle is given in Example 8.10 (page 217). For a refrigeration cycle that uses $C_2H_2F_4$ (R134a) as the refrigerant, see Exercise 6.8 (page 159).

Exercise 8.3 Car AC with CO_2 as working fluid. *A car air condition unit (refrigeration cycle) uses CO_2 as the refrigerant. The cycle operates above the critical point and consists of the following four steps (see a similar subcritical cycle in Figure 8.7 on page 207):*

(a) Adiabatic compression: Superheated vapor at 12° C and 35 bar (state 1) is compressed to 100 bar (state 2 which is supercritical). The compressor efficiency is 75%. (b) Cooling (Q_H) with outside air down to 40° C (state 3, still supercritical). (c) Valve: Expansion back to 35 bar (state 4 at 0° C which is about 50% liquid). (d) Evaporation/superheating by taking heat (Q_C) from the inside of the car (back to state 1).

Data: *pH-diagram for CO_2 (available at the book's home page).*

Problem. *What is the cooling duty (Q_C) when the power consumption in the compressor is $W = 2$ kW, and what is the mass flow of refrigerant? Find the coefficient of performance $\text{COP}_C = Q_C/W$. What is the theoretically maximum COP if the outside air is at 35° C and the inside of the car is at 18° C?*

Commercial refrigerants

The **working fluid** for a cooling cycle is called the **refrigerant.** There are many commercially available refrigerants, which are denoted with R + a mystical number.

- **Ammonia** (also known as refrigerant **R717**) is, in principle, well suited for use in refrigerators. The heat of vaporization on a mass basis is large (1365 kJ/kg at its normal boiling point of −33°C) and this is favorable because the mass flow is then small. Nonetheless, ammonia is no longer used in home refrigerators because it is poisonous and flammable. In large industrial cooling plants, for example, for cooling of fish and meat, ammonia is still used.
- For many years, (hydro)**chlorofluorocarbons** ((H)CFCs; freons), were the most widely used refrigerants. Examples are CCl_2F_2 (**R12**) with a boiling point of -30^oC and $CHClF_2$ (**R22**) with a boiling point of −41°C. These components were considered ideal because they are non-poisonous and inflammable, and were generally considered to be inert. However, in the 1908's it became clear that the CFCs contributed strongly to the ozone layer depletion, and they were phased out worldwide in the 1990's.

- They were replaced by similar chlorine-free components which have no effect on the ozone layer. Currently, the most common refrigerant in air condition (AC) systems is **tetrafluoroethane** (CH_2FCF_3, **R134a**). with a heat of vaporization of 217 kJ/kg at its normal boiling point of -26^oC. However, more recently it has become clear that it is a greenhouse gas, with a global warming potential (GWP) of about 1300 relative to CO_2, and the European Union has decided to ban R134a in AC systems in all new cars as from 2011.
- A possible replacement is **carbon dioxide** (CO_2, **R744**), which, however, has no liquid state below 5 bar / -78^oC where it forms "dry ice". Thus, a CO_2-cycle must operate at high pressures, up to about 100 bar (see Exercise 8.3). Furthermore, CO_2 is supercritical at normal temperatures ($T_c = 304$K), so the "condensation" in subprocess (b) is actually supercritical. The use of CO_2 will require a complete redesign of the air conditioning systems, and the refrigerant producers are working on developing an alternative non-flammable refrigerant with an acceptable GWP.
- In home refrigerators, the most common refrigerant at present is **isobutane (R600a)**, which has a heat of vaporization of 367 kJ/kg at its normal boiling point of -11.6^oC. It has zero ozone depletion potential (ODP) and a negligible global warming potential (GWP), but it must be used with some care because of its flammability. However, the quantities are very small – the amount of butane in a domestic refrigerator equals the content of two cigarette lighters.

8.5 Efficiency

Efficiencies are useful for analyzing and comparing processes. There are however many different "efficiencies" and the use of them is often confusing.

8.5.1 Thermodynamic efficiencies

It seems reasonable that the efficiency should express the ratio between real and ideal, and should be 1 for an ideal process. This is the basis for the following general definition of the thermodynamic efficiency:

- For a process that operates between two given states, the thermodynamic efficiency η expresses the (actual) amount of "useful energy" relative to the theoretical obtainable (in a reversible process), and is always a number less than 1.

Here, "useful energy" (or "interesting energy") can typically be work or heat. The thermodynamic efficiency is also called the "second-law efficiency."

With the above general definition, it is usually quite simple to derive expressions for the efficiency for specific cases. We distinguish between processes where we remove or supply energy and have:

1. **Thermodynamic efficiency** for a process where we **take out** "useful energy:"

$$\eta = \frac{\text{actual useful energy out}}{\text{ideal (maximum) useful energy out}} \tag{8.11}$$

We always have $\eta \leq 1$. For the case where "useful energy out"="work out" (for example, a turbine), we have that $\eta = (-W_s)/(-W_s^{\text{rev}}) = W_s/W_s^{\text{rev}}$ (note that

($-W_s^{rev}$) is positive in this case). In Example 8.7, we consider a case (heater) where "useful energy out" = "heat out."

2. **Thermodynamic efficiency** for a process where we **put in** "useful energy:"

$$\eta = \frac{\text{ideal (minimal) useful energy in}}{\text{actual useful energy in}} \tag{8.12}$$

Note that the ratio is here reversed because we focus on what we put in (rather than on what we take out). Again, we always have $\eta \leq 1$. For the case where "useful energy in"="work in" (for example, a compressor), we have that $\eta = W_s^{rev}/W_s$ (note that W_s is positive in this case). In Example 8.20 (page 224), we consider a case (distillation) where "useful energy in" = "heat in."

For some complex processes, where we both put in and take out useful energy, it may be difficult to define a single efficiency.

In order to find the ideal process, we also have to define the surroundings. The value of the thermodynamic efficiency therefore depends on the surroundings of the process. Often our "natural" surroundings at constant temperature T_0 are chosen.

Comment ☺. The everyday equivalent to the second law of thermodynamics is that there is "**no free lunch.**" The everyday equivalents to the thermodynamic efficiencies in (8.11) and (8.12) are (for the case where we focus on what we **take out** (get))

$$\eta = \frac{\text{What you get (actual)}}{\text{What you pay for (theoretical)}}$$

and (for the case where we focus on what we **put in**)

$$\eta = \frac{\text{What you should have paid (theoretical)}}{\text{What you paid (actual)}}$$

8.5.2 Energy "efficiencies"

There are, as suggested above, many other concepts that are called "efficiencies," but which are not thermodynamic efficiencies.

For a process where we want to remove useful energy, the **"energy efficiency"** η_{energy} is used to indicate the fraction of "(useful) energy in" that is converted into "(useful) energy out":

$$\eta_{\text{energy}} = \frac{\text{(useful) energy out}}{\text{(useful) energy in}} \tag{8.13}$$

Note that we divide by "energy in" instead of "maximum energy out," so this is not a thermodynamic efficiency, according to the above definitions, but rather a "conversion" between two energy forms.

An example is a **heat engine** (thermal power plant), where "useful energy out" is work $|W|$ and "useful energy in" is supplied heat Q_H, and we can define the energy efficiency $\eta_{\text{energy}} = |W|/|Q_H|$ (thus particular energy efficiency is often called the *thermal efficiency*). For heat engines, the maximum (reversible) "thermal efficiency" equals the "Carnot efficiency" ($= 1 - T_c/T_H$), and is therefore always less than 1.

Another example is a **refrigerator** where "useful energy out" is cooling $|Q_C|$ and "useful energy in" is work $|W|$. The energy "efficiency" then equals the "coefficient of performance (COP)", that is $\eta_{\text{energy}} = \text{COP}_\text{C} = |Q_C|/|W|$. Note that the ideal (reversible) COP is larger than 1.

From this we see that the "energy efficiency" is not bounded to be between 0 and 1; in some cases it can never become 1 (even for a reversible process) and in other cases it may always be larger than 1.

- **Relationship between thermodynamic efficiency and "energy efficiency."** The thermodynamic efficiency η is the ratio between actual energy efficiency and maximum energy efficiency for a reversible process. For example, for a heat engine, we have that the thermodynamic efficiency is

$$\eta = \frac{\text{thermal "efficiency"}}{\text{Carnot "efficiency"}}$$

- **Carnot "efficiency."** It also follows from the above definitions of thermodynamic efficiency, see (8.11) and (8.12), that the Carnot "efficiency," defined as

$$\eta_{\text{Carnot}} = \frac{\text{ideal (maximum) work out}}{\text{heat in}} = 1 - \frac{T_c}{T_H} \tag{8.14}$$

should not be called an "efficiency," but rather a "maximum conversion" from heat to work. Note that an ideal heat engine with "thermal energy efficiency = Carnot efficiency" (which is less than 1) has a thermodynamic efficiency $\eta = 1.0$!

Example 8.6 (page 206) continued. Refrigerator thermodynamic efficiency. *For the refrigerator in Example 8.6, we found* $\text{COP}_\text{C} = Q_C/W = 3.23$. *The maximum theoretical (reversible)* COP_C *is, in comparison,* $(Q_C/W)^{\text{rev}} = T_C/(T_H - T_C) = 278K/17K = 16.4$. *The thermodynamic efficiency is then*

$$\eta = \frac{(Q_C/W)}{(Q_C/W)^{\text{rev}}} = \frac{3.23}{16.4} = 0.197$$

Example 8.7 Electric heater. *Let us consider an electric heater which uses electric energy* W, *that is, "useful energy in"* $= W$ *(we use the symbol* W *to clearly show that electric energy can be converted 100% into work). The heater supplies heat* Q *to the room, that is, "useful energy out"* $= Q$. *Since all the supplied electric energy* W *is converted to heat, we have from the energy balance that* $Q = W$. *The electric heater therefore seems to have 100% efficiency, but this is only correct if we consider the* **energy efficiency**

$$\eta_{\text{energy}} = \frac{\text{useful energy out}}{\text{energy in}} = \frac{Q}{W} = \frac{W}{W} = 1$$

The thermodynamic efficiency defined by (8.11) is much lower, because we could have used the electricity much more effectively by using it to run a heat pump. Assume that the room temperature is $T_H = 293K$ *(20* ^oC*) while the outdoor temperature is* $T_C = 278K$ *(5* ^oC*). The "maximum useful energy out," which we theoretically could obtain by "pumping" heat from* T_C *to* T_H, *is from (8.4)* $Q_H = W/(1 - \frac{T_C}{T_H}) = W/(1 - \frac{278}{293}) = W/0.051$ *and we have that the* **thermodynamic efficiency** *is only 5.1%:*

$$\eta = \frac{\text{useful energy out}}{\text{ideal (maximum) useful energy out}} = \frac{W}{W/0.051} = 0.051$$

Note that if we instead used the electricity to power a heat pump, then the **energy efficiency** *(COP) would typically be around 4; see Example 8.4 (page 205).*

8.6 Ideal work and exergy

So far, the results in this chapter, including the Carnot factor for the maximum work that can be extracted from heat, have been for *closed systems* (e.g., a cyclic process) where only heat and work are exchanged with the surroundings. Can we extend our results to *open systems*? Yes, for a **steady-state continuous process** the answer is given by (8.19) below! With this result, we can find the maximum work that can be extracted from a process that operates between two given states. For example, we may want to find the maximum work that can be extracted from a process where chemical reactions take place, or the maximum work that can be extracted in a process where we mix two streams, or the maximum work that can be extracted from a stream at high pressure and temperature?

The result can be used to perform an exergy (or availability) analysis, or equivalently, an analysis of lost work. This kind of analysis is important for evaluating the efficiency and sustainability of a process, and is commonly used, for example, in **industrial ecology**.

8.6.1 Ideal work for open system

Consider a continuous steady-state process that takes a given stream from state 1 to state 2. The enthalpy and entropy changes are $\Delta H = H_2 - H_1$ and $\Delta S = S_2 - S_1$. Both of these changes are uniquely given by the initial and final states, and are consequently independent of the specific process. Furthermore, from the first law of thermodynamics (energy balance), we have for a steady-state continuous process, see (4.13),

$$\Delta H = W_s + Q$$

where W_s is work and Q is heat supplied from the surroundings to the system. (Note that we here have omitted terms related to kinetic and potential energy, and also electrochemical work.) However, work and heat are not state functions, and we want to find the **"ideal" work** for a reversible process, that is, the minimum work that we must supply, or equivalently, the maximum work that the system can perform. From the second law of thermodynamics, we have that the total entropy of the system and surroundings always increases, see (7.2):

$$\Delta S_{\text{total}} = \Delta S + \Delta S_{\text{sur}} \geq 0 \tag{8.15}$$

For an ideal (reversible) process, $\Delta S_{\text{total}} = 0$, and together with the first law, this can be used to find the ideal work. However, an unsettled point remains, what are the surroundings? Here we consider our "natural" ("dead") surroundings. The temperature T_0 of the surroundings varies around the world, but unless anything else is said, we choose $T_0 = 298.15$ K (25 °C). With the assumption that **the surroundings have constant temperature** T_0, the entropy change in the surroundings is given by, see (7.6),

$$\Delta S_{\text{sur}} = \Delta S_0 = \frac{-Q}{T_0} \tag{8.16}$$

where Q is supplied heat from the surroundings to the process. Inserted in the second law (8.15), we get

$$Q \leq T_0 \Delta S \tag{8.17}$$

which, inserted in the first law (energy balance) for a steady-state continuous process, $\Delta H = Q + W_s$, gives

$$W_s \geq \Delta H - T_0 \Delta S \tag{8.18}$$

In other words, we have for a **steady-state continuous process** that $W_s \geq W_s^{\text{id}}$, where the ideal (reversible) work with surroundings at constant temperature T_0 is

$$\boxed{W_s^{\text{id}} \triangleq \Delta H - T_0 \Delta S} \quad [\text{J/s; J}] \tag{8.19}$$

- The expression also holds for a batch process operating between given states (feed and product).
- W_s^{id} is the "minimum work that must be supplied to the process," and equivalently $(-W_s^{\text{id}})$ is the "maximum work that can be extracted from the process."
- Enthalpy and entropy are state functions, so the values of ΔH and ΔS can be easily calculated from given stream data for the process that we are analyzing. Note that these values will not depend on the choice of surroundings.

8.6.2 Exergy

To make use of the important result in (8.19), it is convenient to define a new state function, called **exergy (availability)**,

$$B \triangleq E + pV - T_0 S \tag{8.20}$$

where E is the system's total energy. In this chapter, we consider cases where only the internal energy contributes to the energy. That is, *we neglect contributions kinetic and potential energy* (but these contributions may easily be added, if desired). Then $E = U$, and the exergy is

$$B = H - T_0 S \tag{8.21}$$

This is close to the definition of *Gibbs energy*, $G = H - TS$, but the ambient temperature T_0 is used instead of the system's temperature T. For a **steady-state continuous process**, we then have from (8.19) that

$$\Delta B = W_s^{\text{id}} = \Delta H - T_0 \Delta S \tag{8.22}$$

That is, for a process operating between given states, the change in the system's exergy equals the maximum work we can theoretically extract with surroundings at constant temperature T_0. By using a common reference for all components (e.g., the elements), we can also calculate the exergy change for processes with chemical reactions.

8.6.3 Ideal work for chemical reactions

To extract the maximum amount of work, the process needs to be close to reversible. Unfortunately, this is usually difficult for potentially favorable *chemical reactions* with a large negative ΔG (and thus a large negative ΔB, meaning that we can theoretically extract a lot of work). To operate the reaction reversibly, we need to somehow "hold the reaction back". This is possible if we use a **fuel cell** (open system) or **battery** (which is the batch version of fuel cell), where we use the opposing potential of the external electrical circuit to hold the reaction back.

In theory, fuel cells and batteries can obtain 100% thermodynamic efficiency for conversion of the chemical (Gibbs) energy in fuels to work (see page 226). Thus, they are not limited by the Carnot factor which applies if we first transform the chemical energy (heat of reaction) to heat ($Q_H = \Delta H$) and then extract work from the heat. Nevertheless, fuel cells have losses, e.g., due to internal resistance, so the actual efficiency is usually less than 70%.

8.6.4 Lost work

An exergy analysis is used to compare the process' change in exergy ΔB (which is a state function) with the actual net) supplied work W_s (which depends on the given process), and we have that the **lost work** (=the exergy loss) is

Lost work = Actual work − Ideal work = Actual work − Exergy change

or

$$W_{\text{lost}} = W_s - W_s^{\text{id}} = W_s - \Delta B = W_s - \Delta H + T_0 \Delta S \quad (8.23)$$

Let us provide some insight into this expression. By introducing $W_s = \Delta H - Q$ (first law for steady-state process), $-Q = T_0 \Delta S_0$ (entropy change for the surroundings) and $W_s^{\text{id}} = \Delta H - T_0 \Delta S$, the lost work is

$$W_{\text{lost}} = W_s - W_s^{\text{id}} = T_0(\Delta S + \Delta S_0) = T_0 \Delta S_{\text{total}} \geq 0 \quad (8.24)$$

Thus, as expected, the lost work is always positive and it is caused by irreversibilities which make $\Delta S_{\text{total}} > 0$. Or, in other words, any irreversibility in a process carries with it a price in terms of lost work.

Example 8.8 *In a heat exchanger, the heat $Q = 500\ kW$ is transferred from the hot side (stream 1, which is condensing steam at constant temperature $140^o C$) to the cold side (stream 2, which is evaporating water at constant temperature $110^o C$). From the steady-state energy balance, the enthalpy changes for the streams are*

$$\Delta H_1 = -Q = -500kW; \quad \Delta H_2 = Q = 500kW$$

and since $\Delta_{\text{trs}} S = \Delta_{\text{trs}} H / T_{\text{trs}}$ (7.8) for a phase transition at constant temperature, the entropy changes for the streams are

$$\Delta S_1 = \frac{\Delta H_1}{T_1} = \frac{-500}{273.15 + 140} = -1.210kW/K; \Delta S_2 = \frac{\Delta H_2}{T_2} = \frac{500}{273.15 + 110} = 1.305kW/K$$

The exergy changes for the streams are then

$$\Delta B_1 = \Delta H_1 - T_0 \Delta S_1 = -500 - 298.15 \cdot (-1.210) = -139.2kW$$

$$\Delta B_2 = \Delta H_2 - T_0 \Delta S_2 = 500 - 298.15 \cdot 1.305 = 110.9kW$$

For the heat exchanger (overall process) we then have

$$\Delta H = \Delta H_1 + \Delta H_2 = 0$$

$$\Delta S = \Delta S_1 + \Delta S_2 = -1.210 + 1.305 = 0.095kW/K$$

$$\Delta B = \Delta H - T_0 \Delta S = \Delta B_1 + \Delta B_2 = -139.2 + 110.9 = -28.3kW$$

and the lost work is $W_{\text{lost}} = -\Delta B = 28.3\ kW$. The reason for the lost work is irreversibility caused by the temperature difference of $30^o C$ between the hot and cold sides of the heat exchanger.

8.6.5 Equivalent work for process with heat transfer at temperatures different from T_0

This subsection is important for practical analysis, in particular of subprocesses! We have assumed that the surroundings are at constant temperature T_0, where we usually select $T_0 = 298.15$ K. How do we evaluate the ideal work (or exergy or lost work) for a process or subprocess that exchanges heat Q_i with "non-standard surroundings" (e.g., with a utility, a reservoir, another process or another part of the same process) with a temperature $T_i \neq T_0$? For the process we are analyzing, it is reasonable to assume that the heat integration between the "non-standard surroundings" (at T_i) and the standard surroundings (at T_0) is ideal. This corresponds to assuming that the heat supply Q_i at temperature T_i corresponds to a supplied ideal **"equivalent work"** as given by the Carnot factor (8.4),

$$W'_{s,i} = Q_i \left(1 - \frac{T_0}{T_i}\right) \tag{8.25}$$

(because this is the ideal work needed to generate the heat Q_i at temperature T_i, by taking heat from the surroundings at temperature T_0).

Thus, in the above expressions (8.23)-(8.24), we should replace W_s by $W_{\mathrm{s,tot}}$, where

$$W_{\mathrm{s,tot}} = W_s + \sum_i W'_{s,i} \tag{8.26}$$

For example, the expression for the lost work becomes

$$W_{\mathrm{lost}} = \underbrace{W_s + \sum_i W'_{s,i}}_{W_{\mathrm{s,tot}}} - W_s^{\mathrm{id}} \tag{8.27}$$

Here the ideal work (= change in exergy), $W_s^{\mathrm{id}} = \Delta B = \Delta H - T_0 \Delta S$, is as before determined by the inlet and outlet streams of the process (system) we are considering.

8.6.6 Exergy efficiency

For a process where we extract work ($W_s^{\mathrm{id}} < 0$, $W_s < 0$), the exergy efficiency is defined as

$$\eta_{\mathrm{exergy}} = \frac{W_s}{W_s^{\mathrm{id}}} = \frac{1}{1 + \frac{W_{\mathrm{lost}}}{|W_s|}} \tag{8.28}$$

Similarly, for a process where we have to supply work ($W_s^{\mathrm{id}} > 0$, $W_s > 0$), we define

$$\eta_{\mathrm{exergy}} = \frac{W_s^{\mathrm{id}}}{W_s} = 1 - \frac{W_{\mathrm{lost}}}{|W_s|} \tag{8.29}$$

Note that W_s should be interpreted as $W_{\mathrm{s,tot}}$ in (8.26) for cases where the process exchanges heat with non-standard surroundings (with a temperature different from T_0). In almost all cases, the exergy efficiency equals the previously defined thermodynamic efficiencies, see (8.11) and (8.12), and any possible differences are of no practical importance.

Example 8.9 *(a) A system receives 500 kW of energy by heat exchange with a hot utility stream (non-standard surroundings no. 1), which in this case is condensing low pressure (3.8 bar) steam at 140° C. For the hot utility, we then have* $T_1 = 140 + 273.15 = 413.15K$ *and* $Q_1 = 500\ kW$*, and we want to find the corresponding equivalent work supplied to the system.*

Solution. *The Carnot factor is* $\left(1 - \frac{T_0}{T_1}\right) = (1 - \frac{298.15}{413.15}) = 0.278$ *and the equivalent work is*

$$W'_{s,1} = Q_1 \left(1 - \frac{T_0}{T_1}\right) = 500 \cdot 0.278 = 139.2kW$$

(b) Assume now that the system is an evaporating liquid at constant temperature $T_2 = 383.15K$ *(110° C). We want to find the lost work.*

Solution. *From Example 8.8, the exergy change for the system (stream 2) is* $\Delta B_2 = \Delta H_2 - T_0 \Delta S_2 = 500 - 298.15 \cdot 1.305 = 110.9\ kW$*, and the lost work is* $W_{\text{lost}} = W_{\text{s,tot}} - \Delta B = 139.2 - 110.9 = 28.3\ kW$*, which is the same as we found in Example 8.8.*

(c) Finally, we want to find the exergy efficiency of the heat exchanger in terms utilizing the hot utility. We have $W_{\text{s,tot}} = W'_{s,1} = 139.2\ kW > 0$*, and from (8.29) we get* $\eta_{\text{exergy}} = 1 - W_{\text{lost}}/|W_{\text{s,tot}}| = 1 - \frac{28.3}{139.2} = 0.797$.

8.6.7 Summary of exergy analysis

To perform an exergy analysis of a process we need the following information:

1. Enthalpy and entropy of all streams to find $W_s^{\text{id}} = \Delta B = \Delta H - T_0 \Delta S$.
2. Actual work W_s for compressors, expanders, turbines, etc.
3. Heat transfer Q_i to "non-standard" surroundings (or other processes) that are at a temperature $T_i \neq T_0$. This is used to find the equivalent work, $W'_{s,i}$, in (8.25).

The lost work is then from (8.27)

$$W_{\text{lost}} = W_s + W'_s - \Delta B \tag{8.30}$$

Comments on exergy analysis

1. When comparing the overall efficiency of alternative processes, the processes' inlet and outlet streams are often the same, such that $\Delta B = W_s^{\text{id}}$ is fixed. The effectiveness of the processes can then be compared by evaluating the work $W_{\text{s,tot}} = W_s + W'_s$ for the alternative processes.
2. From (8.22) and (8.24), we see that exergy, ideal work, lost work and entropy production are directly linked to each other. One therefore uses the term "exergy analysis" as a generic term which also includes analysis of ideal work, lost work and entropy loss (irreversibility).
3. The term **exergy** was introduced by the German scientist Rant in 1956, and it is the same as the terms

   - Availability
   - Available energy

   Furthermore, the **change in exergy** is the same as

   - Ideal work
   - Maximum extractable (obtainable) work
   - Available work

   and the **exergy loss** (= actual work − exergy change) is the same as the **lost work**.

4. Since the energy is constant (first law), it makes, strictly speaking, no sense to talk about an "energy problem" or "energy shortage." On the other hand, it makes sense to say we have a "exergy problem" because there is too little "useful" energy, such as electric power or oil.
5. An exergy analysis is independent of prices, and is in particular of energy prices. In a way, it can be said that the exergy says something about what the *energy price "ought to be"* in an "ideal" world, where we try to conserve resources. The price p for different energy forms should reflect the exergy content, that is, the price of a specific energy source should be $p_E = kB/E$, where k [\$/J] is the price of 1 J of "pure" exergy (e.g., electricity), and B/E is the exergy fraction in the energy source.
6. An exergy analysis can yield interesting information, but can also be misleading because it does not say anything about the possibility or cost of reducing the exergy loss. Typically, one finds that the exergy loss is large in the reactor, where an irreversible chemical reaction takes place (with a large negative value of $\Delta H - T_0 \Delta S$), but this loss is usually difficult to avoid.
7. Usually, we consider exergy *changes*, $\Delta B = \Delta H - T_0 \Delta S$, and this is sufficient for analyzing individual processes. What about the **"absolute" exergy** B of a system or a stream – does it have any meaning? The answer is "yes," provided one uses our "natural" surroundings as the "reference state." This corresponds to choosing $T_0 = 298$ K and $p_0 = 1$ bar (like we have already done), and for the components, we choose their "natural state" as reference. For example, air is used as reference for N_2, O_2 and Ar; pure water as pure liquid is used as reference for water, etc. Note that this is not the same as the "usual" reference state, which is the elements. The exergy is then zero for the natural surroundings ($B_0 = 0$), and the "absolute" exergy B of a stream is then the maximum work that one can extract from the stream with the natural surroundings as the final state.
   However, the absolute exergy is often of limited practical interest, because it is generally unrealistic to extract all of the "potential" (available) work in a process stream.

8.6.8 Further examples and uses of exergy analysis

Example 8.10 Thermodynamic analysis of refrigerator. *This is a continuation of Example 8.6 (see page 206 and Figure 8.7), and the data for the streams and subprocesses are summarized in Table 8.1. The enthaply and entropy data are obtained from the pH-diagram for ammonia on page 419.*

The values for the exergy B and ideal work $W_s^{\rm id} = \Delta B$ are based on the standard surroundings at $T_0 = 298.15$ K. However, since the process actually operates between surroundings at T_C (refrigerator at 5° C) and T_H (air at 22° C), an equivalent work W_s' (8.25) for the cooling and heating subprocesses must be added to W_s,

$$\text{Subprocess b}: \quad W'_{s,H} = Q_H\left(1 - \frac{T_0}{T_H}\right) = -1453\left(1 - \frac{298.15}{295.15}\right) = 14.8\text{kJ/kg}$$

$$\text{Subprocess d}: \quad W'_{s,C} = Q_C\left(1 - \frac{T_0}{T_C}\right) = 1110\left(1 - \frac{298.15}{278.15}\right) = -79.8\text{kJ/kg}$$

where from energy balances we have $Q_H = H_3 - H_2$ and $Q_C = H_1 - H_4$.

We note that the largest lost work is in the condenser (subprocess b) because of the large temperature difference between the hot vapor and the surrounding temperature. This seems difficult to avoid because the surroundings have a constant temperature at 22° C (T_H), whereas the hot gas exiting the compressor is at 150° C (stream 2). However, if we introduce cooling during the compression (subprocess a) then the temperature of the hot exit gas will be lower than 150° C and the losses in the condenser (subprocess b) can be considerably reduced. This

Table 8.1: Stream and subprocess data for ammonia refrigeration cycle

Stream (state)	Phase	t ^{o}C	p bar	H kJ/kg	S kJ/kg K	$B = H - T_0 S$ kJ/ kg
1	g	−10	3	1350	5.76	−367.3
2	g	150	16	1693	5.98	−89.9
3	l	30	16	240	1.47	−198.3
4	l/g	−10	3	240	1.53	−216.2

Subprocess	Ideal work $\Delta B = W_s^{\text{id}}$ kJ/kg	Equiv. work W_s' kJ/kg	Actual work W_s kJ/kg	Lost work $W_s + W_s' - \Delta B$ kJ/kg
a: Compressor ($1 \to 2$)	277.4	0	343	65.6
b: Condenser ($2 \to 3$)	−108.4	14.8	0	123.2
c: Choke valve ($3 \to 4$)	−17.9	0	0	17.9
d: Evaporator ($4 \to 1$)	−151.1	−79.8	0	71.3
Sum	0	−65	343	278

is a typical example where we apparently have large losses in one subprocess (b), but where these can be reduced by changing another subprocess (a). From this it is clear that one cannot expect to achieve the full improvements in efficiency by analyzing subprocesses individually.

Phase change and subambient temperatures. The exergy change for the evaporation or condensation of a pure component is given by

$$\Delta B = \Delta_{\text{vap}} H - T_0 \Delta_{\text{vap}} \quad \text{where from (7.8)} : \Delta_{\text{vap}} S = \frac{\Delta_{\text{vap}} H}{T_b}$$

Here $\Delta_{\text{vap}} H$ is the enthalpy change for the process and T_b is the boiling point at the given pressure (i.e., $T = T_b$). Thus, the exergy change for phase transition (here boiling) is

$$\Delta B = \Delta_{\text{vap}} H(T_b) \cdot \left(1 - \frac{T_0}{T_b}\right) \tag{8.31}$$

Remark 1 As expected, the exergy change (= the work that can be extracted) is given by the heat of vaporization (or condensation) $\Delta_{\text{vap}} H$ multiplied by the Carnot factor $\left(1 - \frac{T_0}{T}\right)$.

Remark 2 Note that the exergy ΔB changes sign compared to $\Delta_{\text{vap}} H$ as the process temperature $T = T_b$ passes the surrounding's temperature T_0. Thus, a condensing vapor ($\Delta H < 0$) can perform work (relative to T_0) when $T > T_0$, but not when $T < T_0$ (rather, it requires work to provide cooling at subambient temperatures). The reverse applies to evaporation. For example, we found in the previous example (see Table 8.1) a negative exergy change of −151.1 kJ/kg in the evaporator (which means that work can theoretically be extracted) because the temperature is −10°C, which is well below T_0 = 298.15 K (=25°C).

Remark 3 Because of the Carnot factor, it is better to condense at high temperatures. However, there is also a competing effect because $\Delta_{\text{vap}} H(T_b)$ is smaller at high temperatures, and eventually becomes zero at the critical point.

Example 8.11 Maximum (ideal) work from combustion of methane. *What is the maximum amount of (electrical) work that we can extract by combustion of methane? We assume that we have available a mixture of methane and air (which contains oxygen) that burns according to the following chemical reaction*

$$CH_4(g) + 2O_2(g) = CO_2(g) + 2H_2O(g)$$

Note that the water product (H_2O) is here assumed to be in gas form, corresponding to the **lower heating value** *(lower heat of combustion). With surroundings at constant temperature T_0, the ideal work is $W_s^{\rm id} = \Delta H - T_0\Delta S$, where ΔH and ΔS are the changes in enthalpy and entropy for the reaction. Note that if we* assume that the feed and product are ideal gases at $T_0 = 298$K and 1 bar, *then $\Delta H - T_0\Delta S$ is the standard Gibbs energy for the reaction at $T_0 = 298$ K, that is, we have*

$$W_s^{\rm id} = \Delta_{\rm r}G^{\ominus}(298) \tag{8.32}$$

For combustion of methane, we get

$$\begin{aligned} \Delta_{\rm r}G^{\ominus}(298) &= G_{CO_2(g)} + 2G_{H_2O(g)} - G_{CH_4(g)} - 2G_{O_2(g)} \\ &= -394.36 + 2\cdot(-228.57) + (-50.72) - 2\cdot 0 = -800.78 \text{ kJ/mol} \end{aligned}$$

where we have used data for the standard Gibbs energy of formation from page 416. Thus,

$$W_s^{\rm id} = \Delta_{\rm r}G^{\ominus}(298) = -800.78\frac{\text{kJ}}{\text{mol}} = -49.92\frac{\text{MJ}}{\text{kg methane}}$$

Conclusion: *The maximum work that can be extracted by combustion of methane (e.g., in a fuel cel) is 49.92 MJ/kg methane. This is the "lower" value with the water product in gas form, and we will use this value later for evaluating the efficiency of a natural gas power plant.*

Comments on ideal work (exergy) for combustion of fossil fuels*:*

1. **Gibbs energy versus enthalpy of reaction.** *As derived in (8.32), one should use the Gibbs energy when computing the maximum work for chemical reactions. However, if you read literature on power plants, you will find that they use the heat of reaction (combustion) ($\Delta_{\rm r}H^{\ominus}$) rather than the Gibbs energy ($\Delta_{\rm r}G^{\ominus}$). Fortunately, the error is of no practical significance for combustion of fossil fuels. For example, for combustion of methane, the Gibbs energy of reaction, $\Delta_{\rm r}G^{\ominus} = -800.78$ kJ/mol $= -49.92$ MJ/kg, is approximately equal to the heat of reaction, $\Delta_{\rm r}H^{\ominus} = -802.34$ kJ/mol $= -50.02$ MJ/kg (these are the "lower" values with H_2O (g) as the product). The reason for this small difference is that the entropy change for combustion of methane is $\Delta_{\rm r}S^{\ominus}(298) = -5.23$ J/mol K, which is a small value.*
2. **Combustion of coal***: For the combustion of carbon (graphite), $C(s) + O_2(g) = CO_2(g)$ (which is the same as the formation reaction for CO_2), we have $\Delta_{\rm r}G^{\ominus} = -393.51$ kJ/mol which again is close to $\Delta_{\rm r}H^{\ominus} = -394.36$ kJ/mol. Thus, for combustion of fossil fuels in general, the (lower) ideal work is approximately equal to the (lower) heat of combustion.*
3. *Additional work can theoretically be extracted by condensing the water in the product(flue gas), but only about 2%. This follows since maximum "higher" work (with water as a liquid product) is $W_s^{\rm id} = \Delta_{\rm r}G^{\ominus}(298) = -50.99$ kJ/kg methane, which is only about 2% higher than the "lower" value of -49.92 kJ/kg methane.*
4. *In theory (yes, this is really theoretical) some additional work can be extracted in the step where we mixing the flue gas (CO_2 and H_2O) with air (see Exercise 7.8 page 168).*

Example 8.12 Minimum work for LNG cooling process. *We want to calculate the minimum work to produce 1 kg liquefied natural gas (LNG) when it is assumed that the* *surroundings are at (a) $T_0 = 273$ K (arctic) and (b) $T_0 = 298$ K (tropics); the feed is methane gas at 10°C / 1 bar and the product is methane liquid at −164°C/1 bar.*

Solution. *The ideal work equals the change in exergy*

$$W_s^{\rm id} = \Delta B = \Delta H - T_0 \Delta S$$

From the pH-diagram for methane, the change in enthalpy and entropy when going from 10°C/1 bar to -164°C/1 bar is

$$\Delta H = -280kJ/kg - 570kJ/kg = -850\ kJ/kg$$
$$\Delta S = 5.2kJ/kgK - 11.4kJ/kgK = -6.2\ kJ/kgK$$

The ideal (minimum) work for producing liquefied natural gas (LNG) is then
(a) $T_0 = 273K$: $W_s^{\rm id} = \Delta H - T_0\Delta S = -850 + 273 \cdot 6.2 = 843\ kJ/kg$
(b) $T_0 = 298K$: $W_s^{\rm id} = \Delta H - T_0\Delta S = -850 + 298 \cdot 6.2 = 998\ kJ/kg$
(this is a positive number, which means that we must supply work).

Mixing is an irreversible process, and with surroundings at temperature T_0, the lost work is generally given by (8.23),

$$W_{\rm lost} = W_s - W_s^{\rm id} = W_s - \Delta H + T_0 \Delta S$$

where the Δ represents the change from the inlet to the outlet stream. The energy balance for a steady-state process is $\Delta H = Q + W_s$, and since a mixing process is adiabatic ($Q = 0$) and has no work ($W_s = 0$), the energy balance gives $\Delta H = 0$. The lost work in a steady-state mixing process is then

$$W_{\rm lost} = T_0 \Delta S \tag{8.33}$$

Example 8.13 Ideal mixing of pure components. *From (7.12), the mixing entropy when mixing pure components at constant temperature and pressure is*

$$\Delta S = \Delta_{\rm mix} S = -R \sum_i n_i \ln x_i \quad [\rm J/K] \tag{8.34}$$

As an example assume that we mix 0.4 mol/s of component A with 0.6 mol/s of component B. Assuming an ideal mixture, the mixing entropy is

$$\Delta_{\rm mix} S = -R \sum_i n_i \ln x_i = -8.31(0.4 \ln 0.4 + 0.6 \ln 0.6)\ \rm J/K\ s = 5.6 J/K\ s$$

The lost work in an adiabatic steady-state mixing process is then

$$W_{\rm lost} = T_0 \Delta S = 298.15 \cdot 5.6 J/s = 1.67 \rm kJ/s$$

Example 8.14 Lost work by mixing in natural gas power plant. *In a natural gas power plant, two mixing processes take place, and we want to find the associated lost work. This is a continuation of Example 8.11, where we found that the maximum work that can be extracted by combustion of methane is −800.78 kJ/mol methane. We consider the following two mixing processes*

1. *Mixing of methane and air (to produce the feed) ($\Delta S_1 > 0$)*
2. *Mixing of combustion product with air (at the exit of the process) ($\Delta S_2 > 0$)*

Solution. *For the first mixing process, from Example 7.7 (page 167) the entropy change is* $\Delta S_1 = 27.44$ *J/K mol methane, and the lost work is* $W_{\text{lost},1} = T_0 \Delta S_1 = 298 \cdot 27.44 = 8.18$ *kJ/mol methane. Similarly, for the second mixing process from Example 7.8 (page 168) the lost work is* $W_{\text{lost},2} = T_0 \Delta S_2 = 298 \cdot 89.0 = 26.51$ *kJ/mol methane. The total lost work by mixing is 34.7 kJ/mol. This is 4.3% of the total available work of 800.78 kJ/mol. Unfortunately, it is difficult to eliminate these losses. However, in the next example, we consider a mixing process where it may be possible to extract the work.*

Example 8.15 Salt power plant. *It is theoretically possible to extract work during mixing, and this can be utilized in a salt power plant where we mix water (from a river) and seawater. Let us calculate the theoretical work that can be extracted by mixing 1 l of pure water (55.5 mol = 1000 g/18 g/mol) with 1 l of seawater (in theory, the maximum work is obtained by using an infinite amount of seawater, but using about equal amounts is more realistic in practice). The surroundings are assumed to have constant temperature* $T_0 = 298K$. *Assume that the salinity of seawater is 3.3%, that is 1 l seawater contains 33 g/l of salt (NaCl). How much can theoretically be extracted if we mix 1 l of pure water with 1 l of saltwater?*

Solution. *Since the molar mass of NaCl is 58.4 kg/kmol, 33 g/l corresponds to* 33 $kg/m^3/58.4\ kg/kmol = 0.565\ kmol/m^3 =$ 0.565 *mol/l of NaCl. However, when dissolved in water, NaCl splits in two ions,* Na^+ *and* Cl^-. *Thus, 1 l of seawater consists of approximately 0.565 mol/l* Na^+, *0.565 mol/l* Cl^-, *and 55.5 mol water. The corresponding mole fractions are 0.01 (*Na^+*), 0.01 (*Cl^-*) and 0.98 (*H_2O*). After mixing with 55.5 mol (1 l) of pure water, the mole fractions in the product ("brackish water") are approximately 0.005 (*Na^+*), 0.005 (*Cl^-*) and 0.99 (*H_2O*). From (7.14), the entropy change for this mixing 1 l of pure water (55.5 mol* H_2O*) with 1 l of seawater (55.5 mol* H_2O *+ 0.565 mol Na + 0.565 mol Cl) is*

$$\begin{aligned}\Delta S &= R\,[55.5\ln(1.0/0.99) + 55.5\ln(0.98/0.99) + 2\cdot 0.565\ln(0.01/0.005)] \\ &= R\,[0.5578 - 0.5635 + 2\cdot 0.3916] = 0.7776[\text{mol}]\text{R} = 6.46\text{J/K}\end{aligned}$$

That is, the entropy change is 6.46 J/K kg pure water. The first two terms almost cancel, so we note that, in practice, it is only the entropy change of salts (Na, Cl) that matter. The theoretical mixing work per kg pure water is

$$W_s^{\text{id}} = -T_0\Delta S = -298.15 \cdot 3.18 \quad \text{J/kg water} = -1926\text{J/kg water}$$

(which is negative since we theoretically can perform work). With a water flow of 1000 kg/s (1 m^3/s*), we can then theoretically extract 1.93 MW of work. For example, the Mississippi river has an average flow of 20 000* m^3/s *and it is then theoretically possible to produce about 39 GW work (electric power) by mixing it with a similar amount of seawater. This is about 10% of the current US electricity production.*

Note that the energy of 1926 J/kg water corresponds to an osmotic pressure (see page 382) of about 19 bar ($W = V\Delta p = (m/\rho)\Delta p$, *that is,* $\Delta p = W\rho/m = 1926J \cdot 1000kg/m^3/1kg = 19.26\cdot 10^5\,N/m^2 = 19.26$ *bar). A salt power plant utilizes the osmotic pressure of about 19 bar that is set up across a membrane with pure water on the one side and brackish water (product) on the other side. Due to the concentration difference, pure water moves naturally through the membrane from low pressure (1 bar) to high pressure (about 19 bar). This "pressure energy" can then be extracted as work in a water turbine. The seawater must be pumped up to 19 bar, but the energy for this can also be obtained by "exchanging" pressure with a similar amount of brackish water. In practice, efficiencies of up to 50% can be obtained in such a salt power plant, but this requires large membrane areas.*

Separation is the opposite of mixing, and if we have surroundings at constant temperature T_0, then we must *supply* a minimum work $W_s^{\text{id}} = \Delta H - T_0\Delta S$ to separate a mixture. In the examples below, we assume that the process is such that $\Delta H = H_{\text{out}} - H_{\text{in}} = 0$, that is, we concentrate on the contribution from the entropy change.

Example 8.16 Minimum work for separation of ideal mixture into pure components. *For an ideal mixture with the inlet and outlet streams at the same temperature and pressure and no phase change, the entropy for separating the mixture into its pure components is given by (7.12), but with the opposite sign, that is*

$$\Delta S = -\Delta_{\text{mix}} S = R \sum_i n_i \ln x_i \quad [\text{J/K}]$$

Furthermore, $\Delta H = 0$ (no phase change!) and we get

$$W_s^{\text{id}} = -T_0 \Delta_{\text{mix}} S; \quad \Delta_{\text{mix}} S = R \sum n_i \ln x_i \tag{8.35}$$

Here x_i is mole fraction of component i in the feed mixture. For example, the minimum work for separating 1 mol of a mixture of 40 mol-% A and 60 mol-% B into the pure components is $W_s^{\text{id}} = -T_0 \Delta S = -RT_0 \sum x_i \ln x_i = -8.31 \cdot 298.15 \cdot (0.4 \ln 0.4 + 0.6 \ln 0.6)$ J/mol = 1.67 kJ/mol, which is a positive number since work must be supplied for the separation.

Example 8.17 Minimum work for CO_2 removal (postcombustion separation). *We want to find the minimum work for removing CO_2 from the exhaust gas (flue gas) from a natural gas power plant. The temperature is assumed to be 298.15 K. From Example 8.21 (page 230) the following exhaust gas is obtained by combustion of 1 kg (62.5 mol) methane:*

$$1177 \text{ mol } N_2, \quad 169 \text{ mol } O_2, \quad 62.5 \text{ mol } CO_2, \quad 125 \text{ mol } H_2O, \quad \text{Total}: \ 1533.5 \text{ mol}$$

Solution. *We assume ideal gas. Then from (8.35) the minimum work for extracting pure CO_2 from this stream is at $T_0 = 298$ K:*

$$W_s^{\text{id}} = -T_0 \Delta S = -T_0 R \left(62.5 \ln \frac{62.5}{1533.5} + 1471 \ln \frac{1471}{1533.5} \right) = 0.64 \text{ MJ/kg methane}$$

This is a positive number, which means that work must be supplied. This ideal separation work is 1.2% of the ideal work of 50 MJ/kg obtainable by combustion of methane (see Example 8.11, page 219). In addition, we need to compress the CO_2 so that it can be stored or injected into the ground. If we assume that we must compress the CO_2 from 1 bar to 300 bar, then the associated minimum work for reversible isothermal compression of ideal gas at $T_0 = 298K$ is

$$W_s^{\text{id}} = nRT_0 ln(p_2/p_1) = RT_0 62.5 \ln(300/1) = 0.88 \text{ MJ/kg methane}$$

This is another 1.8% of the ideal work of 50 MJ/kg. In total, we loose in the ideal case 1.52 MJ/kg (3.0%) of the original ideal work.

Exercise 8.4* *In the above example, the exhaust gas was separated into two gas products (pure CO_2 and the remaining exhaust gas).*

(a) Find the ideal work for separating the exhaust gas into four pure gas components (CO_2, N_2, O_2, H_2O) (you should find that the work is about 5 times larger in this case).

(b) Find the ideal work for separating the exhaust gas into two gas products (pure CO_2 and a N_2/O_2-mixture) plus pure <u>liquid</u> water.

Exercise 8.5 Minimum work for separation of air (precombustion separation). *An alternative approach is to use pure oxygen instead of air for combustion of methane. The exhaust gas then contains only H_2O and CO_2 (and no N_2), which are easy to separate. Find the minimum separation work for producing from air the necessary amount of pure O_2 needed for combustion of 1 kg methane (you will find that the work is slightly larger than the 0.66 MW/kg needed for the post-combustion separation, but on the other hand, separation of air is probably an easier process to realize in practice).*

Example 8.18 Minimum work and energy for thermal separation process (reversible distillation). *The most common example of a thermal separation process is distillation. In a distillation column (see page 22) the feed stream (F) is separated into a "light" distillate product (D) and a "heavy" bottom product (B) by supplying heat in the bottom (Q_H) and removing heat (Q_C) by cooling in the top. Distillation is the most common method for separating liquid mixtures, and it is often claimed to have low efficiency. However, as shown in the following examples, this is not really true provided the distillation unit is properly heat-integrated.*

In this example, we derive general expressions for the minimum work and minimum energy (heat) for separation by (reversible) distillation. Actually, the process needs not be distillation, as the expressions apply to any thermal separation process where the "separation work" is supplied indirectly by supplying heat (Q_H) at high temperature (T_H) and removing heat (Q_C) at low temperature (T_C) .

Let us first find the minimum separation work, which with surroundings at temperature T_0 is given by, see (8.19),

$$W_s^{\text{id}} = \Delta H - T_0 \Delta S$$

Here $\Delta H = H_D + H_B - H_F$ and $\Delta S = S_D + S_B - S_F$ are computed from the stream data. If both the feed (F) and the products (D, B) are liquids at their boiling points, then it is a good assumption to set $\Delta H \approx 0$. We then have

$$W_s^{\text{id}} \approx -T_0 \Delta S \tag{8.36}$$

The main contribution is then from entropy change, which is negative for a separation process ($\Delta S < 0$) which means that W_s^{id} is positive, that is, work must be supplied to facilitate the separation.

However, no work is supplied to a distillation column, i.e., $W_s = 0$, yet a separation takes place. How is this possible? The answer is that the supply of energy (heat) at a high temperature provides an indirect supply of work, because the energy could have been used to produce work, for example, in an ideal heat engine. Specifically, as shown in (8.25), the work that is indirectly supplied to the column when supplying the heat $Q_H > 0$ at (high) temperature T_H is

$$W'_{s,H} = Q_H \left(1 - \frac{T_0}{T_H}\right) \tag{8.37}$$

Correspondingly, the work that is indirectly performed by the column by removing heat $Q_C < 0$ at (low) temperature T_C is

$$W'_{s,C} = Q_C \left(1 - \frac{T_0}{T_C}\right) \tag{8.38}$$

The energy balance for distillation columns gives $\Delta H = W_s + Q_H + Q_C$, where $W_s = 0$ and we have already assumed that $\Delta H = 0$. The energy balance then gives $Q_C = -Q_H$, that is, the heat Q_C removed in the condenser is the same as the heat Q_H supplied in the reboiler. The net work that is indirectly supplied to the process due to supply of heat (Q_H and Q_C) is then

$$W_{s,\text{tot}} = W'_{s,C} + W'_{s,H} = Q_H T_0 \left(\frac{1}{T_C} - \frac{1}{T_H}\right) \tag{8.39}$$

This expression can now be used to find the minimum energy (heat) $Q_H^{\min}$ that must be supplied to a distillation process, assuming that the separation process itself is reversible. We start by noting that from (8.27) and (8.36), the lost work for the process is

$$W_{\text{lost}} = W_{\text{s,tot}} - W_s^{\text{id}} = W_{\text{s,tot}} + T_0 \Delta S$$

The minimum heat $Q_H^{\min}$ is obtained when $W_{\text{lost}} = 0$. This gives $Q_H^{\min} T_0 \left(\frac{1}{T_C} - \frac{1}{T_H}\right) + T_0 \Delta S = 0$, and we have that the minimum supplied heat for an ideal (reversible) distillation process operating between temperatures T_H and T_C is

$$Q_H^{\min} = \frac{-T_C \Delta S}{1 - \frac{T_C}{T_H}} \tag{8.40}$$

(note that T_0 drops out). Since the distillation process is reversible, the heat transfer is also reversible with no temperature difference between the system (the column) and the surroundings. Thus, we have that T_C is the temperature in the condenser (= boiling temperature of the "light" distillate product), and T_H is the temperature in the reboiler (= boiling temperature of the "heavy" bottom product).

Comment: It is interesting to note that with $T_C = T_0$, $Q_H^{\min}$ is the minimum separation work $W_s^{\text{id}} (= -T_0 \Delta S)$ divided by the Carnot factor $1 - T_C/T_H$.

Example 8.19 Minimum energy for separating methanol-water by thermal separation process (reversible distillation). *Here, we apply the equations derived in the previous example to a specific case. We want to find the minimum energy (heat) needed to separate 1 mol of ideal mixture of 40 mol-% methanol and 60 mol-% water into the pure components using distillation. For the separation, we have available energy (heat) at $T_H = 373K$ and cooling at $T_C = 338$ K, which are the boiling points for the two pure components at 1 bar.*

Solution. *For an ideal mixture we have from (8.35) that*

$$\Delta S = R \sum x_i \ln x_i = R\ (0.4 \ln 0.4 + 0.6 \ln 0.6) = -0.67R = 5.57 \text{ J/mol K}$$

The Carnot factor is $1 - \frac{T_C}{T_H} = 1 - \frac{338}{373} = 0.094$ and from (8.40) the minimum heat for reversible distillation is $Q_H^{\min} = 338 \cdot 5.57/0.094$ J/mol = 20.0 kJ/mol.

Example 8.20 Efficiencies of real distillation process. *Finally, let us consider a real distillation process that separates a feed mixture of 40 mol-% methanol and 60 mol-% water into (almost) pure components. From Example 8.19, the minimum energy (heat) is $Q_H^{\min} = 20.0$ kJ/mol for a reversible process that uses heat at $T_H = 373$ K and cooling at $T_C = 338$ K as the "separation agent" to separate the mixture. For a conventional distillation column (see page 22), which is the most common case of a thermal separation process, the heat required for 1 mol of a liquid feed using an infinite number of separation stages is*[5]

$$Q_H^{\text{min.dist}} = \Delta_{\text{vap}} H \left(x_L + \frac{1}{\alpha - 1}\right) \quad [\text{J/mol}]$$

where x_L is the mole fraction of light component in the feed, $\Delta_{\text{vap}} H$ is the average heat of vaporization and α is the relative volatility between the components. For our methanol-water

[5] In practice, one may get close to this value with a finite number of distillation stages. For more on distillation see, for example, I.J. Halvorsen, S. Skogestad: "Distillation Theory," *Encyclopedia of Separation Science*, D. Wilson (Editor-in-chief), Academic Press, 2000 (available at S. Skogestad's homepage).

mixture, we have $x_L = 0.4$, $\Delta_{\text{vap}}H = 40$ *kJ/mol and* $\alpha = 3.8$ *(estimated from (7.36)). This gives* $Q_H^{\text{min.dist}} = 30.3$ *kJ/mol. In a real column, with a finite number of stages, one should expect about 5-10% higher energy consumption, so let us in the following use for our real distillation process,*

$$Q_H = 1.05 Q_H^{\text{min.dist}} = 31.8 \text{ kJ/mol}$$

By defining "useful energy in" as "heat supply" Q_H*, the* **thermodynamic efficiency** *(8.12) for distillation of this feed mixture is then*

$$\eta = \frac{\text{ideal (minimum) heat supply}}{\text{actual heat supply distillation}} = \frac{Q_H^{\text{min}}}{Q_H} = \frac{20 \text{ kJ/mol}}{31.8 \text{ kJ/mol}} = 0.63$$

Thus, distillation as a separating process has a good thermodynamic efficiency (63% in this case). On the other hand, the **"energy efficiency"** *(8.13) for the distillation process obtained by defining "useful energy out" as "separation work" is only 5.2%:*

$$\eta_{\text{energy}} = \frac{\text{separation work}}{\text{heat supply distillation}} = \frac{W_s^{\text{id}}}{Q_H} = \frac{1.66 \text{ kJ/mol}}{31.8 \text{ kJ/mol}} = 0.052$$

Here, we have used that the ideal separation work for our feed mixture is $W_s^{\text{id}} = -T_0\Delta S = -T_0 R(0.4 \ln 0.4 + 0.6 \ln 0.6) = 1.66$ *kJ/mol. The low energy efficiency of 5.2% seems to indicate that distillation is a poor separation process, but this is misleading. Indeed, distillation uses a lot of heat (*Q_H *is large) and would be inefficient if we made no use of the heat removed by cooling (*Q_C*). However, if the distillation column is ideally heat integrated, for example using a heat pump between the reboiler (*Q_H*) and condenser (*Q_C*), then the thermodynamic efficiency of a conventional distillation process is 63% for this specific mixture.*

Comment. *We derive, as expected, the same thermodynamic efficiency of 63% if we instead of heat supply consider work, and set the "actual work" as the "total equivalent work" (*$W_{\text{s,tot}}$*):*

$$\eta = \frac{\text{minimum work}}{\text{actual work}} = \frac{W_s^{\text{id}}}{W_{\text{s,tot}}} = \frac{-T_0\Delta S}{Q_H T_0\left(\frac{1}{T_C} - \frac{1}{T_H}\right)} = \frac{1.66 \text{ kJ/mol}}{2.63 \text{ kJ/mol}} = 0.63$$

The lost work $W_{\text{lost}} = W_{\text{s,tot}} - W_s^{\text{id}}$ *= 2.63 kJ/mol - 1.66 kJ/mol = 0.97 kJ/mol is due to irreversibilities inside the distillation column caused by mixing of streams with different composition and temperature.*

8.7 Gas power plant

Here, you will learn to design a natural gas power plant "by hand," which also gives a very good review of subjects from this and previous chapters.

The purpose of a power plant is to extract work (produce electricity), that is, we have that "useful energy out" = "(net) work out," and the thermodynamic efficiency is

$$\eta = \frac{\text{(net) work out}}{\text{ideal (maximum) work out}} = \frac{|\text{W}_\text{s}|}{|\text{W}_\text{s}^{\text{id}}|}$$

We found in Example 8.11 (page 219) that, with surroundings at $T_0 = 298$K, the maximum work that can be extracted by combustion is $|W_s^{\text{id}}| = |\Delta_r G^\ominus(298)|$.[6] In

[6] The efficiency of a thermal power plant is usually based on $\Delta_r H^\ominus(298)$ (heat of combustion) rather on $\Delta_r G^\ominus(298)$, but this does not matter much because we happen to have $\Delta_r G^\ominus(298) \approx \Delta_r H^\ominus(298)$ for combustion; see comment on page 219. For our natural gas power plant, we use $\Delta_r G^\ominus(298) \approx \Delta_r H^\ominus(298) \approx -50$ MJ/kg, which is the value for methane.

theory, this work can be obtained in a reversible **fuel cell** that directly extracts electricity. However, at present there exists no combustion fuel cell for fossil fuels.

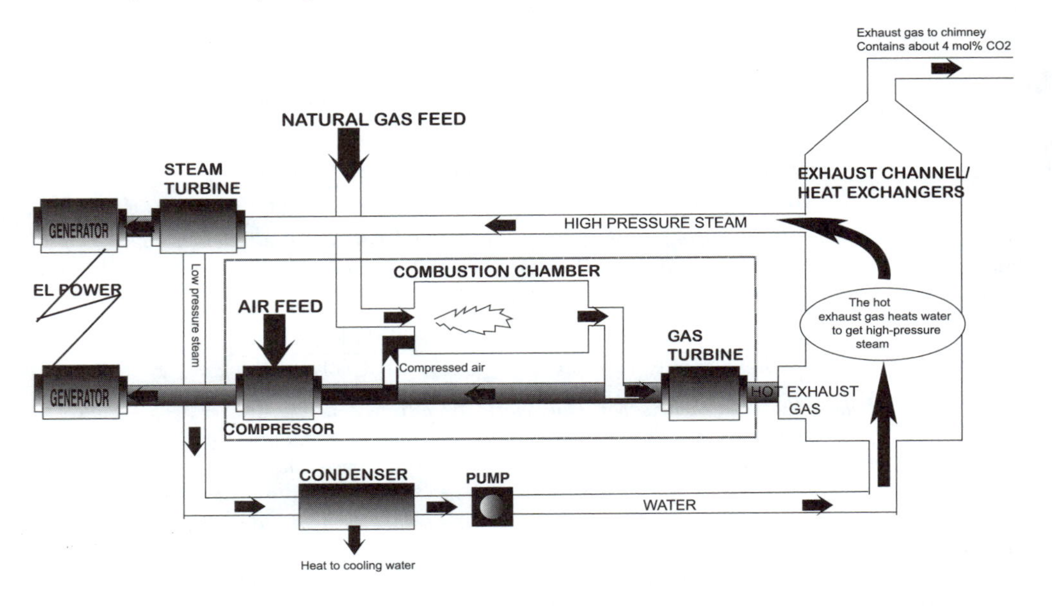

Figure 8.8: Schematic diagram of a natural gas power plant

In practice, one uses a *thermal power plant* where the chemical energy ($\Delta_r G$) is first converted to heat at high temperature and pressure, and then work is extracted using turbines. A schematic diagram of "combined cycle" natural gas power plant is shown in Figure 8.8. It consists of:

- **The combustion gas turbine part** (inside dashed box in Figure 8.8). Most of the work (electricity) is extracted by expanding the combustion gas, at high temperature and high pressure, in a gas turbine.
- **The steam turbine part.** Figure 8.8 is called a **combined cycle** power plant, because following the combustion gas turbine, there is a steam turbine part (**Rankine cycle**) where additional work is extracted by using the hot exhaust gas (combustion product from gas turbine) to generate high pressure steam which is expanded in a steam turbine to produce additional work.

In such a combined cycle natural gas power plant, it is possible to obtain a total thermodynamic efficiency η of more than 60%.

8.7.1 The combustion gas turbine process

A *gas turbine* (section) consists of (see also Figure 8.9)

- A compressor, where air is compressed
- A combustion chamber, where air and fuel are mixed and burned
- A turbine, where the combustion gas is expanded and work is extracted

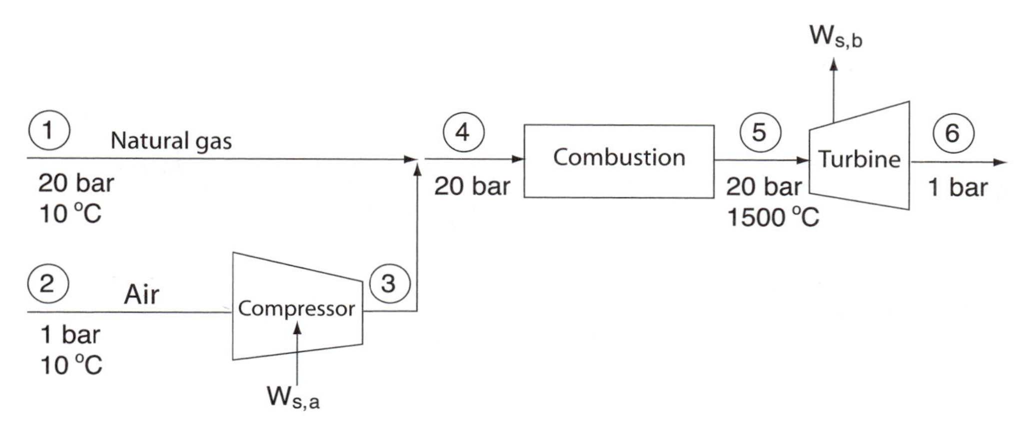

Figure 8.9: Flow sheet for the gas turbine section in Example 8.21.

This is very similar to an airplane jet engine, but instead of moving the plane, electricity is produced. Here are some typical numbers for a current natural gas power plant:

- Fuel (for example, methane) and air are compressed to 10-30 bar and react (burn) in a combustion chamber at temperatures up to about 1500 ^oC. A high temperature is desired for the turbine but material problems provide a practical limit of about 1500 ^oC. Thus, to limit the temperature rise, the amount of air is about 2.5 to 3 times the stoichiometric value.
- Work is removed from the combustion gas by expanding it to atmospheric pressure in a *turbine*. About half of the work in the turbine is used to run the compressor, while the remaining is used to run a generator that produces electricity.
- The exhaust gas from the turbine is typically at about 450-650 ^oC, and the work that is removed in a single turbine is typically from 10 MW to 330 MW.
- The efficiency $|W|/|W_s^{\rm id}|$ for the gas turbine process itself is about 35-42%.
- The largest pollution problem, besides CO_2, is emissions of NO_x, which are about 15-25 ppm with current combustion technology. Sulfur, i.e., emissions of SO_2, is a problem for the combustion of coal, but not in the combustion of natural gas because it does not normally contain sulfur or can be removed.

8.7.2 The steam turbine process

In the second part of a combined cycle power plant, the steam turbine section, the remaining heat in the combustion gas is used to produce additional work. This is done by installing heat exchangers in the exhaust gas duct to generate high pressure steam, which subsequently is expanded in one or more steam turbines using a Rankine cycle. Some typical numbers for the steam turbine process are (see also Figure 8.10):

- The exhaust gas from the gas turbine at about 645 ^oC is cooled to about 90-95 ^oC before leaving the plant through the chimney. A lower chimney temperature would be desirable, but it may result in condensation of water which may result in corrosion.

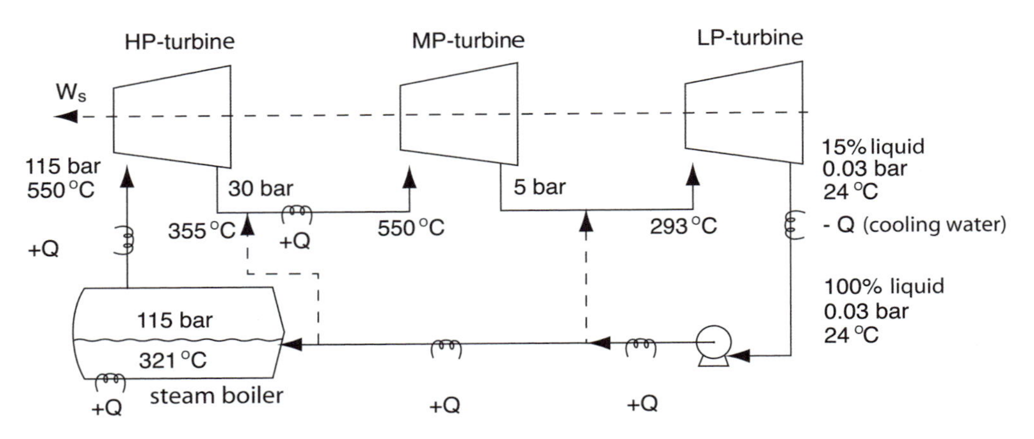

Figure 8.10: Flow sheet for a steam turbine process. The heat "$+Q$" in five places is supplied by cooling the combustion gas (stream 6 in Figure 8.9). The numbers are typical for a power plant with current technology.

- The exhaust gas (hot side) is cooled in countercurrent heat exchangers by heating water and generating steam (cold side). Most of the heat is removed in the steam boiler, where water evaporates to saturated steam at about 321 ^{o}C and 115 bar.
- It is optimal to have many pressure levels for steam, for example, three levels: high pressure (HP) steam at 110-115 bar, medium pressure (MP) steam at 25-30 bar and low pressure steam (LP) at 4-5 bar. The high pressure steam and medium pressure steam are superheated to about 550 ^{o}C (more superheating is favorable) before work is extracted in the turbines. The low pressure steam goes directly from the exit of the medium pressure turbine to the low-pressure turbine (with no superheating) and is at about 293 ^{o}C.
- From a thermodynamic point of view, it is desirable to adjust the pressure levels and amount of steam such that the temperature difference between the hot and cold side is small throughout the heat exchangers in the exhaust gas duct, but this is difficult to accomplish since the maximum pressure is about 115 bar. This means that the evaporation must take place at temperatures of 321 ^{o}C or lower.
- For large plants ($>$ 100 MW), the thermodynamic efficiency of the steam turbines is about 88% for the high and medium pressure turbines, and about 91% for the low pressure turbine. For smaller plants (about 10 MW), the efficiencies are about 5%-points lower.
- The outlet pressure of the low-pressure turbine is typically 0.03–0.07 bar. A low pressure is desired in order to extract the maximum amount of work in the turbine, but it is limited by the temperature of the cooling water. There is a condenser after the low-pressure turbine, where the steam (vapor) is condensed, and the low pressure then equals the vapor pressure of water at the condensing temperature. For example, the vapor pressure of water at 24 ^{o}C is about 0.03 bar, and at 40 ^{o}C is about 0.07 bar. Note that the heat of condensation in the steam is "lost" because it is available at a low temperature (24 – 40°C) – it is then useless for producing work and can hardly be used even as a heating source.
- The steam (vapor) that leaves the low-pressure turbine usually contains some water

(droplets), and another limitation is that the water (liquid) content must not exceed 13-16% to avoid damage to the turbine.

A modern natural gas power plant is a complex plant, but it is still relatively "simple" compared to a large-scale chemical process. An ammonia plant or a methanol plant has, for example, in addition to the main process, a steam turbine process for utilizing the heat in the exhaust gas used to heat the steam reformer.

In the examples below, simple hand calculations, based on the assumption of ideal gas and constant heat capacity, are used to establish mass and energy balances for a natural gas power plant. These calculations are reasonably accurate, but more importantly, provide excellent insight. For more detailed calculations, thermodynamic diagrams or computer programs should be used – in particular, this applies to the steam turbine process, where the pressures are relatively high and the temperatures low, so that non-ideality is expected.

Example 8.21 Gas power plant: The gas turbine section *(see Figure 8.9). Assume that*

- *Natural gas (stream 1) is available as methane at 20 bar and 10 °C*
- *Air (stream 2) is available at 1 bar and 10 °C*

The air is compressed to 20 bar (stream 3) in a compressor with (isentropic) efficiency 80% and mixed with natural gas. This stream (stream 4) is burned in a combustion chamber. The combustion gas at maximum 1500 °C (stream 5) is expanded to 1 bar (stream 6) in a turbine with (isentropic) efficiency 85%. Furthermore, assume that

- *The lower heat of combustion (298 K, 1 bar, $H_2O(g)$ product) for methane is −802 kJ/mol or −50.0 MJ/kg , see (4.34), and for combustion of methane, you can assume that the maximum theoretical work is $W_s^{\text{id}} = 50$ MJ/kg (see Example 8.11, page 219).*
- *Assume ideal gas and constant heat capacities C_p: 30 J/mol K (N_2, O_2), 36 J/mol K (H_2O), 36 J/mol K (CH_4), 37 J/mol K (CO_2).*
- *Assume that the air is 21% O_2 and 79% N_2.*

Task: *(a) What ratio of air and natural gas gives a combustion temperature of 1500 °C? Calculate the composition of the combustion gas.*

(b) What is the temperature after the expansion in the turbine (stream 6)?

(c) How much net work is produced (turbine minus compressor) [J/kg methane] and what is the efficiency W/W_s^{id}?

Solution. *Basis: 1 mol air and x mol natural gas (CH_4).*

(a) **Compression of air from 1 to 20 bar.** *Ideal isentropic compression: $T_3'/T_2 = (20/1)^{8.31/30} = 2.29$, that is, $T_3' = 2.29 \cdot 283\ K = 649\ K$, and $W_a' = C_p(T_3' - T_2) = 30 \cdot (649 - 283) = 10980$ J. Actual: $W_a = W_a'/0.8 = 13725$ J, i.e., $T_3 = T_2 + W_a/C_p = 283 + 13725/30 = 740.5$ K.*

Combustion *occurs according to the reaction*

$$CH_4 + 2O_2 = CO_2 + 2H_2O(g)$$

Conversion is assumed complete with respect to methane, that is, the extent of reaction is x [mol]. From the energy balance, we have that $H_{\text{out}} - H_{\text{in}} = 0$, where $H_{\text{in}} = H_1 + H_3$, and $H_{\text{out}} = H_5$, where $T_5 = 1773$ K (1500 °C). We obtain $H_{\text{out}} - H_{\text{in}}$ by considering the following subprocesses ("method 2"):

1. *Heat methane from $T_1 = 283K$ to $T_5 = 1773K$: $\Delta_1 H = x36(1773 - 283) = 53640x$ [J].*
2. *Heat air from $T_2 = 740.5K$ to $T_5 = 1773K$: $\Delta_2 H = 30(1773 - 740.5) = 30975$ J.*

3. *React at* $T_4 = 1773K$. *Since* C_p *is assumed constant we have* $\Delta_r C_p^\ominus = 37 + 2 \cdot 36 - 36 - 2 \cdot 30 = 13$ *J/molK, and from (B.56), the heat of reaction at temperature* T_5 *is* $\Delta_r H^\ominus(T_5) = \Delta_r H^\ominus(298) + \Delta_r C_p^\ominus \cdot (T_5 - 298) = -802000 + 13 \cdot (1773 - 298) = -783000$ *J/mol. We then have:* $\Delta_3 H = -783000x$ *[J].*

We set the sum to 0 and get $53640x + 30975 - 783000x = 0$, *which gives* $x = 30975/729360 = 0.0425$ *mol methane. To burn this,* $2 \cdot 0.0425 = 0.0849$ *mol* O_2 *is consumed, while the air (1 mol) contains 0.21 mol* O_2. *Thus, the ratio between the actual and stoichiometric amount of air is* $0.21/0.0849 = 2.47$. *The total amount in stream 5 is 1.0425 mol, and the fraction of nitrogen is* $0.79 \cdot 1$ *mol/*1.0425 *mol* $= 0.7578$ *and similarly for the other components. The composition of the combustion gas (stream 5) is then:*

$$75.78\% N_2, \quad 12.00\% O_2, \quad 4.07\% CO_2, \quad 8.15\% H_2O$$

(b) **Expansion in turbine from 20 bar to 1 bar.** *The mean heat capacity in stream 5 is* $C_p = \sum_i x_i C_{pi} = 30.7$ *J/ mol K. Ideally, we then have for isentropic expansion:* $T_6'/T_5 = (1/20)^{8.31/30.7} = 0.444$, *that is,* $T_6' = 0.444 \cdot 1773 = 788K$. *The corresponding ideal work is* $W_b' = |1.0425 \cdot 30.7(788 - 1773)| = 31525$ *J/ mol air and the actual work performed is* $W_b = 0.85W_b' = 26796$ *J/ mol air. The actual exit temperature of the turbine is* $T_6 = 1773 - 26796/(1.0425 \cdot 30.7) = 936$ *K (663 °C).*

(c) **Net work.** *The work generated in the turbine (26796 J/mol air) is about twice the work that must be supplied for compressing the air (13725 J/mol air). The net performed work is:* $W_{\text{gasturbine}} = W_b - W_a = 26796 - 13735 = 13061$ *J/mol air. On a methane basis,* $W_{\text{gasturbine}} = 13061/0.0425 = 307500$ *J/ mol methane, that is, with a methane molar mass of 16 g/mol, the net performed shaft work is*

$$W_{\text{gasturbine}} = 19200 \text{ kJ/kg methane}$$

and we find that the thermodynamic efficiency in the gas turbine process by itself is $W_{\text{gas turbine}}/W_s^{\text{id}} = 19200/50000 = 0.384$ *(38.4%).*

Example 8.22 Steam turbine process. *Consider a relatively simple (older) steam turbine process, as shown in Figure 8.11, where*

- *Superheated high-pressure (HP) steam at 100 bar and 529 °C is produced by exchanging heat with the combustion gas (stream 6 in Figure 8.5).*
- *This high-pressure steam (stream 1) expands adiabatically to get 28.5 bar medium-pressure (MP) steam (stream 2) in a high-pressure turbine with 77% (isentropic) efficiency*
- *The MP-steam is further expanded (without intermediate superheating) to 0.096 bar (stream 3) in a low-pressure turbine with 82% (isentropic) efficiency.*

Note there is no medium-pressure turbine and low-pressure steam in this process. In both turbines, work is extracted.

Data: *At 45 °C, the vapor pressure of the water is 0.096 bar. Assume that the heat of vaporization for water is 2400 kJ/kg at 45 °C. In order to simplify the calculation, it is assumed that the steam (gas-phase water) is an ideal gas with constant heat capacity 2 kJ/kg = 36 J/mol.*

(a) Assume that the hot combustion gas, which comes from the combustion of methane in the gas turbine in Example 8.21, is cooled from 663 °C to 160 °C. How much high-pressure steam is produced [kg steam/kg methane]?

Choose 1 kg steam as basis in the following calculations:

(b) How much work is produced in the high-pressure turbine?

(c) What is the temperature of the medium-pressure steam?

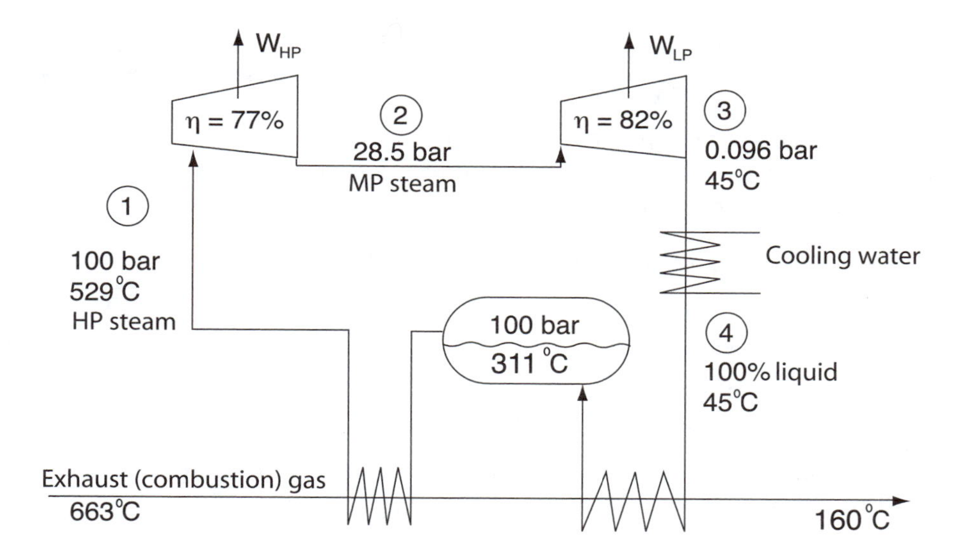

Figure 8.11: Flow sheet for the steam turbine process in Example 8.22.

(d) How much work is produced in the low-pressure turbine?

(e) If your calculations are correct, you will find that the temperature of outlet steam from the LP turbine (stream 3) is lower than the saturation temperature of 45 °C. This implies that some of the steam condenses to form liquid (water). How much water is generated (note that the moisture content should not exceed about 10%)?

(f) How much work is removed in the two steam turbines [J/kg vapor]? Find also the pump work for taking the liquid from 0.096 bar to 100 bar (the pump is not shown in the flow sheet).

(g) Calculate the energy efficiency (8.13) for the steam turbine process alone and compare it with the theoretical energy efficiency (the Carnot factor) when the "heat reservoir" is the combustion gas (where the temperature changes from 663 °C to 160 °C), and it is assumed that cooling is available at 10 °C ("cold reservoir").

Now look at the entire combined cycle gas power plant:

(h) Calculate the total produced work [J/ kg methane], including the net work produced in the gas turbine (Example 8.21). What is the total thermodynamic efficiency?

*(i) Finally, plot the cooling curve for the combustion gas and the heating curve for the steam (vapor) in a figure with enthalpy on the x-axis and temperature on the y-axis. This plot shows the temperature profile if we used a single heat exchanger. From the plot, find the smallest temperature difference (the "pinch point") in the heat exchanger. (If the curves cross each other (*crossover*), then we have negative temperature difference and the process is infeasible).*

Solution. *(a)* **Amount of steam.** *The amount of steam is given by the energy balances. We choose 1 kg (*$= 1/(16 \cdot 10^{-3}) = 62.5$ *mol) methane feed as basis. We found in Example 8.21 that 0.0425 mol methane and 1 mol air give 1.0425 mol combustion gas. By scaling we then have that*

- *1 kg methane gives* $\frac{62.5}{0.0425} \cdot 1.0425$ *mol = 1533 mol combustion gas.*

The heat capacity of the combustion gas is $C_p = 30.7$ *J/mol K. The heat released, when 1533 mol of combustion gas is cooled from 663 °C to 160 °C, is*

$$Q_{\text{steam turbine}} = nC_p(T_{H1} - T_{H2}) = 1533 \; mol \cdot 30.7 J/mol \; K \cdot (663 - 160) \; K = 23672 \; kJ$$

This amount of heat is used to take water (liquid) at 45 °C (stream 4) to superheated vapor at 529 °C (stream 1). This can be divided into two subprocesses (this may not be what occurs in practice, but because enthalpy is a state variable the result is the same):

1. *Evaporate water at 45 °C:* $\Delta_{\rm vap}H = 2400$ *kJ/kg.*
2. *Heat steam from 45 °C to 529 °C:* $C_p(T_1 - T_4) = 2\ kJ/kg\ K \cdot (529 - 45)K = 968\ kJ/kg.$

The total heat required for producing vapor for the steam turbines is then

$$Q_{\rm steam\ turbine} = m_{\rm steam} \cdot \underbrace{(2400 + 968)}_{\rm 3368\ kJ/kg\ steam}$$

and the amount of steam becomes

$$m_{\rm steam} = \frac{Q_{\rm steam\ turbine}}{3368} = \frac{23672 \text{ kJ/kg methane}}{3368 \text{ kJ/kg steam}} = 7.03 \frac{\text{kg steam}}{\text{kg methane}}$$

(b) **HP turbine.** *We choose 1 kg steam as basis in the following. Ideal (reversible) adiabatic expansion of HP steam in the high pressure turbine:*

$$\frac{T_2'}{T_1} = \left(\frac{p_2}{p_1}\right)^{R/C_p} = \left(\frac{28.5}{100}\right)^{8.31/36} = 0.748$$

that is, $T_2' = 802 \cdot 0.748 = 600\ K$. *Ideal work:* $W_{HP}' = |C_p(T_2' - T_1)| = |2 \cdot (600 - 802)| = 404$ *kJ/kg. Actual work:* $W_{HP} = W_{HP}' \cdot 0.77 = 311.1\ kJ/kg.$

(c) **MP steam.** *Actual temperature of medium-pressure steam:* $W_{HP} = |C_p(T_2 - T_1)|$ *gives* $T_2 = 646.5\ K\ (373\ ^oC)$

(d) **LP turbine.** *Ideal (reversible) adiabatic expansion of MP steam in low-pressure turbine:*

$$\frac{T_3'}{T_2} = \left(\frac{p_3}{p_2}\right)^{R/C_p} = \left(\frac{0.096}{28.5}\right)^{8.31/36} = 0.268$$

that is, $T_3' = 646.5 \cdot 0.268 = 173.2\ K$. *Ideal work:* $W_{LP}' = C_p(T_2 - T_3') = 2 \cdot (646.5 - 173.2) = 946.6\ kJ/kg$. *Actual work:* $W_{LP} = W_{LP}' \cdot 0.82 = 776.2\ kJ/kg.$

(e) **Condensing steam (stream 3) from LP turbine.** *The "actual" temperature of the gas in stream 3 is:* $T_3 = T_2 - W_{LP}/C_p = 646.5 - 776.2/2 = 258.4\ K\ (-15\ ^oC)$. *However, this is lower than the condensation (saturation) temperature at 0.0996 bar which is 45 °C, so in practice stream 3 is a mixture of gas (saturated steam) and liquid at 45 °C. From an energy balance, the liquid fraction in stream 3 is*

$$\frac{C_p \Delta T}{\Delta_{\rm vap} H} = \frac{2 \cdot 60 \text{ kJ/kg steam}}{2400 \text{ kJ/kg condensed}} = 0.050 \text{ kg condensed/kg steam}$$

that is, there is about 5% water (l) (more exact calculations with steam tables or a steam enthalpy diagram (see page 420) give that the water (liquid) content is about 10%).

(f) **Work from steam turbines.** *The work produced in the two steam turbines is*

$$W_{\rm steam\ turbines} = W_{HP} + W_{LP} = (311.1 + 776.2) \frac{\rm kJ}{\rm kg\ steam} = 1087 \frac{\rm kJ}{\rm kg\ steam}$$

From (6.14), the pump work for 1 kg of water is

$$W_{\rm pump}^{\rm rev} = \frac{p_2 - p_1}{\rho} = \frac{(100 - 0.096) \cdot 10^5\ N/m^2}{1000\ kg/m^3} \approx 10 \frac{\rm kJ}{\rm kg\ steam}$$

This is less than 1% of the work from the steam turbines and can therefore be neglected.

(g) **Energy efficiency for the steam turbine process alone.** *The heat supplied is, as found in subtask (a),* $Q_H = Q_{\text{steam turbine}} = 3368$ *kJ/kg steam. The energy efficiency (8.13) in the steam turbine process alone is then*

$$\frac{W_{\text{steam turbines}}}{Q_{\text{steam turbine}}} = \frac{1087}{3368} = 0.323 \quad (32.3\%)$$

Note that even with 100% efficiency in the steam turbines, the "energy efficiency" would only be $(404 + 877.6)/3368 = 0.38$. *This is mainly because most of the condensation heat (which is removed in the cooling water) cannot be removed as work. Let us compare the energy efficiency of 32.3% with the theoretical (maximum) value that is given by the Carnot "efficiency" (factor). We assume that the heat capacity of the combustion gas is constant, and we then get from (8.7) in Example 8.2 that we need to use the logarithmic mean value for* T_H *when calculating the Carnot factor. We have* $T_{H1} = 936K$ *(663* ^oC*),* $T_{H2} = 433$ *K (160* ^oC*) and* $T_C = 283$ *K (10* ^oC*). We get*

$$T_4 = T_{H,\log} = (936 - 433)/\ln(936/433) = 652.5 \text{ K}$$

and the **Carnot factor** *is*

$$1 - \frac{T_C}{T_H} = 1 - \frac{283K}{652.5K} = 0.56 \quad (56.6\%)$$

Compared to this maximum achievable value, the thermodynamic efficiency in the steam turbine process is then $0.323/0.566 = 0.57$ *(57%). Still, this quite far from 100%, which shows, as expected, that there is potential for improvement in the steam turbine section.*

(h) **Total work and thermodynamic efficiency for the entire process (gas turbine plus steam turbines).** *We choose here again 1 kg methane feed as basis. From subtask (a), the amount of steam is 7.03 kg steam/kg methane, so this corresponds to*

$$W_{\text{steam turbines}} = 1087 \ \frac{\text{kJ}}{\text{kg steam}} \cdot 7.03 \ \frac{\text{kg steam}}{\text{kg methane}} = 7640 \ \frac{\text{kJ}}{\text{kg methane}}$$

In comparison, we found in Example 8.21 that the net work in the gas turbine is $W_{\text{gas turbine}} = 19200$ *kJ/ kg methane. The total work from the gas turbine and steam turbines is then, for this specific example process,*

$$W_{\text{tot}} = W_{\text{gas turbine}} + W_{\text{steam turbines}} = 26840 \ \frac{\text{kJ}}{\text{kg methane}}$$

that is, total thermodynamic efficiency for the process is

$$\frac{W_{\text{tot}}}{W_s^{\text{id}}} = \frac{26840}{50000} = 0.537 \quad (53.7\%)$$

With an improved steam turbine process, where we remove more energy by cooling the combustion gas further (we cannot get it much lower than 45 ^oC *in our process because of "temperature crossover"; see below), and utilize the energy better by having hotter HP steam, by superheating the MP steam and by condensing at lower temperature (see Figure 8.10), the total efficiency can be increased (try yourself!).*

(i) **Check of temperature crossover.** *We choose 1 kg methane feed as basis. As shown in Figure 8.12, the cooling curve for the combustion gas is a straight line with negative slope equal to the inverse of the heat capacity, that is*

$$C_{p,h} = m_h c_{p,h} = 1533 \ mol \cdot 30.7 \frac{J}{mol \ K} = 47063 \ J/K$$

The heating curve consists of three parts:

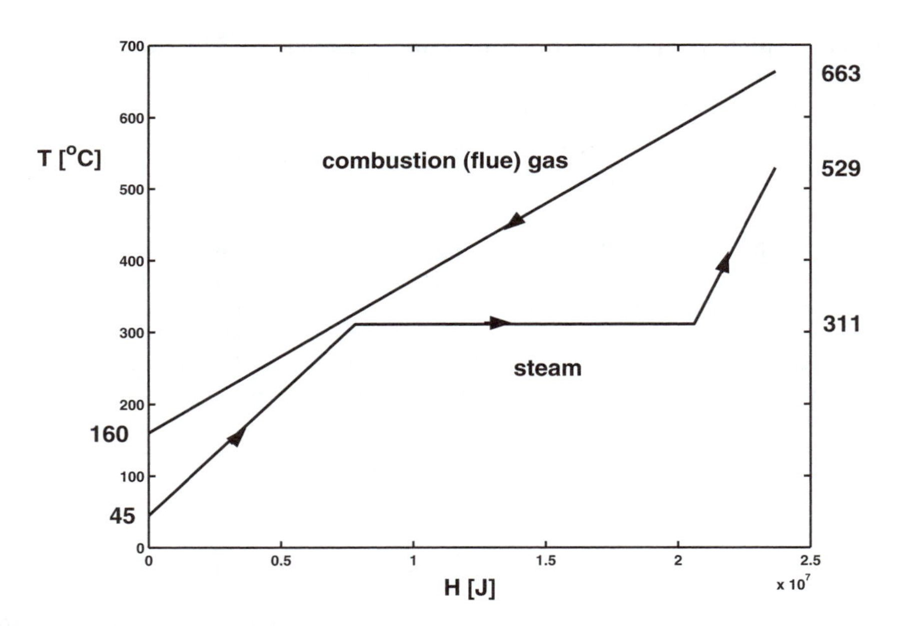

Figure 8.12: Heating and cooling curves for the steam turbine process in Example 8.22.

1. *Straight line for heating of the liquid with positive slope* ($C_{p,c}(l) = m_c c_p(l) = 7.03\ kg \cdot 4180\ J/kg\ K = 29385\ J/K$).
2. *Straight flat line: Evaporation at 100 bar and 311 °C.*
3. *Straight line for superheating the steam with positive slope* ($C_{p,c}(g) = m_c c_p(g) = 7.03 \cdot 2000 = 14060\ J/K$).

The smallest temperature difference between the hot and cold sides (the "pinch") is where the "cold" liquid reaches 311 °C. The corresponding gas temperature is $160 + 29385 \cdot (311 - 45)/47063 = 326.1$ *°C, that is, the temperature difference is 15.1 °C. This is quite small but still acceptable.*

The numbers calculated in the above examples are a bit difficult to compare because we have used a different basis in each subtask. Therefore, we summarize everything for a basis feed of **1 kg methane** in Table 8.2.

8.7.3 Remarks: Gas power plant

Remark 1 Corrected efficiency for case study. Above, we assumed that methane (feed) is provided at 20 bar, but for a thermodynamic analysis it is more correct to assume a standard pressure 1 bar. The reversible work for an isothermal compression of 1 kg methane from 1 bar to 20 bar at 10 °C is $W_s' = nRT_0 \ln(p_2/p_1) = 62.5 \cdot 8.31 \cdot 283 \cdot \ln 20 = 0.44$ MJ. The thermodynamic efficiency, when this work is subtracted, is reduced from 53.7% to 52.8%:

$$\eta = \frac{26.84 - 0.44}{50.00} = 0.528 \quad (52.8\%)$$

Remark 2 Improved process. From Table 8.2, the thermodynamic efficiency of the process in Examples 8.21 and 8.22 is 53.7%. A similar example, also based on ideal gas and

Table 8.2: Summary of results for natural gas power plant (Examples 8.21 and 8.22)

Natural gas feed (methane)	1.00	kg	=	62.5	mol
Air feed	41.68	kg	=	1471.7	mol
Pressure gas turbine	20	bar			
Inlet temp. gas turbine	1500	^{o}C			
Outlet temp. gas turbine	663	^{o}C			
Outlet temp. combustion gas	160	^{o}C			
Outlet temp. LP steam turbine	45	^{o}C			
Air compression (η =80%)	−20.20	MJ			
Gas turbine (η =85%)	39.40	MJ			
Net gas turbine	19.20	MJ			
HP steam turbine (η =77%)	2.19	MJ			
LP steam turbine (η =82%)	5.53	MJ			
Sum steam turbines	7.64	MJ			
Net work produced	26.84	MJ			
Efficiency: 26.84/50.00 =	0.537				

neglecting pressure drops and heat losses, but with better efficiencies for the turbines (90%) and a more complicated steam turbine process with three pressure levels and cooling to 24 ^{o}C, is given as an extra exercise at the book's homepage. Here, a thermodynamic efficiency for the process of about 62% is achieved. This can be further improved by increasing the combustion pressure (inlet pressure to the gas turbine) – computations with the turbine pressure increased from 20 bar to 30 bar using the flowsheet simulator Hysys (where ideal gas is not assumed) gave an overall thermodynamic efficiency of more than 64%.

Remark 3 Heat from natural gas. If the objective is to use natural gas as a heat source, for example, in a home by simply burning the gas, then a high combustion temperature is not required and the pressure is usually 1 bar. In such cases, an energy efficiency up to 100% can be achieved, but the thermodynamic efficiency is much lower – typically about 10% (see Example 8.7 page 211). The reason for this is that one could theoretically (i) convert the chemical energy into electric energy (for example, in a gas power plant), and could then (ii) use the electricity to run a heat pump that "pumps" heat from the outdoor air into the house; see the next remark for details.

Remark 4 Be careful about energy efficiencies. You have probably understood by now that I (the author) am sceptical when it comes to the use of energy "efficiencies". The reason is that almost any energy "efficiency" can be achieved by allowing for heat pumps. For example, take our simple gas power plant where we achieved an thermodynamic (exergy) efficiency of above 50% for production of electricity. If we wanted to maximize the energy "efficiency" of this plant, we could use this work (electricity) to

1. Cold location (e.g., Norway): Run the compressor in a heat pump system to produce heat
2. Hot location (e.g. Florida): Run the compressor in a cooling cycle to produce cold

In either case, we could achieve an overall energy efficiency from feed to energy product of well above 100%. For example, assume that thermodynamic efficiency of the power plant is 50% and that the coefficient of performance is 5 (COP=5) for either the (1) heat pump system or (2) the cooling cycle (both of which are realizable in practice). The overall *energy efficiency* from feedstock (natural gas) to energy product is then $0.5 \cdot 5 = 2.5$ (250%)!

8.8 Summary

Heat engine. It is possible to extract work from high-temperature (T_H) heat, but the second law of thermodynamics says that this can only happen if some of the heat is "lost" by cooling at a lower temperature (T_C). The maximum fraction of the heat supplied (Q_H) that can be extracted as work is given by the Carnot factor

$$\left(\frac{|W|}{|Q_H|}\right)^{\text{rev}} = 1 - \frac{T_C}{T_H}$$

To maximize the work, the temperature difference $T_H - T_C$ should be as <u>large</u> as possible. The thermodynamic efficiency η, for a heat (thermal power) engine, is the ratio between the actual value of W/Q_H and the Carnot factor.

Heat pump/refrigerator. It is possible to transfer heat from low (T_C) to high (T_H) temperature, but the second law of thermodynamics says that this can only happen if we supply work. For a heat pump, the maximum ratio of heat to work (the coefficient of performance) is

$$\text{COP}_{\text{C}}{}^{\text{rev}} = \left(\frac{|Q_H|}{|W|}\right)^{\text{rev}} = \frac{T_H}{T_H - T_C}$$

which is the inverse of the Carnot factor. For a refrigerator, the focus is on the cooling, and it is more interesting to look at the maximum cooling,

$$\text{COP}_{\text{H}}{}^{\text{rev}} = \left(\frac{|Q_C|}{|W|}\right)^{\text{rev}} = \frac{T_C}{T_H - T_C}$$

For both a heat pump and a refrigerator, it is favorable that the temperature difference $T_H - T_C$ is as <u>small</u> as possible.

Exergy and lost work. More generally, the ideal work for a reversible continuous process with surroundings at constant temperature T_0 is

$$W_s^{\text{id}} = \Delta H - T_0 \Delta S$$

We can introduce the state function, exergy, defined as

$$B = H - T_0 S$$

and we then have $W_s^{\text{id}} = \Delta B$ for a process that operates between two states (in and out). The lost work can be found by comparing W_s^{id} with the actual work, and we have

$$W_{\text{lost}} = W_s - W_s^{\text{id}} = W_s - \Delta B = T_0 \Delta S_{\text{total}}$$

9

Mechanical energy balance

The mechanical energy balance (Bernoulli's equation) is used to compute pressure changes for flow systems. Here, we derive Bernoulli's equation from macroscopical energy considerations, but it can alternatively, as shown in fluid mechanics, be derived from Newton's second law[1].

9.1 The "regular" energy balance

We consider a process with one inflow and one outflow, see Figure 9.1. The energy balance for a steady state continuous process between states 1 (inflow) and 2 (outflow) can, as shown in (4.39), be written as

$$\boxed{H_2 + m\alpha_2 \frac{v_2^2}{2} + mgz_2 = H_1 + m\alpha_1 \frac{v_1^2}{2} + mgz_1 + Q + W_s} \quad [J; J/s] \qquad (9.1)$$

where we have included terms for kinetic and potential energy.

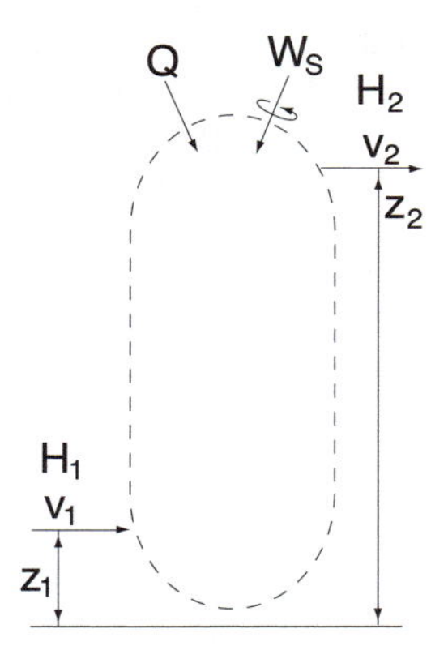

Figure 9.1: Steady-state energy balance for flow process from point 1 to point 2

The velocities v_1 and v_2 are mean velocities defined by

$$v_i \text{ [m/s]} = \frac{\dot{V}_i}{A_i} = \frac{\dot{m}_i}{\rho_i A_i} \qquad (9.2)$$

[1] Most of the figures and empirical data in this chapter are taken from A. Lydersen, *Kjemiteknikk*, Tapir, 1972.

where $\dot{V}_i$ [m^3/s] is the volumetric flow and A_i [m^2] is the cross section area (for example, the cross section of a pipe). The factor α in (9.1) corrects for the velocity not being the same over the entire cross section, that is, it corrects for the mean of kinetic energy not necessarily being equal to $m\frac{v^2}{2}$, where v is the mean velocity defined in (9.2). For turbulent flow (the most common), the velocity profile is almost flat and $\alpha \approx 1$, while for laminar flow in pipe $\alpha = 2$, see Figure 9.2.

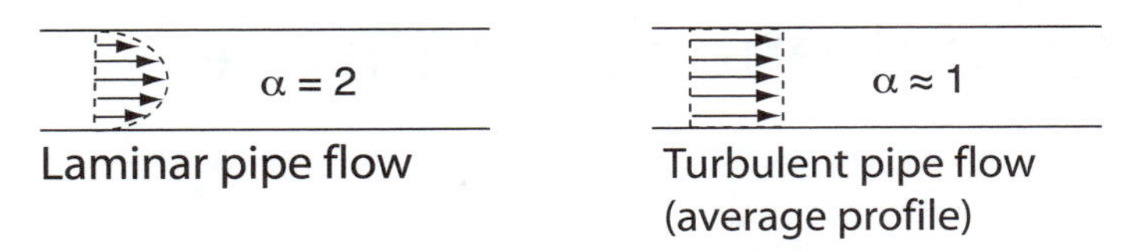

Figure 9.2: Velocity profiles for pipe flow

9.2 Mechanical energy

By "mechanical energy," we mean the energy terms related to motion and position as described by Newton's laws of physics. Mechanical energy can in theory be directly converted to work, and one can reversibly go between the different mechanical energy forms. In the "regular" energy balance, we have previously found that changes in mechanical energy, such as kinetic energy and potential energy, are usually small compared to changes in internal (thermal) energy. On the other hand, in the mechanical energy balance, we exclude internal energy and focus on the mechanical energy terms and the conversion between them. The executive summary of this chapter is: Mechanical energy is preserved if we have a reversible process without friction.

Comment: One might wonder when one is supposed to use the "regular" energy balance (9.1) with kinetic and potential energy included, and when one should use the mechanical energy balance (9.5) (which we will derive, and which also has kinetic and potential energy included). The answer is usually:

- The "regular" energy balance is used when one is interested in temperature changes.
- The mechanical energy balance is used when one is interested in pressure and/or velocity changes.

9.3 Reversible shaft work and friction

The mechanical energy balance can be derived from Newton's second law. However, we choose a different approach: First, we derive an expression for the reversible shaft work, then introduce the concept of friction, combine this with the "regular" energy balance and end up with the mechanical energy balance. Let us first derive the expression for the reversible shaft work and its relation to friction.

itya steady state continuous process, where the inflow has enthalpy H_1, temperature T_1, pressure p_1, velocity v_1 and height z_1, while the outflow has enthalpy H_2, temperature T_2, pressure p_2, velocity v_2 and height z_2; see Figure 9.1. What is the maximum shaft work that can be extracted? Or equivalently: What is the minimum shaft work W_s that must be supplied? In order to find this, we need to consider a reversible process (without friction loss). We have already in (6.4) derived that, if we neglect kinetic and potential energy, the maximum useful shaft work is

$$W_s^{\text{rev}} = \int_{p_1}^{p_2} V dp = m \int_{p_1}^{p_2} \frac{dp}{\rho} \quad [J]$$

where we have introduced the density $\rho = m/V$ [kg/m^3]. But, in addition, also kinetic and potential energy can be converted into shaft work (and *vice versa*), that is, we have the following generalization

$$W_s^{\text{rev}} = m \int_{p_1}^{p_2} \frac{dp}{\rho} + m\left(\alpha_2 \frac{v_2^2}{2} - \alpha_1 \frac{v_1^2}{2}\right) + mg(z_2 - z_1) \quad [J] \tag{9.3}$$

where v is the mean velocity defined in (9.2).

The actual supplied work is larger than that given in (9.3). We write

$$W_s = W_s^{\text{rev}} + \Phi \quad [J] \tag{9.4}$$

where $\Phi > 0$ is the friction loss or the lost work. *The friction loss Φ expresses how much mechanical energy is converted into (useless) thermal energy (friction heat).* Φ is positive ($\Phi > 0$) for all real processes. Despite persistent attempts during the last centuries, no one has been able to create a self-sustaining machine (*perpetuum mobile*) with $\Phi \leq 0$ and this is, of course, impossible according to the second law of thermodynamics. For an (idealized) reversible process, we have $\Phi = 0$.

9.4 The mechanical energy balance

Inserting (9.3) into (9.4) gives the "mechanical energy balance" for a steady state continuous flow process operating between states 1 and 2:

$$\boxed{m\alpha_2 \frac{v_2^2}{2} + mgz_2 + m \int_{p_1}^{p_2} \frac{dp}{\rho} + \Phi = m\alpha_1 \frac{v_1^2}{2} + mgz_1 + W_s} \quad [J; J/s] \tag{9.5}$$

The "mechanical energy" includes kinetic energy ($m\frac{v^2}{2}$), potential energy (mgz) and "potential pressure energy" ($m \int \frac{dp}{\rho}$) in the flow system (stream). Note that mechanical energy is *not* a conserved quantity and the friction term Φ expresses the inevitable loss of mechanical energy that we have in any real process.

On mass basis, the mechanical energy balance (9.5) can be written

$$\alpha_2 \frac{v_2^2}{2} + gz_2 + \int_{p_1}^{p_2} \frac{dp}{\rho} + \frac{\Delta p_f}{\rho} = \alpha_1 \frac{v_1^2}{2} + gz_1 + \frac{W_s}{m} \quad [J/kg] \tag{9.6}$$

where we used the common convention of writing the friction term Φ/m as

$$\frac{\Phi}{m} = \frac{\Delta p_f}{\rho} \quad [J/kg]$$

where Δp_f [N/m^2] is known as the friction pressure drop. The work W_s is usually given in [J/s] and m is then in [kg/s].

Note that the velocities v_1 and v_2 must also satisfy the steady-state mass balance $m_1 = m_2$ [kg/s] (see also (4.40)), that is,

$$m_1 = \rho_1 v_1 A_1 = \rho_2 v_2 A_2 = m_2 \quad [kg/s] \tag{9.7}$$

where A_i [m^2] is the cross section area, for example of a pipe. This special form of the mass balance is called the **continuity equation**. For an incompressible fluid (most liquids), ρ is constant and the continuity equation is

$$v_1 A_1 = v_2 A_2 \quad [m^3/s] \tag{9.8}$$

and we see that the velocity is uniquely determined by the cross section area.

9.4.1 The Bernoulli equation for incompressible flow

For an **incompressible fluid** (most liquids), ρ is constant, and also for gases with a small pressure drop, ρ can be assumed approximately constant. In this case, the potential pressure energy term becomes

$$\int_{p_1}^{p_2} \frac{dp}{\rho} = \frac{p_1 - p_2}{\rho}$$

and the mechanical energy balance (9.6) can be written in the following form, which is known as the **generalized Bernoulli equation**:

$$\frac{p_2}{\rho} + g z_2 + \alpha_2 \frac{v_2^2}{2} + \frac{\Delta p_f}{\rho} = \frac{p_1}{\rho} + g z_1 + \alpha_1 \frac{v_1^2}{2} + \frac{W_s}{m} \quad [J/kg] \tag{9.9}$$

Equivalently, (9.9) can be written in pressure form

$$\boxed{\underbrace{p_2 + \rho g z_2 + \rho \alpha_2 \frac{v_2^2}{2}}_{\text{out}} + \underbrace{\Delta p_f}_{\text{friction loss}} = \underbrace{p_1 + \rho g z_1 + \rho \alpha_1 \frac{v_1^2}{2}}_{\text{in}} + \underbrace{\Delta p_{\text{lift}}}_{\text{supplied}}} \quad [\frac{\text{J}}{\text{m}^3} = \frac{\text{N}}{\text{m}^2}] \tag{9.10}$$

or equivalently

$$\Delta(p + \rho g z + \rho \alpha \frac{v^2}{2}) = \Delta p_{\text{lift}} - \Delta p_f \quad [\text{N/m}^2] \tag{9.11}$$

where the "pressure head" $\Delta p_{\text{lift}} = W_s \rho / m$ is the pressure increase from supplied mechanical work (using a pump, blower or compressor). Here, $(p + \rho g z)$ is often called the "static pressure" (because it does not depend on the moving flow) and $\rho \alpha v^2/2$ the "dynamic pressure" (because of the moving flow). According to Bernoulli's equation, the change in the "total pressure" (static plus dynamic) is then the pressure head minus the friction pressure drop.

If we further (1) neglect friction ($\Delta p_f = 0$), (2) assume that $\alpha = 1$ (that is, no averaging of velocity is required) and (3) assume no mechanical work ($W_s = 0$), then the "original" **Bernoulli equation** follows from (9.10):

$$\boxed{p + \rho g z + \rho \frac{v^2}{2} = \text{constant}} \quad [\text{N/m}^2] \tag{9.12}$$

In words, the "total pressure" ("static" plus "dynamic") is constant for frictionless flow without mechanical work. This equation can alternatively be derived from Newton's second law (see fluid mechanics) and was presented by Daniel Bernoulli in 1738, more than one hundred years before the first law of thermodynamics (energy balance). An important implication of (9.10) and (9.12) is that the pressure goes down when the velocity increases, for example in a restriction. A simple experiment that illustrates this effect is to hold two sheets of paper such that they are aligned next to each other. When we blow air between them from the top, the sheets will be move together because of the lower pressure that is created by the flow (velocity) between the sheets, see Figure 9.3(a).

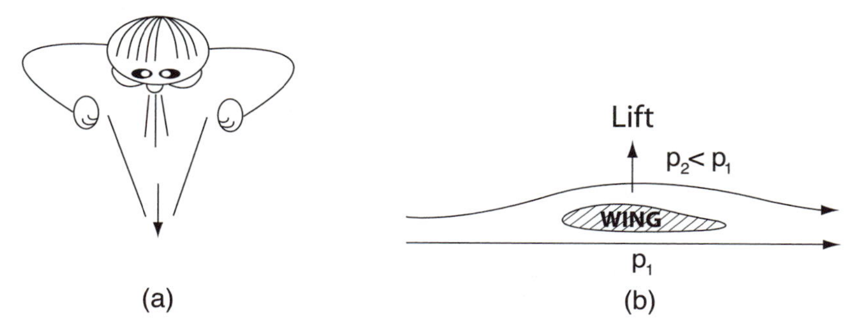

Figure 9.3: Practical consequences of lower pressure caused by a velocity increase: (a) Two sheets are drawn together when you blow between them. (b) An airplane lifts.

Another practical example is the "lift" that is generated by an airplane's wings. Here, the wings are designed such that the air must travel further to pass over the wing than under the wing. Thus, the velocity over the wing is larger, and according to Bernoulli's equation (9.12), the pressure is lower, that is, the pressure difference over the wing results in a force that lifts the plane (see Figure 9.3(b)).

Example 9.1 Pressure drop in restriction. *Water flows in a pipe with velocity 2 m/s. The pipe is restricted (temporarily) by a choke, where the cross section area is 25%. Calculate how much lower the pressure is in the restriction when the friction loss is neglected.*

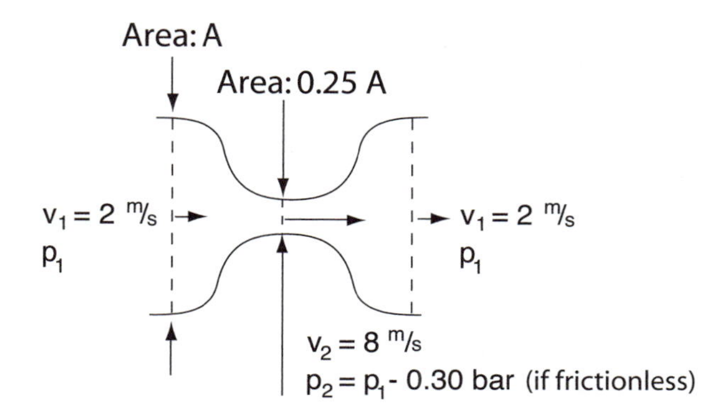

Figure 9.4: Pressure drop in restriction

We assume turbulent flow such that $\alpha = 1$. Furthermore, changes in potential energy can be neglected and the Bernoulli equation (9.12) gives for frictionless flow

$$p_1 + \rho\frac{v_1^2}{2} = p_2 + \rho\frac{v_2^2}{2} \tag{9.13}$$

From the mass balance (9.8), we have that

$$v_2 = v_1\frac{A_1}{A_2} = 2 \text{ m/s} \cdot \frac{1}{0.25} = 8 \text{ m/s}$$

and from (9.13), we get

$$p_2 - p_1 = \rho\left(\frac{v_1^2}{2} - \frac{v_2^2}{2}\right) = 1000\left(\frac{2^2}{2} - \frac{8^2}{2}\right) = -30000 \ N/m^2 = -0.30 \text{ bar}$$

Comments Bernoulli.

1. The assumption $\alpha = 1$, used when deriving (9.12), applies to turbulent flow in a pipe. But the assumption $\alpha = 1$ also applies to other flow regimes if one follows an imaginary "streamline" (drop a microscopic particle and follow its movement), since the velocity along a streamline is a "point variable" and averaging is not needed. This was Bernoulli's original approach.

2. In chokes and orifices for measurement purposes, the pressure drop Δp in the restriction is used as an indirect measurement of the velocity, and we have for incompressible flow (see Example 9.1):

$$v_1 = \sqrt{2\left(\frac{A_1^2}{A_2^2} - 1\right)}\sqrt{\frac{\Delta p}{\rho}} \quad [\text{m/s}]$$

The volumetric flow rate is then $V_1 = v_1 A_1$ [m^3/s], and the mass flow rate is $m = \rho V_1 = \rho v_1 A_1$ [kg/s].

3. Bernoulli's equation, (9.10) and (9.12), is derived for incompressible flow (constant density ρ). For small density changes (including gases), the following approximate extension of the Bernoulli equation is often used:

$$\underbrace{p_2 + \rho_2 g z_2 + \rho_2\alpha_2\frac{v_2^2}{2}}_{\text{out}} + \underbrace{\Delta p_f}_{\text{friction loss}} = \underbrace{p_1 + \rho_1 g z_1 + \rho_1\alpha_1\frac{v_1^2}{2}}_{\text{in}} + \underbrace{\Delta p_{\text{lift}}}_{\text{supplied}} \quad [\frac{\text{N}}{\text{m}^2}] \tag{9.14}$$

The difference from (9.10) is that ρ_1 and ρ_2 may be different.

4. Bernoulli's equation (9.10) can be written in "head form" by dividing both sides by ρg.

$$\Delta(\frac{p}{\rho g} + \frac{v^2}{2g} + z) = \frac{W_s}{mg} - \frac{\Delta p_f}{\rho g} \tag{9.15}$$

The "pressure head" is $p/(\rho g)$, the "velocity head" is $v^2/2g$, the "pumping head" is $\Delta p_{\text{lift}}/(\rho g) = W_s/(mg)$ and the friction head is $\Delta p_f/(\rho g)$.

9.4.2 The friction term for flow in pipes

To use the mechanical energy balance, we must know the friction term $\Phi = m\Delta p_f/\rho$ [J; J/s]. For a compressor or pump, this is rather simple; here, $\Phi = W_s - W_s^{\text{rev}}$ can be calculated from the efficiency of the equipment.

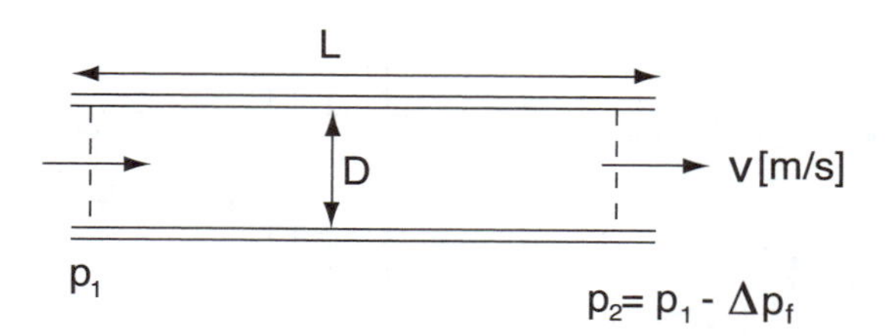

Figure 9.5: The friction pressure drop for pipe flow is $\Delta p_f = 4f\frac{L}{D}\rho\frac{v^2}{2}$ [N/m^2]

For other equipment and pipe fittings, the friction term is usually found by using (more or less) empirical expressions for the friction pressure drop Δp_f. For example, for flow in horizontal, straight pipes with diameter D [m], the friction pressure drop over the pipe length L [m] (see Figure 9.5) is:

$$\Delta p_f = 4f\frac{L}{D}\rho\frac{v^2}{2} \quad [\text{N/m}^2] \tag{9.16}$$

where the friction factor f is given in Figure 9.6 as a function of the non-dimensional Reynolds number

$$Re = \frac{\rho v D}{\mu} \tag{9.17}$$

where μ [kg/m s] is the **viscosity** of the fluid. The viscosity of water is about 10^{-3} kg/m s (= 1 cP) and for gases at 1 bar about 10^{-5} kg/m s (= 0.01 cP). A typical value for the friction factor in turbulent flow is $f = 0.005$.

Note that there are different curves in Figure 9.6 depending on whether we have laminar ("smooth") or turbulent ("chaotic") flow. The flow pattern changes from laminar to turbulent flow when the Re number exceeds approximately 2300 (laminar flow may be possible at higher Re numbers, but if we hit the pipe with, for example, a hammer, it will switch to turbulent, and the pressure drop will rise). For laminar flow in pipes, we have that $f = 16/Re$ (this is shown theoretically in fluid mechanics) and the friction pressure drop is proportional to the velocity,

$$\text{Laminar flow in pipes}: \quad \Delta p_f \sim v$$

For turbulent flow, at high Re numbers, the friction factor f is independent of the Re number (see Figure 9.6), and the friction pressure drop is proportional to the square of the velocity,

$$\text{Turbulent flow in pipes (high Re number)}: \quad \Delta p_f \sim v^2$$

In most practical cases, we have high Reynolds numbers and turbulent flow. For turbulent flow, the friction factor f is also a function of the roughness of the pipe surface, ϵ, and as expected, the friction increases when the surface is more rough; see Figure 9.6.

Equation (9.16) gives the friction pressure drop in horizontal, straight pipes. In addition, there is friction pressure drop in fittings (for example, bends and restrictions) and valves. We may represent these with (9.16) using an equivalent pipe length, or

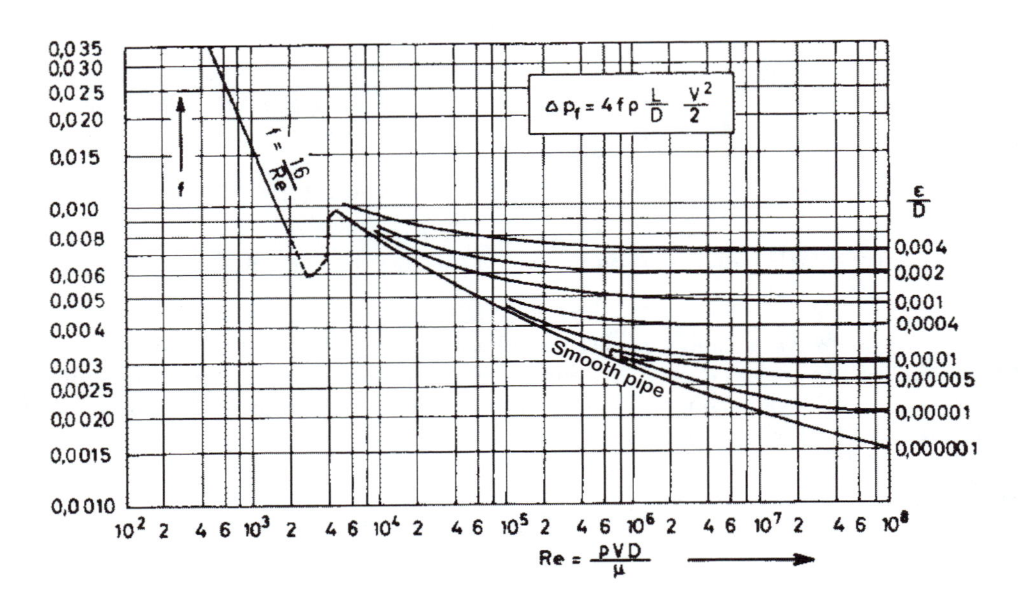

Figure 9.6: Friction factor f for flow in pipes as a function of Reynolds number in laminar and turbulent flow regions. In the turbulent region (Re> about 2300), the error limit is about ±10%. The roughness ϵ typically varies from 1 mm for cement pipes to 0.045 mm for regular steel pipes and to 0.0015 mm for smooth surface-treated pipes.

alternatively the "number of lost velocity heights" n, that is, as the factor n in the equation

$$\Delta p_f = n\rho\frac{v^2}{2} \quad [N/m^2] \tag{9.18}$$

Some typical values for n are given in Table 9.1. We see that n is about 0.5 to 1.5 for a bend or for a large change in cross section area. For valves, n of course depends on the valve design and opening.

Example 9.2 *A typical* **valve equation** *for liquid flow is given in (1.8),*

$$q = C_d f(z) A\sqrt{\Delta p/\rho}$$

where q $[m^3/s]$ is the volumetric flowrate and C_d [dimensionless] is the valve constant (relative capacity coefficient). The velocity is $v = q/A$ $[m/s]$, so for a fully open valve ($f(z) = 1$), (1.8) may be rearranged to give

$$\Delta p = (2C_d^2)\cdot \rho v^2/2$$

and by comparing with (9.18) it follows that $n = 2C_d^2$ or

$$C_d = \sqrt{n/2}$$

For example, from Table 9.1 we see that a typical value for a fully open valve is $n = 2$, which corresponds to $C_d = 1$ (in US engineering units, this corresponds to $C'_d = 20$, which according to Liptak[2] *is a typical value for control valves).*

A crude **rule of thumb** is that the total friction pressure drop is about 0.1 bar for each "major" equipment, for example, a reactor or a heat exchanger. It is possible to reduce the pressure drop, but this requires larger equipment volumes to lower the velocity, and in addition lower velocity means less effective heat and mass transfer. Reducing the pressure drop is particularly important at low pressures, that is, at 1 bar or lower, where the friction pressure drop $\Delta p_f/p$ is relatively larger compared to the total pressure.

Example 9.3 Pressure drop in heat exchanger. *120 m^3/h of seawater is used as a coolant in a heat exchanger (see Figure 5.2 page 130). The seawater flows on the tube side through 62 parallel tubes, each 6m long with an internal diameter of 18 mm. The density of the seawater is 1030 kg/m^3 and the viscosity is 1.13 cP = 0.00113 kg/m s. At the inlet of the tubes (inside the heat exchanger), there is a pressure drop corresponding to 0.5 velocity heights ($n = 0.5$) and at the exit there is a pressure drop corresponding to $n = 1$. For a steel tube, the surface roughness is $\epsilon = 0.045$ mm. Calculate the pressure drop for the heat exchanger on the pipe side.*

Solution. *The mean velocity in the heat exchanger tubes is*

$$v = \frac{120}{3600 \cdot 62 \cdot \frac{\pi}{4} \cdot 0.018^2} = 2.11 \quad m/s$$

The sum of the pressure drops at the inlet and exit to each tube is

$$\Delta p_f = (0.5 + 1)\rho\frac{v^2}{2} = 1.5 \cdot 1030 \cdot \frac{2.11^2}{2} = 3439 \ N/m^2 = 0.034 \ bar$$

The Reynolds number in the pipes is

$$Re = \frac{\rho v D}{\mu} = \frac{1030 \cdot 2.11 \cdot 0.018}{0.00113} = 34600$$

The relative roughness is $\epsilon/D = 0.045/18 = 0.0025$, and from Figure 9.6, the friction factor is $f = 0.0075$. The pressure drop in the pipes is then

$$\Delta p_f = 4f\frac{L}{D}\rho\frac{v^2}{2} = 4 \cdot 0.0075 \cdot \frac{6}{0.018} \cdot 1030 \cdot \frac{2.11^2}{2} = 22928 \ N/m^2 = 0.229 \ bar$$

The total (friction) pressure drop through the heat exchanger is then $0.034 + 0.229 = 0.263$ bar.

(Note that we have not included the pressure drop at the inlet and exit of the heat exchanger. This is because the pressure drop is a function of the velocity in the pipes to and from the heat exchanger, which we do not know.)

[2] B.L. Liptak (Editor), *Instrument Engineers' Handbook*, Volume II (Process control and optimization), 4th Edition, CRC (Taylor & Francis), p. 1051 (2006)

Table 9.1: Friction pressure drop in fittings, valves, etc. given both as equivalent pipe length divided by the pipe diameter, L_e/D, and as the factor n in the equation $\Delta p_f = n\rho v^2/2$

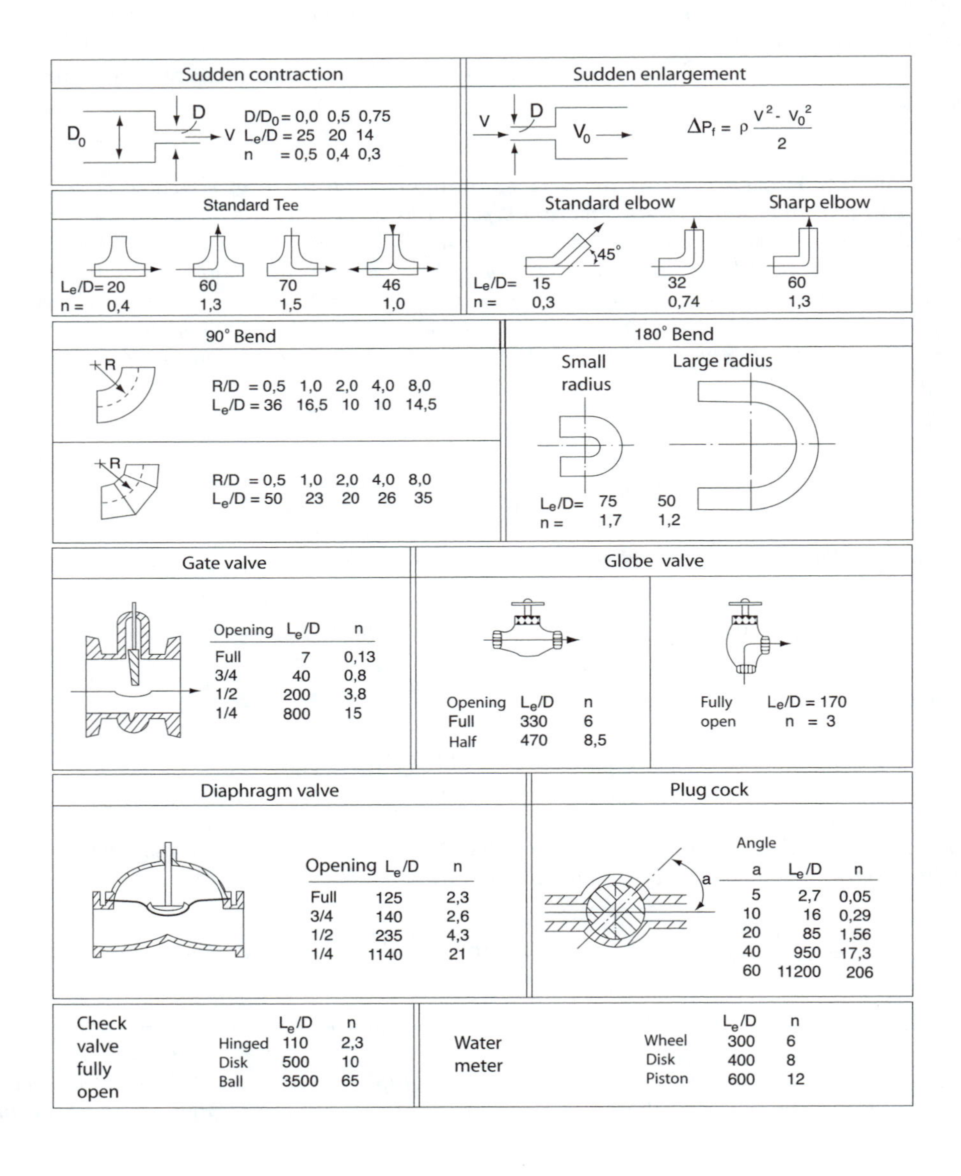

9.5 Compressible flow in pipe (gases)

For the case of with no shaft work ($W_s = 0$), the mechanical energy balance (9.6) from point 1 to 2 in a continuous process is

$$\alpha_2 \frac{v_2^2}{2} + gz_2 + \int_{p_1}^{p_2} \frac{dp}{\rho} + \frac{\Delta p_f}{\rho} = \alpha_1 \frac{v_1^2}{2} + gz_1 \quad [J/kg] \tag{9.19}$$

Differentiating (9.19) gives

$$vdv + gdz + \frac{dp}{\rho} + \frac{dp_f}{\rho} = 0 \tag{9.20}$$

where we have *assumed* turbulent flow ($\alpha \approx 1$) and used $d(v^2)/2 = vdv$. For pipe flow, the differential friction pressure drop over a small pipe length dL is from (9.16)

$$dp_f = 4f \frac{dL}{D} \rho \frac{v^2}{2} \quad [\mathrm{N/m^2}] \tag{9.21}$$

Because the density ρ is small for gases, the contributions from potential and kinetic energy can usually be neglected compared to pressure changes[3] in the mechanical energy balance. We then get from (9.20),

$$dp = -dp_f = -4f \frac{dL}{D} \rho \frac{v^2}{2} \quad [N/m^2] \tag{9.22}$$

that is, all the pressure drop is caused by friction. From the continuity equation (mass balance) (9.7), the term ρv^2 can be written

$$\rho v^2 = \frac{1}{\rho} \frac{m^2}{A^2}$$

where for an **ideal gas**

$$\rho = \frac{pM}{RT}$$

[3] The material balance (continuity equation) in differential form is

$$d(\rho v A) = 0$$

and for constant cross section A, we get $vd\rho + \rho dv = 0$, that is, $vdv = -v^2 d\rho/\rho$. Using the ideal gas law, we then get for the case with constant temperature that the kinetic contribution is

$$vdv = -v^2 \frac{d\rho}{\rho} = -v^2 \frac{M}{RT} \frac{dp}{\rho}$$

Let us compare the "kinetic contribution" vdv with the "pressure contribution" dp/ρ for a gas with molar mass 20 g/mol and temperature 280 K. We have that the contribution from kinetic energy is less than 1% of the pressure contribution when $|v^2 \frac{M}{RT}| < 0.01$ or $v < 0.1\sqrt{\frac{RT}{M}} = \sqrt{\frac{8.31 \cdot 280}{20 \cdot 10^{-3}}} = 0.1 \cdot 341 = 34$ m/s. Usually, the gas velocity is much lower than this. More generally, we only need to include kinetic energy for gases when we start approaching the speed of sound $c_s = \sqrt{\frac{\gamma RT}{M}}$, where $\gamma = C_p/C_V$.

where M [kg/mol] is the molar mass. We then get

$$dp = -2f \frac{1}{D} \frac{RT}{pM} \frac{m^2}{A^2} dL \quad [N/m^2] \tag{9.23}$$

which applies generally for turbulent pipe flow for an ideal gas when we *neglect kinetic and potential energy.* To integrate (9.23), we multiply both sides by p, so that the left side becomes pdp, and if we assume isothermal flow (T constant) and A constant, the integration from 1 (entrance of the pipe) to 2 (exit) gives

$$\boxed{p_1^2 - p_2^2 = 4f \frac{L}{D} \frac{RT}{M} \frac{m^2}{A^2}} \tag{9.24}$$

This applies to isothermal flow of ideal gas in a pipe with constant cross section and with neglected kinetic and potential energy (this is actually an useful formula even though we have introduced many simplifications).

Note that the above expressions apply to pipe flow, where it is reasonable to neglect changes in kinetic energy. The expressions do not apply at high flowrates, above about 50 m/s.

Example 9.4 Gas pipeline. *A 1150 km long pipeline at the sea bottom transports 2500 million Nm^3 of gas per year.*[4] *The internal diameter of the pipeline is 500 mm (0.5 m) and the pipe material roughness is $\epsilon = 0.045$ mm. The pressure at the pipeline inlet is 180 bar, and it is allowed to drop to 100 bar, before it needs to be raised back again to 180 bar in a compressor booster station.*

(a) Show that the contributions from kinetic and potential energy can be neglected (the height variations are maximum 100m) in the mechanical energy balance.

(b) How many compressor booster stations are needed if the final delivery pressure must be at least 100 bar?

Data: *Assume molar mass 20 g/mol, viscosity 10^{-5} kg/m s, ideal gas, constant temperature of 280 K and 8500 hours of operation per year.*

Solution. *The area of the pipe is $A = (\pi/4)D^2 = (\pi/4)0.5^2 = 0.1963\ m^2$. The molar flow rate is*

$$n = \dot{n} = \frac{2500 \cdot 10^6\ Nm^3/y}{8500\ h/y \cdot 3600\ s/h} \cdot \frac{1}{22.414\ kmol/Nm^3} = 3.645\ \text{kmol/s}$$

that is, the mass flow is

$$m = \dot{m} = \dot{n}M = 3.645 \cdot 10^3\ mol/s \cdot 20 \cdot 10^{-3}\ kg/mol = 72.9\ \text{kg/s}$$

At the inlet (1), we have

$$p_1 = 180\ \text{bar} \Rightarrow \rho_1 = \frac{p_1 M}{RT} = \frac{180 \cdot 10^5\ N/m^2 \cdot 20 \cdot 10^{-3}\ kg/mol}{8.31\ J/mol\ K \cdot 280K} = 154.7\ kg/m^3$$

$$v_1 = \frac{m}{\rho_1 A} = \frac{72.9\ kg/s}{154.7 kg/m^3 \cdot 0.1963 m^2} = 2.40\ m/s$$

At the "exit" (2), before the booster station, we have

$$p_2 = 100 bar \Rightarrow \rho_2 = 86.0\ kg/m^3; \quad v_2 = 4.32\ m/s$$

[4] A Normal cubic meter (Nm^3) is defined at 1 atm and 0 °C – and 22.414 Nm^3 corresponds to 1 kmol, see page 14.

(a) The change in kinetic energy is

$$\frac{v_2^2}{2} - \frac{v_1^2}{2} = \frac{4.32^2}{2} - \frac{2.40^2}{2} = 9.0 - 2.9 = 6.1\ J/kg$$

The change in potential energy is at most

$$gz_{max} = 10 \cdot 100 = 1000\ J/kg$$

The change in pressure energy is given by $\int dp/\rho$, *and integrating with constant temperature and using* $\rho = \frac{PM}{RT}$ *(ideal gas) gives*

$$\int_1^2 \frac{dp}{\rho} = \frac{RT}{M}\int_1^2 \frac{dp}{p} = \frac{RT}{M}\ln\frac{p_2}{p_1} = \frac{8.31 \cdot 280}{20 \cdot 10^{-3}}\ln\frac{100}{180} = -68383\ \mathrm{J/kg}$$

and we see that we can safely neglect the kinetic (contribution less than 0.01%) and potential energy (contribution at most 1.5%).
(b) We want to find the pipe length L that gives a pressure drop from 180 bar to 100 bar. Since the temperature is constant, we have from (9.24)

$$L = \frac{p_1^2 - p_2^2}{4f\frac{RT}{M}\frac{m^2}{A^2}}D \tag{9.25}$$

From the mass balance, $m = \rho VA$, *and the Re-number is*

$$Re = \frac{\rho v D}{\mu} = \frac{mD}{A\mu}$$

Note that we have the same Re-number throughout the pipe

$$Re = \frac{72.9\ kg/s \cdot 0.5\ m}{0.1963\ m^2 \cdot 10^{-5}}\ kg/(m\ s) = 1.86 \cdot 10^7$$

The flow is turbulent since $Re > 2300$. *We further have* $\epsilon/D = 0.0001$, *and from Figure 9.6 (244), we obtain*

$$f = 0.003$$

which gives

$$L = \frac{(180^2 - 100^2)10^{10} \cdot 0.5}{4 \cdot 0.003 \cdot \frac{8.31 \cdot 280}{20 \cdot 10^{-3}}\left(\frac{72.9}{0.1963}\right)^2} = 582 \cdot 10^3\ \mathrm{m}$$

that is, 582 km. With one booster station, we may reach $582 \cdot 2 = 1164$ *km, which is sufficient since the total pipeline is 1150 Km.*
Comment: One source of error in the above calculations is the assumption of ideal gas. Since the pressure is high and the temperature is low, non-ideality is expected.

9.6 A remark on friction

The friction Φ expresses the loss in mechanical energy. But what happens with this friction energy, that is, where does it go?

Consider a steady-state process with inflow (1) and outflow (2). Comparing the mechanical energy balance (9.5) with the "regular" energy balance in (9.1) and using

(6.5) together with the definition of enthalpy, we find that the friction loss can be expressed as

$$\Phi = U_2 - U_1 - Q + \int_{V_1}^{V_2} pdV \tag{9.26}$$

where $V = m/\rho$ [m^3; m^3/s] is the volume or volumetric flow of the fluid. For an incompressible fluid (most liquids), V is constant, that is, $dV = 0$ and the last term equals 0, and we arrive at the important conclusion that the friction energy Φ is either converted into internal energy $(U_2 - U_1)$ or transferred to the surroundings as heat $(-Q)$. This is consistent with our intuition that friction is the conversion of mechanical energy into "thermal energy," and confirms that the concept "heat of friction" is reasonable (by the way, if you would like to see a picture of the author in a frictionless Superman costume then click on the picture on my homepage ☺). For a compressible fluid (gases), we see from (9.26) that friction can also contribute to increasing the volume (expansion gas), which is a form of "useless" mechanical work.

9.7 Summary

The *"regular" energy balance* for a steady-state continuous process between the state 1 (inflow) and state 2 (outflow) is

$$H_2 + m\alpha_2 \frac{v_2^2}{2} + mgz_2 = H_1 + m\alpha_1 \frac{v_1^2}{2} + mgz_1 + Q + W_s \quad [J; J/s]$$

In the "regular" energy balance, the enthalpy term $H = U + pV$, which is primarily a function of temperature, usually dominates. Therefore, the "regular" energy balance is normally used when one is interested in temperature changes. If one is interested in pressure changes, the *mechanical energy balance* is more useful:

$$m\alpha_2 \frac{v_2^2}{2} + mgz_2 + m\int_{p_1}^{p_2} \frac{dp}{\rho} + \Phi = m\alpha_1 \frac{v_1^2}{2} + mgz_1 + W_s \quad [J; J/s]$$

Mechanical energy is not a conserved quantity and the friction term Φ [J; J/s] expresses how much mechanical energy is converted into thermal energy (internal energy U and heat Q). For an incompressible fluid (most liquids), ρ is constant and the mechanical energy balance can be written in the following form, which is called the *generalized Bernoulli equation*:

$$\underbrace{\Delta(p + \rho g z + \rho\alpha \frac{v^2}{2})}_{\text{out}-\text{in}} = \underbrace{\Delta p_{\text{lift}}}_{\text{supplied}} - \underbrace{\Delta p_f}_{\text{friction loss}} \quad [\text{N/m}^2]$$

The friction pressure drop Δp_f can be determined from empirical correlations. If we further (1) neglect friction ($\Delta p_f = 0$), (2) assume $\alpha = 1$ (that is, averaging the velocity is unnecessary) and (3) assume no mechanical work ($\Delta p_{\text{lift}} = 0$), we get the "original" *Bernoulli equation*:

$$p + \rho g z + \rho \frac{v^2}{2} = \text{constant} \quad [N/m^2]$$

An important implication of this equation is that the pressure drops when the velocity increases, for example in a restriction.

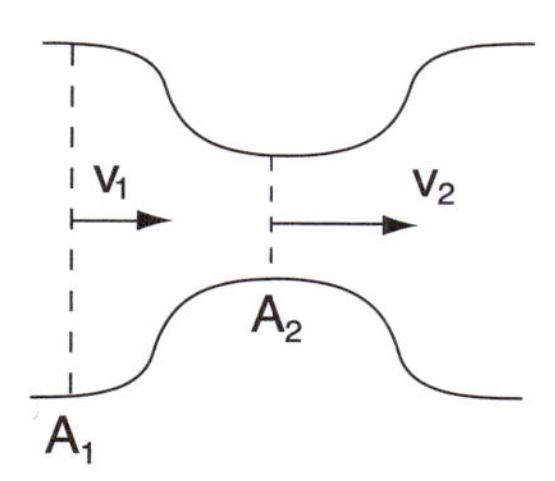

Figure 9.7: Restriction in a pipe

Exercise 9.1 *A gas with temperature* $T_1 = 400K$ *and pressure* $p_1 = 2$ *bar flows in a pipe with velocity* $v_1 = 10$ *m/s. The gas passes through a restriction (2), as shown in Figure 9.7. It can be assumed that the flow is lossless and adiabatic. The gas is assumed ideal with molar mass 35 g/mol and constant heat capacity* $c_p = 35$ *J/K mol. There is no accumulation.*

(a) Explain why the entropy is constant.

(b) Formulate the energy balance from 1 to 2 with the kinetic energy included.

(c) The gas velocity in the restriction is $v_2 = 50$ *m/s. Use the energy balance to find the temperature* T_2.

(d) What is the pressure p_2 *in the restriction (remember that the entropy is constant)?*

(e) Formulate the mass balance from 1 to 2, and express this with density and velocity. Find the area in the restriction relative to the area before the restriction, A_2/A_1.

10

Chemical reaction engineering

In Chapter 3, we assumed the extent of reaction or the conversion to be given, but except for the cases where we assumed chemical equilibrium, we didn't say anything about how it could be determined. To do this, we need a kinetic expression for the reaction rate and details about the reactor. This is the field of chemical reaction engineering considered in this chapter.

10.1 Reaction kinetics

Reaction kinetics involves the study of the rate of chemical reactions. In some cases, the reactions are very fast and it is reasonable to assume that we achieve equilibrium. However, in most cases, equilibrium is not reached and to increase the rate of reaction r, one can increase the temperature or use a catalyst. In this chapter, we use reaction kinetics for two problems:

1. *Design:* Calculate how large the reactor must be (necessary residence time) in order to achieve the desired conversion.
2. *Simulation:* Calculate the conversion for a given reactor (with given residence time).

In industrial reactors, the residence time may vary from milliseconds to hours.

10.1.1 Reaction rate

The reaction rate r_j [mol/m^3, s] for the *reaction* j is defined as

$$\boxed{r_j = \frac{\text{mol of component formed by reaction } j \text{ per unit of time and per unit of volume}}{\text{stoichiometric coefficient for the component in reaction } j}} \tag{10.1}$$

Comments to the definition:

- Note that r here is per unit of volume, so it is the "local" rate, and may vary throughout the reactor.
- Other units for r can be used, for example per mol in the reactor [mol(reacted) / mol(reactor), s], but this is not used here. For heterogeneous catalytic reactions, r is often given per mass or unit area of catalyst.
- Since we divide by the stoichiometric coefficient, we get the same value for the rate r_j no matter which component in the reaction we consider.

- We can also consider the rate of reaction for an individual component. The rate of reaction for an arbitrary component A in reaction j is

$$r_{A,j} = \nu_{A,j} r_j \tag{10.2}$$

where $\nu_{A,j}$ is the stoichiometric coefficient for component A in reaction j. For example, for the reaction $A \rightarrow B$, we have (we omit the subscript j since there is only one reaction)

$$r = -r_A = r_B$$

For example, for the reaction $N_2 + 3H_2 \rightarrow 2NH_3$, we have $r = -r_{N_2} = \frac{-r_{H_2}}{3} = \frac{r_{NH_3}}{2}$.

- The amount of component A generated by chemical reactions can, as shown in Chapter 3, be written as

$$G_A = \sum_j \nu_{A,j} \xi_j = -n_{A,0} X_A$$

where ξ_j is the extent of reaction for reaction j and X_A is the (total) conversion. The extent of reaction ξ_j is directly given by the reaction rate r_j. If we have a reactor with volume V and constant composition and temperature throughout the reactor (so that r_j is constant), then

$$\xi_j = r_j V \tag{10.3}$$

The amount of component A that is generated in all reactions is then

$$G_A = \sum_j \nu_{A,j} \xi_j = \sum_j \nu_{A,j} r_j V = r_A V \tag{10.4}$$

where

$$r_A = \sum_j \nu_{A,j} r_j \quad [\text{mol A/m}^3\text{, s}] \tag{10.5}$$

is the overall reaction rate for generation of component A. Note that when using reaction rates, we do not need to limit ourselves to independent reactions, as recommended in Chapter 3. For example, component A may be formed by several alternative reaction mechanisms, and we should sum up the contributions from all possible reactions; see (10.5).

- **Matrix formulation.** For many components and reactions, we can collect the amounts generated into the vector $\underline{G} = \begin{bmatrix} G_A \\ G_B \\ \vdots \end{bmatrix}$, and from (3.21) we have that

$$\underline{G} = N^T \underbrace{\underline{r} V}_{\underline{\xi}} \tag{10.6}$$

where N is the stoichiometric matrix (see page 90), $\underline{r} = \begin{bmatrix} r_1 \\ r_2 \\ \vdots \end{bmatrix}$ is the vector of reaction rates for the reactions and V is the reactor volume.

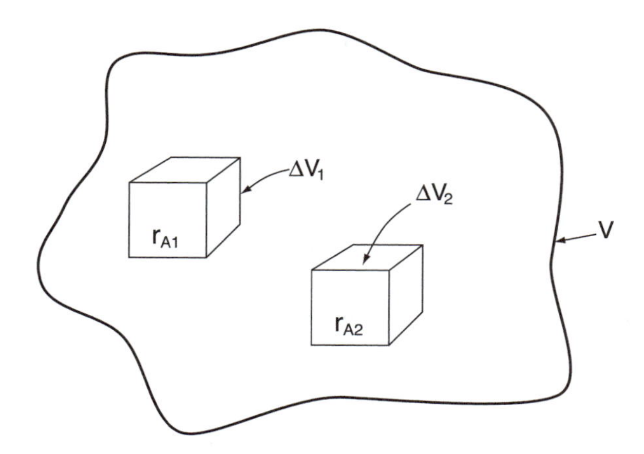

Figure 10.1: Large volume consisting of many small elements of volume

- In this chapter, we limit ourselves for simplicity to cases with only one reaction.
- When deriving (10.4) and (10.6), we assumed that the composition is the same throughput reactor. More generally, the reactor can, as shown in Figure 10.1, consist of many small elements, each with volume ΔV [m^3]. In such a small element of volume, the amount of component A generated

$$\Delta G_A = r_A \Delta V \quad [\text{molA/s}]$$

where r_A is the "local" reaction rate for component A. If the entire reactor consists of many such small elements, the total amount of component A generated is $G_A = \sum_i \Delta G_{Ai} = \sum_i r_{Ai}\Delta V_i$. For differentially small elements of volume dV_i, we can replace the summation by integration and get, for an arbitrary component A, that the amount generated in the reactor i is

$$\boxed{G_A = \int_0^V r_A dV} \quad [\text{mol A/s}] \tag{10.7}$$

Here, r_A is a function of concentration and temperature and the value will vary with the location in the reactor. For the case of ideal mixing (continuous stirred tank reactor, CSTR), the compositions and temperature are uniform such that r_A is constant throughout the reactor

10.1.2 Rate equation

The rate equation gives the dependency of the reaction rate r on temperature and composition. Here, we assume that this dependency can be separated such that the reaction rate can be written

$$r = k(T) \cdot f(\text{composition}) \tag{10.8}$$

For examle, $r = k \cdot c_A$. Here, k is the **reaction rate "constant,"** which is not really a "constant" since it depends on temperature; see below. The specific form of the function f (concentration) depends on the **reaction mechanism**; see below.

10.1.3 Reaction rate constant

The reaction rate constant k is assumed to depend on temperature only, and one normally uses the semi-empirical **Arrhenius equation**

$$\boxed{k(T) = A \cdot e^{-E/RT}} \tag{10.9}$$

where A is the frequency factor and E [J/mol] is the activation energy for the reaction. The numerical value of A is often very large and it is more convenient to rewrite (10.9) with the rate constant $k(T_0)$ at temperature T_0 as a basis, $k(T_0) = A \cdot e^{-E/RT_0}$. We then have

$$k(T) = k(T_0)e^{-\frac{E}{R}\left(\frac{1}{T}-\frac{1}{T_0}\right)} \tag{10.10}$$

Taking the logarithm of (10.9) gives

$$\ln k = \ln A - \frac{E}{R} \cdot \frac{1}{T}$$

Thus, with an Arrhenius temperature dependency, $\ln k$ as a function of $1/T$ is a straight line with slope $-E/R$. This can be used to determine the activation energy E from experimental data. The activation energy is a measure of the reaction's sensitivity to temperature – the higher the value of E, the more k changes with the temperature.

Exercise 10.1 *A common (but rather questionable) "rule of thumb" states that the reaction rate (and consequently, the value of k) is doubled when the temperature increases by 10K. What does this presume about the value of the activation energy E when we assume the temperature is 300K?*

10.1.4 Order of reaction and reaction mechanism

The dependency of the reaction rate on composition, f in (10.8), is determined by the reaction mechanism. For example, for a *first-order irreversible reaction*

$$A \rightarrow B$$

the reaction rate (with elementary kinetics) is

$$r = kc_A \quad [\text{mol/m}^3,\ \text{s}]$$

That is, $f = c_A$, where c_A [mol/m^3] is the concentration of A. In this case, the reaction rate constant k has unit [1/s]. For a *second-order irreversible reaction*

$$2A \rightarrow B$$

we have (with elementary kinetics)

$$r = kc_A^2 \quad [\text{mol/m}^3,\ \text{s}]$$

where the reaction rate constant k has unit [m^3/mol A, s].

By "**elementary kinetics**" we mean that the reaction equation tells what actually happens, so that all the components given in the reaction need to come together (collide) simultaneously for the reaction to occur. The likelihood for a collision is found by multiplying the concentrations of the components, as given by the stoichiometry. For example, for the forward reaction $A + 2B \rightarrow C$ the reaction rate with elementary kinetics is $r = kc_A c_B^2$.

By the term "**reaction order**," we mean the power to which the concentration is raised in the expression for f. Consider, for example, the reaction

$$A \rightarrow B$$

with reaction rate

$$r = kc_A^\alpha$$

Here, α is the reaction order. With $\alpha = 0$, we have a zeroth order reaction and the reaction rate is independent of concentration c_A. Correspondingly, a first-order reaction has $\alpha = 1$, and a second-order reaction has $\alpha = 2$. The reaction $A + 2B \rightarrow C$ with

$$r = kc_A c_B^2$$

is said to be first-order with respect to A and second-order with respect to B, and the overall reaction order is $1 + 2 = 3$. Note that the order does not need to be an integer.

For a **reversible reaction**, we also need to consider the reverse reaction. For example, for the reaction

$$2A \rightleftharpoons B$$

the reaction rate can be given by

$$r = k_1 c_A^2 - k_2 c_B$$

For **gas phase reactions**, partial pressure rather than concentration is usually used. For example, for the gas phase reaction $A + B \rightarrow C$, we may write

$$r = kp_A p_B$$

where p_i [bar] is the partial pressure of component i and k here has unit [mol A/m^3, s bar^2].

More generally, instead of concentration or partial pressure, the activity a of the components is used in the reaction rate, that is, we write $r = k(T) \cdot f(a_i)$.

For cases with **elementary reaction kinetics** the stoichiometric coefficients appear directly as exponents in the rate expressions. For example, for the gas phase reaction

$$aA + bB \leftrightharpoons cC + dD$$

with elementary reaction kinetics we have

$$r = k_1 p_A^a p_B^b - k_2 p_C^c p_D^d$$

However, in general, the reaction mechanisms are not so simple. For example, for a heterogeneous catalytic reaction, the rate is often proportional to the fraction θ of active surface sites occupied (**Langmuir kinetics**). Here, θ cannot exceed 1 and the resulting reaction rates are in the form (for example, for the reaction $A + B \rightarrow P$)

$$r = k\theta_A\theta_B = k\frac{b_A p_A \cdot b_B p_B}{(1 + b_A p_A + b_B p_B + b_C p_C)^2}$$

In addition, other factors may be important and affect the resulting reaction rate expression, for example, "resistance" related to diffusion of the components. The reaction rate and the reaction order are normally determined empirically, based on experimental data (see also page 267).

Example 10.1 Kinetics for ammonia synthesis. *Consider the gas phase reaction for formation of ammonia over iron catalyst,*

$$N_2 + 3H_2 \leftrightharpoons 2NH_3$$

If we assume elementary kinetics, we would get $r = k_1 p_{N_2} p_{H_2}^3 - k_2 p_{NH_3}^2$, *but experimentally it has been found that the kinetics are better described using the Temkin-Pyzhev equation,*

$$r = k_1 \frac{p_{N_2} p_{H_2}^{1.5}}{p_{NH_3}} - k_2 \frac{p_{NH_3}}{p_{H_2}^{1.5}} \tag{10.11}$$

10.1.5 Consistency between kinetics and thermodynamics

If a reaction runs for a sufficiently long time (for example, if we leave a batch system to itself, or use a sufficiently large reactor for a continuous process), then we will reach the state of chemical equilibrium where the net reaction rate is zero, $r = 0$. For example, consider the reaction

$$aA + bB \leftrightharpoons cC + dD \tag{10.12}$$

and assume that the reaction rate depends only on the components that appear in the reaction,

$$r = k_1 f_1(a_A, a_B, a_C, a_D) - k_2 f_2(a_A, a_B, a_C, a_D) \tag{10.13}$$

(where a is the activity, and k_1 and k_2 are reaction rate constants). At equilibrium, we have $r = 0$ and we derive

$$\frac{k_1}{k_2} = \frac{f_2(a_A, a_B, a_C, a_D)}{f_1(a_A, a_B, a_C, a_D)} \tag{10.14}$$

However, chemical equilibrium is also ruled by thermodynamics, and from (B.46) (page 386) we must at equilibrium have that

$$K = \frac{a_C^c \cdot a_D^d}{a_A^a \cdot a_B^b} \tag{10.15}$$

where K is the thermodynamic equilibrium constant, which depends on temperature only. In order to have consistency between (10.14) and (10.15), we must have

$$\frac{k_1}{k_2} = K \tag{10.16}$$

and also

$$\frac{f_2(a_A, a_B, a_C, a_D)}{f_1(a_A, a_B, a_C, a_D)} = \frac{a_C^c \cdot a_D^d}{a_A^a \cdot a_B^b} \tag{10.17}$$

This line of thinking led Guldberg and Waage in 1864 to propose the law of mass actions. They postulated (10.17) by assuming elementary reactions, and then from (10.14) derived the thermodynamic equilibrium condition (10.15).

Example 10.1 continued. *For the reaction for formation of ammonia, (10.11), we have at equilibrium:*

$$r = 0 \quad \Rightarrow \quad \frac{k_1}{k_2} = \frac{f_2}{f_1} = \frac{p_{NH_3}/p_{H_2}^{1.5}}{p_{N_2}p_{H_2}^{1.5}/p_{NH_3}} = \frac{p_{NH_3}^2}{p_{N_2}p_{H_2}^3}$$

For ideal gas, we have that $a_i = p_i/p^{\ominus}$ (where $p^{\ominus}$=1 bar) and we see the reaction rate (10.11) for ammonia satisfies (10.17).

It is also necessary to have consistency between the activation energy and the heat of reaction $\Delta_r H^{\ominus}(T)$. If we assume Arrhenius temperature dependency

$$k_1 = A_1 e^{-E_1/RT}; \quad k_2 = A_2 e^{-E_2/RT}$$

then $k_1/k_2 = K$ in (10.15) dictates the following relationship

$$E_1 - E_2 = \Delta_r H^{\ominus}(T) \tag{10.18}$$

Proof: $k_1/k_2 = K$ gives $\ln k_1 - \ln k_2 = \ln K$. Differentiating with respect to temperature gives $d\ln k_1/dT - d\ln k_2/dT = d\ln K/dT$. Here, $d\ln k_1/dT = E_1/RT^2$, $d\ln k_2/dT = E_2/RT^2$ (Arrhenius), and from thermodynamics, see (B.53), $d\ln K/dT = \Delta_r H^{\ominus}/RT^2$, and we derive $E_1 - E_2 = \Delta_r H^{\ominus}$.

Example 10.2 *For an exothermic reaction with $\Delta_r H^{\ominus} = -80$ kJ/mol and $E_1 = 40$ kJ/mol, we need, for the reverse reaction, to have that $E_2 = 40$ kJ/mol $-(-80)$ kJ/mol = 120 kJ/mol.*

Remark 1 The derived relationships (10.16)–(10.18) between thermodynamics and kinetics only apply if we assume that the kinetics can be expressed using the components that are included in the reaction. In particular, this applies to elementary reactions. However, for more complex reactions schemes, for example, an equilibrium reaction $A \leftrightharpoons B$ that takes place over several steps, $A \leftrightharpoons Q \leftrightharpoons B$, the intermediate component Q is included in the kinetics, but not in the thermodynamic relationship for $A \leftrightharpoons B$. Thus, the derived relationships (10.16)-(10.18) apply to *individual* reactions, but not to the overall reaction (unless one is able to eliminate Q from the kinetics – which is normally not possible).

Remark 2 Since the equilibrium constant K is always finite, there are strictly speaking no "irreversible" reactions (without a reverse reaction, that is, with $k_2 = 0$), but in practice many reactions are shifted to the left or they are slow such that we never achieve equilibrium and there is then little purpose in including the reverse reaction.

10.1.6 Catalysis

A catalyst is a component that is not included in the overall reaction's stoichiometry (and is therefore not consumed in the reaction), but which nevertheless increases the reaction rate r. This usually occurs because the catalyst lowers the activation energy E of the reaction.

A well-known example from everyday life is the catalyst used to "clean" the exhaust gas of a car. This catalyst contains both platinum and rhodium. Platinum is the active component for the oxidation $CO + 0.5O_2 \rightarrow CO_2$, while rhodium is the active component for the reduction $2NO \rightarrow N_2 + O_2$.

Most industrially important reactions take place using catalysts:

- Ammonia: The synthesis step where ammonia is formed from nitrogen and hydrogen uses an iron catalyst.
- Methanol: The synthesis step where methanol is formed from CO and hydrogen uses a catalyst where the most important active component is copper.
- Formaldehyde: The reaction where formaldehyde is formed from methanol and oxygen may use a silver catalyst (silver net).
- Nitric acid: The reaction where ammonia reacts with oxygen and forms NO uses a platinum catalyst (platinum net).

In all of the above reactions, including the car catalyst, the reactants and products are in gas phase, while the catalyst is in solid phase. The reaction takes place on "active" sites on the catalyst's surface. This is called **heterogeneous catalysis** because the catalyst and the reactant/product are in different phases. This is often very practical because there is then no problem with separating the catalyst from the product. There are also many industrial examples of **homogeneous catalysis**, where the catalyst is in the same phase (gas or liquid) as the reactant. Examples are acids or bases that catalyze many liquid-phase reactions. In biological reactions, **enzymes** act as catalysts. These are large molecules with molar mass over 6000 g/mol that usually have a three-dimensional structure, which favor certain molecules and hinder others.

Some reactions are **autocatalytic**. This occurs when the forward reaction rate depends on one of the products. For example, we may for the overall reaction $A \rightarrow B$, have a reaction rate $r = kc_Ac_B$. The reason for this may be that the "actual" reaction is $A + B \rightarrow 2B$. An important example of autocatalysis is the biochemical reaction for growth of cells, where B is the cell mass. For an autocatalytic reaction, one needs some product to initiate the reaction, and autocatalytic reactions "strange" behavior, for example, the reaction rate increases with conversion, whereas the opposite is the case for normal reactions.

10.2 Reactor calculations and reactor design

Here, we formulate the component mass balance for several reactor types: (1) batch reactor, (2) continuous stirred tank reactor (CSTR) and (3) continuous plug flow reactor. The resulting equations can be used for analyzing what takes place in a given reactor (simulation), or for calculating the necessary reactor volume (design). We *assume isothermal conditions* in this chapter, which means we do not need the energy balance.

It is important to emphasize that there is actually nothing new in the section – we just formulate the mass balances and introduce the reaction rate r. Let us first consider a simple example that generalizes the reactor types (1) and (2) mentioned above.

Example 10.3 *Consider the reaction $A \rightarrow 2B$ and assume the reaction rate is $r = kc_A$ [mol/m^3, s]. From the stoichiometry, $r_A = (-1)r$ and $r_B = 2r$, see (10.2), and from the general dynamic balance equation (2.8), the component mass balances for a reactor with volume V and perfect mixing become*

$$\frac{dn_A}{dt} = \dot{n}_{A,in} - \dot{n}_{A,out} + (-1)rV \quad \text{[mol A/s]}$$

$$\frac{dn_B}{dt} = \dot{n}_{B,in} - \dot{n}_{B,out} + 2rV \quad [\text{mol B/s}]$$

Let us now concentrate on component A. We consider two special cases:

1. *Closed reactor (batch reactor; "beaker"). Here*

$$\dot{n}_{A,in} = 0; \quad \dot{n}_{A,out} = 0$$

   and the mass balance is

$$\frac{dn_A}{dt} = r_A V \tag{10.19}$$

2. *Steady-state continuous stirred tank reactor (CSTR). Here*

$$\frac{dn_A}{dt} = 0$$

   and if we introduce the standard notation $\dot{n}_A = \dot{n}_{A,out}$ and $\dot{n}_{A0} = \dot{n}_{A,in}$, the mass balance is

$$\dot{n}_A = \dot{n}_{A0} + r_A V \tag{10.20}$$

Introducing the conversion X_A. Below, we consider the simplified case with only one reaction, and it is sufficient to formulate the balance for reactant A. For practical calculations, it is useful to introduce the conversion X_A of component A. From (3.9), we then have for a batch reactor and a continuous steady-state reactor (but not for the general case in Example 10.3):

$$n_A = n_{A0}(1 - X_A) \tag{10.21}$$

where n_{A0} and n_A [mol A or mol A/s] are the amounts of A in the reactor feed and product, respectively (in the continuous case, we sometimes use dots for n_A to show more clearly that it is a rate [mol A/s]). For calculations, we also need to express r_A as a function of X_A, and next are some examples that show how this can be done.

Example 10.4 *Consider a reaction $A \rightarrow P$ with reaction rate $r = -r_A = kc_A$ that takes place in a reactor with constant volume V.* **Task:** *Determine the function $r_A(X_A)$.*

Solution. *Introducing $c_A = n_A/V$ in (10.21) we get $c_A = c_{A0}(1 - X_A)$, and we derive*

$$-r_A = kc_{A0}(1 - X_A) \tag{10.22}$$

Here, we have assumed that the volume V is constant, but more generally, one uses $V(X_A)$.

Example 10.5 *Consider a reaction $A + B \rightarrow P$ with reaction rate $r = -r_A = kc_Ac_B$.* **Task:** *Determine the function $r_A(X_A)$.*

Solution. *Assuming constant volume, we have from (10.21) that $c_A = c_{A0}(1 - X_A)$, and from the stoichiometry $c_B = c_{B0} - c_{A0}X_A$, that is, we find*

$$-r_A = kc_{A0}c_{B0}\left(1 - (1 + \frac{c_{A0}}{c_{B0}})X_A + \frac{c_{A0}}{c_{B0}}X_A^2\right)$$

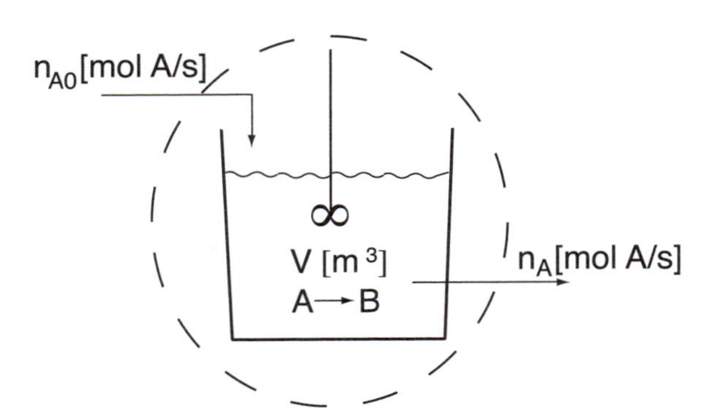

Figure 10.2: Continuous stirred tank reactor (CSTR)

10.2.1 Ideal continuously stirred tank reactor (CSTR)

Let us consider a continuously stirred tank reactor (CSTR) with perfect (ideal) mixing (see Figure 10.2). Note that the term "ideal continuously stirred tank" means that the stirring is perfect (*ideal mixing*); it does not have anything to do with the thermodynamic assumption of "ideal mixture." The CSTR reactor is well known in industry and is an important idealized reactor type. We assume steady-state conditions, that is, there is no accumulation in the system. The mass balance then is "Out = In + Generated," which for component A becomes

$$\boxed{\dot{n}_A = \dot{n}_{A0} + r_A V} \quad [\text{mol A/s}] \tag{10.23}$$

This simple mass balance is called the "design equation" for a CSTR. The equation can be solved with respect to the volume,

$$V = \frac{\dot{n}_{A0} - \dot{n}_A}{-r_A} \tag{10.24}$$

We can alternatively introduce the conversion from (10.21) and get

$$V = \dot{n}_{A0} \frac{X_A}{(-r_A)} \quad [\text{m}^3] \tag{10.25}$$

Note that we have assumed perfect mixing. This means that the concentration is the same throughout the reactor and that the concentration in the outlet stream, c_A, is the same as the concentration in the reactor.

10.2.2 Batch reactor

Let us here consider a batch reactor with no inflow or outflow (closed system) (see Figure 10.3). A well known example is a reaction in a beaker. The component mass balance for a closed system is, also see (2.8),

$$\frac{d}{dt}\text{Inventory} = \underbrace{\text{(Net) formed in reactions}}_{\text{Per unit of time}} \quad [\text{mol A/s}] \tag{10.26}$$

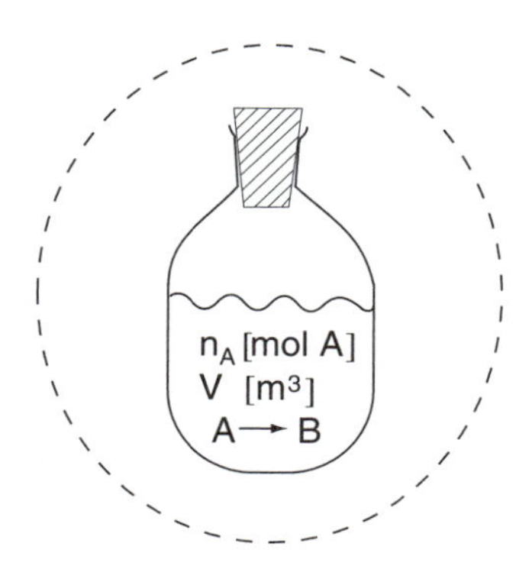

Figure 10.3: Batch reactor (reaction in a beaker)

For component A, the inventory is n_A [mol A], and if we assume perfect mixing, then the amount generated in the reaction is $G_A = r_A V$ [mol A /s]. We get

$$\boxed{\frac{dn_A}{dt} = r_A V} \quad \text{[mol A/s]} \tag{10.27}$$

This simple mass balance is called design equation for a batch reactor in differential form, and no assumption of constant volume is made. We can introduce the conversion X_A, using $n_A = n_{A0}(1 - X_A)$, and since n_{A0} is constant, the mass balance (10.27) then gives $n_{A0} d(X_A)/dt = -r_A V$. Separation of variables gives

$$dt = n_{A0} \cdot \frac{dX_A}{-r_A \cdot V}$$

and integrating from the time $t = 0$, where $X_A = 0$, to time t, gives the reaction time for a batch reactor

$$t = n_{A0} \int_0^{X_A(t)} \frac{dX_A}{-r_A \cdot V} \quad \text{[s]} \tag{10.28}$$

This is the design equation for a batch reactor in integrated form.

By introducing the concentration $c_A = n_A/V$ [mol A/m^3], we can write component balance (10.27) in the form $d(c_A V)/dt = r_A V$. Assuming **constant volume**, we get for a batch reactor

$$\frac{dc_A}{dt} = r_A \quad \text{[mol A/s m}^3\text{]} \tag{10.29}$$

The use of (10.29) to compute the concentration c_A as a function of time for various reaction orders is shown in Chapter 10.2.5 (page 267).

Remark. Warning (and a kick to chemistry teachers). In most chemistry and physical chemistry textbooks, (10.29) is used to define the reaction rate r_A, but this is **completely wrong** – see (10.1) for the correct definition. As derived above, (10.29) is the component balance for a batch reactor with constant volume (which indeed is the type of reactors chemists mostly work with), and it is only in this case that $r_A = dc_A/dt$. For other reactor types, completely different expressions apply; for example, for a steady state continuous stirred tank reactor, we have $dc_A/dt = 0$, and if we used (10.29) as the definition of the reaction rate, this would give $r_A = 0$ (which of course is wrong), whereas the correct answer from (10.23) is $r_A = (n_A - n_{A0})/V$.

10.2.3 Plug flow reactor (PFR)

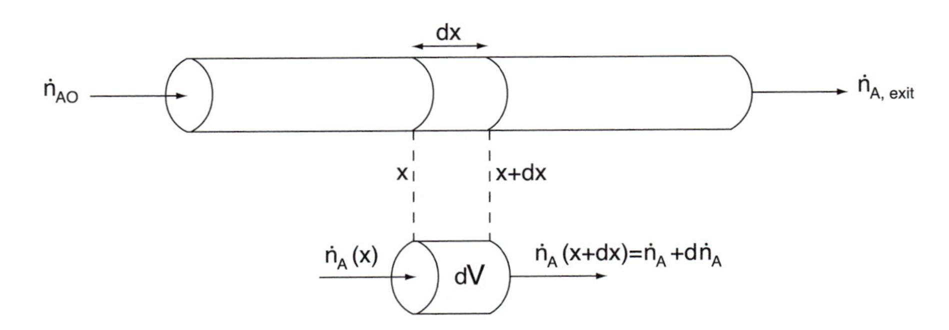

Figure 10.4: Plug flow reactor consisting of many small volume elements

Consider a reactor shaped in the form of a long pipe and assume that the mass moves as a *plug* (without backmixing) through the reactor (see Figure 10.4). This is called a tubular reactor and the idealized case without backmixing is called a plug flow reactor (PFR).

We can imagine that the reactor is divided into a large (infinite) number of small volumes dV. Within each of these small volumes we can assume perfect mixing (as for a CSTR), and the component mass balance over a small volume is (see Figure 10.4 and note that x is reactor length, and not conversion X)

$$\begin{aligned} \text{Mol A in :} &\quad \dot{n}_A \\ \text{Mol A out :} &\quad \dot{n}_A + d\dot{n}_A \\ \text{Mol A generated :} &\quad r_A \cdot dV \end{aligned}$$

The mass balance "Out = In + Generated" then gives

$$\boxed{d\dot{n}_A = r_A dV} \quad [\text{mol A/s}] \tag{10.30}$$

which is called the design equation in differential form for a PFR. Introducing the conversion from (10.21), $\dot{n}_A = \dot{n}_{A0}(1 - X_A)$, gives upon differentiation $d\dot{n}_A = -\dot{n}_{A0}dX_A$, and (10.30) becomes $-\dot{n}_{A0}dX_A = r_A dV$. Separating variables gives

$$dV = \dot{n}_{A0}dX_A/(-r_A)$$

Integration from the inlet, where $V = 0$ and $X_A = 0$, to the exit with volume V, gives (note that $V = \int_0^V dV$):

$$V = \dot{n}_{A0} \int_0^{X_A} \frac{dX_A}{-r_A} \tag{10.31}$$

which is called the design equation for a PFR in integral form.

10.2.4 Comparison of reactor types

Here, we use the mass balance equations derived above to calculate the required reactor volume to obtain a given conversion. We start with an example where we compare the two idealized continuous reactor types, CSTR and PFR.

Example 10.6 Calculation of reactor volume for CSTR and PFR. *Consider a first-order reaction $A \longrightarrow B$ with reaction rate $r = -r_A = kc_A$. We assume that the reaction temperature is constant (isothermal), and at this temperature the reaction rate constant is $k = 0.036 s^{-1}$. The volumetric feed flow rate is $\dot{V}_0$ l/s and has concentration $c_{A0} = 5.5$ mol/l. The desired conversion in the reactor is $X_A = 0.8$. Calculate the required volume for (a) a CSTR and (b) a plug flow reactor (PFR).*

Solution. *For this first-order reaction, we have from (10.22) that*

$$-r_A = kc_A = kc_{A0}(1 - X_A)$$

Furthermore, the feed flow rate of component A is

$$\dot{n}_{A0} = c_{A0}\dot{V}_0$$

(a) **Stirred tank reactor.** *From the "design equation" (10.25) (that is, the component balance) for a CSTR, we have*

$$V = \frac{\dot{n}_{A0} X_A}{-r_A} = \frac{c_{A0}\dot{V}_0 X_A}{kc_{A0}(1 - X_A)} = \frac{\dot{V}_0}{k}\frac{X_A}{(1 - X_A)} \tag{10.32}$$

We note that the feed concentration c_{A0} drops out for the case with a first-order reaction in a CSTR. With numbers inserted, we then get

$$V = \frac{4 \text{l/s}}{0.036\ s^{-1}} \cdot \frac{0.8}{1 - 0.8} = 444\ l$$

(b) **Plug flow reactor.** *From the integral form "design equation" (10.31) (that is, the integrated mass balance) for a PFR, we have*

$$V = \dot{n}_{A0} \int_0^{X_A} \frac{dX_A}{-r_A} = \frac{\dot{n}_{A0}}{c_{A0}} \int_0^{X_A} \frac{dX_A}{k(1 - X_A)} = \dot{V}_0 \int_0^{X_A} \frac{dX_A}{k(1 - X_A)}$$

This can be integrated analytically using $\int dx/(1 - x) = -\ln(1 - x)$, and we get

$$V = \frac{\dot{V}_0}{k} \ln \frac{1}{1 - X_A} \tag{10.33}$$

We see again that the feed concentration drops out. With numbers inserted, we get

$$V = \frac{4 \text{ l/s}}{0.036\ s^{-1}} \cdot \ln 5 = 178\ l$$

which is significantly smaller than for a CSTR.

Plug flow reactor versus CSTR

In the above simple example, the integral for PFR could be solved analytically, but for other cases it is often simpler and more instructive with a graphical solution where we plot $\frac{1}{-r_A}$ as function of X_A (Levenspiel plot). From Figure 10.5 we see that a PFR is better than a CSTR as long as $\frac{1}{-r_A}$ increases when X_A increases, that is, as long as the reaction rate $-r_A$ decreases when the conversion X_A increases. This gives the following rule:

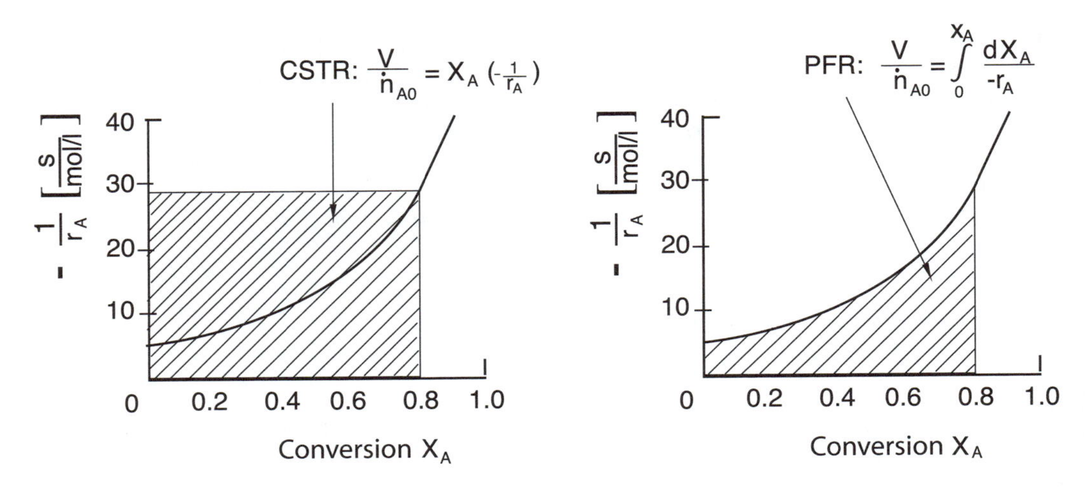

Figure 10.5: Graphical determination of reactor volume (Levenspiel plot)

> For an isothermal reactor with one reaction, a plug flow reactor (PFR) requires a smaller reactor volume than a continuous stirred tank reactor (CSTR) if the reaction rate decreases with increasing conversion, that is, as long as the reaction order is positive.

This is also easy to understand: With a positive reaction order, the reaction goes faster when the concentration of the reactants is high. It is then unfavorable with a continuous stirred tank reactor where the feed is mixed directly into the reactor such that all of the reaction takes place at the "low" exit concentration.

For reactions with negative reaction order, for example, an autocatalytic reaction (for example, biological cell growth), a CSTR is better than a PFR.

For cases with several reactions, a CSTR reactor can be more favorable than a PFR even for "normal" reactions with positive reaction order. For example, if the main reaction is of first order ($r_1 = k_1 c_A$) and we have a second-order undesired side reaction ($r_2 = k_2 c_A^2$), then it is favorable for the selectivity to keep c_A low (since $r_1/r_2 = k_1 k_2 c_A^{-1}$ has negative "order").

All the discussion in this section is for isothermal reactions. For an adiabatic reactor with exothermic reaction and a "cold" feed, a CSTR may be better because the reaction is faster at high temperature. The reaction may have difficulties "getting started" in a plug flow reactor since the feed is cold, while on the other hand, in a CSTR the feed is mixed directly into the hot reactor. In practice, this problem is solved for the plug flow reactor by using the hot product stream from the reactor to pre-heat the feed (for example, see Figure 4.15, page 121).

Similarity between plug flow reactor and CSTR

From the derivation of the design equations for a plug flow reactor, note that n identical continuous stirred tank reactors (CSTR) in series, each with volume V/n, will behave like a PFR reactor with volume V when $n \to \infty$.

Similarity between batch reactor and plug flow reactor

By introducing $\dot{n}_{A0} = c_{A0}\dot{V}_0$, where $\dot{V}_0$ [m^3/s] is the feed flow to the reactor, into the integrated design equation (10.31) for the plug flow reactor, we find that the required residence time in the PFR is

$$\tau_0 = \frac{V}{\dot{V}_0} = c_{A0}\int_0^{X_A} \frac{dX_A}{-r_A} \tag{10.34}$$

Compare this with the required batch reaction time t in (10.28), which, with constant reactor volume V, becomes

$$t = c_{A0}\int_0^{X_A(t)} \frac{dX_A}{-r_A} \tag{10.35}$$

where $c_{A0} = n_{A0}/V$ is the feed concentration (when the reaction starts). We see that (10.34) and (10.35) are identical, that is, the required residence time τ_0 in a plug flow reactor is the same as the reaction time t in a batch reactor. Intuitively, this is reasonable, as a small volume element (plug) that moves through a plug flow reactor, is like a small batch reactor with the same reaction time.

This can also be seen by comparing the mass balance equations for the batch reactor in (10.27) and the plug flow reactor in (10.30). They are identical if we introduce, in the latter, $dx = dV/V$, where x is the non-dimensional reactor length from 0 to 1:

$$\frac{dn_A}{dt} = r_A V \quad \text{and} \quad \frac{d\dot{n}_A}{dx} = r_A V \quad [mol\ A/s]$$

10.2.5 Finding the reaction order and rate constant

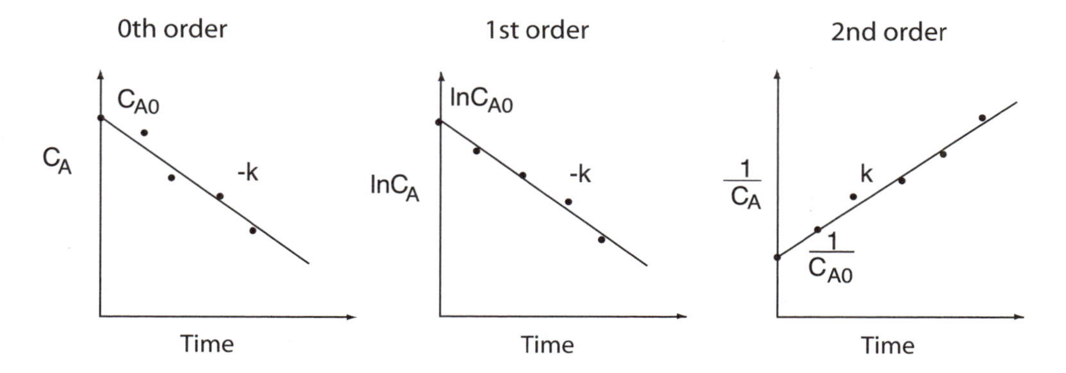

Figure 10.6: Experimental determination of the reaction rate constant k

Here, we show how to determine the reaction order α and the reaction rate constant k based on experimental concentration data from an *isothermal batch reactor* ("reaction in beaker"). If we only have one reaction (A $\rightarrow$ product) and assume constant volume, the mass balance for component A in a batch reactor becomes, see (10.29),

$$\frac{dc_A}{dt} = r_A \quad [\text{mol A/s m}^3] \tag{10.36}$$

Let us assume the reaction rate can be written in the following simple form

$$-r_A = kc_A^{\alpha} \tag{10.37}$$

where k is constant. We have several cases for the reaction order α:

1. $\alpha = 0$ (0th order reaction). Then $dc_A/dt = -k$ or $dc_A = -kdt$, which, when integrated from c_{A0} at time 0 to c_A at time t, gives

$$c_A = c_{A0} - kt \tag{10.38}$$

   That is, for a 0th order reaction, we get a straight line with slope $-k$ if we plot c_A against time (see Figure 10.6).
2. $\alpha = 1$ (1st order reaction). Then $dc_A/dt = -kc_A$ or $\frac{dc_A}{c_A} = -kdt$, which, when integrated from c_{A0} at time 0 to c_A at time t, gives $\ln\frac{c_A}{c_{A0}} = -kt$, or

$$\ln c_A = \ln c_{A0} - kt \tag{10.39}$$

   That is, for a 1st order reaction, we get a straight line with slope $-k$ if we plot $\ln c_A$ against time (see Figure 10.6).
3. $\alpha = 2$ (2nd order reaction). Then $dc_A/dt = -kc_A^2$ or $\frac{dc_A}{c_A^2} = -kdt$, which, when integrated from c_{A0} at time 0 to c_A at time t, gives

$$\frac{1}{c_A} = \frac{1}{c_{A0}} + kt \tag{10.40}$$

   That is, for a 2nd order reaction we get a straight line with slope k if we plot $1/c_A$ against time (see Figure 10.6).
4. More generally, for a reaction order $\alpha \neq 1$, we have $dc_A/dt = -kc_A^{\alpha}$ or $dc_A/c_A^{\alpha} = -kdt$, which, when integrated from c_{A0} at time 0 to c_A at time t, gives[1]

$$c_A^{(1-\alpha)} = c_{A0}^{(1-\alpha)} + (\alpha - 1)kt \tag{10.41}$$

   That is, for a reaction of order $\alpha \neq 1$, we get a straight line with slope $(\alpha - 1)k$ if we plot $c_A^{1-\alpha}$ against time. Thus, by raising c_A to different powers and checking if we get a straight line as a function of t, the order α can be determined, even for cases where α is not an integer.

The method described above for determining the reaction order is called the "integral method." The main assumption is that the reaction rate can be written in the form $-r_A = kc_A^{\alpha}$ in (10.37) where k is constant. In addition to isothermal reaction conditions, this requires that either (1) only one reactant (component A) is included in the kinetics, or (2) we have $c_{A0} = c_{B0}$ for the reaction $A + B \rightarrow$ products, or (3) the other components that effect the reaction kinetics (and enter into the rate expression) are available in such large quantities that their concentrations do not vary with time.

If these conditions are not satisfied, then it is recommended to formulate (guess) a rate expression ($-r_A$), and to compute (estimate) the parameters in the rate equation

[1] Use $\int x^n dx = \frac{1}{n+1}x^{n+1}$, i.e., (set $n = -\alpha$) $\int \frac{1}{x^{\alpha}}dx = \frac{-1}{\alpha-1}x^{-\alpha+1}$.

using numerical regression. Consider, for example, the reaction $A+B \rightarrow P$ and assume that the reaction rate is $r = -r_A = kc_A^\alpha c_B^\beta$. At a given temperature, there are three unknown parameters: k, α and β. They can be obtained numerically by seeking the values that give the smallest deviation between measured and calculated values for, for example, $c_A(t)$ and $c_B(t)$.

Half time. The half time in a batch reactor is when $c_A(t_{1/2}) = c_{A0}/2$ for the reactant A. An alternative method for finding the reaction order is to consider the relationship between the half time and initial concentration c_{A0}. With our rate expression $-r_A = kc_A^\alpha$ we derive ($\alpha \neq 1$):

$$t_{1/2} = \frac{c_{A0}^{1-\alpha}\left(0.5^{1-\alpha} - 1\right)}{k(1-\alpha)} \tag{10.42}$$

We note that the half time increases with c_{A0} for $\alpha < 1$, and decreases with c_{A0} for $\alpha > 1$. For a first-order reaction ($\alpha = 1$), we derive

$$t_{1/2} = \ln 2/k \tag{10.43}$$

which is independent of c_{A0}. Alternatively, if the reaction order is known, we may use these expressions for the half time to find the rate constant k.

Exercise 10.2 *Derive the formula (10.42) for $t_{1/2}$.*

10.2.6 Choice of reaction conditions

The optimal choice of reaction conditions depends on many factors, including the reaction kinetics, thermodynamics (heat of reaction and equilibrium conditions), reactor type, costs for separation and recycle, and so on. We do not consider all of them here but give some simple guidelines for the optimal choice of temperature and pressure in the reactor.

Pressure. Let us first consider the pressure in a gas phase reactor (or possibly the concentration in a liquid phase reactor). High reaction rates (and reduction of reactor volume) are favored by high pressure. However, if the reaction is close to equilibrium, then the reverse reaction may be important and it is not obvious that a high pressure is favorable. According to Le Chatelier's principle, reversible reactions with mole number reduction (for example, $N_2 + 3H_2 = 2NH_3$) are favored by high pressure, while reversible reactions with mole number increase (for example, $CH_4+H_2O = CO+3H_2$) are favored by low pressure.

Temperature and conversion. A *high* temperature increases the reaction rates, which results in a reduction in the required reactor volume. However, a high reaction temperature does not necessarily give a high conversion, because of equilibrium (thermodynamic) limitations. In fact, according to Le Chatelier's principle, the equilibrium conversion for exothermic reactions (for example, $N_2+3H_2 = 2NH_3$ with $\Delta_r H^\ominus(298) = -92$ kJ/mol) is favored by *low* temperature, whereas it for reversible endothermic reactions (for example, $CH_4+H_2O = CO+3H_2$ with $\Delta_r H^\ominus(298) = +206$ kJ/mol) is favored by *high* temperature.

Temperature and heat integration. Another issue is heat integration. Normally, we need to supply heat for an endothermic reaction and remove heat (cool) for an exothermic reaction. In general, heat is more valuable at high temperature (as expressed, for example, by the Carnot factor). Therefore, in order to extract "high-grade" (high-temperature) heat from an exothermic reaction, the reaction should take place at high temperature (which by the way is the opposite of what is favored by the equilibrium). Correspondingly, in order to be able to use "low-grade" (low-temperature) heat for an endothermic reaction, the reaction should take place at low temperature (which again is the opposite of what is favored by the equilibrium).

Other factors. However, there are many other factors to consider. The most important is often side reactions which result in formation for byproducts. The reaction conditions are often chosen to suppress side reactions, and this can drastically change the above recommendations. The optimal reaction conditions are determined from a trade-off between these conflicting factors, and should generally be based on an economic criterion.

Exercises

Exercise 10.3* *An irreversible liquid-phase reaction $A \rightarrow P$ with reaction rate $(-r_A) = kc_A$ takes place in an isothermal ideal continuous stirred tank reactor (CSTR). The density of the liquid can be assumed constant. The volumetric feed rate is 0.1 m^3/min, $c_{A0} = 10\ mol/m^3$ and the reactor volume is $1m^3$. The temperature is 50 ^oC and the reaction rate constant at this temperature is 0.003 s^{-1}.*

(a) What is the conversion in the reactor?

(b) The reaction is instead performed in an isothermal plug flow reactor (PFR) with volume 1 m^3. What is the conversion now?

Exercise 10.4 *A liquid-phase reaction $A \rightarrow P$ is performed in an isothermal ideal continuous stirred tank reactor (CSTR). The reaction rate is $(-r_A) = kc_A$. The reaction rate constant is given at two temperatures: $k = 0.015\ s^{-1}$ (50 ^oC) and $k = 0.024\ s^{-1}$ (60 ^oC). The reaction takes place with surplus solvent such that the density can be assumed constant. The volumetric feed rate is $72m^3/h$ and $c_{A0} = 10\ mol/m^3$.*

(a) Formulate the mass balance for component B.

(b) The reaction is performed at 65 ^oC. What reactor volume is needed to achieve a conversion of 94% ?

(c) The reaction is endothermic and the reactor's heat supply is lost due to an error such that the temperature drops to 35 ^oC. What is the conversion now?

Exercise 10.5* *Component A dimerizes in an irreversible liquid-phase reaction $2A \rightarrow P$ with reaction rate $r_P = kc_A^2$. The reaction takes place in a continuous stirred tank reactor (CSTR). The volumetric feed rate is 0.1 m^3/s and $c_{A0} = 2\ mol/l$. The rate constant at 363 K is $k = 0.0003$ l/mol s. The activation energy is $E = 96$ kJ/mol.*

(a) The reactor operates at 386 K. Find the rate constant at this temperature.

(b) Formulate the mass balance for component A.

(c) Given that the conversion of A is 66%, what is the reactor volume?

(d) Will the conversion increase if you replace the CSTR reactor with a plug flow reactor (PFR) with the same volume (give your reasons)?

Exercise 10.6* *(a) The irreversible liquid-phase reaction $A \rightarrow P$ with reaction rate $-r_A = kc_A$ takes place in an isothermal batch reactor with constant volume. After five minutes, the conversion of A is 30%. How long does it take before the conversion is 50%?*

Data: $T = 323\ K$, $c_{A0} = 1\ mol/l$, $V_{\text{batch}} = 100l$.

(b) The reaction is instead performed in a plug flow reactor (PFR) with volume 1 m^3 and feed rate 0.1 m^3/min. What is the conversion now?

Exercise 10.7 *The wastewater from a plant contains a harmful substance G, which is removed by converting it chemically to the harmless substance H. You have studied the reaction in an ideal batch reactor at 50 ° and have obtained the following data:*

time (min)	c_G *[mol/m^3]*
0	13.53
18	7.73
36	4.40
54	2.47
72	1.39
90	0.77
108	0.44
126	0.25

The reactor volume and density can be assumed constant.

(a) Determine the reaction order and rate constant k at 50 °C, k_{50}.

(b) Given that the activation energy for the reaction is 100 kJ/mol: What is the reaction rate constant at 33 °C, k_{33}?

(c) Industrially, the reaction is performed in a CSTR at 33 °C. 1 m^3/h wastewater containing 10 mol/m^3 of the substance G is treated. The authorities require that the outlet concentration of G must not exceed 1 mol/m^3. How large must the reactor be to meet the requirement?

(d) The factory is expanded, and the amount of wastewater that must be treated is doubled (the inlet concentration of G is still 10 mol/m^3). The permit from the authorities, however, does not allow one to increase the total discharge of G (that is, we have to reduce the exit concentration from 1 to 0.5 mol/m^3). Suggest two alternative solutions for obtaining sufficient treating capacity. Discuss (please include calculations) which solution you would choose.

11

Process dynamics

In a dynamic system, the values of the variables change with time, and in this chapter we quantify the well-known fact that "things take time." We also consider dynamic modeling, dynamic responses (analysis), dynamic simulation (numerical calculation) and process control.

11.1 Introduction

Some reasons for considering a system's dynamics and obtaining dynamic models are:

1. To describe the time behavior of a batch process.
2. To describe the transient response of a continuous process (e.g., dynamic change from one steady state to another).
3. To understand the dynamics of the process (analysis), for example, as expressed by the time constant.
4. To develop a "training simulator" for operator training.
5. For "what occurs if" studies, for example, as a tool in a HAZOP analysis ("what happens if this valve is closed?").
6. For optimization and control (control structure, tuning of controllers, model-based control).

Note that when it comes to dynamics, there is no difference between a model for a batch process a continuous process.

The dynamic models we consider in this chapter are given in the form of differential equations,

$$\frac{dy}{dt} = f(y, u) \tag{11.1}$$

where u is the independent variable and y the dependent variable, as seen from a cause-and-effect relationship. With a dynamic model, it is possible, given the system's initial state ($y(t_0) = y_0$) and given the value of all of the independent variables ($u(t)$ for $t > t_0$), to compute ("simulate") the value of the dependent variables as a function of time ($y(t)$ for $t > t_0$).

Up to now, we have studied steady-state behavior, where time t was not a variable. The steady-state model $f(y, u) = 0$ gives the relationship between the variables u and y for the special case when $dy/dt = 0$ ("the system is at rest").

The basis for a dynamic model can be

1. Fundamental: From balance equations + physics/chemistry; see the next section
2. Empirical (regression-based): From experimental data (measurements)

Often we use a combination, where the parameters of a fundamental model are obtained from experimental measurement data.

Comment on notation. The dot notation ($\dot{X}$) is used other places in this book to indicate rate variables (e.g., $\dot{m}$ [kg/s] denotes the mass flow rate). However, in other fields and books, particularly in control engineering, the dot notation indicates time derivative (that is $\dot{m} \equiv dm/dt$). Since we work, in this chapter, with both time derivatives and rates, we here choose to avoid the dot notation altogether. The following special symbols are instead used for rates (amount of stream per unit of time):

- Molar flow rate: $F \equiv \dot{n}$ [mol/s]
- Mass flow rate: $w \equiv \dot{m}$ [kg/s]
- Volumetric flow rate: $q \equiv \dot{V}$ [m^3/s]

11.2 Modeling: Dynamic balances

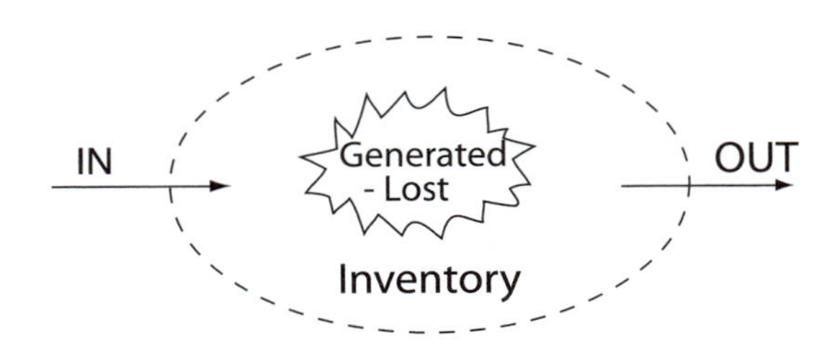

Figure 11.1: The balance principle

Here, we show how dynamic models can be derived from the balance equations for total mass, energy and component mass (mole). This gives, at the same time, an overview and a review of the material presented in previous chapters. Consider a system with a well-defined boundary ("control volume"), see Figure 11.1. The starting point for a fundamental model is the balance equations (see Chapter 2).

$$\boxed{\underbrace{\text{Change Inventory}}_{\text{accumulated in the system}} = \underbrace{\text{In} - \text{Out}}_{\text{through the system's boundary}} + \underbrace{\text{Generated} - \text{Loss}}_{\text{internally in the system}}}$$

In this chapter, the terms "change," "in," "out," "generated" and "loss" are always per unit of time. Mathematically, the general balance equation *per unit of time* is (see (2.8) on page 42):

$$\boxed{\frac{dB}{dt} = B_{\text{in}} - B_{\text{out}} + B_{\text{generated}} - B_{\text{loss}}} \quad [\frac{\text{kg}}{\text{s}}, \frac{\text{mol}}{\text{s}}, \frac{\text{J}}{\text{s}}, \ldots] \tag{11.2}$$

Here B is the inventory of the quantity that we are considering (inside the system's boundary), $\frac{dB}{dt}$ is the change in the inventory per unit of time, $B_{\text{in}} - B_{\text{out}}$ is net supplied through the system's boundary (with mass flows or through the wall) and $B_{\text{generated}} - B_{\text{loss}}$ is net supplied internally in the system. For conserved quantities (*mass* and *energy*), we have $B_{\text{generated}} = 0$ and $B_{\text{loss}} = 0$. *Component mass (mol)* is not conserved, so we have to include a term for "net generated in chemical reactions," which represents the sum of "generated" and "lost." Similarly, *momentum* (mechanical energy) is not conserved and we have to include a friction term.

In principle, the balance equations are easy to formulate, but we need to decide:

1. Which control volume (where do we draw the boundary for the quantity we are balancing)?
2. Which balance (which quantity are we considering, for example, mass or energy)?

The answer to the last question is typically:

- Interested in mass, volume or pressure: *mass balance*
- Interested in concentration: *component balance*
- Interested in temperature: *energy balance*
- Interested in the interaction between flow and pressure: *Mechanical energy balance* (= *momentum balance* = Bernoulli = Newton's second law) (in some of the examples below, we use the static momentum balance where the term for acceleration is neglected).

11.2.1 Dynamic total mass balance

The total mass balance per unit of time is

$$\boxed{\frac{dm}{dt} = w_{\text{in}} - w_{\text{out}}} \qquad [\text{kg/s}] \tag{11.3}$$

where m [kg] is the system's mass ("inventory of mass inside the control volume"), dm/dt [kg/s] is the change in mass inventory per unit of time and $w_{\text{in}} - w_{\text{out}}$ [kg/s] are the mass flow rates for for the entering and exiting streams (bulk flow). By introducing the density, we get

$$\frac{d(\rho V)}{dt} = \rho_{\text{in}} q_{\text{in}} - \rho_{\text{out}} q_{\text{out}} \quad [\text{kg/s}]$$

where V [m^3] is the system's volume, q_{in} [m^3/s] and q_{out} [m^3/s] are the volumetric flow rates and ρ, ρ_{in} and ρ_{out} [kg/m^3] are the (average) densities.

For liquid-phase systems, it can often be assumed that the density ρ is constant (that is, $\rho = \rho_{\text{in}} = \rho_{\text{out}} =$ constant), and the mass balance becomes a "volume balance"

$$\text{Constant density}: \quad \frac{dV}{dt} = q_{\text{in}} - q_{\text{out}} \quad [\text{m}^3/\text{s}] \tag{11.4}$$

Quotation marks are here used to show that volume is generally *not* a conserved quantity. In practice, it is often the liquid level (or height h [m]) that is of interest. The relationship between volume and level is $V = Ah$ for a tank with constant cross section area A [m^2], and more generally $V = \int A(h)dh$ when A varies with height. We then get

$$\frac{dV}{dt} = A\frac{dh}{dt} + h\frac{\partial A}{\partial h}\frac{dh}{dt}$$

where the last term is zero for a constant cross section area A (since $\partial A/\partial h = 0$).

Note that the total number of moles in the system is generally not a conserved quantity, that is, the total mole balance is

$$\frac{dn}{dt} = F_{\text{in}} - F_{\text{out}} + G \quad [\text{mol/s}] \tag{11.5}$$

where G [mol/s] is the net generated number of moles in chemical reactions.

11.2.2 Dynamic component balance

The dynamic **component balance** can, for an arbitrary component A, be written

$$\boxed{\frac{dn_A}{dt} = F_{A,\text{in}} - F_{A,\text{out}} + G_A} \quad [\text{mol A/s}] \tag{11.6}$$

(we normally use mole basis, but the component balance can also be written on weight basis [kg A/s]). Here, n_A [mol A] is the inventory (amount) of component A inside the system's boundary, $F_{A,\text{in}} - F_{A,\text{out}}$ [mol A/s] are the molar flow rates of A in the streams (bulk flow) and G_A [mol A/s] is net generated in the chemical reactions. This can, from (3.7), be calculated from

$$G_A = \sum_j \nu_{A,j}\xi_j \quad [\text{mol A/s}]$$

where $\nu_{A,j}$ is the stoichiometric coefficient for component A in reaction j, and ξ_j [mol/s] is the extent of reaction for reaction j. Instead of the extent of reaction, one can alternatively use the reaction rate, and from (10.7), write

$$G_A = \int_0^V \underbrace{\sum_j \nu_{A,j} r_j}_{r_A} dV \quad [\text{mol A/s}] \tag{11.7}$$

where r_j [mol/ m^3 s] is the reaction rate for reaction j. Note that we in the dynamic case usually do *not* restrict ourselves to independent reactions because this makes it more difficult to introduce the reaction rate. The reaction rate is a function of concentration and composition, and generally varies with the position in the reactor (and therefore the integral in (11.7)).

For example, for a first-order reaction $A \rightarrow B$, we can have that

$$r = k(T)c_A \quad [\text{mol A/s m}^3]$$

Here, we have $r_A = -r$, where the sign is negative because A is consumed in the reaction and the stoichiometric coefficient is $\nu_A = -1$. We often assume that the temperature dependency of the reaction rate constant k follows Arrhenius' equation

$$k(T) = Ae^{-E/RT}$$

where A is a constant and E [J/mol] is the activation energy. We also introduce

$$\bar{c}_A = n_A/V; \quad c_{A,\text{in}} = F_{A,\text{in}}/q_{\text{in}}; \quad c_{A,\text{out}} = F_{A,\text{out}}/q_{\text{out}}$$

where $\bar{c}_A$ [mol/m^3] is the average concentration of A in the reactor. Similarly, the average reaction rate is defined $\bar{r}_A = (\int r_A dV)/V$. Then $G_A = \bar{r}_A V$ and the component balance can be written

$$\boxed{\frac{d(\bar{c}_A V)}{dt} = c_{A,\text{in}} q_{\text{in}} - c_{A,\text{out}} q_{\text{out}} + \bar{r}_A V} \quad [\text{mol A/s}] \tag{11.8}$$

Here we have used concentration c, but we may alternatively use mole fraction or weight fraction.

Example 11.1 Ideal continuous stirred tank reactor (CSTR). *Here we have perfect mixing and we do not need to use average values, that is,* $\bar{c}_A = c_A$ *and* $\bar{r}_A = r_A$*. Furthermore, we have that* $c_{A,\text{out}} = c_A$ *and the component balance (11.8) is*

$$\frac{d(c_A V)}{dt} = c_{A,\text{in}} q_{\text{in}} - c_A q_{\text{out}} + r_A V \tag{11.9}$$

If we, in addition, **assume constant density** ρ*, we can introduce the "volume balance" (11.4) such that the left side of (11.9) is*

$$\frac{d(c_A V)}{dt} = c_A \frac{dV}{dt} + V \frac{dc_A}{dt} = c_A (q_{\text{in}} - q_{\text{out}}) + V \frac{dc_A}{dt}$$

The "out term" in (11.9) then drops out and the component balance for a CSTR becomes

$$\boxed{V \frac{dc_A}{dt} = (c_{A,\text{in}} - c_A) q_{\text{in}} + r_A V} \quad [mol\ A/s] \tag{11.10}$$

Note that, with the assumption of constant density, this equation applies even if the reactor volume V *varies.*

With a little practice, the balance (11.10) may be set up directly: "The concentration change in a CSTR is driven by the inflow having a different composition plus the contribution for chemical reaction." However, it is generally recommended to start from equation (11.6).

11.2.3 Dynamic energy balance

The general energy balance (4.10) over a time period Δt with $\Delta U = U_f - U_0$ gives, as $\Delta t \to 0$, the dynamic energy balance:

$$\boxed{\frac{dU}{dt} = H_{\text{in}} - H_{\text{out}} + Q + W_s - p_{\text{ex}} \frac{dV}{dt}} \quad [\text{J/s}] \tag{11.11}$$

Here, U [J] is the internal energy for the system (inside the control volume), while $H_{\text{in}} - H_{\text{out}}$ is the sum of internal energy in the streams plus the flow work that the streams perform on the system as they are "pushed" in or out of the system. The term $-p_{\text{ex}} \frac{dV}{dt}$ is the work supplied to the system when its volume changes; it is negligible for most systems. Q [J/s] is supplied heat (through the system's wall), while W_s [J/s] is supplied useful mechanical work (usually shaft work, for example, from a compressor, pump or turbine). Note that there is no term of the kind "heat generated in chemical reaction" because the heat of reaction is indirectly included in the internal energy, and thus in the terms dU/dt, H_{in}, and H_{out}.

"Complete" general energy balance. Note that I, as before, have been a bit lazy when writing the energy balance in the "general" form in (11.11). When necessary, terms for kinetic and potential energy must be added to U and H, and other work terms such as electrochemical work W_{el} must be included. Thus, as stated in the "energy balance reading rule" on page 4.4:

- Shaft work W_s [J/s] really means $W_s + W_{\text{el}}+$ other work forms.
- Internal energy U of the system [J] really means $E = U + E_K + E_P$ + other energy forms. Here E_K is kinetic energy and E_P is potential energy of the system.
- Enthalpy H of the in- and outstreams [J/s] really means $H + E_K + E_P$ + other energy forms. For a stream, $E_K = w\alpha v^2/2$ and $E_P = wgz$, see page 125, where w [kg/s] is the flow rate.

Energy balance in enthalpy

We usually prefer to work with enthalpy, and introducing $U = H - pV$ in (11.11), gives

$$\boxed{\frac{dH}{dt} = H_{\text{in}} - H_{\text{out}} + Q + W_s \underbrace{-(p_{\text{ex}} - p)\frac{dV}{dt} + V\frac{dp}{dt}}_{\text{pressure-volume changes}}} \quad [\text{J/s}] \qquad (11.12)$$

Here, $H = mh$ [J] is the enthalpy of the system (inside the control volume), where m [kg] is system mass and h [J/kg] is its specific enthalpy.

Comments:

1. The term "pressure-volume changes" in (11.12) and (11.13) is often negligible.
   - The term is exactly zero (also for gases) for cases with constant pressure and volume.
   - The term is exactly zero (also for gases) for cases where the pressure is constant and equal to the surrounding's pressure ($p = p_{\text{ex}}$=constant).
   - Even with varying pressure, the term is approximately zero for liquids and solids, because the volume V is relatively small for such systems.

   However, the term "pressure-volume changes" can be considerable for gases with varying pressure, for example, for a gas pipeline.
2. We have $\frac{dH}{dt} = m\frac{dh}{dt} + h\frac{dm}{dt}$ and by introducing the mass balance (11.3), the energy balance on "mass flow basis" becomes

$$m\frac{dh}{dt} = w_{\text{in}}(h_{\text{in}} - h) - w_{\text{out}}(h_{\text{out}} - h) + Q + W_s \underbrace{-(p_{\text{ex}} - p)\frac{dV}{dt} + V\frac{dp}{dt}}_{\text{pressure-volume changes}} \quad [\text{J/s}] \quad (11.13)$$

3. All enthalpies must refer to a common reference state. If we use, for example, the elements at 298 K and 1 bar as the reference, the enthalpy H (or h) is the sum of (1) chemical formation energy, (2) "latent" phase transition energy (if the phase differs from the standard state), (3) thermal energy ("sensitive heat c_p"), (4) mixing energy and (5) pressure-correction energy; see page 364.
4. Enthalpy $H(T, p, f, n_j)$ [J/kg] is generally a function of temperature T, pressure p, phase distribution f (where f is fraction of light phase) and composition (n_j). The time derivative

of the enthalpy in (11.12) can then be written

$$\frac{dH}{dt} = \underbrace{\frac{\partial H}{\partial T}}_{C_p}\frac{dT}{dt} + \frac{\partial H}{\partial p}\frac{dp}{dt} + \underbrace{\frac{\partial H}{\partial f}}_{\Delta_{\text{trs}} H}\frac{df}{dt} + \sum_j \frac{\partial H}{\partial n_j}\frac{dn_j}{dt} \tag{11.14}$$

This expression may be useful in some cases, but for numerical calculations it is generally recommended to work directly with H (or U) as the internal variable ("state") rather than T; see page 316 on solving the resulting differential-algebraic equations (DAE).

11.2.4 Energy balance in temperature

Here, we want to derive a differential equation in temperature, $dT/dt = \cdots$. This gives insight and is useful for some calculations. The expressions for dT/dt presented below depend on the following assumptions:

- The enthalpy's dependency of pressure is neglected, which is reasonable in most cases.
- The phase distribution in the system and in each stream does not change, which is reasonable in most cases.
- The enthalpy's dependency of composition is neglected, which is reasonable in many cases, for example, if each stream's composition is constant (actually, this assumption is not made for the case with chemical reaction in case III).

This means that the three last terms in (11.14) drop out, and the specific enthalpies in (11.13) are only a function of temperature, that is,

$$h(T) = h(T_{\text{ref}}) + \int_{T_{\text{ref}}}^{T} c_p(T)dT \tag{11.15}$$

Here $h(T_{\text{ref}})$ is constant, because the composition and phase distribution is constant. When we put everything into the energy balance (11.13), the contribution from the reference-terms ($h(T_{\text{ref}}), h_{\text{in}}(T_{\text{ref}}), h_{\text{out}}(T_{\text{ref}})$) will appear as terms for heat of phase change (e.g., heat of vaporization) or heat of reaction. Let us next consider three cases.

I. No reaction and no phase transition

For the case with *no reaction and no phase transition*, the reference-terms drop out and (11.13) becomes

$$mc_p(T)\frac{dT}{dt} = w_{\text{in}}\int_{T}^{T_{\text{in}}} c_p(T)dT - w_{\text{out}}\int_{T}^{T_{\text{out}}} c_p(T)dT + Q + W_s \underbrace{-(p_{\text{ex}} - p)\frac{dV}{dt} + V\frac{dp}{dt}}_{\text{pressure}-\text{volume changes}}$$

If we, in addition, assume that the *heat capacity is constant* (independent of temperature), the energy balance becomes

$$mc_p\frac{dT}{dt} = w_{\text{in}}c_p(T_{\text{in}} - T) - w_{\text{out}}c_p(T_{\text{out}} - T) + Q + W_s \underbrace{-(p_{\text{ex}} - p)\frac{dV}{dt} + V\frac{dp}{dt}}_{\text{pressure}-\text{volume changes}} \tag{11.16}$$

This is further simplified for an ideal stirred tank (CSTR), where we have $T_{\text{out}} = T$.

II. With phase transition

Let us consider a somewhat more complex case with phase transition, where we cannot use (11.16), because the reference terms $h(T_{\text{ref}})$ do not drop out of the energy balance.

Example 11.2 Phase transition: Energy balance for evaporator.

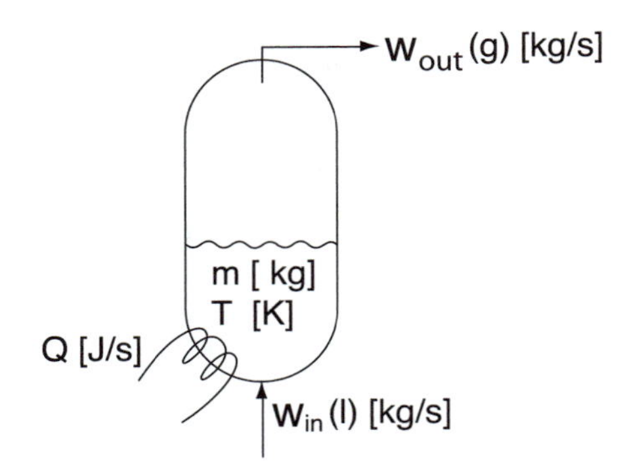

Figure 11.2: Evaporator for water

We consider an evaporator for water as shown in Figure 11.2. We neglect the mass of gas compared to the mass of liquid in the system (inside the evaporator). The mass balance is

$$\frac{dm}{dt} = w_{\text{in}} - w_{\text{out}} \quad [\text{kg/s}]$$

Since we assume only liquid in the tank, we can neglect the terms with "pressure-volume changes" (even when the pressure varies). We also have no shaft work ($W_s = 0$). The energy balance (11.13) then becomes

$$m\frac{dh}{dt} = w_{\text{in}}(h_{\text{in}} - h) - w_{\text{out}}(h_{\text{out}} - h) + Q \quad [\text{J/s}]$$

The enthalpy h [J/kg] of the liquid in the tank is only a function of temperature (because the remaining terms in (11.14) can be neglected or are zero). Thus, we have $dh/dt = c_{pL}dT/dt$, where we use c_{pL} [J/K kg] with subscript L to indicate that it is a liquid.

The inflow and the mass in the evaporator have the same composition and phase (liquid). We then have

$$h_{\text{in}}(T_{\text{in}}) - h(T) = \int_T^{T_{\text{in}}} c_{pL}(T)dT \quad [\text{J/kg}]$$

We assume perfect mixing such that $T = T_{\text{out}}$. Since the outlet stream is in gas phase, we then get

$$h_{\text{out}}(T) - h(T) = \Delta_{\text{vap}}h(T) \quad [\text{J/kg}]$$

where $\Delta_{\text{vap}}h(T)$ is the heat of vaporization for water at T (which takes into account the change in reference due to the phase transition). The energy balance (11.13) for the evaporator then becomes

$$\boxed{mc_{pL}(T)\frac{dT}{dt} = w_{\text{in}}\int_T^{T_{\text{in}}} c_{pL}(T)dT - w_{\text{out}}\Delta_{\text{vap}}h(T) + Q} \quad [\text{J/s}] \tag{11.17}$$

Note that (11.17) also applies when the mass m in the tank varies with time, because the mass balance $\frac{dm}{dt} = w_{\text{in}} - w_{\text{out}}$ was used when deriving (11.13).

Comments.

1. *The heat of vaporization is often given at a temperature T_{ref} (for example, at the normal boiling point at 1 atm). The heat of vaporization at T can then be found by adding the following subprocesses: (1) Cooling the liquid from T to T_{ref}, (2) evaporation at T_{ref}, and (3) heating the gas from T_{ref} to T. We then get*

$$\Delta_{\text{vap}}h(T) = \Delta_{\text{vap}}h(T_{\text{ref}}) + \int_{T_{\text{ref}}}^{T} (c_{pV} - c_{pL})dT$$

   where $\Delta_{\text{vap}}h(T_{\text{ref}})$ is the heat of vaporization at temperature T_{ref}, and c_{pV} is the heat capacity of the steam.
2. *Temperature and pressure are related by the equilibrium vapor pressure: $p = p^{\text{sat}}(T)$ (see page 180).*

With a little practice, it is possible to formulate energy balances of this kind directly: We imagine "standing in the tank" (the system) and use the temperature and phase here as the reference. Then we consider what can be the source of changes in the system's temperature. In the example with the evaporator, (11.17) can be derived as follows:

> "The temperature change in the tank (left side) is driven by the inflow having a different temperature than the tank (first term right side), and by enthalpy being removed by evaporation (second term) and by heat being supplied (third term)."

The term for the outlet stream drops out since it has the same temperature as the tank. More generally, it is recommended to start from the basic equations.

Exercise 11.1 *Derive the energy balance for a flash tank with inventory n [mol], feed F [mol/s], vapor product D [mol/s] and liquid product B [mol/s] (make a flow sheet). Show that it becomes*

$$nC_{pL}\frac{dT}{dt} = FC_{pL}(T_F - T) + D \cdot \Delta_{\text{vap}}H(T)$$

What are the units for the quantities in the equation? Which assumptions have been made when deriving this?

III. With chemical reaction

For cases with chemical reaction, it is usually most convenient to use a molar basis. We return to (11.12) and introduce $H(T, p, n_j) = \sum_j n_j \bar{H}_{m,j}(T, p)$. Here, $\bar{H}_{m,j}$ [J/mol] is the "partial molar enthalpy" for component j in the mixture. For cases with negligible heat (enthalpy) of mixing, we have that $\bar{H}_{m,j} = H_{m,j}$, where $H_{m,j}$ is the molar enthalpy of pure component j in its actual phase. With this as a starting point, let us derive the general energy balance in terms of temperature (dT/dt) for a continuous stirred tank reactor (CSTR).

Example 11.3 Energy balance with temperature for CSTR. *We consider an ideal continuous stirred tank reactor (CSTR) where a chemical reaction takes place (Figure 11.3). Let us, as an example, consider the reaction $2A \rightarrow B$, but the derivation below is general and applies to any reaction. The reaction rate is $r(T, c_A)$ [mol/s m^3], and if we take into consideration the stoichiometry, the component balances are:*

$$\frac{dn_A}{dt} = F_{A,\text{in}} - F_{A,\text{out}} + \underbrace{\nu_A r V}_{G_A} \quad [\text{mol A/s}]$$

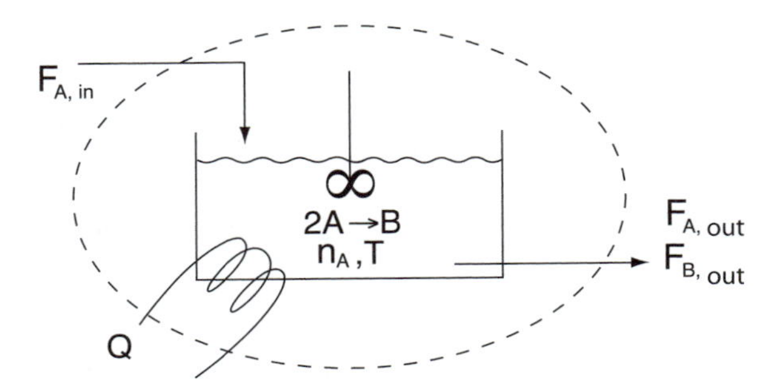

Figure 11.3: Continuous stirred tank reactor (CSTR) with heating

$$\frac{dn_B}{dt} = F_{B,\text{in}} - F_{B,\text{out}} + \underbrace{\nu_B r V}_{G_B} \quad [\text{mol B/s}]$$

where the stoichiometric coefficients in our example are $\nu_A = -2$ *and* $\nu_B = 1$. *We assume no shaft work and neglect the "pressure-volume contribution." The energy balance is then*

$$\frac{dH}{dt} = H_{\text{in}} - H_{\text{out}} + Q \quad [\text{J/s}] \tag{11.18}$$

If we neglect the enthalpy of mixing, the enthalpy can be written

$$H(T, p, n_A, n_B) = n_A H_{m,A}(T, p) + n_B H_{m,B}(T, p) \quad [\text{J}]$$

where $H_{m,j}(T, p)$ *[J/mol] is the molar enthalpy for component* j. *Here, we choose the elements in their standard states at 298.15 K and* $p^\ominus = 1$ *bar as the reference. If we neglect the pressure's influence on the enthalpy, we then have*

$$H_{m,j}(T, p) = H_j^\ominus(T) + \Delta H_{\text{trs}} \quad [\text{J/mol}]$$

where $H_j^\ominus(T) = \Delta_{\text{f}} H_j^\ominus(T)$ *[J/mol] is the standard enthalpy of formation for generating component* i *in its standard state at* T *and 1 bar from the elements at 298 K and 1 bar, and* ΔH_{trs} *is the enthalpy change for the change in reference from the standard state to actual state (phase). If we assume that there is no phase change, we can then write*

$$\begin{aligned} H &= n_A H_A^\ominus(T) + n_B H_B^\ominus(T) \quad [\text{J}] \\ H_{\text{in}} &= F_{A,\text{in}} H_A^\ominus(T_{\text{in}}) + F_{B,\text{in}} H_B^\ominus(T_{\text{in}}) \quad [\text{J/s}] \\ H_{\text{out}} &= F_{A,\text{out}} H_A^\ominus(T_{\text{out}}) + F_{B,\text{out}} H_B^\ominus(T_{\text{out}}) \quad [\text{J/s}] \end{aligned}$$

Inserting into the energy balance (11.18) gives

$$n_A \overbrace{\frac{dH_A^\ominus(T)}{dT}}^{C_{p,m,A}(T)} \frac{dT}{dt} + n_B \overbrace{\frac{dH_B^\ominus(T)}{dT}}^{C_{p,m,B}(T)} \frac{dT}{dt} + H_A^\ominus(T)\frac{dn_A}{dT} + H_B^\ominus(T)\frac{dn_B}{dT}$$

$$= F_{A,\text{in}} H_A^\ominus(T_{\text{in}}) + F_{B,\text{in}} H_B^\ominus(T_{\text{in}}) - F_{A,\text{out}} H_A^\ominus(T_{\text{out}}) - F_{B,\text{out}} H_B^\ominus(T_{\text{out}}) + Q$$

We assume perfect mixing such that $T = T_{\text{out}}$. *By inserting the expressions for* dn_A/dt *and* dn_B/dt *from the mass balance, and rearranging the terms (a bit of a work is needed here...), we finally derive the energy balance in "temperature form":*

$$\boxed{n C_{p,m} \frac{dT}{dt} = F_{\text{in}} \int_T^{T_{\text{in}}} C_{p,m,\text{in}}(T) dT + \left(-\Delta_{\text{r}} H^\ominus(T)\right) rV + Q} \tag{11.19}$$

(For cases with many reactions, the term $\left(-\Delta_r H^{\ominus}(T)\right) rV$ *is replaced by* $\sum_j \left(-\Delta_r H_j^{\ominus}(T)\right) r_j V$*).*
For our reaction $2A \rightarrow B$*, we have*

$$\Delta_r H^{\ominus}(T) = \sum_j \nu_j H_j^{\ominus} = H_B^{\ominus} - 2H_A^{\ominus} \quad [J/K\ mol] \tag{11.20}$$

Furthermore,

$$n = n_A + n_B \quad [\text{mol}]$$

$$F_{\text{in}} = F_{A,\text{in}} + F_{B,\text{in}} \quad [\text{mol/s}]$$

and the molar heat capacities for the reactor (system) and feed are

$$C_{p,m} = \frac{n_A}{n} C_{p,m,A}(T) + \frac{n_B}{n} C_{p,m,B}(T) \quad [\text{J/K mol}]$$

$$C_{p,m,\text{in}}(T) = \frac{F_{A,\text{in}}}{F_{\text{in}}} C_{p,m,A}(T) + \frac{F_{B,\text{in}}}{F_{\text{in}}} C_{p,m,B}(T) \quad [\text{J/K mol}]$$

Let us summarize the assumptions that have been made when deriving (11.19):

1. *All streams have the same phase.*
2. *Perfect mixing such that* $T = T_{\text{out}}$*.*
3. *Heat of mixing is neglected.*
4. *The pressure's influence on the enthalpy is neglected.*

Note that (11.19) applies to the case with varying composition in the reactor and a varying amount of n ("holdup") in the reactor. For a more detailed example with dynamic simulation, see page 311.

With a little experience, it is again possible to directly formulate the energy balance (11.19) in temperature form for a continuous stirred tank reactor:

> "The temperature change in the reactor (left side) is driven by the difference between the feed and reactor temperatures (first term left side), by the heat of reaction (second term) and by the supplied heat (third term)."

Comments.

1. We note that the "heat of reaction" appears as a separate term when we choose to write the energy balance in the "temperature form" in (11.19).
2. The energy balance in temperature form (11.19) gives interesting insights and is useful in many situations. However, it is usually simpler for numerical calculations (dynamic simulation) to stay with the original form (11.11) or (11.12) with U or H as the state (differential) variables. See page 316 for solving the resulting DAE equations.

11.2.5 Steady-state balances

The dynamic balances derived above are all in the form $dy/dt = f(y, u)$. We usually assume that the system is initially "at rest" (steady-state) with $dy/dt = 0$. The steady-state (nominal) values for u and y are here indicated by using superscript *, and we have that $f(y^*, u^*) = 0$.

11.3 Dynamic analysis and time response

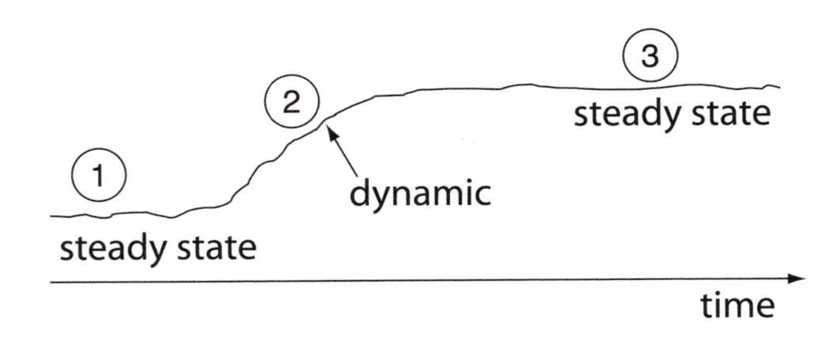

Figure 11.4: Dynamic response in output y to step change in input u

We want to understand what happens when we get an imbalance from the steady-state, such that the system's states change with time. For this purpose, let us consider the following incident (see Figure 11.4):

1. The system is initially "at rest" (steady state).
2. A change in one of the system's independent variables ("input" u) occurs, for example, a change in external conditions or a parameter change, such that we get an imbalance and the system's dependent variables (states and "outputs" y) change with time.
3. After a while (actually when $t \to \infty$), the system will eventually approach a new equilibrium state, where it is again "at rest" (new steady state).

Some examples are

- If we, on a winter's day, turn on more heat in a room, the temperature will start rising. The change is largest in the beginning, and "eventually" the temperature will approach a new steady state value (where again the system is at rest).
- If we push the accelerator ("gas") pedal of a car, then the car's speed will increase. The change is largest in the beginning, and "eventually" the speed will reach a new steady-state value (where again there is a balance between the forward force from the engine and the resistance force from the air).
- In a chemical reactor we have a continuous supply of reactant. If we increase ("disturb") the concentration of the reactant, the product concentration will also increase. The change is largest in the beginning, and "eventually" the product concentration will approach a new steady state value.

In all these cases, we go from one steady state to another, and a steady-state model is sufficient to calculate the initial and final states. However, we need a dynamic model to say something about the dynamic response and to quantify what we mean by "eventually." By the term **response**, we mean the time response for the dependent variable (output) y when we change the independent variable (input) u. In the three cases mentioned above we have

- Room: $u = Q$ (heating), $y = T$
- Car: $u = w$ (fuel flow), $y = v$ (speed)
- Reactor: $u = cin$, $y = c_{\text{out}}$

Four important responses are (see Figure 11.5):

Step response. This is the response in the dependent variable y to a *step change* (persistent change) in the independent variable u. Mathematically, the change in u is

$$u(t) = \left\{ \begin{array}{ll} u_0 & t \leq t_0 \\ u_\infty = u_0 + \Delta u & t > t_0 \end{array} \right\}$$

where Δu is the magnitude of the step. A step response was considered in the three cases above.

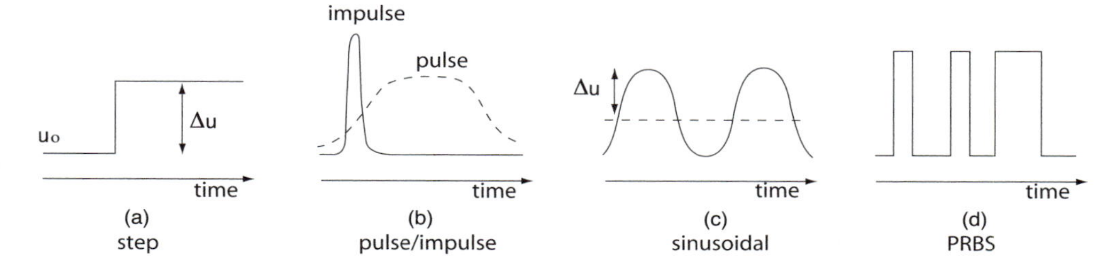

Figure 11.5: Time signals for input $u(t)$

Impulse response. A **pulse** is a temporary change of the independent variable u, and if the duration is very short (negligible) compared to the system's dynamics, we have an **impulse**. The impulse response is the resulting response in y. For a process engineer, an example of an impulse is to "throw a bucket" of something into a tank. For a chemist or a medical doctor, an injection with a needle gives an impulse.

For a flow system, the so-called **residence time distribution (RTD)** is actually the concentration impulse response of a non-reacting component.

Frequency response (sinusoidal input). This is the resulting response in y to a persistent sinusoidal variation in the independent variable u,

$$u(t) = u_0 + \Delta u \cdot \sin(\omega t)$$

For small changes, we can assume that the system is linear, and the output signal is also sinusoidal with the same frequency ω:

$$y(t) = y_0 + \Delta y \cdot \sin(\omega t + \phi)$$

The frequency response is characterized by two parameters: The gain $\Delta y/\Delta u$, and the phase shift, ϕ. Both depend on the frequency ω [rad/s], and by varying the frequency ω, we get information on how the system reacts to quick (ω large) and slow (ω small) input variations. Frequency analysis is an important tool in control engineering.

PRBS response. This is the response in y when the independent variable u changes at "random" times between two given values (PRBS = pseudo-random binary sequence). This may give a good "dynamic distribution" and is sometimes an effective method for obtaining experimental data that can be used for estimating (="identify" in control engineering) parameters in an empirical dynamic model for the relationship between u and y.

The step response is very popular in process engineering because it is simple to perform, understand and analyze. In the following, we study the step response in more detail.

11.3.1 Step response and time constant

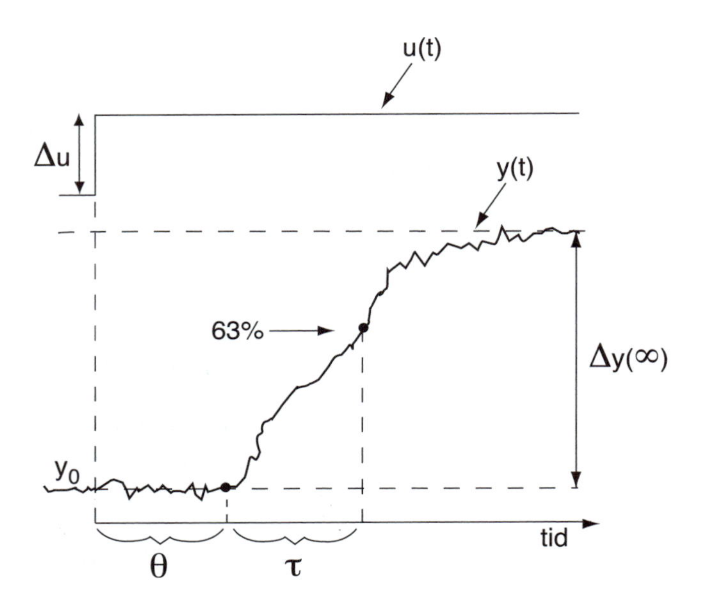

Figure 11.6: Experimental step response

We consider a system that is initially "at rest," that is, at steady state with $dy/dt = 0$. A *step-change* then occurs in the independent variable u, which takes the system away from its initial steady state. We assume that the system is stable such that it eventually approaches a new steady state. The resulting **step response** in $y(t)$ is often characterized by the following three parameters (see Figure 11.6):

(Steady state) Gain $k = \frac{\Delta y(\infty)}{\Delta u}$.

(Effective) Delay θ – the time it takes before y "takes off" in the "right" direction. Thus, $\Delta y(\theta) \approx 0$.

Time constant τ – additional time it takes to reach 63% of the total change in y (that is, $\Delta y(\tau + \theta) = 0.63 \Delta y(\infty)$).

Here

- $\Delta u = u(\infty) - u(t_0)$ – magnitude of step change in u
- t_0 – time when step change in u occurs (often $t_0 = 0$ is chosen)

- $\Delta y(t) = y(t) - y(t_0)$ – the resulting change in y
- $y(t_0) = y_0$ – initial (given) steady state
- $y(\infty)$ – final (new) steady state

The value of $\Delta y(\infty) = y(\infty) - y(t_0)$, and thereby of the steady state gain k, can be determined from a steady state model, if one is available.

The cause of the delay (time delay) θ may be a transport delay (for example a pipe) or a delay in a measurement, but in most cases it represents the contribution from many separate dynamic terms that, altogether, give a response that resembles a delay (hence the term "effective" delay).

The time constant τ characterizes the system's dominant "inertia" against changes. It is defined as the additional time (after the time delay) it takes the variable to reach 63% (more precisely, a fraction $1 - e^{-1} = 1 - 0.3679 \approx 0.63$, see below) of its total change. Why do we not let the time constant be the time it takes to reach all (100%) of its change? Because it generally take an infinitely long time for the system to reach exactly its final state, so this would not give a meaningful value.

The values of the parameters k, τ and θ are independent of the size of the step (independent of the value of Δu), provided the step Δu is sufficiently small such that we remain in the "linear region." On page 301, we show how we can derive a linear model.

11.3.2 Step response for first-order system

The basis for the definition of τ given above is the simplest case with one linear differential equation (first-order system). Here, we study this system in more detail. A first-order system can be written in the following standard form

$$\boxed{\tau \frac{dy}{dt} = -y + ku}, \quad y(t_0) = y_0 \tag{11.21}$$

where

- u is the independent variable (input)
- y is the dependent variable (output)
- τ is the time constant
- k is the gain

We now assume that

1. The system is "at rest" at time t_0 with $dy/dt = 0$, that is, for $t \leq t_0$ we have $u = u_0$ and $y_0 = ku_0$.
2. The independent variable u changes from u_0 to a constant value $u = u_0 + \Delta u$ at time t_0.

As proven below, the solution ("step response") can then be written as

$$\boxed{y(t) = y_0 + \left(1 - e^{-t/\tau}\right) k\Delta u} \tag{11.22}$$

or

$$\underbrace{\Delta y(t)}_{y(t)-y_0)} = \underbrace{\Delta y(\infty)}_{y(\infty)-y_0} \left(1 - e^{-t/\tau}\right) \tag{11.23}$$

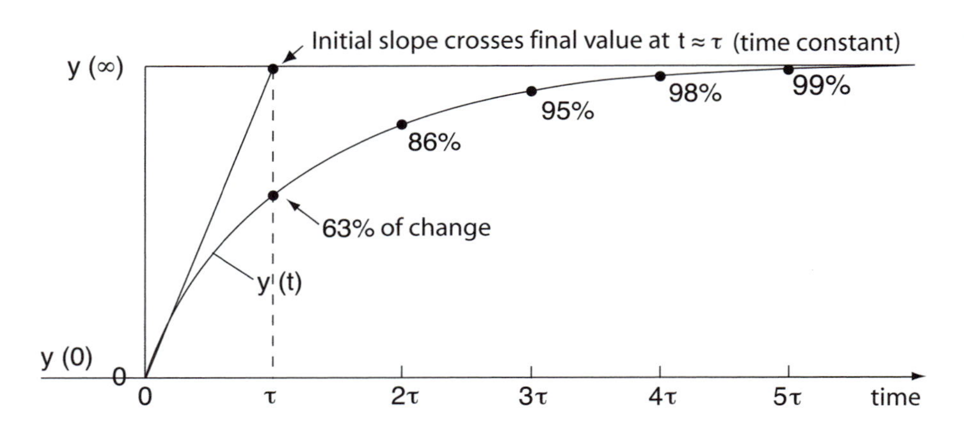

Figure 11.7: Step response for first-order system

(you should try to remember this one). k is the steady state gain, and when $t \to \infty$ we have $e^{-t/\tau} \to 0$ and the system approaches a new steady state where $\Delta y(\infty) = k\Delta u$. Notet that the exponential term $1 - e^{-t/\tau}$ describes how fast the system approaches its new steady state, and as a function of the non-dimensional time t/τ we have:

t/τ	$1 - e^{-t/\tau}$	Value	Comment
0	$1 - e^{0} =$	0	
0.1	$1 - e^{-0.1} =$	0.095	
0.5	$1 - e^{-0.5} =$	0.393	
1	$1 - e^{-1} =$	0.632	63% of change is reached after time $t = \tau$
2	$1 - e^{-2} =$	0.865	
3	$1 - e^{-3} =$	0.950	
4	$1 - e^{-4} =$	0.982	98% of change is reached after time $t = 4\tau$
5	$1 - e^{-5} =$	0.993	
∞	$1 - e^{-\infty} =$	1	

The time response is plotted in Figure 11.7. We note that at time $t = \tau$ (the time constant), we have reached 63% of the total change, and after four time constants, we have reached 98% of the change (and we have for all practical purposes arrived at the new steady state). Note also from Figure 11.7 that the initial slope of the response (at time $t = 0$) goes through to the point $(\tau, y(\infty))$. This can be shown mathematically from (11.23):

$$\frac{dy}{dt} = (y(\infty) - y_0)\,\frac{1}{\tau}e^{-t/\tau} \quad \Rightarrow \quad \left(\frac{dy}{dt}\right)_{t=0} = \frac{y(\infty) - y_0}{\tau} \tag{11.24}$$

This means that the response $y(t)$ would reach the final value $y(\infty)$ at time τ if it continued unaltered (in a straight line) with its initial slope.

Comments.

1. As seen from the proof below, (11.23) applies also to cases where the system is not initially at rest. This is not the case for (11.22).

2. For cases where τ is negative, the system is unstable, and we get that $y(t)$ goes to infinity when t goes to infinity.

3. From (11.24) and $\Delta y(\infty) = k\Delta u$, we derive that

$$\frac{1}{\Delta u}\left(\frac{dy}{dt}\right)_{t=0} = \frac{k}{\tau} \tag{11.25}$$

This means that the initial slope k' of the "normalized" response $\Delta y(t)/\Delta u$ is equal to the ratio k/τ, i.e., $k' \triangleq k/\tau$.

Proof: Step response for a first-order system

Consider a first-order system in standard form, (11.21),

$$\tau\frac{dy}{dt} = -y + ku; \quad y(0) = y_0 \tag{11.26}$$

where both τ and ku are constant. There are many ways of solving the linear differential equation (11.26). We can for example use separation of variables and derive

$$\frac{dy}{y - ku} = -\frac{dt}{\tau}$$

Integration gives

$$\int_{y_0}^{y} \frac{dy}{y - ku} = \int_0^t -\frac{dt}{\tau} \quad \Rightarrow \quad \ln\frac{y - ku}{y_0 - ku} = -\frac{t}{\tau}$$

and we get the general solution

$$y(t) = ku + e^{-t/\tau}(y_0 - ku)$$

We subtract y_0 from both sides and get

$$y(t) - y_0 = \left(1 - e^{-t/\tau}\right)(ku - y_0) \tag{11.27}$$

Since $e^{-t/\tau} \to 0$ as $t \to \infty$, we have that $y(\infty) = ku$, and by introducing deviation variables

$$\Delta y(t) \triangleq y(t) - y(0) \tag{11.28}$$

we find that (11.27) can be written in the following general form

$$\Delta y(t) = \Delta y(\infty)\left(1 - e^{-t/\tau}\right) \tag{11.29}$$

We have so far not assumed that the system is "at rest" at $t = t_0$, but let us do this now. We then have at $t = t_0$ that $dy/dt = 0$, which gives

$$y_0 = ku_0$$

and (11.27) gives for a system that is initially at rest:

$$\underbrace{\Delta y(t)}_{y(t)-y_0} = \left(1 - e^{-t/\tau}\right) k \underbrace{\Delta u}_{u-u_0} \tag{11.30}$$

Example 11.4 Concentration response in continuous stirred tank

We consider the concentration response for component A in a continuous stirred tank without chemical reaction (see Figure 11.8). We assume constant liquid density ρ and constant volume V. The system is assumed to be at rest at $t = 0$. We want to find the step response for $t > 0$ given the following data

$$V = 5m^3; \quad q = 1m^3/h$$

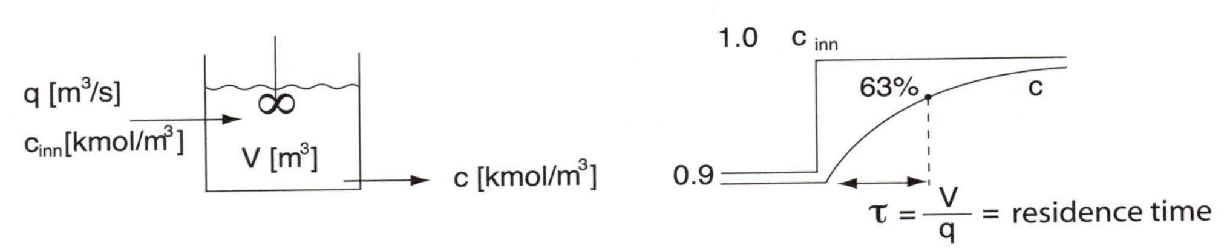

Figure 11.8: Continuous stirred tank without reaction

$$c_{A,\text{in}} = \left\{ \begin{array}{ll} c_0 = 0.9\ kmol/m^3 & t \leq 0 \\ c_\infty = 1.0\ kmol/m^3 & t > 0 \end{array} \right\}$$

Solution. *With constant density and constant volume, the mass balance gives that the volumetric inlet and outlet flow rates are equal, $q_{\text{in}} = q_{\text{out}} = q$. We further assume perfect mixing in the tank such that $c_{A,\text{out}} = c_A$. The component balance for A in the tank is then [mol A/s]*

$$\frac{d}{dt}(c_A V) = q c_{A,\text{in}} - q c_A \tag{11.31}$$

With constant volume V this gives

$$\frac{V}{q}\frac{dc_A}{dt} = -c_A + c_{A,\text{in}} \tag{11.32}$$

This is in standard form (11.21) with

$$u = c_{A,\text{in}}; \quad y = c_A$$

and

$$k = 1; \quad \boxed{\tau = \frac{V}{q}} \quad \frac{[m^3]}{[m^3/s]} = [s]$$

Here, V/q [s] is the residence time for mass in the tank, that is, the time constant in this case equals the residence time. From (11.22), the solution of (11.32) (the step response) is given by

$$c_A(t) = c_0 + \left(1 - e^{-t/\tau}\right)\Delta c_{A,\text{in}} \tag{11.33}$$

where $\Delta c_{A,\text{in}} = c_\infty - c_0 = 0.1\ kmol/m^3$. At time $t = 0$, we then have that $c_A(0) = c_0 = 0.9\ kmol/m^3$, and concentration rises such that it is, at time $t = \tau = 5\ h$ (the residence time), $c_A = 0.9 + \left(1 - e^{-1}\right) \cdot 0.1 = 0.963\ kmol/m^3$, and at time $t = \infty$, $c_A(\infty) = 0.9 + 0.1 = 1\ kmol/m^3$ (as expected).

11.3.3 Additional examples of step responses for first-order systems

Here, we consider some relatively simple examples with only one differential equation which give first-order step responses (Figure 11.7).

Example 11.5 Temperature dynamics in continuous stirred tank. *Consider the continuous process in Figure 11.9 where a liquid stream of 1 kg/s (constant) flows through a mixing tank with constant volume 1.2 m³. The density of the liquid is 1000 kg/m³ (constant) and the heat capacity is 4 kJ/kg K. Perfect mixing in the tank is assumed.*

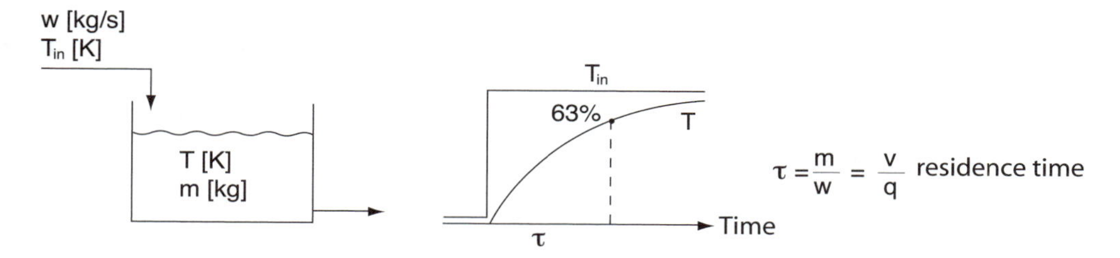

Figure 11.9: Temperature dynamics in continuous stirred tank without reaction

The process is initially operated at steady state such that the inlet temperature $T_{\rm in}$ *is* $50^o C$ *and the outlet temperature* $T_{\rm out} = T$ *is* $50^o C$ *(so we assume no heat loss). Suddenly, the temperature of the inflow is changed to 60 °C (step change). The outlet temperature will also "eventually" reach 60 °C.* **The question is***: What is the time constant, that is, how long does it take before the temperature in the tank (and outlet stream) has increased by* $0.63 \cdot 10 = 6.3^o C$ *to 56.3 °C?*

Solution. *Since the mass in the tank is constant, the mass balance gives* $w_{\rm out} = w_{\rm in} = w = 1$ *kg/s. The energy balance (11.12) for the tank is (liquid)*

$$\frac{dH}{dt} = H_{\rm in} - H_{\rm out} \quad [J/s]$$

With the assumption of constant heat capacity c_p*, this gives*

$$mc_p \frac{dT}{dt} = wc_p(T_{\rm in} - T)$$

or equivalently

$$\frac{m}{w}\frac{dT}{dt} = -T + 1 \cdot T_{\rm in}$$

With $y = T$ *and* $u = T_{\rm in}$ *we see that this is in standard form (11.21) with*

$$\tau = \frac{m}{w} = \frac{\rho V}{w} = \frac{1000 \cdot 1.2}{1} = 1200 \ s; \quad k = 1$$

In other words, it will take $\tau = 1200$ *s = 20 min (the residence time m/w) before the outlet stream's temperature reaches 56.3 °C (and it will take an infinitely long time before it reaches 60 °C).*

Note that the time constant also for this example equals the residence time*. This is true for changes in both concentration and temperature for a continuous stirred tank without reaction or heating.*

Example 11.6 Temperature dynamics in continuous stirred tank with heat exchange.

Consider the same example as above, where the inlet temperature is changed from 50 °C (initial steady state) to 60 °C, but we have heating (see Figure 11.10) such that the temperature in the tank is 70 °C (initial steady state). We consider the response and determine the time constant for the following two cases:

1. *An electric heater is used such that the supplied heat* Q *is independent of the temperature* T *in the tank.*
2. *We have a heat exchanger with condensing stream on the hot side. The supplied heat is* $Q = UA(T_h - T)$ *where* T_h *(hot side temperature) is constant at 110 °C.*

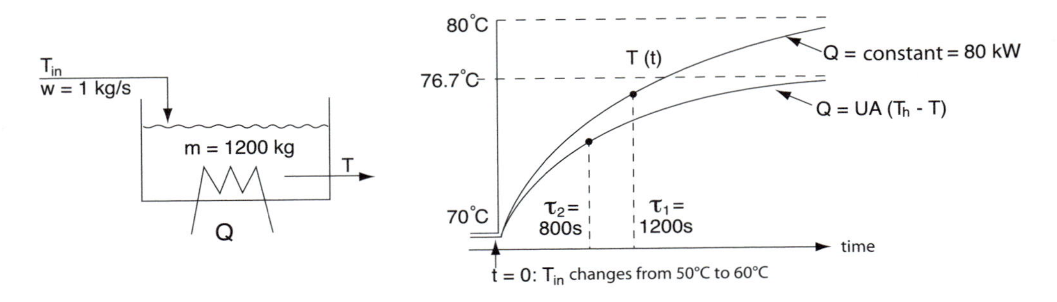

Figure 11.10: Continuous stirred tank with heating

Solution. *The energy balance (11.12) becomes [J/s]*

$$mc_p \frac{dT}{dt} = wc_p(T_{\text{in}} - T) + Q$$

At the initial steady state ($dT/dt = 0$), we have (before the change in T_{in})

$$Q = -wc_p(T_{\text{in}} - T) = -1 \text{ kg/s} \cdot 4000 \ J/kg \ K \cdot (50 - 70)K = 80000 J/s = 80 \ kW$$

1. *For the case when Q is independent of T, transformation to the standard form (11.21) gives that the time constant is $\tau = m/w = 1200$ s (residence time), and that the gain from T_{in} to T is $k = 1$, that is, the steady-state temperature rise in the tank is 10 ^oC, that is, it will eventually rise to 80 ^oC.*
2. *For the case where Q depends on T, the energy balance becomes*

$$mc_p \frac{dT}{dt} = wc_p(T_{\text{in}} - T) + UA(T_h - T) \tag{11.34}$$

and transformation to the standard form (11.21) gives

$$\tau = \frac{mc_p}{wc_p + UA}; \quad k = \frac{wc_p}{wc_p + UA}$$

The time constant τ and the gain k are both smaller than in case 1. The reason is that the heat exchanger counteracts some of the temperature change ("negative feedback"). For numerical calculations, we need to know the value of UA. We have $UA = Q/(T_h - T)$, and from the initial steady state data, we find $UA = 80 \cdot 10^3/(110 - 70) = 2000 \ W/K$. The time constant and the gain are then

$$\tau = \frac{mc_p}{wc_p + UA} = \frac{1200 \cdot 4000}{1 \cdot 4000 + 2000} = 800 \ s; \quad k = \frac{4000}{4000 + 2000} = 0.67$$

that is, the temperature in the tank only increases by 6.7 ^oC to 76.7 ^oC – while in case 1 with an electric heater it increased by 10 ^oC.

Although k and τ are different, we note that $k' = k/\tau = 1/1200$ is the same in both cases, and since from (11.25) $lim_{t \to 0} \Delta T'(t) = (k/\tau) \cdot \Delta T_{\text{in}}$, this means that the initial responses are the same (see also Figure 11.10). This is reasonable also from physical considerations, since the "counteracting" negative feedback effect from the heat exchanger only comes in after the tank temperature T starts increasing which leads to a reduction in $Q = UA(T_h - T)$.

Example 11.7 Dynamics of cooking plate. *Let us consider a cooking plate with mass $m = 1$ kg and specific heat capacity $c_p = 0.5$ kJ/kg K. The cooking plate is heated by electric power and the supplied heat is $Q_1 = 2000$ W. The heat loss from the cooking plate is $UA(T - T_o)$ where T is the cooking plate's temperature, $T_o = 290K$ is the temperature of the surroundings, $A = 0.04m^2$ and U is the overall heat transfer coefficient. If we leave the plate unattended, then we find that $T \rightarrow 1000K$ when $t \rightarrow \infty$. What is the time constant for the cooking plate (defined as the time it takes to obtain 63% of the final temperature change)?*

Solution. *This is a closed system without mass flows and shaft work, and since the cooking plate is solid, we can neglect energy related to pressure-volume changes. The energy balance (11.12) around the cooking plate (the system) gives*

$$\frac{dH}{dt} = Q$$

Here, there are two contributions to the supplied heat Q, from electric power and from heat loss, that is,

$$Q = Q_1 - UA(T - T_o)$$

The enthalpy of the cooking plate is a function of temperature, that is, $dH/dt = mc_p dT/dt$. The energy balance becomes

$$mc_p \frac{dT}{dt} = Q_1 - UA(T - T_o) \tag{11.35}$$

In order to determine the overall heat transfer coefficient U, we use the steady state temperature $T^* = 1000K$. At steady state, the energy balance is $0 = Q_1 - UA(T^* - T_o)$ and we find

$$U = \frac{Q_1}{A(T^* - T_o)} = \frac{2000}{0.04(1000 - 290)} = 70.4 \quad [W/m^2\ K]$$

We assume that the overall heat transfer coefficient U is constant during the heating. The dynamic energy balance (11.35) is then a linear first-order differential equation which can be written in standard form

$$\tau \frac{dT}{dt} = -T + ku \tag{11.36}$$

where

$$\tau = \frac{mc_p}{UA} = 177.5 \ s$$

and

$$ku = \underbrace{\frac{1}{UA}}_{k_1} \underbrace{Q_1}_{u_1} + \underbrace{1}_{k_2} \cdot \underbrace{T_0}_{u_2}$$

In other words, we find that it takes time $t = \tau = 177.5$ s (about 3 min) to obtain 63% of the final change of the cooking plate's temperature.

Example 11.8 Response of thermocouple sensor in coffee cup. *Temperature is often measured with a thermocouple sensor based on the fact that electric properties are affected by temperature. We have a thermocouple and a coffee cup and perform the following experiments:*

1. *Initially, we hold the thermocouple sensor in the air (such that it measures the air temperature).*
2. *We put the thermocouple into the coffee (and keep it there for some time so that the thermocouple's temperature is almost the same as the coffee's temperature).*
3. *We remove it from the coffee (the temperature will decrease and eventually approach the temperature of air – actually, it may temporarily be lower than the air temperature because of the heat required for evaporation of remaining coffee drops).*

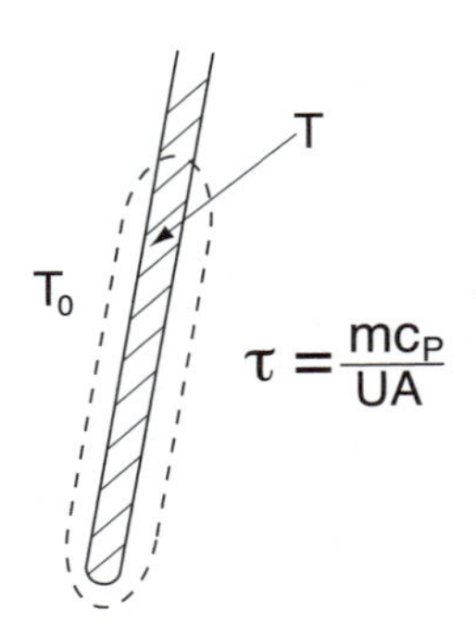

Figure 11.11: Thermocouple

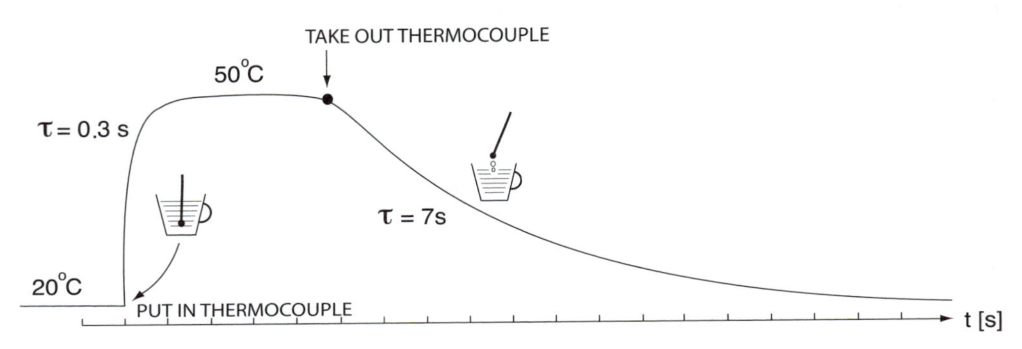

Figure 11.12: Coffee cup experiment

Task 1. What happens? *Sketch the expected temperature response.*

Solution: *The result of an actual experiment performed by the author is shown in Figure 11.12. We see that the response is similar to a standard first-order response. However, it is striking that the response is much quicker when we put the sensor into the coffee (time constant about 0.3 s) than when we remove it (time constant about 7s).*

Task 2. Can you explain this? *Formulate a dynamic model and find an analytical expression for the time constant.*

Solution: *Since we want to find the response in temperature, we need to formulate an energy balance, and since it is the thermocouple's temperature, the energy balance should be around the thermocouple. The general energy balance is given in (11.12). Since there are no streams, we have that $H_{\rm in} - H_{\rm out} = 0$. There is also no shaft work ($W_s = 0$), and the contribution from "pressure-volume changes" can be neglected. The energy balance (11.12) around the thermocouple is then simply*

$$\frac{dH}{dt} = Q$$

Here, $dH/dt = mc_p dT/dt$ where m is the mass of the thermocouple and T its temperature. The supplied heat to the thermocouple from the surroundings is

$$Q = UA(T_o - T)$$

The energy balance then becomes

$$mc_p \frac{dT}{dt} = UA(T_o - T) \tag{11.37}$$

where

- T – *temperature of thermocouple [K]*
- T_o – *temperature of surroundings (coffee or air) [K]*
- m – *mass of thermocouple [kg]*
- $c_p(T)$ – *specific heat capacity of thermocouple [J/kg K]*
- A – *area of thermocouple [m²]*
- U – *heat transfer coefficient from surroundings to thermocouple [W/m² K]*

(11.37) can be rewritten as

$$\frac{mc_p}{UA}\frac{dT}{dt} = T_o - T$$

With $y = T$ and $u = T_o$ this is in standard form (11.21) with

$$\tau = \frac{mc_p}{UA}; \quad k = 1 \tag{11.38}$$

(note that we get the same expression for the time constant as for the cooking plate in Example 11.7). At steady state, $dT/dt = 0$, and we have as expected that $T = T_o$. Thus, following a step in the surrounding's temperature T_o, the thermocouple's temperature T should exponentially (with time constant τ) approach T_o, and this is indeed confirmed by the experiment.

Some comments on coffee cup experiment

1. *The time constant is independent of the temperatures T and T_o (this is not immediately obvious for someone who does not know any process dynamics).*
2. *The time constant τ is constant if c_p and U are constant (this seems to be a reasonable assumption during each of the two experiments).*
3. *The time constant was observed to be $7s/0.3s = 23$ times larger when the thermocouple was removed from the coffee. Since $\tau = \frac{mc_p}{UA}$ where mc_P/A is constant, this must be (provided our theory is correct) because U is about 23 times higher when the thermocouple is in the coffee than when it is in air. This seems reasonable because heat transfer is usually much better to liquid than to gas.*
4. *In general, we desire a fast measurement, that is, we want the time constant τ to be small for the sensor. This is obtained by reducing the thermocouple's heat capacity mc_p [J/K], and making a design such that UA [W/K] is large. In order to protect the thermocouple, it is often placed in a pocket, which is not favorable because it increases the mass m and also reduces U. We can reduce this effect by choosing a pocket material with a small heat capacity mc_p (but at the same time with a good conductivity) and designing the pocket such that the outer area A is as large as possible.*

Final comment on comparison of coffee cup experiment with theory

Being good engineers, we are very eager to compare our experimental results with theoretical calculations. I used a cylindrical thermocouple, that is,

$$\frac{V}{A} = \frac{(\pi/4)D^2L}{\pi DL} = \frac{1}{4}D$$

where $D = 1.6$ mm, $\rho = 2700$ kg/m³ and $c_p = 800$ J/kg K (aluminium). We can from (11.38) calculate the overall heat transfer coefficient U (SI units):

$$U = \frac{V\rho c_p}{A\tau} = \frac{1}{4}\frac{D\rho c_p}{\tau} = \frac{864}{\tau} \quad \text{[using SI units]}$$

Here, I found experimentally $\tau = 0.3s$ (coffee, that is, water) and $\tau = 7s$ (air), which gives $U = 2880$ W/m^2 K (water) and $U = 123$ W/m^2 K (air). Immediately, the value 2880 W/m^2 K seems very high, because it is similar to values we find in heat exchangers with forced convection, and here we have natural convection. Let us compare with theoretical values for **natural convection** *to air and water. For natural convection,*[1] $Nu = 0.5(Gr \cdot Pr)^{0.25}$, *where the non-dimensional groups Nu, Gr and Pr are defined as*

$$Nu = \frac{hD}{k}; \quad Pr = \frac{c_p \mu}{k}; \quad Gr = \frac{g\beta \Delta T D^3}{(\mu/\rho)^2}$$

Inserting and rearranging gives

$$h = 0.5\left(\frac{k^3 c_p \rho^2 g \beta}{\mu}\right)^{0.25} \cdot \left(\frac{\Delta T}{D}\right)^{0.25}$$

where k is the **thermal conductivity**, *β the* **thermal expansion coefficient** *and μ the* **viscosity** *of the fluid. We use the following physical and transport data:*

$$\textbf{Air}: \quad k = 0.027\frac{W}{K\ m}; c_p = 1000\frac{J}{kg\ K}; \mu = 1.8 \cdot 10^{-5}\frac{kg}{m\ s}; \rho = 1.2\frac{kg}{m^3}; \beta = \frac{1}{T} = 0.003\frac{1}{K}$$

$$\textbf{Water}: \quad k = 0.7\frac{W}{K\ m}; c_p = 4200\frac{J}{kg\ K}; \mu = 10^{-3}\frac{kg}{m\ s}; \rho = 1000\frac{kg}{m^3}; \beta = 0.001\frac{1}{K}$$

We then find for natural convection (SI units)

$$\text{Air}: \quad h = 1.31 \cdot \left(\frac{\Delta T}{D}\right)^{0.25}$$

$$\text{Water}: \quad h = 173 \cdot \left(\frac{\Delta T}{D}\right)^{0.25}$$

Note from this that with natural convection, the heat transfer coefficient h to water is more than 100 times higher than to air. If we use $D = 10^{-3}$ m and $\Delta T = 10$ K (mean temperature difference between coffee and air; the exact value is not that important since it is raised to the power 0.25) we get $\left(\frac{\Delta T}{D}\right)^{0.25} = 10$ (SI units) and if we assume $U \approx h$ (that is, we assume that the heat conduction inside the thermocouple is very fast), we estimate theoretically that $U = 13.1$ W/m^2K (air) and $U = 1730$ W/m^2K (water). We see that the theoretical U-value for water (1730 W/m^2 K) is quite close to the experimental (2880 W/m^2 K), while the theoretical U-value for air (13.1 W/m^2 K) is much lower than the experimental (123 W/m^2 K) estimated from the experiment. The reason for this is probably remaining water droplets on the thermocouple which evaporate and improve the heat transfer for the case when we remove the thermocouple from the coffee.

Example 11.9 Mass balance for filling a bathtub without plug. *Here, we consider the dynamics for the volume (level) in a bathtub with no plug, see Figure 11.13. The model can also describe the dynamics of the outflow for a tank or the change in the water level in a lake following a rainfall. We consider a rectangular bathtub with liquid volume $V = Ah$ where A [m^2] is the base of the tub and h [m] is the liquid height. We assume that the density ρ is constant.*

The control volume (boundary) for the system is the whole bathtub, and the inventory of mass is $m = \rho V$ [kg]. Mass is a conserved quantity, and from (11.3) we get that

$$\frac{dm}{dt} = w_{\text{in}} - w_{\text{out}} \quad [kg/s] \tag{11.39}$$

[1] For more details on this, and in general on modeling and balance equations, see: R.B. Bird, W.E. Stewart and E.N. Lightfoot, *Transport Phenomena*, Wiley, 1960.

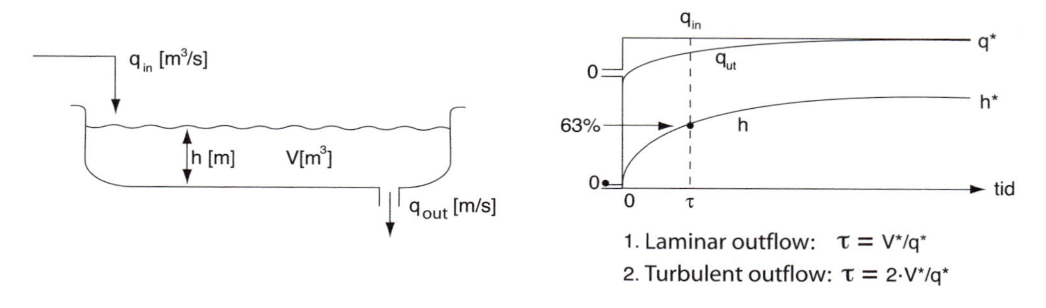

Figure 11.13: Bathtub without plug

and with the assumption of constant density we get the "volume balance"

$$\frac{dV}{dt} = q_{\text{in}} - q_{\text{out}} \quad [m^3/s] \tag{11.40}$$

This equation describes the volume change in a bathtub while it is filled or emptied. With a plug, we have $q_{\text{out}} = 0$, *and the process is a "pure integrator," that is, there is no natural feedback that counteracts the increase in* V. *However, here we consider the case with no plug, and there is a "natural negative feedback," because* q_{out} *is a function of amount of water in the bathtub, that is,* q_{out} *increases when the liquid height* h *increases. We have from the static momentum balance (= mechanical energy balance):*[2]

1. **Laminar flow exit:** $q_{\text{out}} = k_l h$
2. **Turbulent flow exit:** $q_{\text{out}} = k_t \sqrt{h}$

The flow pattern is probably turbulent, but for simplicity let us assume laminar flow.

1. Laminar outflow. *Inserting* $V = Ah$ *into the "volume balance" gives*

$$\frac{d(Ah)}{dt} = A\frac{dh}{dt} = q_{\text{in}} - k_l h \quad [m^3/s] \tag{11.41}$$

This is a first-order differential equation in $h(t)$ *that can be rearranged into the standard form (11.21),*

$$\tau \frac{dh}{dt} = -h + k \cdot q_{\text{in}}$$

Thus, we have $\tau = A/k_l$ *and* $k = 1/k_l$ *and the solution is*

$$h(t) = \frac{1}{k_l}\left(1 - e^{-\frac{k_l t}{A}}\right) q_{\text{in}} \tag{11.42}$$

We find that $h(t)$ *increases with time, most sharply at first, but then the increased level (*h*) results in a larger outflow, and we eventually reach (for* $t \to \infty$*) at a balance point (steady state) where* $q^*_{\text{out}} = q_{\text{in}}$ *and* h *no longer increases. The steady-state value,* h^*, $t = \infty$ *is from (11.42)*

$$h^* = h(\infty) = \frac{q_{\text{in}}}{k_l} \tag{11.43}$$

[2] The outlet stream of the bathtub is driven by the pressure difference $\rho g h$ over the hole where the water exits. At steady state this pressure difference equals the friction pressure drop, i.e., $\Delta p_f = \rho g h$. From fluid mechanics (see page 243) we have $\Delta p_f \sim q$ for laminar flow and $\Delta p_f \sim q^2$ for turbulent flow, and it follows that $q \sim h$ (laminar) and $q \sim \sqrt{h}$ (turbulent).

- *We can alternatively derive (11.43) from the steady state mass balance, $q_{\text{in}} = q_{\text{out}}$ [m³/s]. Here, $q_{\text{out}} = k_l h$ and (11.43) follows.*
- *The time constant is $\tau = A/k_l$. Here, the steady-state flow rate is $q^* = k_l h^* (= q^*_{\text{out}} = q^*_{\text{in}})$, that is, $k_l = q^*/h^*$, and it follows that*

$$\tau = \frac{A}{k_l} = \frac{Ah^*}{q^*} = \frac{V^*}{q^*}$$

which equals the residence time of the bathtub. However, so that you won't think that the time constant always equals the residence time, please note that for turbulent outflow the time constant is twice the residence time; this is shown on page 302.

The following example illustrates that the dynamics of gas systems are usually very fast. This is primarily because of a short residence time, but it is usually further amplified by small relative pressure differences.

Example 11.10 Gas dynamics. *A large gas tank is used to dampen flow rate and pressure variations. Derive the dynamic equations and determine the time constant for the pressure dynamics. We assume for simplicity that the inlet and outlet flow rates of the tank are given by $F_{\text{in}} = c_1(p_{\text{in}} - p)$ [mol/s] and $F_{\text{out}} = c_2(p - p_{\text{out}})$ [mol/s] where the "valve constants" c_1 and c_2 are assumed to be equal ($c_1 = c_2 = c$).*

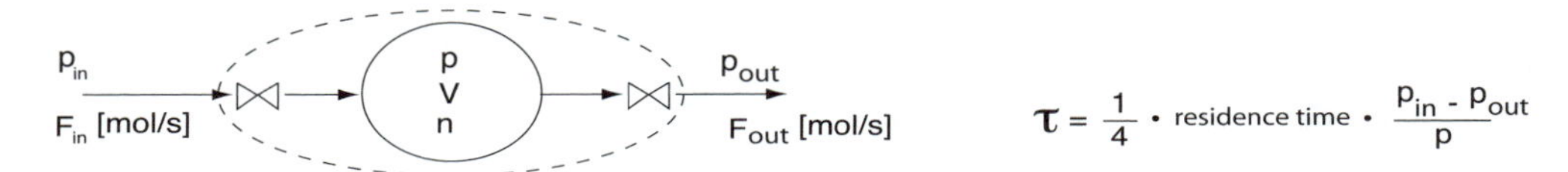

Figure 11.14: Gas dynamics

Solution. *The mass balance is*

$$\frac{dn}{dt} = F_{\text{in}} - F_{\text{out}} \quad \text{[mol/s]}$$

We assume constant volume V and ideal gas,

$$n = \frac{pV}{RT}$$

The mass balance then gives:

$$\frac{V}{RT}\frac{dp}{dt} = c(p_{\text{in}} - p) - c(p - p_{\text{out}})$$

This equation can be used to compute p as a function of p_{in}, p_{out} and time. Rearranged into standard form (11.21), we see that the time constant is

$$\tau = \frac{V}{2cRT} = \frac{n}{2cp} \tag{11.44}$$

From the steady-state mass balance we get $p^* = (p^*_{\text{in}} + p^*_{\text{out}})/2$, so at steady state

$$F^* = F^*_{\text{in}} = F^*_{\text{out}} = c \cdot \frac{p^*_{\text{in}} - p^*_{\text{out}}}{2}$$

Substituting the resulting value for c into (11.44) gives

$$\tau = \frac{n^*}{2cp^*} = \frac{1}{4} \cdot \frac{n^*}{F^*} \cdot \frac{p^*_{\text{in}} - p^*_{\text{out}}}{p^*} \tag{11.45}$$

that is, the time constant is 1/4 of the residence time, n/F, multiplied by the relative pressure difference, $(p_{\text{in}} - p_{\text{out}})/p$. For gas systems, both these terms are usually small, which explains why the pressure dynamics are usually very fast.

For example, with $p^*_{\text{in}} = 10.1$ bar, $p^* = 10$ bar and $p^*_{\text{out}} = 9.9$ bar we get

$$\tau = \frac{1}{4} \cdot \frac{n^*}{F^*} \cdot \frac{10.1 - 9.9}{10} = \frac{1}{4} \cdot \frac{1}{50} \cdot \frac{n^*}{F^*}$$

that is, the time constant for the pressure dynamics in the tank is only 1/200 of the (already small) residence time.

Example 11.11 First-order reaction in batch reactor (or in beaker)

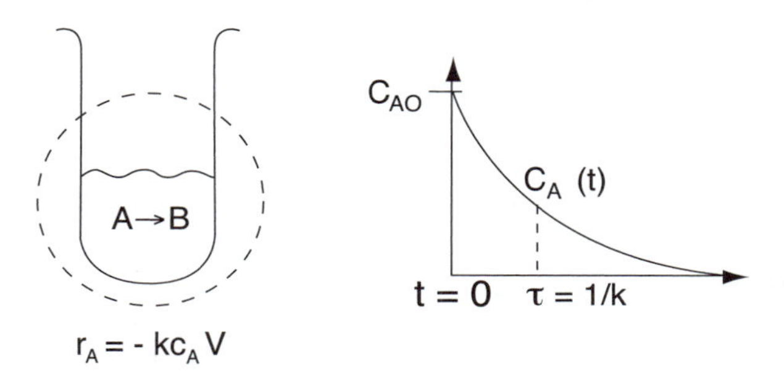

Figure 11.15: Reaction in beaker

Consider a beaker where component A reacts according to the first-order irreversible reaction $A \rightarrow B$. Derive the equation that describes the concentrations dynamics when temperature is assumed constant.

Solution. *There are no inlet and outlet streams, so the component balance for the beaker is*

$$\frac{d(c_A V)}{dt} = r_A V \quad [\text{mol A/s}] \tag{11.46}$$

where r_A is the reaction rate for "generation" of component A, which for a first-order reaction is $r_A = -kc_A$ [mol A/m^3, s], where k [s^{-1}] is constant since the temperature is constant. If we, in addition, neglect changes in the volume, we get

$$\frac{dc_A}{dt} = -kc_A \tag{11.47}$$

which gives a first-order response $c_A(t) = c_A(0)e^{-t/\tau}$ with time constant $\tau = 1/k$ (note that k here is the reaction rate constant and not the gain). We note that $c_A \rightarrow 0$ when $t \rightarrow \infty$, that is, the final steady state has complete conversion of A.

Comment. *This is a batch process, so the system is not initially at steady state. However, this is not a requirement, and (11.47) can be solved when we know the initial concentration $c_A(0)$ at the start of the experiment.*

Exercise 11.2* Evaporator. *Take another look at the evaporator in Example 11.2 (page 280). What is the time constant for the temperature response?*

Exercise 11.3 First-order reaction in CSTR. *Consider a continuous stirred tank reactor (CSTR) where component A decomposes in a first-order irreversible reaction $A \rightarrow B$ with reaction rate $r = (-r_A) = kc_AV$ [mol A/s]. (Note that k here is the reaction rate* *constant and not the process gain). The feed concentration is $c_{A,F}$. Derive the equation that describes the concentration dynamics when temperature is assumed constant. Find the time constant and gain for the response.*

11.3.4 Time response for more complex systems

In the previous section, we considered in detail the step response for systems with only one differential equation which can be written in "standard" form $\tau dy(t)/dt = -y(t) + k\ u(t)$. This gave rise to a first-order response. Although many systems can be written (or approximated) by a first-order response, it must be emphasized that the responses are generally far more complex.

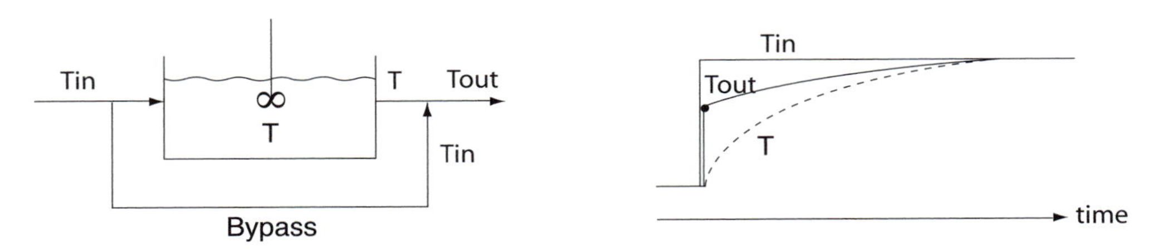

Figure 11.16: Temperature response for stirred tank with *bypass*

- Even for systems with only one linear differential equation, the response can be different from that described above, either because the system is non-linear or because the response has a "direct term," that is, the equation can be written in the form

$$\tau dx(t)/dt = -x(t) + ku(t); \quad y(t) = c \cdot x(t) + d \cdot u(t)$$

  where the $d \neq 0$ gives a "direct term" from u to y (see for example Figure 11.16 which shows the response of a stirred tank with bypass).
- If we have two first-order systems in series, for example two stirred tanks, the total response will be second-order, and if we have n first-order systems in a series, the total response is nth-order. The response for such higher-order systems will usually have a "flatter" initial response (see Figure 11.22, page 309), and is often approximated as an effective time delay.
- We will also have a higher-order response if the model consists of several coupled differential equations, for example, an adiabatic reactor with coupled material and energy balance (see Figure 11.24, page 312).

The analytic expression for the time response of higher-order system is usually rather complicated, and often there is no analytical solution. However, by linearizing the system, as discussed in the next section (Section 11.4), it is possible to use effective mathematical tools for analyzing the system, for example, by computing the system's "poles" (=eigenvalues $= -1/$time constant) and "zeros." The most important tool for analyzing more complex systems is nevertheless "dynamic simulation," that is, numerical solution of the equations. This is discussed in Section 11.5.

Exercise 11.4 *(a) Derive the model for the stirred tank with bypass shown in Figure 11.16 and (b) find an analytical expression for the time response.*

11.4 Linearization

Consider a dynamic model

$$\frac{dy}{dt} = f(y, u) \tag{11.48}$$

This model is **linear** if the function $f(y, u)$ is linear, which means that if we double the change in u (or in y) then the change in f is doubled. In general our models are nonlinear, but we are often interested in studying the response of small changes in u, and we can then use a linearized model. The most important use of linearized models is in control engineering, where the objective of the control is indeed to keep y close to its desired value (that is, Δy is indeed small) such that the assumption of linear model often holds well.

Let y^* and u^* denote the values of y and u at the operating point $*$ (or along the nominal trajectory $y^*(t)$) where we linearize the model. This is often a steady-state point but does not need to be. A first-order **Taylor-series expansion** ("tangent approximation") of the function $f(y, u)$, where we neglect the second-order (with $\Delta u^2, \Delta y^2, \Delta u \Delta y$) and higher-order terms, gives a linearized approximation

$$f(y, u) \approx \underbrace{f(y^*, u^*)}_{f^*} + \underbrace{\left(\frac{\partial f}{\partial u}\right)^* \Delta u + \left(\frac{\partial f}{\partial y}\right)^* \Delta y}_{\Delta f} \tag{11.49}$$

where $\Delta u = u - u^*$ and $\Delta y = y - y^*$ represent the deviations from the nominal operating point. The approximation is exact for small values of Δu and Δy. Further, we have that

$$\frac{d\Delta y}{dt} = \frac{d(y - y^*)}{dt} = \frac{dy}{dt} - \underbrace{\frac{dy^*}{dt}}_{f^*}$$

For the non-linear model (11.48) we have then derived a **linearized model in deviation variables**,

$$\frac{d\Delta y}{dt} = \Delta f = \underbrace{\left(\frac{\partial f}{\partial y}\right)^*}_{a} \Delta y + \underbrace{\left(\frac{\partial f}{\partial u}\right)^*}_{b} \Delta u \tag{11.50}$$

where the coefficients a and b denote the local derivatives with respect to y and u, respectively. Comparing this with the standard form for first-order systems in (11.21),

$$\tau \frac{d\Delta y}{dt} = -\Delta y + k \Delta u$$

we find

$$\tau = -\frac{1}{a}; \quad k = -\frac{b}{a}$$

Thus, linearized models can be used to determine the time constant τ.

Example 11.12 Linearized model for turbulent outflow of tank. *This is a continuation of Example 11.9 (page 296) where we considered laminar outflow of a bathtub. For case 2 with turbulent outflow, $q_{\text{out}} = k_t\sqrt{h}$, the "volume balance" (11.41) for filling the bathtub becomes*

$$A\frac{dh}{dt} = q_{\text{in}} - k_t\sqrt{h} = f(h, q_{\text{in}}) \quad [m^3/s] \tag{11.51}$$

Here, the function f is non-linear in h. Linearizing f and introducing deviation variables gives, see (11.50),

$$A\frac{d\Delta h}{dt} = \Delta f = \Delta q_{\text{in}} - k_t\frac{1}{2\sqrt{h^*}}\Delta h$$

Comparison with the standard form with $y = \Delta h$ and $u = \Delta q_{\text{in}}$ gives $\tau = 2\sqrt{h}^* A/k_t$, where from (11.51), $k_t = q^*/\sqrt{h^*}$ and q^* is the steady state flow. Further rearrangement of the expression for the time constant gives

$$\tau = 2\frac{\sqrt{h}^* A}{k_t} = 2\frac{h^* A}{q^*} = 2 \cdot \frac{V^*}{q^*}$$

That is, the time constant is two times the residence time (while it was equal to the residence time with laminar outflow). In other words, we can, by comparing the experimental time constant with the residence time, predict whether the outflow is laminar or turbulent. Also note that the steady state gain $k = \Delta h(\infty)/\Delta q_{\text{in}} = 2h^*/q^*$ for turbulent flow is twice that of laminar flow.

Comment. *Note that the initial response for $h(t)$ (expressed by the slope $k' = k/\tau$) is the same for both cases, $k' = k/\tau = 1/A$. This is reasonable since the outlet flow (where the difference between turbulent and laminar flow lies) is only affected after the level starts changing.*

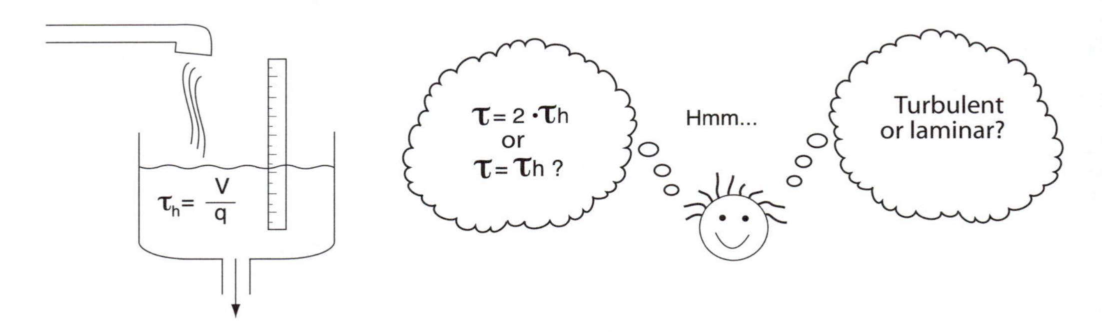

Figure 11.17: Student anxious to check the outflow from a sink

Exercise 11.5 Experiment at home. *You should check whether the outflow from your sink is laminar or turbulent by comparing the time constant τ of the dynamic response in sink level with the residence (holdup) time $\tau_h = V/q$:*

1. *With the plug out, adjust the inflow such that the level is at a steady state where the sink is a little more than half full.*
2. *Reduce the inflow and record the level response (use a ruler and read off the level at regular intervals). From this experiment estimate the time constant τ (when 63% of the steady-state change is reached). This assumes that the area A is reasonably constant in the region between the two steady-state levels.*

3. *Temporarily lead the water somewhere else (but keep the same flow), for example, into a bucket, such that the sink is emptied. Put in the plug and let again the water flow into the sink. Measure the time it takes to fill the tank to its previous level. This is the residence time* $\tau_h = V/q$.
4. *If* $\tau \approx \tau_h$, *the outflow is laminar, and if* $\tau \approx 2\tau_h$, *it is turbulent. (Note that it is possible, but not very likely, that you get a transition from turbulent to laminar flow when* q *is reduced).*
5. *Another way of checking whether the flow is laminar or turbulent is to find the residence time* τ_h *for two different steady state levels (see point 3); if the flow is laminar, then* $\tau_h = A/k_l$ *is independent (!) of the level* h, *but if the flow is turbulent, then* $\tau_h = \sqrt{h}A/k_t$ *increases with the square root of the level.*

Multivariable and higher-order systems. We have above assumed that we have a scalar model with one input variable u and one output variable y. It is, however, easy to generalize the linearization to the multi-dimensional case where the coefficients (derivatives) $A = \partial f/\partial y$ and $B = \partial f/\partial u$ become matrices. The model in deviation variables is then

$$\frac{d\Delta y}{dt} = A\Delta y + B\Delta u$$

Δu**:** vector of independent variables (inputs or disturbances)

Δy**:** vector of dependent state variables (often denoted x)

(Note that we, for simplicity, have not introduced separate symbols for vectors, but we could for clarity have written $\underline{u}$ and $\underline{y}$).

The concept of time constant is less clear in the multivariable case, but we can instead compute the eigenvalues λ_i of the matrix A:

- We find that the "time constants" $\tau_i = -1/\lambda_i(A)$ appear in the linearized time response which contains the term e^{-t/τ_i}. For the scalar case with only one equation ($A = a =$ scalar), the eigenvalue of A equals a, and we find $\tau = -1/a$.
- The system is (locally) stable if and only if all eigenvalues of A have a negative real part (i.e., the eigenvalues are in the left-half complex plane).

11.5 Dynamic simulation with examples

By the expression "dynamic simulation," we mean "numerical solution (integration) of the system's differential equations as a function of time."

We consider a dynamic system described by the differential equations

$$\frac{dy}{dt} = f(y, u)$$

where

1. The initial state $y(t_0) = y_0$ is known (we need one for every differential equation).
2. The independent variables $u(t)$ are known for $t > t_0$.

Using "dynamic simulation," we compute numerically $y(t)$ for $t > t_0$ by integrating the above equation,

$$y(t) = y_0 + \int_{t_0}^{t} f(y(t), u(t))dt$$

(strictly speaking, this should be $y(t) = y_0 + \int_{t_0}^{t} f(y(\tau), u(\tau))d\tau$ but we are a bit sloppy to simplify the notation).

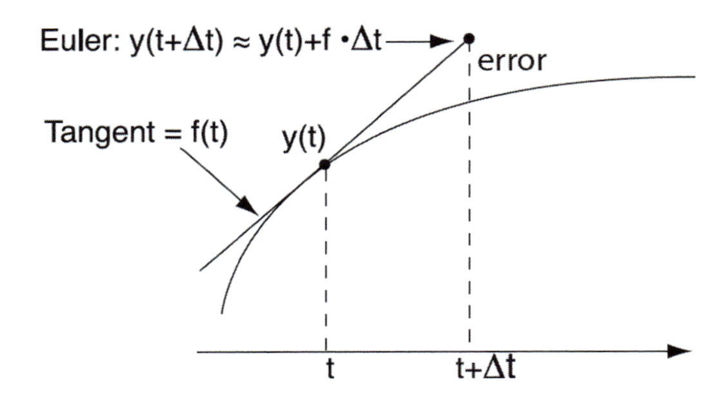

Figure 11.18: Euler integration

The simplest method is **Euler integration**, see Figure 11.18, where we assume that the derivative $f(y, u)$ is piecewise constant over a time interval Δt. If we are at time t, then the value of y at time $t + \Delta t$ is

$$y(t + \Delta t) = y(t) + \int_{t}^{t+\Delta t} f(y, u)dt \approx y(t) + f(y, u)\Delta t$$

where $f(y, u)$ is the local derivative (tangent) at time t. We repeat this at time $t + \Delta t$ and so on, as explained next.

Algorithm for Euler integration:

1. Start at $t = t_0$ with a known initial state y_0).
2. Compute the derivative $f = f(y(t), u(t))$ at time t.
3. Euler approximation: Assume the derivative f is constant over the period Δt and compute $y(t + \Delta t) \approx y(t) + f \cdot \Delta t$.
4. Stop if $t \geq t_{\text{final}}$; otherwise set $t := t + \Delta t$ and $y(t) := y(t + \Delta t)$ and go to step 2.

The algorithm is best understood by considering an example.[3]

Example 11.13 Euler integration: Concentration response for tank.
Consider the continuous stirred tank in Figure 11.19 with the following given data:

- *$V = 5\ m^3$ = constant*
- *q= 1 m^3/min (assumed constant)*

[3] The unit for time (t) is minutes [min] in almost all examples in this chapter.

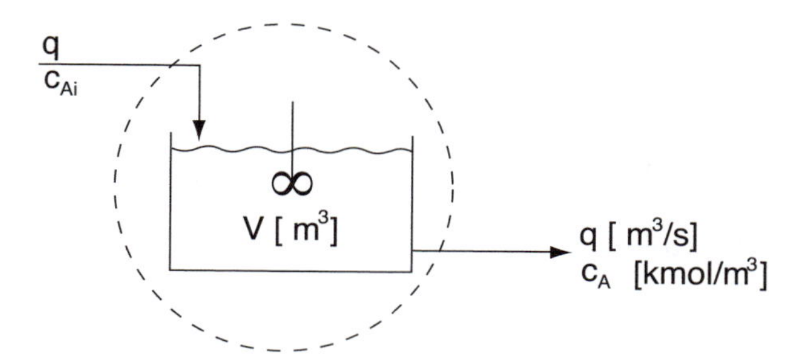

Figure 11.19: Stirred tank without reaction

- *Step change in* $c_{Ai} = \begin{cases} 0.9 \text{ kmol/m}^3, & t \leq 0 \\ 1.0 \text{ kmol/m}^3, & t > 0 \end{cases}$
- *Ideal mixing and constant density is assumed.*

The component balance $d(c_A V)/dt = qc_{Ai} - qc_A$ *[kmol A/s] is rearranged to:*

$$\frac{dc_A}{dt} = \frac{q}{V}(c_{Ai} - c_A) = f(c_A)$$

For $t \leq 0$*, we assume the system is at steady state and the component balance gives* $c_A = c_{Ai} = 0.9$ *kmol/m*^{3} *(the initial value for* c_A*). The exact solution of the differential equation for* $t \geq 0$ *is from (11.22)*

$$c_A(t) = 0.9 + 0.1 \cdot (1 - e^{-t/5})$$

where the time constant is $\tau = V/q = 5$ *min (residence time).*

Let us compare this with Euler integration using $\Delta t = 0.1$ *min, which is a relatively small step compared to the time constant of 5 min. The steps of the algorithm are:*

1. *At* $t = t_0 = 0$ *set* $c_A(t) = 0.9$.
2. *With* $c_{Ai} = 1$ *(constant), we have for* $t \geq 0$*:*

$$f(c_A) = \frac{q}{V}(c_{Ai} - c_A) = 0.2(1 - c_A)$$

3. *Euler approximation: Value of* c_A *at time* $t + \Delta t$ *is:*

$$c_A(t + \Delta t) \approx c_A(t) + f(c_A) \cdot \Delta t = c_A + 0.2(1 - c_A) \cdot 0.1 = 0.98c_A(t) + 0.02$$

4. *Set the value for* $c_A(t + \Delta t)$ *to* $c_A(t)$ *and go to step 2.*

We then get:

t	$c_A(t)$	$f(c_A) = 0.2(1 - c_A)$	$f \cdot \Delta t$	Euler $c_A(t + \Delta t) \approx c_A(t) + f \cdot \Delta t$	Exact $c_A(t + \Delta t)$
0^+	0.9	0.02	0.002	0.902	0.90198
0.1	0.902	0.0196	0.00196	0.90396	0.90392
0.2	0.90396	0.0192	0.00192	0.90588	0.90582
0.3	0.90588	0.0188	0.00188	0.90776	0.90768
⋮	⋮	⋮	⋮	⋮	⋮
5.0	0.9636	0.0073	0.00073	0.9643	0.9628
⋮	⋮	⋮	⋮	⋮	⋮
20.0	0.9982	0.0004	0.00004	0.9982	0.9982

We see, as expected, that Euler integration gives a numerical error; see also Figure 11.20. This error can be reduced by reducing the step length Δt, but this increases the computational effort and if it becomes too small it may conflict with the accuracy of the computer. On the other hand, if Δt gets too large, the Euler integration may go unstable.

There are many possible improvements to Euler integration

- Higher-order method: Include more terms in the Taylor-series expansion for y (Euler assumes $y \approx y_0 + f\Delta t$).
- Introduce step length control (adjusting Δt during integration).
- Use an implicit solution that avoids the possible instability, for example, implicit Euler:

$$y(t + \Delta t) \approx y(t) + f\left(y(t + \Delta t), u(t + \Delta t)\right) \cdot \Delta t$$

  which has to be solved with respect to $y(t + \Delta t)$.

Examples of MATLAB routines which include improvements of this kind are `ode45` and `ode15s` (the latter is recommended for most problems).

Euler integration with MATLAB

We continue Example 11.13. First, we write the following MATLAB routine to compute the derivative $dc_A/dt = f(c_A)$ (and save it in the file `conctank.m`):

```
function DYDT=f(t,y)
% This is file conctank.m
% Concentration response of tank with no reaction
% inlet: Time t and state vector y
% OUTPUT: derivatives DYDT
%
% Usage with odeeuler: [T,Y]=odeeuler(@conctank,[0 10],0.9,0.1)
% Usage with ode15s:   [T,Y]=  ode15s(@conctank,[0 10],0.9)
% Plot results:        plot(T,Y)
%
% I. Data (parameters and independent variables)
V=5;     % tank volume
q=1;     % volumetric flow rate
tau=V/q; % residence time
cai=1;   % inlet concentration for t>0
% II. Extract present value of states
ca=y;    % tank concentration
% III. Evaluate derivatives of states
f= (cai - ca)/tau;
DYDT=f;
```

We note that the routine that calculates the derivative (in this case `conctank.m`) generally contains the following parts:

I. Data (given values for parameters and independent variables u).

II. Extract variables from the state vector `y` (in this example there is only one state so the "vector" `y` has only one element). It is recommended that the variables be given physical names in order to enhance the readability of the code, for example `ca=y`.

III. Evaluate the derivative, that is, compute function `f` (which s returned to the MATLAB integration routine).

In addition, we need a program that computes the numerical solution ("performs the integration"). Below is a simple program for **Euler integration** which is saved in the file `odeeuler.m`:

```
function [tout,yout]=odeeuler(odefile,tspan,y0,H)
% This is the function odeeuler.m
% Simple integration routine written by SiS in 1998
% Usage:        [T,Y]=odeeuler(@F,TSPAN,Y0,H)
% for example:  [T,Y]=odeeuler(@conctank,[0 10],0.9,0.1)
%
%    T - solution time vector.
%    Y - solution state (output) vector.
%    F - filename with diff.eqns. (see also help ode15s).
%    TSPAN = [initial_time final_time}
%    Y0 - initial state vector
%    H - integration step size
%
t0=tspan(1); tfinal=tspan(2);
% Initialize
tout=t0; yout=y0; neq=length(y0); t=t0; y=y0;
% Integrate
while t < tfinal,
  t=t+H;
  f=feval(odefile,t,y);
        for i=1:neq,
        y(i)=y(i)+H*f(i);
        end
  tout=[tout;t]; yout=[yout; y];
end
```

We can now use MATLAB to compute the concentration response using Euler integration:

```
>>  [T,Y]=odeeuler(@conctank,[0 1],0.9,0.1)

T =
         0
    0.1000
    0.2000
    0.3000
    0.4000
    0.5000
    0.6000
    0.7000
    0.8000
    0.9000
    1.0000
    1.1000

Y =
    0.9000
    0.9020
    0.9040
    0.9059
    0.9078
    0.9096
    0.9114
    0.9132
    0.9149
    0.9166
    0.9183
    0.9199
```

```
>>  [T,Y]=odeeuler(@conctank,[0 20],0.9,0.1);  % semicolon avoids output to the screen
% The result is compared with the more exact solution with ode15s:
>>  [T1,Y1]=ode15s(@conctank,[0 20],0.9);
>>  plot(T,Y,T1,Y1,'--')  % see plot in Figure
```

Figure 11.20 compares the results of the Euler integration with a more accurate and effective integration method (`ode15s` in MATLAB). The difference is small in this case.

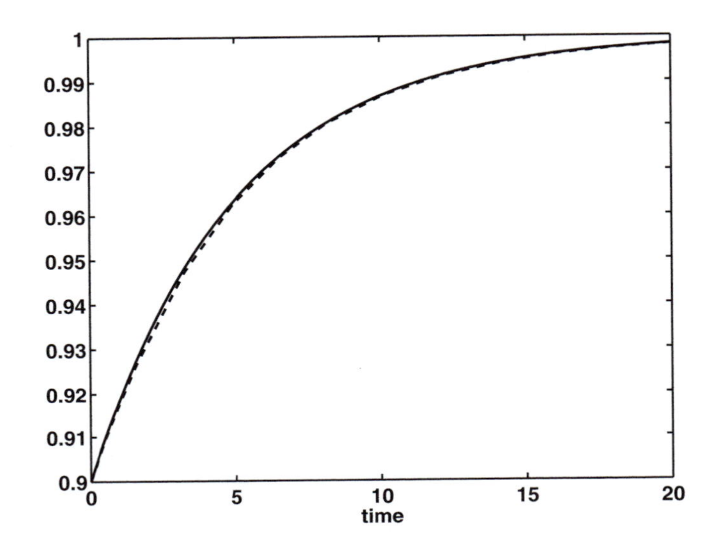

Figure 11.20: Concentration response with `odeeuler` (solid) and `ode15s` (dashed) for a tank.

Example 11.14 Three tanks in series. *This is an extension of Example 11.13, where we had a feed of 1 m^3/min to a large tank of $V_1 = 5\ m^3$. We add two smaller tanks with volume $V_2 = V_3 = 1.5m^3$ (Figure 11.21).*

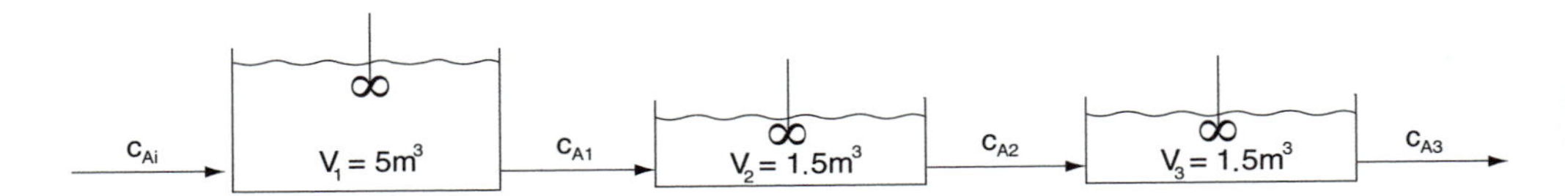

Figure 11.21: Three tanks in a series

The component balance for the "old" tank (tank 1) gives

$$\frac{dc_A}{dt} = \frac{q}{V_1}(c_{Ai} - c_{A1}) = f_1(c_{A1}, c_{Ai})$$

The component balances for the new tanks 2 and 3 give

$$\frac{dc_{A2}}{dt} = \frac{q}{V_2}(c_{A1} - c_{A2}) = f_2(c_{A1}, c_{A2})$$

$$\frac{dc_{A3}}{dt} = \frac{q}{V_3}(c_{A2} - c_{A3}) = f_3(c_{A2}, c_{A3})$$

For $t \leq 0$, steady-state conditions are assumed and the component balances give $c_{A1} = c_{A2} = c_{A3} = c_{Ai} = 0.9$ kmol/m^3 (which is the initial value for the three states). The dynamic

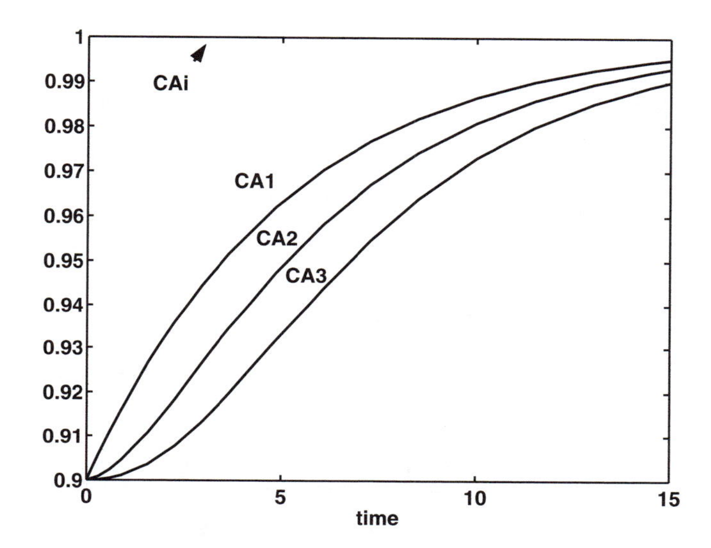

Figure 11.22: Concentration response for three tanks in series

response is shown in Figure 11.22. Note the typical second-order response for c_{A2}, which starts "flat," that is, the first derivative is initially zero. For c_{A3}, the initial response is even "flatter" since the second derivative is also initially zero.

Here, we used the following routine to compute the derivatives (saved in file `conctank3.m`*):*

```
function DYDT=f(t,y)
% This is file conctank3.m
% INPUT: Time t and state vector y
% OUTPUT: derivatives DYDT
% Usage with ode15s:   [T,Y]=  ode15s(@conctank3,[0 15],[0.9 0.9 0.9])
%
% I. Data (parameters and independent variables)
V1=5;        % volume tank 1
V2=1.5;      % volume tank 2
V3=1.5;      % volume tank 3
q=1;         % volumetric flow rate
tau1=V1/q; tau2=V2/q; tau3=V3/q; % residence time
cai=1;       % inlet concentration for t>0
% II. Extract present value of states
ca1=y(1);    % concentration big tank 1
ca2=y(2);    % concentration small tank 2
ca3=y(3);    % concentration small tank 3
% III. Evaluate derivatives of states
f1= (cai - ca1)/tau1;
```

```
f2= (ca1 - ca2)/tau2;
f3= (ca2 - ca3)/tau3;
DYDT=[f1; f2; f3];
```

Example 11.15 Isothermal continuous stirred tank reactor (CSTR).

In an isothermal continuous stirred tank reactor (CSTR) with constant volume V, two reactions take place

$$A \rightarrow B; \quad r_1 = k_1 c_A$$

$$B \rightarrow C; \quad r_2 = k_2 c_B$$

Data: $c_{AF} = 10\ kmol/m^3$ *(feed concentration),* $c_{BF} = 0\ kmol/m^3$, $c_{CF} = 0\ kmol/m^3$, $V = 0.9\ m^3$, $q = 0.1\ m^3/min$, $k_1 = 1\ min^{-1}$, $k_2 = 1\ min^{-1}$.

Task: *Plot the responses of c_A and c_B to a step increase in q of 20%.*

Solution. *Component balances for A, B and C give*

$$\begin{aligned}
\frac{d}{dt}(c_A V) &= q c_{AF} - q c_A - k_1 c_A V \\
\frac{d}{dt}(c_B V) &= 0 - q c_B + k_1 c_A V - k_2 c_B V \\
\frac{d}{dt}(c_C V) &= 0 - q c_C + k_2 c_B V
\end{aligned}$$

The steady-state concentrations are found by setting the time derivatives to 0. We find

$$\begin{aligned}
c_A^* &= \frac{q c_{AF}}{q + k_1 V} = 1 \text{ kmol/m}^3 \\
c_B^* &= \frac{k_1 V}{q + k_2 V} c_A^* = 0.9 \text{ kmol/m}^3 \\
c_C^* &= \frac{k_2 V}{q} c_B^* = 8.1 \text{ kmol/m}^3
\end{aligned}$$

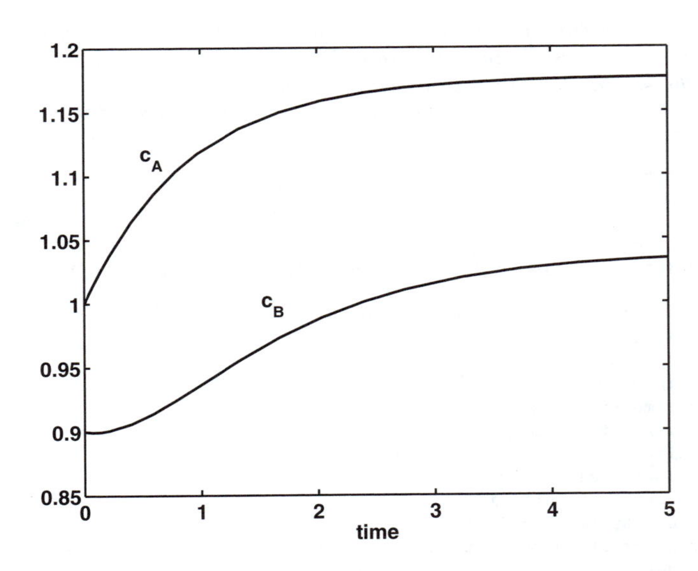

Figure 11.23: Concentration response for isothermal CSTR after a step increase in flowrate

The resulting time response is shown in Figure 11.23. We observe, as expected, a typical "first-order" response for c_A with time constant $V/(q+k_1V) = (0.9/(0.12+1\cdot 0.9))$ min $=$ 0.88 min. The response for c_B is however not a typical second-order response – we have a so-called inverse response *where c_B initially drops (it is not so easy to see) and then reverses and ends up with a steady state increase from 0.90 to 1.034 kmol/m^3. The reason is that an increase in the feed rate initially lowers c_B because of the dilution effect. However, in the long run the lower residence time results in less B being consumed in the reaction $B \rightarrow C$.*

Here, we used the following MATLAB routine in order to compute the derivative of the three concentrations (state variables):

```
function DYDT=f(t,y)
% This is file cstr3.m
% INPUT: Time t and state vector y
% OUTPUT: derivatives DYDT
% Usage with ode15s:   [T,Y]=  ode15s(@cstr3,[0 5],[1.0 0.9 8.1])
%
% I. Data (parameters and independent variables)
cAF=10; cBF=0; cCf=0;  % inlet concentrations
V = 0.9;               % reactor volume (constant)
q = 0.1*1.2;           % 20% increase in q
k1 = 1; k2 = 1;        % rate constants
% II. Extract present value of states
cA=y(1);
cB=y(2);
cC=y(3);
% III. Evaluate derivatives of states
f1= (q*cAF - q*cA - k1*cA*V)           /V;
f2= (q*cBF - q*cB + k1*cA*V - k2*cB*V) /V;
f3= (q*cCf - q*cC           + k2*cB*V) /V;
DYDT=[f1; f2; f3;];
```

Let us now take a look at some more complicated examples where the temperature varies and we also need to use the energy balance.

Example 11.16 Exothermic CSTR with cooling.

In a continuous stirred tank reactor (CSTR) with constant volume V and cooling, we have the exothermic reversible reaction $A \leftrightharpoons B$. The component balances for A and B give

$$V\frac{dc_A}{dt} = qc_{AF} - qc_A - rV \quad [mol\ A/min]$$

$$V\frac{dc_B}{dt} = qc_{BF} - qc_B + rV \quad [mol\ B/min]$$

where the reaction rate is $r = k_1c_A - k_2c_B$ [mol/m^3 min]. The energy balance (11.19) gives

$$\rho V c_p \frac{dT}{dt} = \rho q c_p(T_F - T) + rV(-\Delta_r H^{\ominus}(T)) + Q \quad [J/min]$$

where the "supplied" heat by cooling is $Q = UA(T_c - T)$. The reactor feed $c_{AF} = 10$ kmol/m^3, $c_{BF} = 0$ and $T_F = 300K$, and the cooling temperature is $T_c = 430K$. We assume that the heat capacity and heat of reaction $\Delta_r H^o$ are independent of temperature.

The remaining data are as given in the MATLAB file `cstrT.m` *(see below). By using a long simulation time (10000 min) in MATLAB, the steady state values in the reactor are numerically determined to be*

$$c_A^* = 2.274\ kmol/m^3;\ c_B^* = 7.726\ kmol/m^3;\ T^* = 444.0\ K$$

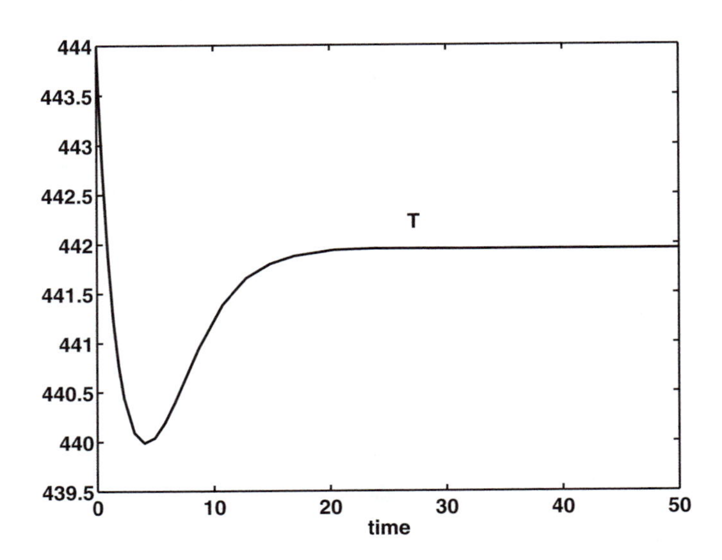

Figure 11.24: Temperature response for exothermic CSTR to a 20% feedrate increase

Increasing the feed rate q by 20% (from 0.10 to 0.12 m^3/min) (by editing the file `cstrT.m`*; try yourself!) gives a drop in the steady-state temperature from 444.0 K to 441.9 K. The dynamic response is shown in Figure 11.24, and we note that we have a rather strange response. The temperature first drops (because we supply more cold feed), but then it rises because more reactant is converted and the reaction is exothermic. This is not an inverse response because the response does not cross its original value.*

If we increase the feed rate q by 50% (to 0.15 m^3/min; try yourself!), we find that the temperature drop is so large that the reaction "extinguishes" (that is, the reactor becomes unstable), and the temperature drops all the way down to 348.7 K, which is much lower than the "cooling" temperature (try yourself!).

Further simulations. *(1) With a very large (infinite) reactor volume, we approach chemical equilibrium and the steady-state reactor temperature is 453.4 K (independent of q, but you may need to run for a very long time). (2) Removing the cooling gives an equilibrium temperature of 461.0 K ("adiabatic temperature rise"). (3) If the cooling is removed at normal conditions the reactor temperature is 453.5 K (rather than 444.0 K). (Try this and other changes yourself! It is easy with the MATLAB program; and if you think it is too much work to write it yourself then you can get it from the author's home page).*

```
function DYDT=f(t,y)
% This is file cstrT.m
% INPUT: Time t and state vector y

% OUTPUT: derivatives DYDT
% Usage with ode15s:   [T,Y]=  ode15s(@cstrT,[0 50],[2274 7726 444.0])
% Plot:                plot(T,Y(:,3))

% All in SI units except time which is in minutes.
% I. Data (parameters and independent variables)
cAF=10000;             % feed concentration of A [mol/m3]
cBF=0;                 % feed concentration of B [mol/m3]
TF=300;                % feed temperature [K]
V = 0.9;               % reactor volume [m3]
```

```
q = 0.10;                % volumetric flow rate [m3/min] (CAN CHANGE, e.g., to 0.12)
dhr=  -80e3;             % Heat of reaction [J/mol]
cp= 4.e3;                % specific heat capacity [J/kg K]
rho = 1000;              % mean density [kg/m3]
U = 1000*60;             % overall heat transfer coefficient [J/min,m2,K]
A = 5;                   % heat transfer area [m2]
Tc = 430;                % cooling temperature [K]
k1_400 = 0.1;            % rx1: rate constant at 400K [1/min]
k2_400 = 0.001;          % rx2: rate constant at 400K [1/min]
E1 = 60e3;               % rx1: activation energy [J/mol]
E2 = E1 - dhr;           % rx2: activation energy [J/mol]
R = 8.31;                % gas constant [J/K mol]
% II. Extract present value of states
cA=y(1);
cB=y(2);
T=y(3);
% IIIa. Intermediate calculations
k1 = k1_400 * exp(-(E1/R) * (1/T - 1/400));
k2 = k2_400 * exp(-(E2/R) * (1/T - 1/400));
r = k1*cA - k2*cB;
Q = U*A*(Tc-T);
% IIIb. Evaluate derivatives of states
Vdcadt = q*cAF - q*cA - r*V;                     % [mol A/min]
Vdcbdt = q*cBF - q*cB + r*V;                     % [mol B/min]
mcpdTdt = rho*q*cp*(TF-T) + r*V*(-dhr) + Q;      % [J/min]

f1 = Vdcadt/V;
f2 = Vdcbdt/V;
f3 = mcpdTdt / (rho*cp*V);

DYDT=[f1; f2; f3];
```

In the above example, we assumed that the heat capacities and the heat of reaction were independent of temperature. For the more general cases, it is recommended that the energy balance is written in its original form with U (or H) as a state, and that T is found numerically from the implicit algebraic equation $U = U_0(T, p, n_i)$, as described for the flash tank in Example 11.18.

Exercise 11.6 Second-order reaction in CSTR *Consider a continuous stirred tank reactor (CSTR) where component A decomposes in a second-order irreversible reaction* $2A \rightarrow B$ *with reaction rate* $r = r_B = kc_A^2 V$ *[kmol/s]. The following steady state data are given:* $V^* = 30$ *m*^{3} *(constant),* $q^* = 0.5$ *m*^{3}*/s,* $c_{AF}^* = 4$ *kmol/m*^{3} *,* $c_{BF}^* = 0$ *kmol/m*^{3} *(feed),* $c_B^* = 1$ *kmol/m*^{3} *(product and tank).*

(a) Derive the equations that describe the concentration dynamics when the temperature and volume are assumed constant.

(b) Use the steady state data to determine c_A^* *and the reaction rate constant* k*.*

(c) Linearize the model and determine an expression for the time constant for the concentration response for component A in the nominal working point.

(d) Sketch the expected response $c_A(t)$ *(in product/tank) when we at* $t = 0$ *throw in some catalyst such that* k *is doubled. (For calculations by hand you can for example use Euler integration of the balance for component A with* $\Delta t = 5$ *s). What is the new steady state value of* c_A*? What is the time constant?*

Distillation examples

Example 11.17 Dynamics of distillation column *(Figure 11.25). In a distillation column, the components are separated based on their difference in volatility, and multiple* *stages and countercurrent flow are used to enhance this. Here, we look at a very simple distillation column with only three equilibrium stages (a reboiler, a feed stage, and a stage above the feed) plus a total condenser. We separate a binary mixture with a constant relative volatility of 4.78. The MATLAB file* `dist.m` *given below should be self-explainable. In order to find the steady state column profile, we simulate for a very long time. We then find the following mole fractions of the lightest component on the four stages (including the total condenser):*

```
xss =
0.0998    0.3160    0.6536    0.9002
```

That is, we have about 10 mol% light component in the bottom product and about 90 mol% in the top product. The subsequent response to a step in the feed rate F by 20%, with constant reflux (L) and boilup (V), is shown in Figure 11.25. We note that the responses are close to first-order, in spite of the fact that we have four coupled differential equations.

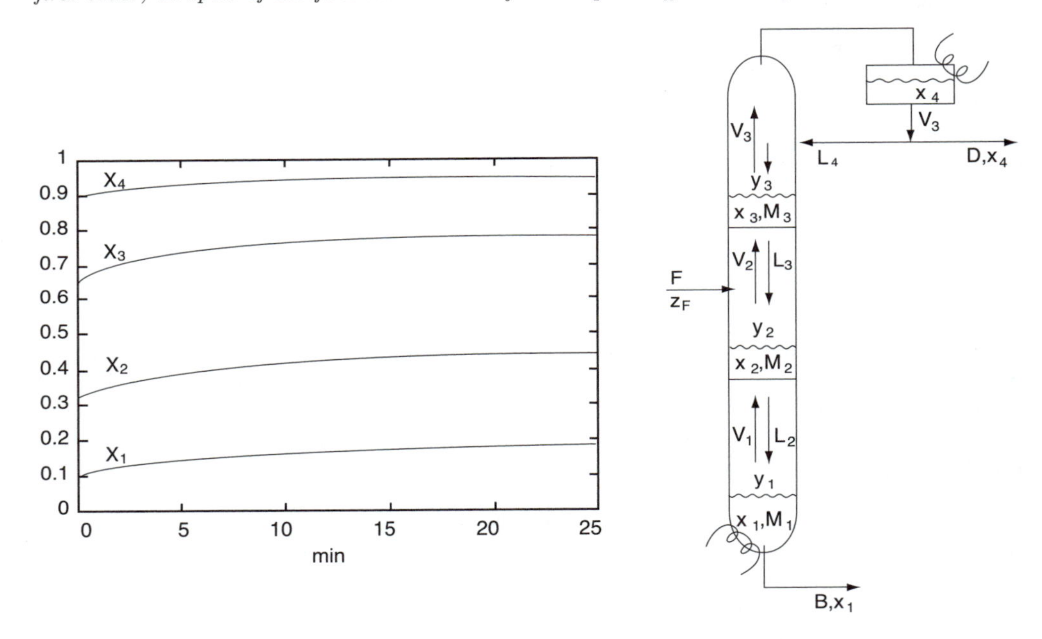

Figure 11.25: Concentration response for distillation column

```
function DXDT = f(t,x)
% This is the file dist.m
% Distillation column with reboiler (stage 1), a feed stage  (stage 2),
%  ... a stage above this (stage 3) and a total condenser (stage 4)
% Assumptions; Binary mixture with constant alfa and constant molar flows
%              Molar holdup on all stages is 1 kmol (M=1)
% States x : vector of liquid mole fractions of light component on the stages
% Usage:
%  x0 = [0.5, 0.5, 0.5, 0.5]; % initial states (not steady-state)
%  [T,X] = ode15s(@dist,[0 1000],x0) % First simulate to t=1000 (steady-state)
%  xss = X(length(X),:) % Save the steady-state mole fractions
%  [T,X] = ode15s(@dist,[0 20],xss) % Run new simulation (e.g change F=1.2)

% I. Data (parameters and independent variables)
```

```
% Assume constant relative volatility
alfa = 4.78;

% Feed rate [kmol/min] and feed composition  (may change this)
F=1.0; zF=0.5;

% Flows in the column [kmol/min] (feed liquid; constant molar flows)
V=3.55; V1=V; V2=V; V3=V;
L=3.05; L4=L; L3=L; L2=L+F;
% Assume constant condenser and reboiler holdup (perfect level control):
D=V3-L4; B=L2-V1;

% II. Extract present value of states
%   ..... Not needed here since x is the state which is already a good name

% IIIa. Intermediate calculations
% Vapor-liquid equilibrium (constant relative volatility)
y(1) = alfa*x(1)/(1+(alfa-1)*x(1));
y(2) = alfa*x(2)/(1+(alfa-1)*x(2));
y(3) = alfa*x(3)/(1+(alfa-1)*x(3));
y(4)=x(4);  % total condenser

% IIIb. Evaluate derivatives of states
% Component balances (assume constant stage holdups M1=M2=M3=M4=1 [kmol])
DXDT(1) = L2*x(2)-V1*y(1)-B*x(1);
DXDT(2) = L3*x(3)+V1*y(1)-L2*x(2)-V2*y(2)+F*zF;
DXDT(3) = L4*x(4)+V2*y(2)-L3*x(3)-V3*y(3);
DXDT(4) = V3*y(3)-L4*x(4)-D*x(4);

% Change vector DXDT to a column vector (MATLAB requires this..).
DXDT=DXDT';
```

The above routine does not make use of MATLAB's vector calculation features. However, below is given an excerpt from a more general routine which uses vectors. Note that we use element-by-element operators `*.` and `./` to multiply and divide vectors. This code also allows for variable stage holdup $M(i)$, which is important if the model is to be used for control purposes. It is simple to change the number of stages `NT` in the column.

```
% From code for general distillation column dynamics
% Vapor-liquid equilibria
i=1:NT-1;     y(i)=alpha*x(i)./(1+(alpha-1)*x(i));

% Need algebraic for computing L(i) (e.g., Francis weir)
% and V(i) (e.g., constant molar flows or ``valve" equation)
% ..... but these are not given here.

% Column mass balances
i=2:NT-1;
dMdt(i) = L(i+1)          - L(i)         + V(i-1)           - V(i);
dMxdt(i)= L(i+1).*x(i+1)  - L(i).*x(i)   + V(i-1).*y(i-1)   - V(i).*y(i);

% Correction for feed at the feed stage
% The feed is assumed to be mixed into the feed stage
dMdt(NF) = dMdt(NF)  + F;
dMxdt(NF)= dMxdt(NF) + F*zF;

% Reboiler (assumed to be an equilibrium stage)
dMdt(1) = L(2)        - V(1)        - B;
dMxdt(1)= L(2)*x(2)   - V(1)*y(1)   - B*x(1);
```

```
% Total condenser (no equilibrium stage)
dMdt(NT) = V(NT-1)          - LT        - D;
dMxdt(NT)= V(NT-1)*y(NT-1)  - LT*x(NT)  - D*x(NT);

% Compute the derivative for the mole fractions from d(Mx) = x dM + M dx
i=1:NT;
dxdt(i) = (dMxdt(i) - x(i).*dMdt(i) )./M(i);

% Output
DYDT=[dxdt';dMdt'];
```

Systems with algebraic equations (DAE system)

In the above examples, all the model equations were differential equations. Actually, we had some algebraic expressions, e.g., the reaction rate constant as function of temperature, but these were explicit in the (dynamic) state variables y_1, i.e., $y_2 = f(y_1, u)$, such that they could easily be evaluated (using `IIIa. Intermediate calculations` in the MATLAB code).

However, more generally, one will in addition to the differential equations

$$\frac{dy_1}{dt} = f_1(y_1, y_2, u) \tag{11.52}$$

also have "implicit" algebraic equations of the form

$$0 = f_2(y_1, y_2, u) \tag{11.53}$$

where y_2 are the extra algebraic variables. Three approaches of dealing with systems with both differential and algebraic equations (DAE systems) are:

1. **Eliminate the algebraic variables** y_2 by substituting relationships for them into the differential equations (which is actually what we do with the "intermediate calculations" in the above examples). This approach does not generally work for all the equations, but it should be used to some extent to reduce the number of variables. However, you should avoid that things get too complicated, because otherwise the code becomes difficult to read and you will make errors.
2. **Use a separate "equation solver" for the algebraic equations** $f_2 = 0$, which is "inside" an ordinary differential equation (ODE) solver (integrator). This approach is common, but may be ineffecient in terms of computing time.
3. **Use a DAE-solver that solves the differential and algebraic equation simultaneously.** The equation set is then written in the form

   $$M\frac{dy}{dt} = f(y, u)$$

   where the "*mass matrix*" M is a square matrix that tells the solver which equations are algebraic. Usually, M is a diagonal matrix with 1's on the diagonal for differential equations and 0's for algebraic equations. This is a general approach and complicated "fixes" are avoided.

To illustrate the three approaches, consider a **dynamic flash**. We derive differential equations (and dynamic states) from the dynamic balances for component mass and energy. The energy balance has internal energy U (or enthalpy H) as the "natural" differential variable (dynamic state):

$$dU/dt = Fh_F(T, \ldots) - Gh_G(T, \ldots) - Lh_L(T, \ldots)$$

where, as indicated, the enthalpies h_F, h_G and h_L are (explicit) functions of temperature T. In addition, we have the algebraic equations, which mostly are associated with the vapor-liquid equilibrium (VLE). Again, these algebraic equations do not depend explicitly on U, but rather on temperature T, etc. As discussed above, there are three approaches to overcome this:

1. In simple cases, we can **eliminate** U as a state variable by substituting its dependency on T and other variables into dU/dt, and rewrite the energy balance with T as a state ($dT/dt = \cdots$). This approach was used in all the previous examples, but generally it will not work, or at least be very cumbersome, see (11.14). For the flash example, it will not work because U depends on the phase distribution f and we lack an expression for df/dt in (11.14).
2. In general, with internal energy U and the component holdups as state variables, we can solve an **"UV-flash"** to compute the temperature, pressure and phase distribution. Here, we make use of the fact that the total volume V of the flash tank is fixed. The UV flash must performed as a **separate "intermediate" calculation**, which requires a separate solver, in addition to the solver for the differential equations (integrator). This approach may require a long computation time because of the nested loops.
3. The recommended approach, used in the example below, is to use a **DAE solver** (`ode15s` in our case) to solve the flash equations and the differential equations **simultaneously**. However, also here we should use "intermediate calculations" (elimination; approach 1) to reduce the number of algebraic equations, for example, for computimg physical properties. For each remaining algebraic equation, we need an associated (algebraic) state variable, which should be chosen such that algebraic equations in the "intermediate calculations" depend explicitly on the state variables. In many cases, we need the temperature T in the "intermediate calculations," so it is recommended to choose T as a (algebraic) state variables. In summary, it recommended to include <u>both</u> U and T in the state vector, by using the following DAE set

$$\begin{aligned} \frac{dU}{dt} &= f_1(T, n_i, \ldots) \\ 0 &= \underbrace{U - U_0(T, n_i, \ldots)}_{f_2} \end{aligned}$$

This corresponds to including U in the differential variables y_1 and T in the algebraic variables y_2 in (11.52)-(11.53).

We next consider an example with a flash tank where we use the simultaneous DAE approach.

Example 11.18 Adiabatic flash. *We have an adiabatic flash tank with feed stream F [mol/s], vapor product G [mol/s] and liquid product L [mol/s]. The feed consists of methanol*

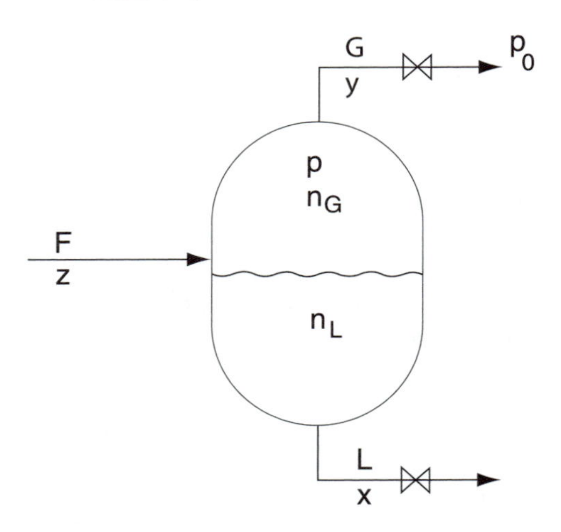

Figure 11.26: Adiabatic flash tank

(1) and ethanol (2). The mole fractions of the light component in the three streams are denoted z (feed), y (vapor) and x (liquid), respectively. The feed is assumed to be liquid and the pressure p_0 downstream the tank (which must be lower than the bubble point pressure of the feed to get flashing) is assumed given; see Figure 11.26. The vapor/liquid equilibrium (VLE) is assumed to be ideal and follow Raoult's law.

For a system with N_c components, we can generally set up N_c mass balances (we choose to use 1 total mass balance, and $N_c - 1$ component balances) and 1 energy balance (11.11);

$$\begin{aligned} \frac{dn}{dt} &= F - G - L \\ \frac{dn_i}{dt} &= Fz_i - Gy_i - Lx_i \quad (i = 1, \ldots, N_c - 1) \\ \frac{dU}{dt} &= Fh_F - Gh_G - Lh_L \end{aligned}$$

Here n [mol] is the total holdup in the tank (in both phases), n_i [mol i] is the holdup of component i in the tank (in both phases) and U [J] is the internal energy in the tank. We have assumed, in the energy balance, that the volume of the tank V_{tot} is constant, such that $p_{\text{ex}} \frac{dV_{\text{tot}}}{dt} = 0$. This gives $N_c + 1$ differential equations, corresponding to $N_c + 1$ dynamic state variables (n, n_i, U). However, in addition we generally have a large number of algebraic equations, which may require us to add algebraic state variables, at least if the algebraic equations are implicit.

First, we have the following algebraic relationships for mass and energy holdups

$$\begin{aligned} n &= n_G + n_L \\ n_i &= n_G\, y_i + n_L\, x_i \quad (i = 1, \ldots N_c - 1) \\ U &= n_G\, h_G + n_L\, h_L - pV_{\text{tot}} \end{aligned}$$

where n_G and n_L is the amount of gas and liquid (the phase distribution) in the tank. Furthermore, we have algebraic expressions for $h_G(T, p, x_i)$ and $h_V(T, p, y_i)$ [J/mol], for G (e.g., valve equation $G = k_g(p - p_0)$), for L (e.g., level control equation $L = k_L(V - V_0)$), for the VLE (from which we can compute p and y_i from T and x_i), for the gas holdup n_G (e.g., ideal gas law), etc. For details, see the MATLAB code below, which should be self-explainable.

It is possible to set up all of these equations as one large equation set (and solve with a DAE solver), but this gives many algebraic variables. In practice, we want to reduce the number of algebraic equations (and the corresponding number of state variables) by inserting any explicit algebraic relationships into the differential equations, as we did earlier using "intermediate computations." There are many ways of doing this, and it will depend on which variables we select as the algebraic state variables.

For the present flash example, we select T and V_L (liquid volume) as the algebraic state variables, in addition to the three dynamic state variables n, n_1 and U. With this choice, all the algebraic equations are explicit in the state variables, except for two algebraic equations for n and U (which are implicit in V_L and T). For details see the MATLAB file `flash.m` *below. Note that the mass matrix M has 0's on the last two diagonal entries, to signal that the last two equations are algebraic rather than differential.*

The steady-state solution is, as before, found by simulating the dynamic response for a long time (unfortunately, it is not allowed in MATLAB to set the mass matrix $M = 0$, which in principle should have been OK). We find at steady state

$$n = 100.4e3 \text{ mol}, \quad n_1 = 48.1e3 \text{ mol}, \quad U = 5.02e8 \text{ J}, \quad T = 344.5 \text{ K}, \quad V_L = 5.02\ m^3$$

We start from this steady-state when performing further simulations. The liquid feed rate is 0.1 m^3/s (2012 mol/s), so the residence time in the flash tank is about 50 s. The dynamic response in the flash tank temperature T to a step increase in the feed temperature T_f from 400K to 440 K is shown in Figure 11.27. This corresponds to an increase in feed enthalpy. Note that the feed is liquid, and the feed pressure (20 bar) is above the bubble point. There is a fast initial temperature increase from 344.5 K to about 345.1 K, related to a fast pressure increase (from $p = 1.03$ bar to 1.05 bar), followed by a slow temperature increase towards 345.3 K at the new steady state, related to the composition change in the liquid phase (from $x_1 = 0.479$ to 0.467 at the new steady state). The increase in vapor flow is from $G = 302$ mol/min to 533 mol/s at the new steady state. The MATLAB file used for this simulation is given below.

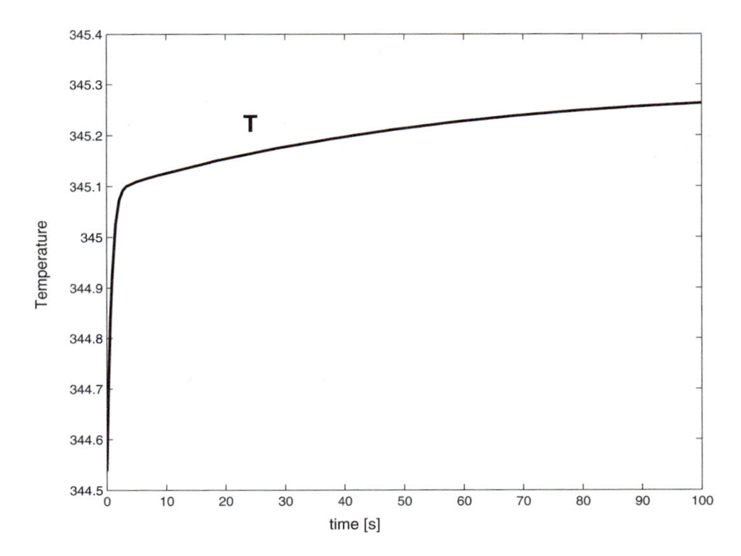

Figure 11.27: Adiabatic flash: temperature response after step in the feed enthalpy

```
function F=f(t,y)
% This is file flash.m
% INPUT: Time t and state vector y
% OUTPUT: Right hand side of DAE set: M dy/dt = f(y)
% States n=y(1); n1=y(2); U=y(3); T=y(4); VL=y(5);
% Usage with ode15s:
%    options = odeset('mass',diag([1 1 1 0 0]));
%    y0 = [100.4e3 48.1e3 5.02e8 344.5 50.17]
%    [t,y]= ode15s(@flash,[0 100],y0,options)
% Plot temperature:   plot(t,y(:,4))

% All in SI units
% I. Data (parameters and independent variables)
% Data for: 1-methanol, 2-ethanol
A1=8.08097; B1=1582.271; C1=239.726;    % Antoine psat1 [mmHg] w/ T [C] (T-range: 15C - 84C)
A2=8.11220; B2=1592.864; C2=226.184;    % Antoine psat2 [mmHg] w/ T [C] (T-range: 20C - 93C)
cpl1=80; cpl2=131; cpv1=44; cpv2=65;    % heat capacity        [J/K mol]
T0=298.15; hvap01=38000; hvap02=43000; % heat of vap. at T0   [J/mol]
Vl1= 40.7e-6; Vl2=58.7e-6;              % liquid molar volumes [m3/mol]
R=8.13;                                 % J/mol K

% Feed data
q=0.1;                                  % m3/s
z1=0.5;                                 % mol1/mol
Tf=1.1*400;                             % K  (increase from 400K to 440K)
p0=1e5;                                 % N/m2  (=1 bar downstream pressure)
Vf = z1*Vl1 + (1-z1)*Vl2;               % m3/mol (molar volume feed)
F = q/Vf;                               % mol/s (feed rate)

% Total tank volume
Vtot = 10;                              % m3

% Valve constant and controller gain
kg=0.1; kl=100e3;                       % note that the P-controller gain kl is large

% II. Extract present value of states
n=y(1);                                 % total holdup in tank (both phases)       [mol]
n1=y(2);                                % component 1 holdup in tank (both phases) [mol1]
U=y(3);                                 % total internal energy (both phases)      [J]
T=y(4);                                 % temperature (same in both phases)        [K]
VL=y(5);                                % liquid volume                            [m3]

% IIIa. Intermediate calculations
% VLE
x1 = n1/n;
p1s=10.^(A1-B1/(T-273.15+C1))/750e-5;  % psat1 from Antoine                     [N/m2]
p2s=10.^(A2-B2/(T-273.15+C2))/750e-5;  % psat2 from Antoine                     [N/m2]
p1 = x1*p1s;                           % partial pressure component 1           [N/m2]
p2 = (1-x1)*p2s;                       % partial pressure component 2           [N/m2]
p = p1+p2;                             % pressure = sum of partial pressures    [N/m2]
y1 =  p1/p;                            % vapor fraction component 1             [mol1/mol]
Vm  = x1*Vl1 + (1-x1)*Vl2;             % molar volume (liquid phase)            [m3/mol]

% Phase distribution
VG = Vtot - VL;                        % gas volume                             [m3]
nL = VL / Vm;                          % liquid holdup                          [mol]
nG = p*VG / (R*T);                     % gas holdup  (ideal gas law)            [mol]

% Enthalpies [J/mol] (Ref.state: pure liquid at T0)
hF = [z1*cpl1 + (1-z1)*cpl2] * (Tf-T0);
hL = [x1*cpl1 + (1-x1)*cpl2] * (T -T0);
hG = [y1*cpv1 + (1-y1)*cpv2] * (T -T0) + y1*hvap01 + (1-y1)*hvap02;
```

```
% Vapor and liquid flow
G = kg*(p-p0);              % simple valve equation for outflow of gas                 [mol/s]
VLs = Vtot/2;               % Setpoint level (volume): keep 50% liquid in tank         [m3]
L = kl*(VL-VLs);            % Level controller with proportional gain kl               [mol/s]

% IIIb. Evaluate right hand side of DAE-set: M dy/dt = f(y)
f1 = F - G - L;                     % =dn/dt  Dynamic: Overall mass balance
f2 = F*z1 - G*y1 - L*x1;            % =dn1/dt Dynamic: Component 1 mass balance
f3 = F*hF - G*hG - L*hL;            % =dU/dt  Dynamic: Energy balance
f4 = U + p*Vtot - hL*nL - hG*nG;    % = 0     Algebraic: Internal energy U
f5 = n - nG - nL;                   % = 0     Algebraic: Total holdup n
F  = [f1; f2; f3; f4; f5];
```

Finding the steady-state. *Above, the steady state was found by simulating the dynamic response for a long time. Alternatively, one may find the steady-state directly, for example, using the function* `fmincon` *in MATLAB:*

```
yss = fmincon('1',y0,[],[],[],[],[],[],@flashss)
```

where the file `flashss.m` *is identical to* `flash.m` *(above) except that the first line is changed to* `function [c,ceq]=f(y)` *and the following line is added at the end:* `c=[]; ceq=F;`.

Remark 1 Removing or adding algebraic state variables. In the MATLAB code given above, we have two algebraic state variables (T and V_L), but actually we can get rid of V_L as a state variable if we do a little work. This follows because the holdup equations are simple, so we can combine them (including $n = n_G + n_L$ which MATLAB solved using `f5` in the code above), and derive an explicit expression for V_L as a function of n, p, V_{tot} and V_m (try yourself, it is easy!). In the MATLAB code above, we then replace the line `VG = Vtot - VL` by the following two lines of "intermediate calculations":

```
VG = (Vtot-n*Vm)/(1 - (p*Vm)/(R*T));
VL = Vtot - VG;
```

We now have only 4 state variables, so we delete the algebraic equation `f5 = n - nG - nL` at the end. Of course, we also need to change the mass matrix and the call to `ode15s`. The final result is of course the same as before.

The main problem when we reduce the number of state variables is that the equations get a bit more messy and it is easy to make mistakes. For this reason, we often choose to *add* "unnecessary" state variables in the problem. This also makes plotting the results easier, as MATLAB stores all the state variables. For example, if we want to plot pressure, then we can simply add a "dummy" state variable (`pdummy=y(6);`) together with a "dummy" algebraic equation (`f6= p - pdummy;`).

Remark 2 Fixing pressure and index problem. In the model of the flash tank given in the MATLAB code above, we let the pressure vary dynamically, but from the very quick initial rise in temperature in Figure 11.27 it follows that the pressure dynamics are very fast. In such cases it might be tempting to say that the pressure is fixed by introducing the algebraic equation $p = p_0$ (`f6 = p-p0;`) and an additional state variable G (and omitting the valve equation for G). This is in principle OK, but it turns out that the integration routine `ode15s` is unable to solve this – we get an error message: "`This DAE appears to be of index greater than 1.`" An "index problem" is often an indication of a non-physical assumption (in this case, it is not physically possible to keep the pressure p constant), and the problem can often be avoided by rewriting the equations, and/or avoiding non-physical assumptions.

11.6 Process control

Automatic **feedback control** is widely used in the process industry, and the instrumentation and control system typically represents 30% of the investments in a plant. For each process variable y that one wants to control one needs

- a measurement of the process variable (y),
- an independent manipulated variable u (usually a valve) that influences y.

We use the following notation

- CV = controlled variable (y, "output")
- MV = manipulated variable (u, "input", independent variable)
- DV = disturbance variable (d, independent variable that we cannot influence)

The MV should have a "direct" and large effect on the CV (with fast dynamics and a small delay or inverse response). The idea of control is to adjust the MV (u) such that the CV (y) is kept close to its desired setpoint y_s, in spite of disturbances d, that is, we want a small control error,

$$e(t) = y(t) - y_s$$

We use **negative feedback**, where the sign of the control action is opposite the sign of the process. This implies that the MV (u) is adjusted such that it *counteracts* changes in the CV (y). A well-known feedback controller from daily life is the on/off controller used in thermostats, where the heat is the MV and temperature is the CV. The on/off controller is simple, but it gives large MV changes (between max and min), and fluctuations in the CV (temperature) are unavoidable. This is undesirable, so in the process industry one normally uses the *proportional-integral-derivative (PID) controller* with algorithm

$$u(t) = u_0 - K_c \left(e(t) + \frac{1}{\tau_I} \int_0^t e(t)dt + \tau_D \frac{de(t)}{dt} \right) \tag{11.54}$$

We see that the MV-change away from its nominal value ($u - u_0$) is a weighted sum of the present value of the error e (the P-term), the integral of the error e (the I-term) and the derivative of the error e (the D-term). The PID controller has three adjustable parameters:

- Gain K_c
- Integral time τ_I [s]
- Derivative time τ_D [s]

The proportional term is usually the most important, and a large value of K_c results in a faster initial response. The integral action causes the MV to change until the error $e(t)$ is zero, that is, we get no steady state off-set. A small value of the integral time τ_I [s] results in the controller returning faster to steady state. For this reason, the integral time is often called the "reset time." The derivative term can give faster responses for some processes, but it often gives "nervous control" with large sensitivity to measurement noise. For this reason, a PI controller (with $\tau_D = 0$) is most common.

There are also other variants of the PID controller, for example, the cascade form, but the differences are usually small for practical purposes. One should, however, note that the vendors use different names and definitions for the three PID parameters. For example, some vendors use the integral gain $K_I = K_c/\tau_I$ and the derivative gain $K_D = K_c\tau_D$. Others use the "proportional band" $100/K_c$, and "reset rate" $1/\tau_I$.

The main problem with negative feedback is that we can get instability if we over-react (if K_c is too large or τ_I is too small) such that we get variations that grow over time.

On-line tuning. Finding good control parameters ("tunings") is not as simple as one may believe. A common (and serious) mistake is to use the wrong sign for K_c, which usually causes the system to drift to an operating point with a fully open or fully closed valve. Tuning is often performed "on-line" using trial-and-error. One usually starts with a controller with a low gain (K_c) and with no integral action ($\tau_I = \infty$). K_c is then gradually increased until either (a) the control performance to disturbances and set-point changes is acceptable, (b) the MV change is too large or (c) the system starts oscillating. If the system starts oscillating, then K_c is reduced by approximately a factor 2 or more. Next, one gradually reduces the integral time τ_I until (a) the settling time (back to the set-point) is acceptable or (b) the system starts oscillating. If the system oscillates, then τ_I is increased by a factor of approximately 2 or more compared to the value that gave oscillations. If the response is too slow then one may try introducing derivative time τ_D, which can be increased until (a) the MV changes become too nervous or (b) the system starts oscillating. If the system starts oscillating, then τ_D is reduced with approximately a factor 2 or more compared to the value that gave oscillations.

Model-based tuning for fast response. Alternatively, model-based tuning is used. The response (without control) from the MV (u) to the CV (y) is recorded and then approximated as a first-order response with a delay, that is, one obtains the model parameters k, τ and θ (see page 286). The following SIMC[4] PI-tunings are recommended

$$K_c = \frac{1}{k}\frac{\tau}{\tau_c + \theta}; \quad \tau_I = \min\{\tau,\ 4(\tau_c + \theta)\} \tag{11.55}$$

Here, the "closed-loop" response time τ_c [s] is the only tuning parameter. A smaller τ_c gives a faster response for the CV, but one may get oscillations and the MV-changes are larger. In order to avoid oscillations and have good robustness (with a good margin to instability), it is recommended to choose τ_c larger than the effective delay, that is, $\tau_c \geq \theta$.

If the response is dominant second order, meaning that the response is well approximated by a second-order response with $\tau_2 > \theta$, then a substantial improvement can sometimes be obtained by adding derivative action, provided there is not too much measurement noise. The response is then approximated by a second-order model with parameters k, τ, τ_2 and θ. For a PID controller on *cascade* form, K_c and τ_I are then as given in (11.55) (but note that the parameter values will change because θ is smaller when we use a second-order model) and the derivative time is

$$\tau_D = \tau_2 \tag{11.56}$$

[4] S. Skogestad, "Simple analytic rules for model reduction and PID controller tuning," *J. Process Control*, Vol. 13 (2003), 291–309.

Note that this is for a so-called cascade PID-form. To get the corresponding PID-parameters for the "ideal" PID form in (11.54), compute the factor $\alpha = 1 + \tau_D/\tau_I$, and multiply K_c and τ_I by α, and divide τ_D by α.

Conservative tuning for smooth response. The tuning procedure outlined above is often time consuming, and as a starting point the following minimum ("conservative") gain can be used[5]

$$|K_{c,\min}| = \frac{|u_0|}{|y_{\max}|} \tag{11.57}$$

where $|u_0|$ is the MV change required to counteract the largest expected disturbance and $|y_{\max}|$ is the largest accepted CV deviation. In industry, the variables have often already been scaled such that $|u_0| \approx |y_{\max}|$ (for example equal to 1) and we get $|K_{c,\min}| \approx 1$. Indeed, this is a common factory setting for the gain. In addition, it is crucial that the sign of K_c is chosen correctly – remember that the control is supposed to counteract and not intensify changes in the CV. As a conservative starting point for the integral time, $\tau_I = \tau$ can be chosen, where τ is the dominant time constant for the effect of the MV on the CV.

Example 11.19 Control of exothermic CSTR. *This is a continuation of Example 11.16 (page 311). We want to keep the reactor temperature $y = T$ approximately constant at $y_s = 444K$. We assume that the reactor temperature can be measured and that we can affect $y = T$ by changing the coolant temperature $u = T_c$. The objective is to design a feedback PI-controller with $y = T$ as the controlled variable (CV) and $u = T_c$ as the manipulated variable (MV). We consider, as before, an increase in the feed rate of 20% (from 0.10 to 0.12 m^3/min) – this is the "disturbance" to the process. Without control, we have found that the reactor temperature T will eventually drop to 441.9 K, but with PI control the MV will counteract the disturbance such that CV = T returns to its desired value (setpoint) of 444 K; see Figure 11.28.*

To tune the controller, we obtained first, without control, the response from the cooling temperature (MV, u) to the reactor temperature (CV, y). This response (not shown in Figure 11.28) can be closely approximated as a first-order response (without time delay θ) with gain $k = \Delta y(\infty)/\Delta u \approx 0.5$ and time constant $\tau \approx 7$ min. For example, this is obtained by simulating a small step in T_c (for example, by changing T_c from 430 to `Tc=431` *and setting* `q=0.1` *in the MATLAB code on page 312), but it can also be found analytically by linearizing the model. We chose the closed-loop response time to be $\tau_c = 3$ min (a lower value gives a faster response, but with larger changes in the MV T_c). From (11.55), this gives the PI-settings*

$$K_c = \frac{1}{0.5}\frac{7}{3+0} = 4.7, \quad \tau_I = \min\{7, 12\} = 7 \text{ min}$$

The response with control is shown in Figure 11.28. We see that the temperature $y = T$ returns to its setpoint $T_s = 444$ K after about 9 minutes (about three times τ_c). The simulation was performed by adding the following lines after point II in the MATLAB code on page 312:

```
% PI-CONTROLLER:  u = u0 - Kc*e - (Kc/taui)*eint, where deint/dt = e
%  Note:  (1) The integrated error eint is introduced as an extra state: eint = y(4)
%         (2) The process ''output" (CV) yreg is in this case the reactor temperature T
%         (3) The process ''input"  (MV) u is in this case the cooling temperature Tc
yreg = T; yregs= 444; e=yreg-yregs; u0 = 430;  eint=y(4); Kc=4.7; taui=7;
u = u0 - Kc*e  - (Kc/taui)*eint;
Tc = u;
```

[5] S. Skogestad, "Tuning for smooth PID control with acceptable disturbance rejection," *Ind. Eng. Chem. Res.*, Vol. 45, 7817-7822 (2006).

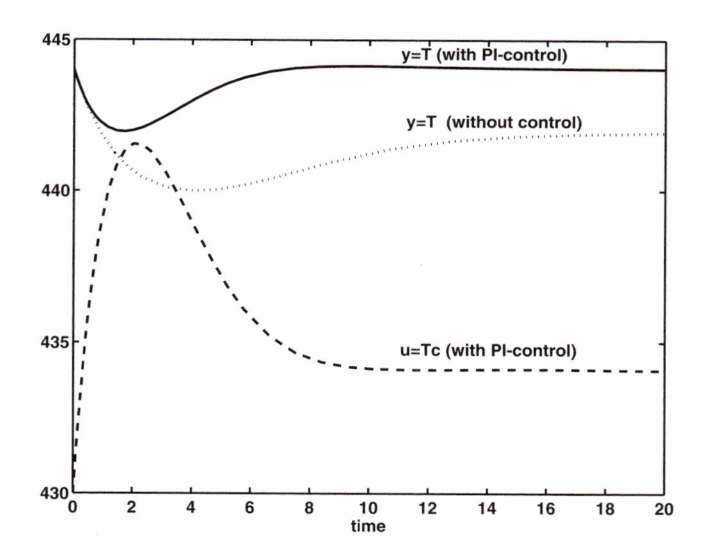

Figure 11.28: Exothermic CSTR with and without control: Temperature response after a 20% increase in feed flow rate

and by changing the last line to: `DYDT=[f1; f2; f3; e];`*. The modified code is saved in the file* `cstrTpi.m` *and can be run by entering:*

```
[T,Y]= ode15s(@cstrTpi,[0 50],[2274 7726 444.0 0]);.
```

11.7 Summary

Typically, the following steps are involved for the derivation and analysis of a dynamic model:

1. Formulate the relevant dynamic balance equations. The main problem is often: Which balance? Which control volume?
2. Use steady state data (obtained at the nominal operating point) to determine any missing parameters in the dynamic model equations.
3. Linearize and analyze the model.
4. Find the dynamic response by solving the dynamic equations ("dynamic simulation").
5. The model can, also, be used to design the control system, for example, to tune a PID controller.

APPENDIX A

Some thermodynamics and physical chemistry

Thermodynamics and physical chemistry are very important subject areas for process engineering calculations. The contents of this appendix could just as well been placed within the main text of the book, but I have chosen to put it here in order to maintain a clearer separation between the different areas and show that thermodynamics is a distinct subject. The reader is recommended to quickly read through the chapter in order to gain an overview, returning later to individual subjects as needed. The presentation is mostly in the form of an overview and you may want to consult other books for more details.

A.1 Concept of mol

The number of moles normally cannot be directly measured, but it is nevertheless a very practical quantity – in particular for systems with chemical reactions and varying composition. The main difference between a chemical engineer and a mechanical engineer is said to be that the eyes of a chemical engineer do not start to wander at the mention of the term *mole*. This section is primarily aimed at non-chemical engineers.

Consider the reaction $N_2 + 3H_2 = 2NH_3$, where 1 molecule of nitrogen reacts with 3 molecules of hydrogen to form 2 molecules of ammonia. In order to describe what happens, one needs to keep track of the number of the various molecules. However, molecules are extremely small so the numbers become extremely large. Therefore, it has been decided to call $N_A = 6.02214 \cdot 10^{23}$ of molecules (or atoms or entities) for a mole (= 1 mol in SI units). Just as a dozen eggs is 12 eggs, 1 mol of eggs is $6.02214 \cdot 10^{23}$ eggs.

N_A is also known as **Avogadro's number** or Avogadro's constant. The number originally came from considering the number of atoms in 1 g of hydrogen, since the atomic mass for hydrogen was set to $M = 1$ g/mol. The current IUPAC definition of 1 mol is that it is the number of atoms in 12 g of carbon-12 (C^{12}), and the atomic mass of hydrogen is then more precisely $M = 1.008$ g/mol.[1]

The hydrogen molecule consists of two hydrogen atoms (H_2) and the molar mass of hydrogen is $M = 2.016$ g/mol. Correspondingly, the atomic mass of oxygen is 16.00 g/mol and the molar mass of oxygen (O_2) is 32.00 g/mol. The molar mass of water (H_2O) is then 18.016 g/mol, that is, $6.02214 \cdot 10^{23}$ molecules of water have a mass of 18.016 g.

[1] The atomic mass of carbon is 12.011 g/mol (rather than 12.000 g/mol for the carbon-12 isotope), because naturally occurring carbon also contains some of carbon-13. There is also a tiny amount of carbon-14, which does not affect the atomic mass at our level of precision, but which is useful for dating objects.

The molar mass for mixtures is simply the molar average,

$$M = \sum_i x_i M_i$$

where M_i is the molar mass and x_i is the mole fraction of substance i. For example, air consists of 78.1 mol% N_2, 21.0% O_2 and 0.9% Ar. The molar mass for the components [g/mol] are 28.01, 32.00 and 39.95, respectively, and the molar mass of air is then

$$0.781 \cdot 28.01 + 0.210 \cdot 32.00 + 0.009 \cdot 39.95 = 28.96 \text{ [g/mol]}$$

Note that I in this book mainly use "molar mass" (M) [g/mol], which has the same numerical value as the closely related quantity "molecular weight" [dimensionless] = relative molar mass (M_r).

Exercise A.1* *Show that 1 liter of water contains 55.5 mol of water molecules.*

A.2 Balancing chemical reactions

A chemical reaction can generally be written in the form $0 = \Sigma_i \nu_i A_i$ where A_i are the chemical components and ν_i the stoichiometric coefficients. To determine the stoichiometric coefficients we use the atom balances. Usually, this balancing is done "by inspection," but alternatively the following systematic procedure can be used:

Step 1. Specify the chemical components that participate in the reaction.

Step 2. Formulate the atomic balances ("atoms in = atoms out").

Step 3. Select the stoichiometric coefficient for one component (select a basis).

Step 4. Determine the remaining stoichiometric coefficients by solving the atom balances.

The main difficulty in this procedure is to specify the components in step 1. If one leaves out one component, there may be no solution, and if one specifies extra components, then the solution is not unique and more than one reaction can be formulated.

Example A.1 Step 1. *Consider a reaction involving components CH_3OH (methanol), CO_2, H_2O and H_2. The chemical reaction on standard form is*

$$0 = \nu_{CH_3OH} CH_3OH + \nu_{CO_2} CO_2 + \nu_{H_2O} H_2O + \nu_{H_2} H_2$$

and we want to find the stoichiometric coefficient ν_i. **Step 2.** *Atoms are conserved in the reaction, so the atom balances for C, H and O give*

$$C: \quad \nu_{CH_3OH} + \nu_{CO_2} = 0$$

$$H: \quad 4\nu_{CH_3OH} + 2\nu_{H_2O} + 2\nu_{H_2} = 0$$

$$O: \quad \nu_{CH_3OH} + 2\nu_{CO_2} + \nu_{H_2O} = 0$$

Step 3. *We select that CO_2 is a reactant and that 1 mol is consumed, that is, $\nu_{CO_2} = -1$.* **Step 4.** *We then have 3 independent equations (the atom balances) in three unknowns, and we find*

$$\nu_{CH_3OH} = 1, \;\; \nu_{H_2O} = 1, \nu_{H_2} = -3$$

The chemical reaction can then be written

$$0 = CH_3OH - CO_2 + H_2O - 3H_2$$

or equivalently

$$CO_2 + 3H_2 = CH_3OH + H_2$$

Extra for interested readers: General method for finding the number of independent chemical reactions (N_r). Assume that we want to balance a single reaction, and we specify N_c components to be included in the reaction (step 1). Assume that for these components we can formulate N_a *independent* atomic balances (step 2). Since for each reaction we can select 1 stoichiometric coefficient (step 3), we then have $N_c - 1$ unknown stoichiometric coefficients, and we need N_a equations to find a unique solution in step 4. There are now three possibilities. (a) If $N_c - 1 = N_a$ or equivalently $N_c - N_a = 1$, then we have as many equations as unknowns, and we can compute the stoichiometric coefficients for the reaction. (b) If $N_c - N_a < 1$, then there are too few components, so we must have left out a component. (c) If $N_c - N_a > 1$ then there are more unknown stoichiometric coefficients than equations, so there are infinitely many solutions, and we can formulate more than one reaction.

To study the last case in more detail, introduce $N_r = N_c - N_a$. If $N_r = 1$, then we have case (a) with a unique solution. In case (c) with $N_r = 2$, we have infinitely many solutions, and we can select one stoichiometric coefficient to be zero. This component is then not included in the reaction (let us call it reaction 1), but we can instead formulate an additional independent reaction (let us call it reaction 2) involving this component. If $N_r = 3$ then we can set the stoichiometric coefficients of two components to zero in reaction 1. We can then specify that the first of these components is only included in reaction 2, and the second only in reaction 3. Again, these 3 resulting reactions are clearly independent. We have then derived the following *general result*:

- **In a system where N_c components are included in reactions, and where we can set up N_a *independent* atomic balances for these components, we have $N_r = N_c - N_a$ *independent reactions*.**

In some cases, it may be difficult to find the number N_a of independent atomic balances. In such cases, one can obtain the atom matrix A, which is simply a "table" that gives the number of atoms in each of the N_c components, and we have that $N_a = \text{rank}(A)$; see page 90 for examples.

A.3 Thermodynamic concepts

Thermodynamics was originally the study of the relationship between heat and mechanical work, but the area of thermodynamics has later been expanded to include, among other things, chemical equilibrium and phase equilibrium.

Central to thermodynamics is the concept of temperature. Thermodynamics concerns itself with systems in internal equilibrium and *time* (t) is therefore not a parameter. Because of the concept of state, we can nevertheless use thermodynamics for actual (irreversible) processes that change over time.

- Make sure that you know the following (see for example the list of concepts on page 18):
  - System (open, closed, isolated, adiabatic)
  - Surroundings
  - Process
  - State
  - Intensive and extensive variables
- **What is temperature T?**
  Microscopically, that is, on the molecular level, temperature is a measure of the intensity of the random (chaotic) motions of the molecules including translation, rotation and vibration.
  Macroscopically, we can observe that a system's state (for example, its volume, pressure or phase) is changed when it comes into contact with another system because of a difference in a property that we call temperature. The change is caused by transferred energy in the form of **heat** Q from high to low temperature until we reach temperature equilibrium.[2]

[2] It was only in the 19th century that the terms temperature (intensive variable) and heat (extensive variable) were clearly distinguished.

The Celsius temperature scale was originally defined by setting water's freezing point to 0^oC and water's boiling point at 1 atm to 100 ^{o}C. Later, it was discovered that there is an absolute lowest temperature, -273.15 ^{o}C, where there is no molecular motion, and from this the Kelvin scale was defined:

$$T[\text{K}] = t[^o\text{C}] + 273.15$$

T is called the "absolute temperature" because $T = 0$ at absolute zero. **In all thermodynamic equations, for example, in the ideal gas law, one must use absolute temperature.**
In the past, temperature was often measured using a mercury thermometer, utilizing the fact that mercury expands when temperature increases. Today, a thermocouple is often used, using the fact that electric conductibility varies with temperature.
For the *thermodynamic definition of temperature*, see (B.7) on page 390.

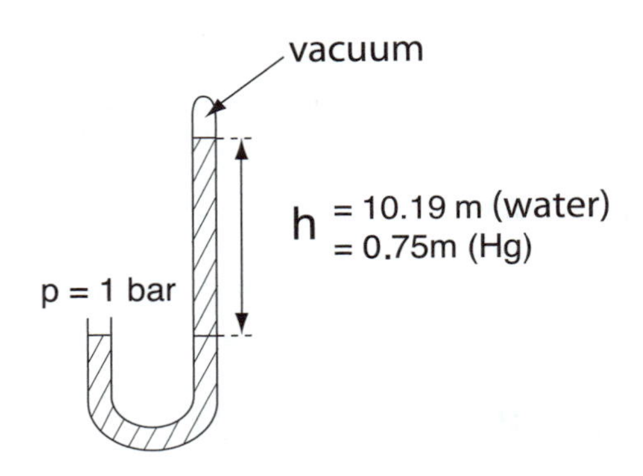

Figure A.1: Measuring pressure

- **What is pressure** p**?** Pressure p [Pa = N/m^2] is defined as force per unit area,

$$p\ [\text{N/m}^2] = \frac{F\ [\text{N}]}{\text{A}\ [\text{m}^2]}$$

A fluid also has an (internal) pressure due to forces between the molecules. A negative pressure is theoretically possible for systems with strong forces of attraction (but not for gases).
Macroscopically, the pressure in a fluid can be measured with a manometer, as shown in Figure A.1, where there is an equilibrium between the hydrostatic pressure from the liquid column with height h and the pressure in the fluid (e.g., outside air). A liquid column with area A and height h has a mass $m = \rho V = \rho h A$ and from Newton's second law the gravitational force of this mass is $F = mg$, where $g \approx 9.81 m/s^2$ is the acceleration of gravity. The **hydrostatic pressure** exerted by the fluid is then

$$p = \frac{F}{A} = \rho g h \tag{A.1}$$

For example, a water column of $h \approx 10.2$ m corresponds to a pressure of

$$p = \rho g h = 1000\ \text{kg/m}^3 \cdot 9.806\ \text{m/s}^2 \cdot 10.2\text{m} = 1.0 \cdot 10^5\ \text{N/m}^2 = 1\ \text{bar}$$

For mercury (Hg), the density is much higher and 1 bar corresponds to a liquid column of approximately $h = 0.75$ m $= 750$ mm.

Example A.2 The barometric formula. *We want to derive a formula for how the air (barometric) pressure p depends on the elevation h (height above sea level), and use this formula to calculate the air pressure at Mount Everest. We assume ideal gas and a constant temperature of $0^o C$.*
Solution. *The expression $p = \rho g h$ in (A.1)* cannot *be used directly to find the variation in air pressure p with height h because the density of air ρ is not constant. However, over a small height difference, dh, we can assume ρ to be constant and the corresponding small pressure drop is*

$$dp = -\rho g dh$$

Note that we have a negative sign because pressure drops as height increases. For an ideal gas, we have from (A.9) that the density is

$$\rho = \frac{pM}{RT}$$

which inserted gives

$$\frac{dp}{p} = -\frac{Mg}{RT} dh$$

To simplify the calculations, we assume that the composition (and thus the molar mass M [kg/mol]), the temperature T [K] and the acceleration of gravity g [m/s^2] are all independent of the height h. Integration of the equation from pressure $p_0 = 1$ atm (at $h = 0$) to p (at height h) then gives the **barometric formula**

$$\ln \frac{p}{p_0} = -\frac{Mg}{RT} h \quad \Rightarrow \quad p = p_0 \cdot \exp\left(-\frac{Mg}{RT} h\right) \tag{A.2}$$

Inserting $M = 29 \cdot 10^{-3}$ kg/mol, $T = 273.15K$ ($0\ ^oC$), $g = 9.81$ m/s^2 and $R = 8.31$ J/mol gives

$$p = p_0 \cdot \exp\left(-1.25 \cdot 10^{-4} \cdot h[\mathrm{m}]\right) \tag{A.3}$$

where $p_0 = 1$ atm $= 1.013$ bar is the normal atmospheric pressure at sea level ($h = 0$). At **Mount Everest**, *we have $h = 8848$ m and we find that the normal pressure is $p = 0.331 p_0 = 0.331$ atm $= 0.335$ bar.*
In practice, the temperature also varies with the height and the result changes, see Example A.11 on page 352.

Exercise A.2* *The boiling point (T_b) depends on pressure, and we want to use (A.3) to find how the boiling point of water depends on elevation.*
(a) Use the Antoine vapor pressure formula from Example 7.13 (page 181) to derive an expression for how the boiling temperature of water depends on elevation h.
(b) Show that the boiling point of water drops about 0.35K per 100 m.
(c) What is the boiling point of water at Mount Everest (8850m) ?

- **What is a reversible process?** A reversible process is a process where there is a complementary process (e.g., the reverse process) that brings the system and its surroundings back to their original states. In a reversible process, there is always a balance between internal (in the system) and external (from the surroundings) forces, such that we can reverse the process if desirable. For a reversible process, we will not be able to see whether a movie film is running forwards or backwards!! A reversible process between two states will, in many cases, take an infinite length of time and may therefore be without any practical interest. Nevertheless, one may sometimes use an *idealized* reversible process between two states to determine changes in the state variables (see below) which will also apply to an actual irreversible process between the same states. **Note. This is very important!**

- **What is a state variable?** This a very important concept! A state variable depends only on the present state of the system and on not how ("the way") has arrived at the state. The system is assumed to be at internal equilibrium in every state.
  This means that if we start with a given amount of matter in a state 1 (e.g., 1 bar, 5m^3), go to state 2 (e.g., 2 bar, 5 m^3) (perhaps by supplying heat), further to state 3 (e.g., 2 bar, 3 m^3) (perhaps by supplying work to compress no volume), and then return to state 1 (1 bar, 5m^3) (perhaps by extracting work while cooling), then the values of all the state variables are unchanged. This is often written in the form (here for internal energy)

$$\oint dU = 0$$

  — Examples of state variables are: entropy S, enthalpy H, internal energy U, volume V, temperature T, pressure p, heat capacity C_p, density ρ, etc. A combination of state variables is also a state variable. For example, $U + pV - TS = H - TS$ is a state variable (it is known as the Gibbs "free" energy and has the symbol G).
  — The following are *not* state variables: Heat Q and work W.

  In order to calculate the change in a state variable from one state to another given state, we often use an *idealized* process, e.g., a reversible process, between the two given states.
- **How many independent state variables are there?** In general, a system in equilibrium has $2+N_c$ independent state variables where N_c is the number of components. For a system with given amounts of the N_c components (e.g., 1 kg of a pure fluid), there are then only 2 independent state variables, e.g., V and p. One should in principle be able to choose any other pair of independent state variables, e.g., p and S, p and H, S and H, S and V, p and T and so on. We can then write the other state variables as a function of these two variables (coordinates), e.g., (here for internal energy U)[3]

$$U = U_1(p, V) = U_2(p, S) = U_3(p, H) = U_4(S, H) = U_5(S, V) = U_6(p, T)$$

  But sometimes we have to be careful in choosing the pair. In partucular, the combination of p and T does not give a unique state for a pure component in a region with several phases. Consider, for example, 1 mol water at 1 atm and 100 ^{o}C, that is, at its normal boiling point. From the given value of p and T, we cannot say anything about the phase distribution.
  This also follows from the **Gibbs phase rule** which states that for a system in equilibrium, the number of independent *intensive* variables taken from the set temperature, pressure and composition that can be specified is

$$F = 2 + N_c - N_P - N_r \tag{A.4}$$

  where N_c is the number of components, N_P is the number of co-existing phases and N_r is the number of independent reactions (see page 89) that are in equilibrium in the system. For example, if there is one component ($N_c = 1$), two phases (e.g., vapor and liquid) ($N_P = 2$) and no reactions ($N_r = 0$) then Gibbs phase rule gives $F == 2 + 1 - 2 - 0 = 1$, so at most one of the intensive variables T and p can be specified independently. Note that Gibbs phase rule does *not* contradict our assertion that there are always $2+N_c$ independent variables for a system in equilibrium, because these independent variables are not restricted to T, p and composition. For example, for our pure-component case with two phases, just mentioned, we can independetly specify the following $2 + N_c = 2 + 1 = 3$ variables: m [kg], h [J/kg] and p [bar]. However, replacing the specification on h with a specification on T does not work inside the two-phase region.

[3] The most favorable choice of independent state variables (the so-called **canonical variables**) depends on the process one is looking at, and many of the apparent mysteries of thermodynamics are relations that arise when switching between different independent variable.

- **What is the critical point?** In the state diagram in Figure A.2a, the thick lines indicate the border of the two-phase region – with liquid at the left (with small volume) and gas at the right (with large volume). Inside this region, we have equilibrium between gas and liquid, and for a pure component, pressure is uniquely given by the temperature $p = p_{\text{sat}}(T)$ (or the other way around, temperature is uniquely given by pressure). At low pressure, the density ρ [kg/m^3] of the liquid is much larger than that of the gas, or equivalently, the (molar) volume (V_m) of the liquid is much smaller than for the gas.
  However, as we increase pressure (and thereby increase the equilibrium temperature in the two-phase region), the gas will be more compressed and its volume decreases. On the other hand, the volume of the liquid is affected far less by the pressure. At sufficiently high pressure, the volume of the gas becomes equal to the density of the liquid. This is the critical point (C in the Figure) where the gas and liquid phases are identical. Above the critical point, there exists no separate phases for gas or liquid.

Example A.3 *The molar volume of water (liquid) at 1.013 bar and 373.15 K (which is the boiling point for water at atmospheric pressure) is* $0.0188 \cdot 10^{-3}$ *m^3/mol* ($V_m = M/\rho$, *where* $M = 18.015 \cdot 10^{-3}$ *kg/mol and* $\rho = 958$ *kg/m^3 for water at 373.15 K). The molar volume of the corresponding water vapor (gas) is about 1600 times larger; for ideal gas* $V_m = \frac{RT}{p} = 8.31 \cdot 373.15\, over 1.013 \cdot 10^5 = 30.61 \cdot 10^{-3}$ *m^3/mol. For water, the critical point is at* $p_c = 220.6$ *bar and* $T_c = 647.1$ *K (374° C) where the volume is* $V_c = 0.0554 \cdot 10^{-3}$ *m^3/mol* ($\rho_c = 325$ *kg/m^3).*
For larger molecules with a larger molar mass M, the critical point is at a lower pressure. For example, for n-hexane the critical point is at 30 bar and 507 K (234° C).

A.4 Thermodynamic diagrams

Consider a system with given amounts of the component, e.g., 1 kg or 1 mol of a pure component. As mentioned above, we then have only two independent state variables. By using the chosen state variables as axes in a coordinate system, the value of any other state variables may be represented by iso-lines in a contour plot. These are the so-called state diagrams or thermodynamic diagrams; see Figure A.2. The following variables are usually shown on thermodynamic diagrams: p, T, H, S and V.

In a pV**-diagram** (Figure A.2a), V is on the x-axis and p on the y-axis, and we can, for example, read off p as function of V for constant values of T (isothermals). An **equation of state** (see below) gives a description in equation form of this type of diagram.

The isotherms in a pV-diagram can be found experimentally by filling a given amount of fluid (e.g., 1 mol) into a cylinder, and then recording how the pressure p depends on volume V. The cylinder needs to be cooled (or heated) to keep the temperature constant during the experiment. We expect the pressure p to increase as we reduce the volume V. For example, this is the case for an ideal gas, where pV remains constant; see for example the isotherm for T_1 for large volumes in Figure A.2a. When the pressure is sufficiently high, we get **condensation** of liquid, and for a pure fluid the pressure will remain constant (at the vapor pressure $p^{sat}(T)$) until all the gas has condensed. At this point, we only have liquid in the cylinder, and a further decrease in volume will give a large increase in pressure. However, in the critical point (C) the gas and liquid phases become identical, and for temperatures above the critical temperature (T_c) it is not possible to observe any condensation when the "gas" is compressed. Here, we write "gas" in quotation marks, because strictly speaking one cannot distinguish between gas and liquid at supercritical conditions. The term (supercritical) fluid is better.

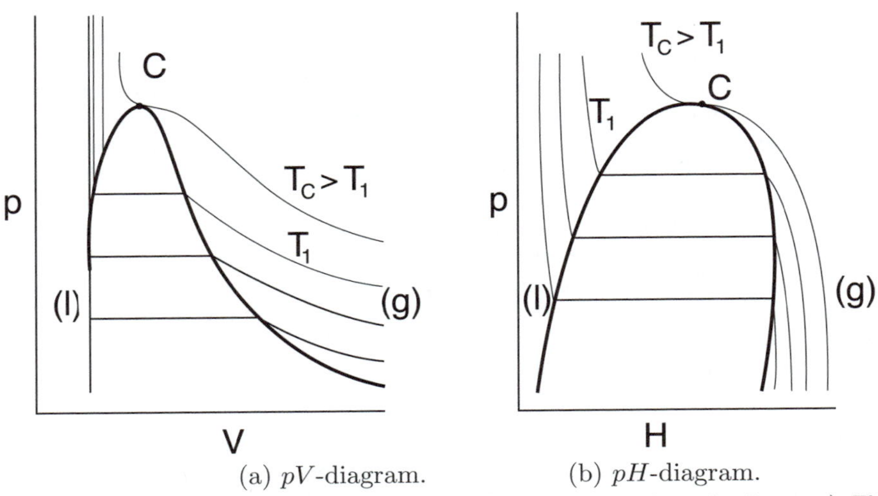

(a) pV-diagram. (b) pH-diagram.

Isothermals are the thin lines (with rising temperature upwards in the diagrams). Thick lines represent phase transition between gas and liquid, with liquid at the left, two-phase in the middle and gas at the right. C is the critical point.

Figure A.2: Typical thermodynamic diagrams for pure component

A pT**-diagram** is impractical because the two-phase region for a pure component is a line (the vapor pressure line), and H, S and V make a jump along this vapor pressure line. Therefore, for process calculations, pH**-diagrams** (see page 115 and Figure A.2b) or HS**-diagrams** (see page 117 and Appendix F) are more common.

A.5 Equations of state

The term "equation of state" usually means an equation that describes how the pressure p in a fluid depends on volume V and temperature T (and, if necessary, composition).

A.5.1 Equation of state for ideal gas

Let us consider a closed system with n mol of gas. Through observations, it has been established that all gases at sufficiently low pressure satisfy the following laws:

Boyle's law: $pV =$ constant (at constant T)

Charles' law: $V =$ constant $\cdot T$ (at constant p)

Avogadro's law $V =$ constant $\cdot n$ (at constant p and T)

By combining the three "laws," the **equation of state for an ideal gas** (the ideal gas law) is derived,

$$\text{Ideal gas law}: \quad pV = \underbrace{\text{constant}}_{R} \cdot nT \tag{A.5}$$

where R [J/mol K] is the gas constant. (A.5) can also be derived from kinetic gas theory or statistical mechanics. From Dalton's law (see below), we have further that the gas constant

R is the same for all gases and also for gas mixtures (of components, atoms and molecules).

For a *mixture* with n_i mol of component i, we can introduce the mole fraction defined by $y_i = n_i/n$, and the **partial pressure** p_i defined by of component i

$$p_i \triangleq y_i p \tag{A.6}$$

Since $\sum y_i = 1$, the total pressure is equal to the sum of the partial pressures,

$$p = p_1 + p_2 + \cdots p_i + \cdots = \sum p_i \tag{A.7}$$

which applies for both ideal and real gases. **Dalton's law** for a gas mixture says that *for an ideal gas, the partial pressure is equal to the pressure the component would have had if it were alone in the given volume* V. From (A.5), this can be written as

$$p_i = \frac{n_i R_i T}{V}$$

Here, we have allowed for the possibility that the constant R_i [J/mol K] may differ for different components. However, if we combine this equation with the definition of partial pressure, $p_i = \frac{n_i}{n} p$, we get $p = nR_iT/V$ and it follows (Why? Try another component!) that the gas constant R_i must be the same for all components.[4] The universal gas constant is $R = 8.3145$ J/mol K.

The ideal gas law (A.5) can be written in many ways, for example, on molar basis

$$\text{Ideal gas law}: \quad pV_m = RT$$

where $V_m = V/n$ [m^3/mol] is the molar volume. Introducing the mass

$$m[\text{kg}] = \text{n}[\text{mol}] \cdot \text{M}[\text{kg/mol}]$$

where M is the molar mass of the gas, we get the ideal gas law on mass basis,

$$\text{Ideal gas law}: \quad pV = m\frac{R}{M}T \tag{A.8}$$

We can also introduce the (mass) density $\rho = m/V$ [kg/m^3] and the ideal gas law becomes

$$\text{Ideal gas law}: \quad p = \rho\frac{R}{M}T \tag{A.9}$$

We note that on a mass basis the gas constant is R/M [J/kg K], which is a function of composition, because the molar mass M depends on composition. Mechanical engineers usually use mass basis, so they must learn a different value of the gas constant (R/M) for every substance.

A.5.2 Deviation from ideal gas

The assumption of ideal gas holds well at low pressures and at high temperature. More generally, the deviation from ideal gas can be quantified by introducing the "compressibility

[4] Dalton's law tells that a given number (e.g., 1 mol = 6.023 10^{23} molecules) of light molecules (e.g., hydrogen) exerts the same pressure as the same number of heavier molecules (e.g., air), which initially seems odd. The reason is that at a given temperature the light molecules move at a higher velocity. For example, the average velocity (*"root mean square"* (rms)) v for translation for air at 298 K is from kinetic gas theory

$$v = \sqrt{\frac{3RT}{M}} = \sqrt{\frac{3 \cdot 8.31 \cdot 298}{29 \cdot 10^{-3}}} = 506\ m/s$$

while the velocity of hydrogen at the same temperature is 1927 m/s.

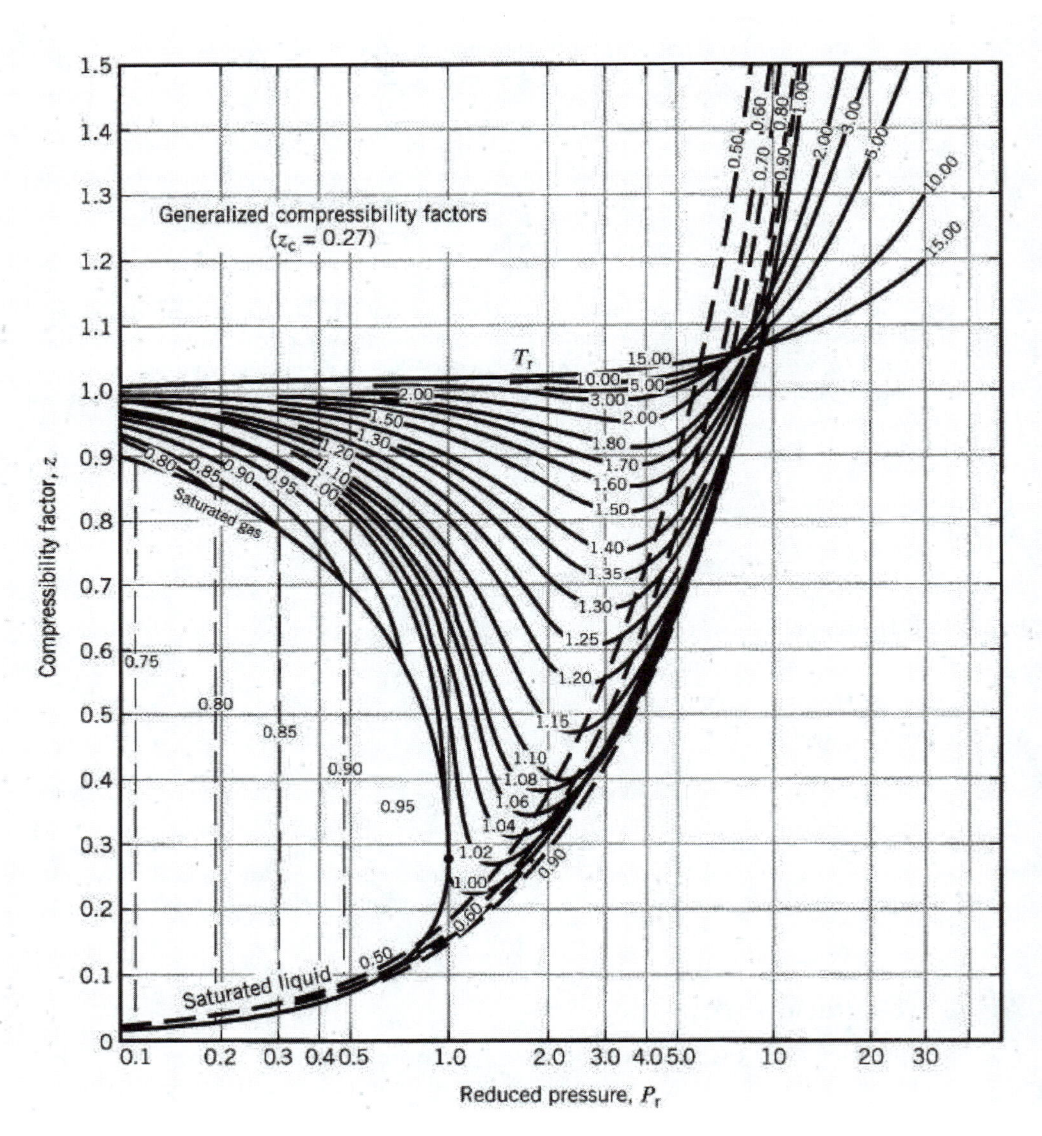

Figure A.3: Generalized compressibility diagram

From: A.L. Lydersen, R.A. Greenkorn, A. Hougen, University of Wisconsin, 1955.

factor" z defined by

$$\text{Real gas}: \quad z = \frac{pV}{nRT} = \frac{pV_m}{RT} \tag{A.10}$$

For an ideal gas, we have $z = 1$. For a "real" gas, z can be expressed by a virial equation, which is a series expansion in pressure or density. The virial expansion in pressure is

$$z = 1 + Bp + Cp^2 + Dp^3 + \cdots \tag{A.11}$$

where the coefficients $B, C, D, \ldots$ are functions of T and composition (for ideal gas, they are, of course, zero). The virial equation has a theoretical basis, as it represents interactions between a molecule and its neighbors, but it is not used much for practical calculations.

Of more practical interest is the generalized compressibility diagram in Figure A.3, which shows z as a function of reduced pressure and temperature,

$$T_r = T/T_c; \quad p_r = p/p_c$$

where T_c is the critical temperature and p_c is the critical pressure for the component.

The basis for the diagram is that most gases (pure components or mixtures) behave similarly if we "normalize" with respect to the critical point by introducing T_r and p_r. (This is sometimes called the "law of corresponding states," but the term "law" is here somewhat misleading since this is only a useful empirical simplification.) From the diagram, we see that at lower pressures the attracting forces dominate and $z < 1$. The deviation from ideal gas is largest close to the critical point (where $z_c = 0.27$ for the fluid given in the diagram). At very high pressures ($p_r > 10$), the repulsive forces between the molecules dominate and z can be much higher than 1. We note that the deviation from ideal gas is small (with $z \approx 1$) at low pressure and also at high temperature. The generalized compressibility diagram is useful because it gives insight into the deviation from ideal gas, but for practical calculations, and in particular for mixtures, we use equations of state, e.g., the SRK or PR equations of state, as described below.

Example A.4 *We will use the generalized diagram in Figure A.3 to calculate the density of ethane at* $p = 103$ *bar and* $T = 333$ *K (60 °C). The critical point for ethane is* $p_c = 48.8$ *bar and* $T_c = 305.4K$*, so the reduced pressure is* $p_r = p/p_c = 2.11$ *and the reduced temperature is* $T_r = T/T_c = 1.09$ *(i.e., we are at a supercritical state which cannot be clearly classified as gas or liquid). From the generalized compressibility diagram, we read off* $z = 0.39$*. The molar volume of the gas is then* $V_m = zRT/p = (0.39 \cdot 8.31 \cdot 333/103 \cdot 10^5)$ $m^3/mol = 0.105 \cdot 10^{-3}$ m^3/mol*, and the density is* $\rho = M/V_m = 30.1 \cdot 10^{-3}/0.105 \cdot 10^{-3}$ $kg/m^3 = 287$ kg/m^3*. To compare, the experimental value is about 296* kg/m^3*.*

A.5.3 Equations of state for real gases

The ideal gas law

$$p = \frac{nRT}{V} = \frac{RT}{V_m}$$

where $V_m = V/n$ [m^3/mol] is the molar volume is applicable at low pressures and high temperatures. Many modified equations have been developed to describe real gases and liquids. The most famous is the van der Waals equation of state.[5] Van der Waals started from the ideal gas law on the form $p = RT/V_m$ and argued as follows: (1) There are attractive forces between the molecules that give a reduction in pressure proportional to $1/V_m^2$, which

[5] Johannes van der Waals (1837–1923) received the 1910 Nobel prize in physics for this equation of state. His Nobel lecture is interesting reading and provides insight into the equation; see the Nobel committee's home page: `http://www.nobel.se/physics/laureates/1910/`.

expresses the likelihood for two molecules being close to each other. (2) When we account for the volume of the molecules, the remaining "free" volume is $V_m - b$ where b is the volume of 1 mol of molecules when packed together. The **van der Waals equation of state** then becomes

$$p = \frac{RT}{V_m - b} - \frac{a}{V_m^2} \tag{A.12}$$

where (1) a is an empirical parameter that describes the attractive forces between the molecules, and (2) b is an empirical parameter that accounts for the volume that the molecules occupy. The gas constant R is as before a universal constant but the parameters a and b depend on composition. At a given p and T, (A.12) can be transformed into a cubic equation in V_m:

$$pV_m^3 - (bp + RT)V_m^2 + aV_m - ab = 0$$

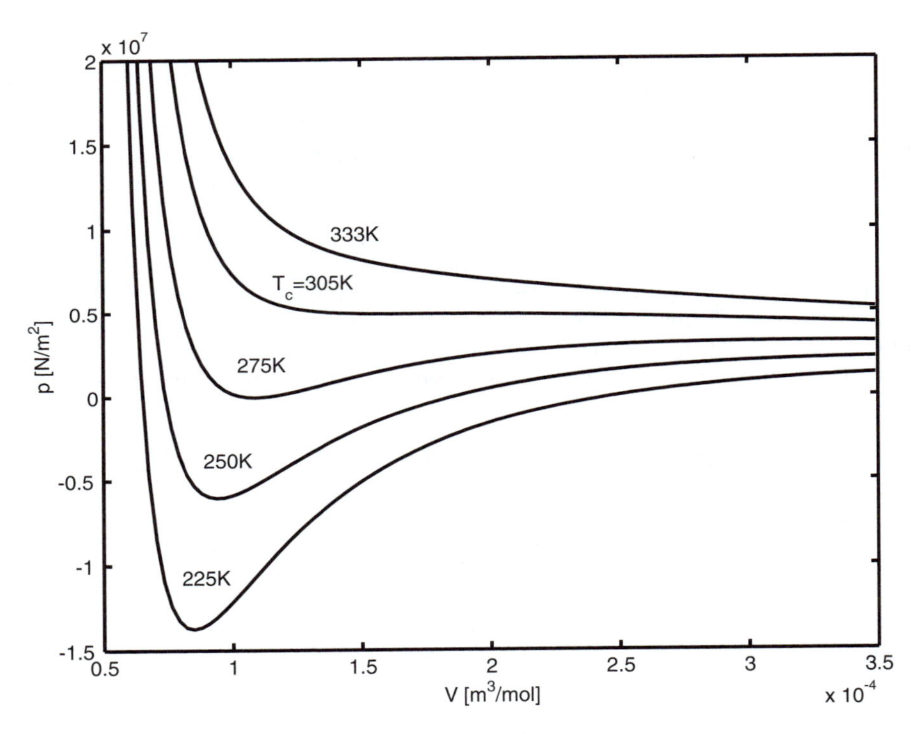

Figure A.4: Isotherms for ethane computed with the SRK equation of state (see Example A.6 for data).

For most cubic equation of states, including the van der Waals equation, the following apply (see Figure A.2a and Figure A.4):

- The pressure goes to infinity when the mole volume V_m approaches the value b because of the repulsion forces represented by the first term in (A.12).
- The attractive forces from the second term in (A.12) reduce the pressure, and it is possible to achieve a negative pressure. This is shown in Figure A.4 which depicts the isotherms for ethane calculated with a cubic equation of state. As mentioned earlier, negative pressures are physically realizable in condensed phases (for example, liquid), but such states are usually unstable (which they indeed are for ethane).

- In the two-phase region, where we have both gas and liquid, the cubic equation has three real, positive solutions ("roots") – the solution with the smallest volume is the liquid volume, the largest is the gas volume, while the middle volume is a non-physical solution.

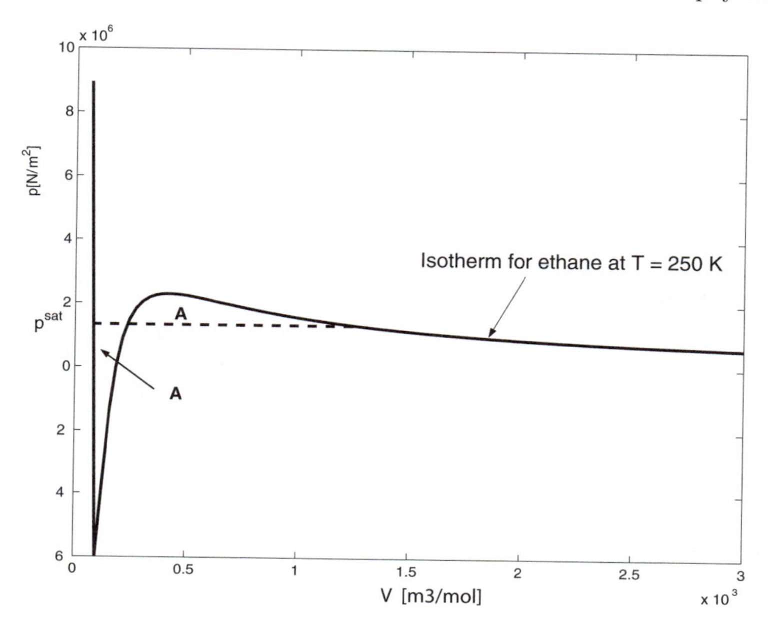

Figure A.5: Maxwell's "equal-area" rule for finding the vapor pressure $p^{sat}(T)$ from an equation of state (in this case with the SRK equation)

- At the critical point (p_c, T_c, V_c), there is no difference between gas and liquid and the three solutions to the cubic equation are identical, that is, the isotherm $T = T_c$ has a turning point (see Figures A.2a and A.4), and at the critical point (p_c, T_c)

$$\frac{dp}{dV} = 0; \quad \frac{d^2p}{dV^2} = 0 \tag{A.13}$$

From this and (A.12), we derive that

$$a = 0.42188 \frac{R^2 T_c^2}{p_c}; \quad b = 0.125 \frac{RT_c}{p_c} \tag{A.14}$$

which can be used to calculate a and b for pure components from critical data (T_c and p_c).

- For a pure component in the two-phase region, the pressure is constant and equal to its vapor pressure $p^{sat}(T)$ (as shown by the horizontal lines in Figure A.2a; they are not shown in Figure A.4). The vapor pressure at a given temperature can easily be found from the isotherm using the Maxwell's "equal-area" rule for pure components: In the two-phase region, the area $-A$ below the horizontal vapor pressure line (on the liquid side) is the same as the area A above the line (on the gas side); see Figure A.5 and Example A.9.

Example A.5 *(continuation of Example A.4). We want to use the van der Waals equation of state to calculate the density of ethane at $p = 103$ bar and $T = 333$ K (which is above the critical point). For ethane, we have $p_c = 48.8$ bar and $T_c = 305.4K$, and (A.14) gives $a = 0.42188 \frac{R^2 T_c^2}{p_c} = 0.557 Nm^4$ and $b = 0.125 \frac{RT_c}{p_c} = 0.0650 \cdot 10^{-3} m^3/mol$. The real solution* *of the resulting cubic equation is $V_m = 0.126 \cdot 10^{-3}$ m^3/mol (see MATLAB code below). The density is then $\rho = M/V = 30.1 \cdot 10^{-3}/0.126 \cdot 10^{-3}$ $kg/m^3 = 239$ kg/m^3, which is a somewhat different from the experimental value of 296 kg/m^3.*

```
% MATLAB code for solving van der Waals equation:
b=0.0650e-3; a=0.557; p=103e5; R=8.31; T=333;
C(1)=p; C(2)=-(b*p+R*T); C(3)=a; C(4) = -a*b
roots(C)
% MATLAB answers: ans =
%  1.0e-003 *
%     0.1039 + 0.1309i
%     0.1039 - 0.1309i
%     0.1259
```

Improved equation of states for practical calculations

The most well-known extensions of the van der Waals equation are the **Soave-Redlich-Kwong (SRK)** and **Peng-Robinson (PR)** cubic equations of state, which are used frequently in practical calculations. Let us take a look at the SRK equation which originates from the modified van der Waals equation (A.12) of **Redlich and Kwong** (RK) (1949):

$$p = \frac{RT}{V_m - b} - \frac{a(T)}{V_m(V_m + b)} \tag{A.15}$$

Redlich and Kwong originally proposed using $a(T) = a_c/\sqrt{T_r}$ where $T_r = T/T_c$. The constants a_c and b can be determined from critical data using (A.13), and we derive

$$a_c = 0.42747\frac{R^2T_c^2}{p_c}; \quad b = 0.08664\frac{RT_c}{p_c} \tag{A.16}$$

At a given p and T, (A.15) can be transformed into a cubic equation in V_m,

$$pV_m^3 - RTV_m^2 + (a - pb^2 - bRT)V_m - ab = 0 \tag{A.17}$$

After its introduction in 1949, the RK equation (A.15) soon gained widespread use for computing densities and other thermodynamic data for real gases.

In theory, an equation of state can also be used for the liquid phase, and if we use the same equation of state for both phases, we can compute the vapor pressure, for example, using Maxwell's "equal-area" rule (see Figure A.5). However, it turns out that the vapor pressures computed from the RK equation are unsatisfactory. For this reason, Giorgio Soave[6] in 1972 published an article where he modified the temperature dependency for a and introduced an extra parameter such that the equation's ability to estimate vapor pressure was greatly improved. **Soave's formula** for a is

$$a(T) = a_c\left[1 + m(1 - \sqrt{T_r})\right]^2 \tag{A.18}$$

where a_c is a constant, $T_r = T/T_c$ and m is the new parameter. Soave proposed estimating m from the correlation $m = 0.48508 + 1.55171\omega - 0.15613\omega^2$, where ω is *acentric factor*.

From (A.18) and (A.16), we can obtain a and b for pure components. For mixtures, Soave proposed the following "mixing rules":

$$a = \sum_i \sum_j x_i x_j \sqrt{a_i a_j}(1 - k_{ij}); \quad b = \sum x_i b_i \tag{A.19}$$

[6] G. Soave, "Equilibrium constants from a modified Redlich-Kwong equation of state," *Chem.Eng.Sci.*, Vol. 27, pp. 1197-1203 (1972).

where the binary interaction parameters k_{ij} need to be determined from experimental data. For an ideal mixture, $k_{ij} = 0$. The combination of (A.15), (A.16), (A.18) and (A.19) is called the **SRK equation of state**.

Example A.6 (continuation of Example A.4). *We want to use the SRK equation to calculate the density of ethane at* $p = 103$ *bar and* $T = 333$ *K. For ethane, we have* $p_c = 48.8$ *bar,* $T_c = 305.4K$ *and* $\omega = 0.099$*, and we find* $a_c = 0.42747\frac{R^2T_c^2}{p_c} = 0.564Nm^4$*,* $b = 0.08664\frac{RT_c}{p_c} = 0.0451 \cdot 10^{-3}m^3/mol$ *and* $m = 0.637$*. This gives* $a(333K) = 0.564[1 + 0.637(1 - \sqrt{333/305.4})]^2 = 0.533Nm^4$*. The cubic equation (A.17) in* V_m *gives the (real) solution* $V_m = 0.118 \cdot 10^{-3}\ m^3/mol$*.*

```
MATLAB: C(1)=p; C(2)=-R*T; C(3)=a-p*b*b-b*R*T; C(4)=-a*b; roots(C).
```

The density is then $\rho = M/V = 30.1 \cdot 10^{-3}/0.118 \cdot 10^{-3}\ kg/m^3 = 255\ kg/m^3$*, which is somewhat different from the experimental value of 296* kg/m^3*. Isotherms for ethane computed with the SRK equation of state are shown in the pV-diagram (Figure A.4).*

Comment. The SRK equation was primarily developed to give good values for the vapor/liquid equilibrium (including vapor pressures) and not to give good values for z or densities. The relatively large deviation in density for supercritical ethane in Example A.6 (255 kg/m^3 versus 296 kg/m^3) is therefore not surprising. The next example shows that vapor pressure prediction for ethane with SRK is very good.

Example A.7 Vapor pressure of ethane from SRK equation of state. *Figure A.5 shows the isotherm for ethane at* $T = 250$ *K computed from the SRK equation of state. From Maxwell's equal-area rule, we find that the vapor pressure at this temperature is* $p^{sat} = 1.32 \cdot 10^6\ N/m^2 = 13.2$ *bar (the vapor pressure is not computed this way in practice). This is very close to the value of 13.1 bar computed from the experimental vapor pressures formula given in Poling, Prausnitz and O'Connell,* The properties of gases and liquids, *5*th *Ed., McGraw-Hill (2001).*

Exercise A.3* *The experimental compressibility factor of saturated ammonia at* $T = 325$ *K and* $p = 21.2$ *bar is* $z = 0.81$*. Check this using (a) the thermodynamic diagram for ammonia on page 419 (you will need to extrapolate outside of the diagram), (b) the generalized compressibility diagram in Figure A.3, (c) van der Waals equation and (d) the Redlich-Kwong equation.* Data: $T_c = 405.7K$, $p_c = 111.3$ *bar.*

Example A.8 3-component mixture. *Consider a mixture of 50 mol-% methane (1), 40 mol-% ethane (2) and 10 mol-%* CO_2 *(3). (a) Determine the SRK parameters a and b for the mixture at 200 K (we need to know the temperature because a depends on* T_r*). (b) What is the density of the gas mixture at 200 K and 4 bar? (c) A 20 l gas tank contains 10 mol of this mixture. What is the density and pressure at 200 K?*

Component data:

Methane: $T_c = 190.6$ *K,* $p_c = 46.0$ *bar,* $\omega = 0.008$.

Ethane: $T_c = 305.4$ *K,* $p_c = 48.8$ *bar,* $\omega = 0.099$.

CO_2*:* $T_c = 304.2$ *K,* $p_c = 73.8$ *bar,* $\omega = 0.239$.

Binary interaction coefficients:

$CH_4 - C_2H_6$: $k_{12} = 0$; $CH_4 - CO_2$: $k_{13} = 0.0973$; $C_2H_6 - CO_2$: $k_{23} = 0.1346$

Solution. *(a) We choose to use volume in l (liter) and pressure in bar. The gas constant in these units is* $R = 0.0813$ *l, bar/K, mole.*

$m_1 = 0.48508 + 1.55171 \cdot 0.0008 - 0.15613 \cdot 0.008^2 = 0.4975$

$m_2 = 0.6372, \quad m_3 = 0.8470$

$a_1 = 0.42747\frac{0.08314^2 \cdot 190.6^2}{46} \cdot [1 + 0.4975(1 - \sqrt{200/190.6})]^2 = 2.2778\ bar\ (l/mol)^2$

$a_2 = 7.0858$ *bar* $(l/mol)^2$, $a_3 = 4.9560$ *bar* $(l/mol)^2$
$b_1 = 0.08664 \frac{0.08314 \cdot 190.6}{46} = 0.02983\ l/mol$
$b_2 = 0.04502\ l/mol$, $b_3 = 0.02969\ l/mol$
$b = \sum_i x_i b_i = 0.5b_1 + 0.4b_2 + 0.1b_3 = \mathbf{0.03589\ l/mol}$
$a(200K) = x_1^2 a_1 + x_2^2 a_2 + x_3^2 a_3 + 2[x_1 x_2 \sqrt{a_1 a_2}(1 - k_{12}) + x_1 x_3 \sqrt{a_1 a_3}(1 - k_{13})$
$+ x_2 x_3 \sqrt{a_2 a_3}(1 - k_{23})] = \mathbf{4.0734\ bar\ (l/mol)^2}$

(b) With given $p = 4$ *bar and* $T = 200K$*, the cubic equation (A.17) in* V_m *has three solutions (3.94, 0.160 and 0.058 l/mol). The largest value* $V_m = 3.94\ l/mol$ *is for gas. The density is then* $\rho = M/V_m = 24.4/3.94 = 6.2\ g/l = 6.2\ kg/m^3$.

(c) We have $V_m = 20l/10mol = 2\ l/mol$*. The molar mass of the mixture is* $M = 0.5 \cdot 16 + 0.4 \cdot 30 + 0.1 \cdot 44 = 24.4\ g/mol$*. The density is* $\rho = M/V_m = 24.4/2 = 12.2g/l = 12.2\ kg/m^3$*. The pressure in the container at* $T = 200\ K$ *is from (A.15)* $p = \frac{RT}{V_m - b} - \frac{a(T)}{V_m(V_m + b)} = \frac{0.0813 \cdot 200}{2 - 0.03589} - \frac{4.0734}{2 \cdot 2.03589} = 7.28\ bar$.

Comments on the SRK equation. The SRK equation gives very good values of vapor pressure for non-polar components, and it also predicts well the phase equilibria for mixtures of such components, for example, for hydrocarbons at high pressures (as found in reservoir conditions). The SRK equation can also be used for mixtures with supercritical components, for example, nitrogen dissolved in heavier hydrocarbons. After its introduction in 1973, the SRK equation quickly became very popular in industry, and almost revolutionized the use of simulations for the design of processes with non-polar components (for example, hydrocarbons, air separation and many others). Those who wish to take a closer look at this can read more about it in a paper that I published while working in industry in 1983.[7]

A drawback with the SRK equation is that it gives a critical compressibility $z_c = 0.33$ for all pure components. This is a rather high value; for example, z_c equals 0.288 for methane, 0.259 for n-octane, 0.290 for nitrogen, 0.271 for benzene and 0.25 for NO. Because of this **Peng and Robinson** in 1976 proposed another modification of the Redlich-Kwong equations which gives $z_c = 0.307$ for pure components. Today, both the SRK and PR equations are used commonly for practical computations and the difference in their predictions is usually small (the main difference seems to be that SRK is more popular in Europe, while PR is more popular in North America).

Use of equation of state for computing "everything"

An equation of state gives a relationship between pressure, volume and temperature. However, more generally, an equation of state can be used to compute the deviation from ideal gas – both for gas and liquid – and we have already discussed that it can be used to compute vapor pressures and vapor/liquid-equilibrium. Since an ideal gas is exactly described if we have data for the heat capacity for the ideal gas, $C_p'(T)$, it follows that we can, with an equation of state (e.g., SRK in (A.15–A.19) or PR), compute "all" of thermodynamic quantities of interest (including enthalpy (see page 356), heat of mixing, entropy, fugacity coefficient, activity, vapor pressure, phase distribution, composition in gas and liquid phases, density, etc.). For this, we need only the following data for each component (believe it or not!):

- Critical temperature T_c
- Critical pressure p_c
- Acentric factor ω (or value of m)
- Ideal gas heat capacity $C_p'(T)$ (usually given as a polynomial in T)

For mixtures, we also need

[7] S. Skogestad, "Experience in Norsk Hydro with cubic equations of state," *Fluid phase equilibria*, Vol. 13, pp. 179-188 (1983).

- The interaction parameter k_{ij} for each binary combination of components i and j in the mixture.

For most non-polar mixtures, we get good results by setting $k_{ij} = 0$. With the SRK and PR equations, one usually finds that the largest deviation compared to experimental data is for the liquid density (so one often uses a separate for liquid density), while the other quantities usually have good accuracy.

How is it possible that we can calculate "everything" from only one equation? First, we need the ideal-gas heat capacities ($C_p'(T)$) for all pure components. These are found in reference tables; e.g., Poling, Prausnitz and O'Connell, *The properties of gases and liquids*, 5th Ed., McGraw-Hill (2001). This gives an accurate description of the energy for an ideal gas as a function of temperature and composition. In addition, we need an equation of state (for example, SRK or PR) which gives the p, V, T relationship for a real gas, and can be used, for example, using relationships such as (A.53) on page 357 to compute the deviation from ideal-gas enthalpy, etc., and from this obtain the enthalpy, entropy, density, and so on for the real gas. The liquid phase and its properties can also be computed given that the equation of state describes both gas and liquid phase. We can also compute the vapor-liquid equilibrium, for example, with Maxwell's equal-area rule, as shown in Figure A.5. The heat of vaporization can also be obtained, for example, using the Clapeyron equation $\frac{\Delta_{\rm vap} H}{T\Delta_{\rm vap} V} = \frac{dp^{\rm sat}}{dT}$, see (7.28).

Example A.9 Derivation of Maxwell's "equal-area" rule for a pure component. *At a given temperature T, the vapor pressure $p^{sat}(T)$ of a liquid (l) is the resulting equilibrium pressure in the gas phase (g) (the vapor). At equilibrium, the pressure and temperature in the two phases are equal ($p_g = p_l = p^{sat}(T)$, $T_g = T_l = T$). In addition, the Gibbs energy in the two phases is equal (see page 179): $G_g = G_l$. This gives the equilibrium condition*

$$\Delta G = G_g - G_l = \int_{\rm liquid}^{\rm gas} dG = \int_{\rm liquid}^{\rm gas} V dp = 0$$

Here, the relationship $dG = V dp$ follows from the fundamental equation $dG = -SdT + V dp$ (see (B.66) for pure component) since we choose to integrate along an isotherm ($dT = 0$). Some further manipulation is needed to obtain the desired condition. Integration by parts gives

$$\int_{\rm liquid}^{\rm gas} V dp = pV|_{\rm liquid}^{\rm gas} - \int_{\rm liquid}^{\rm gas} pdV$$

Since $p_g = p_l = p^{sat}(T)$, the first term on the right side can be written

$$pV|_{\rm liquid}^{\rm gas} = (pV)_g - (pV)_l = p^{sat}(V_g - V_l) = \int_{\rm liquid}^{\rm gas} p^{sat} dV$$

and the equilibrium condition becomes

$$\Delta G = \int_{\rm liquid}^{\rm gas} (p^{sat} - p) dV = 0$$

In other words, in a pV-diagram the areas above and below the line p^{sat} must be equal (see Figure A.5).

A.6 Work, heat and energy

Work, heat and energy are very central concepts and you should familiarize yourself with them. It may take some effort – and you should not worry if you have some difficulties, because science struggled with this for centuries.

- **What is work W ?** Work is the "organized" energy *transfer* when a body (system) is moved under the influence of a force. Work is usually easy to observe and therefore relatively easy to understand. For example, to lift a 20 kg rock up 2 m from the ground, the supplied work (force · displacement) is $W = Fl = mgl \approx 20$ kg $\cdot 10$ m s$^{-2} \cdot 2$ m $= 400$ J (see Figure A.6).

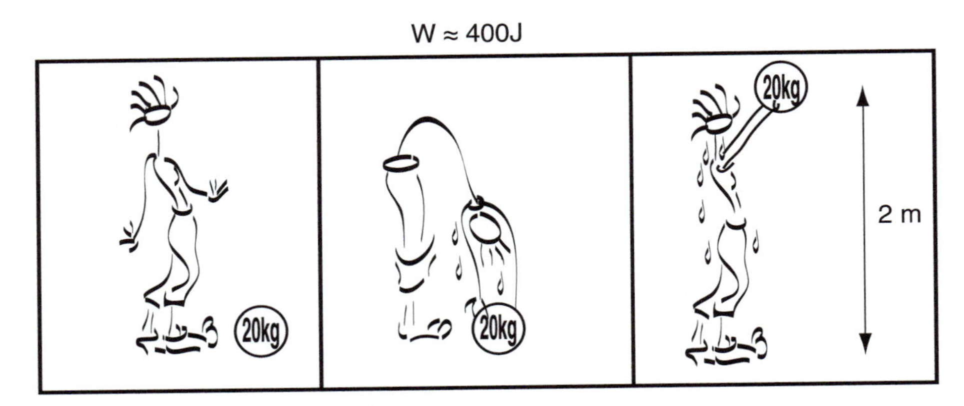

Figure A.6: Example of 400 J of mechanical work

There are many work forms:

— **Expansion work** $W_{\Delta V}$ (often called pV-Work) is the work for a system's volume change.
— **Flow work** W_{flow} is the work performed by a stream that enters or exits the system.
— **Shaft work** W_s is the mechanical work related to pressure changes from/to movable machinery such as compressor, pump or turbine.
— Further, we have the **electrochemical work** W_{el} in a battery or fuel cell
— and **other forms of work**, W_{other}, for example, the surface work needed to change the surface area, or electromagnetic work.
— **The total work** is

$$W = W_{\text{flow}} + \underbrace{W_{\Delta V} + W_s + W_{\text{el}} + W_{\text{other}}}_{W_n} \tag{A.20}$$

where W_n indicates the "useful" *non-flow* work. Usually, we only consider cases where the first three terms contribute, that is,

$$W = W_{\text{flow}} + \underbrace{W_{\Delta V} + W_s}_{W_n}$$

- **What is energy E ?** The classic definition of energy is "ability to perform work." However, this definition dates back to the time when the true relationship between heat and work was yet not established, and is in fact a very poor definition. For example, energy that exists as thermal energy at ambient temperatures is useless as a source of work (see the second law of thermodynamics). A better definition would be that energy is "ability to perform work <u>or</u> release heat," and that energy is a conserved quantity (see page 40). Energy can have many forms, some of which are

  — **Kinetic energy** E_K: macroscopic motion where the center of mass moves (example: the velocity of a gas in a pipe).

— **Rotational energy** E_R: macroscopic motion where the center of mass is at rest (example: liquid that rotates (*swirl*) in a tank).
— **Potential energy** E_P: potential energy compared to a reference level (example: a tank with liquid placed above ground level).
— **Electric energy** E_{el} (example: energy stored in a capacitor).
— **Surface energy** E_s.
— **Internal energy** U: energy of the molecules, including their kinetic, rotational, electric, potential and chemical energy. Sometimes, it is practical to consider internal energy as the sum of *thermal energy* (microscopic kinetic energy from the random (non-organized) motion of the molecules), *chemical bonding energy* and *latent energy* (the last one is "liberated" during phase transition). Internal energy is a state variable that includes the (thermodynamic) energy in a system when we neglect energy related to magnetic, electric and other fields. Internal energy is usually the most important form of energy for "our" systems.
— **The total energy** for the system is the sum of all of these energy forms

$$E = U + E_K + E_R + E_P + E_{\text{el}} + E_S + E_{\text{other}} \tag{A.21}$$

- **What is heat** Q **?** Heat is the "disorganized" transfer of energy that results when systems with different temperatures are contacted. The physical mechanism is the transfer of molecular thermal (kinetic, rotational, etc.) energy as molecules collide or get close to each other.
  Thus, for a closed system, there are two mechanisms for transferring energy between the system and the surroundings:

  Work W – organized energy transfer

  Heat Q – disorganized (or chaotic or thermal) energy transfer

  Note that heat Q is not included in the system's energy E, because heat is the energy transfer between systems whereas the energy E is within a system. In everyday speech, it is common to use the term "heat" to mean "thermal energy," but this is thermodynamically incorrect. On the other hand, it is correct to say that heat is the transfer of thermal energy from one system to another.
- **What is the relationship between energy, work and heat?** The first law of thermodynamics (which is a special case of the *energy balance*) states that energy is a conserved quantity. More precisely, for a **closed system**, the change (increase) in a system's energy is equal to the sum of supplied work and supplied heat

$$\Delta E = Q + W \tag{A.22}$$

Here, $\Delta E = E_f - E_0$ where E_0 is the energy in the initial state (at the time t_0), and E_f is the energy in the final state (at the time t_f). For our systems, internal energy changes are often the main contributions, that is, $\Delta E \approx \Delta U$, and we derive the **the first law thermodynamics** (the energy balance) for a closed system

$$\Delta U = Q + W \tag{A.23}$$

An *isolated* system is a closed system with no exchange of heat or work with the surroundings, that is, $Q = 0$ and $W = 0$, and we get

$$\text{Isolated system}: \quad \Delta U = 0 \tag{A.24}$$

The generalization of the energy balance to open systems is discussed in Chapter 4.1 (page 95).

A.7 Volume change work for closed system

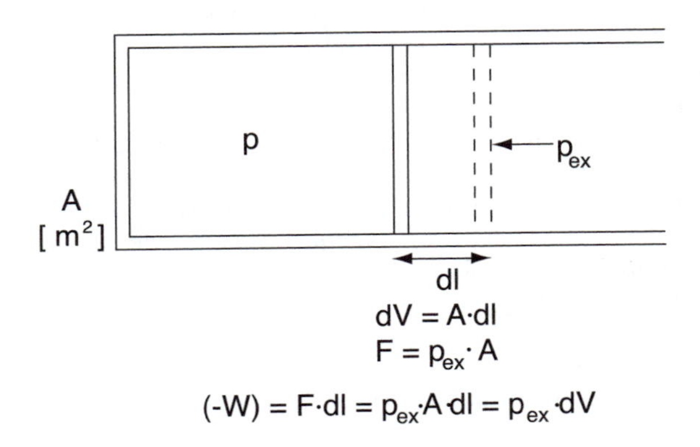

Figure A.7: Work during moving of piston

Consider a closed system with pressure p and surroundings with pressure $p_{\rm ex}$ (*external*). We want to find the expansion work $W = W_{\Delta V}$ related to a volume increase.

We assume that the expansion takes place in a cylinder with a movable piston with area A [m^2]; see Figure A.7. For a small (differential) expansion where the piston moves the length dl [m], the change in volume is $dV = Adl$. The work the system performs on the surroundings when the pistol is moved is force · displacement, where the force is $p_{\rm ex}A$ [N]. That is, performed work is

$$(-dW_{\Delta V}) = p_{\rm ex}Adl = p_{\rm ex}dV \tag{A.25}$$

Note that:

1. The negative sign is due to the convention of W being *supplied* work (from the surroundings). During expansion, work is performed by the system, so the supplied work is negative.
2. One must *always* use the surrounding's pressure $p_{\rm ex}$ when calculating the expansion work. This is because work is an energy exchange with the surroundings. The special case with $p = p_{\rm ex}$ applies only when the expansion process is reversible.

The total performed work for the whole expansion from the initial state (with volume V_0) to final state (with volume V_f) is found by integration,

$$\boxed{(-W_{\Delta V}) = \int_{V_0}^{V_f} p_{\rm ex}dV} \tag{A.26}$$

For a *reversible process*, we have a balance between the forces such that $p = p_{\rm ex}$ and the reversible expansion work is:

$$\boxed{W_{\Delta V}^{\rm rev} = -\int_{V_0}^{V_f} pdV} \tag{A.27}$$

For an *irreversible* (spontaneous, natural) process, there will be a difference between the pressures p and $p_{\rm ex}$. For example, during an expansion with $\Delta V > 0$, we need $p > p_{\rm ex}$. The performed work on the surroundings is then $p_{\rm ex}\Delta V$, while the work that we extract from the system (and which would have been performed on the surroundings during a reversible process) is only $p\Delta V$. The following question is appropriate:

- What happens to the rest of the work, $|(p - p_{\rm ex})|\Delta V$?

The answer is that it is converted to heat by friction. This heat will, depending on how the process takes place, be supplied to the system (as an increase in internal energy) and/or be transferred to the surroundings (as heat Q).

Reversible volume change work for ideal gas.

For an ideal gas, the reversible expansion (volume change) work is

$$W_{\Delta V}^{\text{rev}} = -\int_{V_0}^{V_f} pdV = -\int_{V_0}^{V_f} \frac{nRT}{V} dV \tag{A.28}$$

However, work is not a state function, so in order to calculate the exact value of the integral, we need to specify the process ("the path") in more detail. For the simple case with **constant temperature** T_0 (isothermal process), we get

$$W_{\Delta V}^{\text{rev}} = -nRT_0 \int_{V_0}^{V_f} \frac{dV}{V} = nRT_0 \ln \frac{V_0}{V_f} = nRT_0 \ln \frac{p_f}{p_0} \tag{A.29}$$

where $V_0/V_f = p_f/p_0$ since $pV =$ constant for an isothermal process.

A.8 Internal energy

Let us summarize some fundamental properties about internal energy.

Internal energy. There exists a state function called internal energy U. In a *closed system*, the change in U equals the sum of supplied heat and work,

$$\boxed{dU = \delta Q + \delta W} \tag{A.30}$$

Equation (A.30) is the differential form of the first law of thermodynamics $\Delta U = Q + W$, and if you take a close look, you will see that we have used an "ordinary" differential for U (dU) and a "curly" differential for Q and W (δQ and δW). This is a standard way to show that U is a state function, while Q and W are *not* state functions.

Since U is a state function, it is uniquely determined by the system's state (e.g., for an ideal gas with constant composition, U is only a function of temperature T). Actually, the fact that internal energy is a state function has not been proven, so it should be considered to be a **postulate**[8]. (But the statement is very reasonable as here so far there have been no cases where it has not been valid).

Constant volume. In order to get a clearer understanding of what internal energy is, let us consider a closed system with constant volume where no work is supplied such that $W = 0$. The energy balance (the first law of thermodynamics) gives

$$\Delta U = Q \quad \text{(closed system with constant volume)} \tag{A.31}$$

That is, for a closed system with constant volume, the change in internal energy is equal to the supplied heat (see Figure A.8a). This also gives a practical way of measuring the relationship between internal energy and temperature: We simply measure the temperature as a function of the supplied heat (internal energy) in a closed container with constant volume (*bomb calorimetry*). This is used below to define heat capacity for a system with no phase transition; see (A.36).

[8] A postulate or axiom is a statement that is not proved; there is perhaps a small difference in that an axiom is considered self-evident, which is not necessarily the case for a postulate.

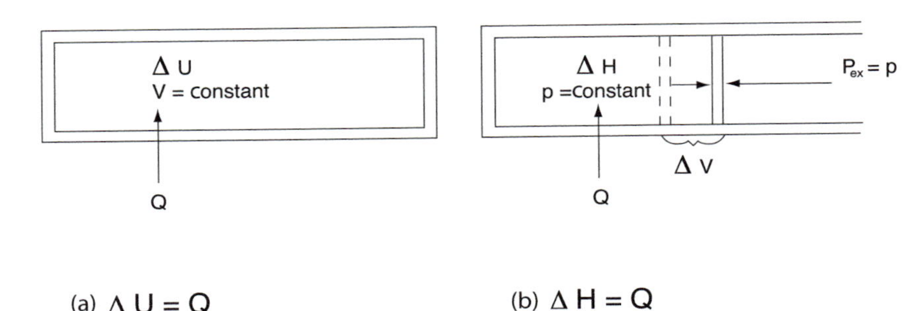

Figure A.8: Supplied heat to a closed system is equal to (a) the change in internal energy when the volume is constant, (b) the change in enthalpy when the pressure is constant

A.9 Enthalpy

The definition of enthalpy is

$$H = U + pV \quad [\text{J}] \tag{A.32}$$

where p is the system pressure [N/m^2] and V is the system volume [m^3]. (Let us check if the term pV really has the unit [J]: p [N/m^2] $\cdot V$ [m^3] gives the unit [Nm] which indeed is the same as [J]).

Enthalpy is a state function since U, p and V all are state functions. The molar ($n = 1$ mol) enthalpy is

$$H_m = U_m + pV_m \quad [\text{J/mol}] \tag{A.33}$$

where V_m is the mole volume [m^3/mol]. For an *ideal gas*, we have $pV_m = RT$ and

$$H_m = U_m + RT \quad \text{J/mol} \tag{A.34}$$

Since internal energy U_m is only a function of temperature for an ideal gas, it then follows that enthalpy is only a function of temperature for an ideal gas (at constant composition).

For most "condensed" phases, such as solids and liquids, we have $H_m \approx U_m$ because the molar volume V_m is small. For example, for water, the density is $\rho = 1000$ kg/m^3, the molar mass is $M = 18$ g/mol and the molar volume is $V_m = M/\rho = 18 \cdot 10^{-3}/1000 = 18 \cdot 10^{-6}$ m^3/mol which is more than a factor 1000 less than a typical mole volume for a gas (for example, the molar volume for an ideal gas at 298 K and 1 bar is $24.4 \cdot 10^{-3}$ m^3/mol).

Constant pressure. In order to get a clearer understanding of what enthalpy is, let us consider a closed system with constant pressure. The volume of the system varies as a function of temperature, and here we consider the case of reversible expansion (pressure-volume) work (that is, $p_{\text{ex}} = p =$ constant). We then have $W_{\Delta V}^{\text{rev}} = -\int_{V_0}^{V_f} pdV = -p(V_f - V_0) = -p\Delta V$ and the energy balance (A.23) becomes

$$\Delta U = Q^{\text{rev}} - p\Delta V$$

But from the definition of enthalpy, we have at constant pressure that

$$\Delta H = \Delta U + p\Delta V$$

The energy balance can therefore in this case be written

$$\Delta H = Q^{\text{rev}} \qquad \text{(closed system with constant pressure)} \tag{A.35}$$

where superscript ^{rev} denotes that the pV work is reversible. We therefore conclude that for a closed system with constant pressure and reversible pressure-volume work, the change in the system's enthalpy equals the supplied heat (see Figure A.8b).

Comment. (A.35) can be used to experimentally measure enthalpy, but an unfortunate consequence of (A.35) is that many believe that the use of enthalpy is limited to processes with constant pressure. However, this is completely wrong for open systems with inlet and outlet flows. Here, the enthalpy represents the sum of the flow's internal energy and its accompanying flow work, and the pressure may very well vary (see page 99).

Exercise A.4 *Consider a process where we heat an ideal gas with heat capacity $C_{p,m} = 30$ J/mol K from 20 ^oC to 80 ^oC and at the same time increase its pressure from 6 bar to 8 bar. Suggest (a) a closed process, and (b) an open (continuous) process that takes the system between the two states (note that there may be many possible processes). (c) What is the difference in the enthalpy change, the work and the supplied heat [J/mol] for the two processes? (Enthalpy is a state function, so the enthalpy change is the same in both cases, but the work and heat will differ).*

A.10 Heat capacity

The heat capacity C_p [J/K] equals the amount of heat Q [J] that must be supplied to increase the system's temperature with 1 K at *constant pressure* for a closed system, with no phase transition or reaction. The corresponding expansion is assumed to be reversible, such that the same heat amount is liberated when the temperature is lowered by 1 K. Similarly, the heat capacity C_V [J/K] equals the amount of heat (Q) that must be supplied in order to increase the temperature 1 K at *constant volume* in a closed system with no phase transition or reaction. The mathematical definitions are

$$C_p \triangleq \left(\frac{\partial Q^{\text{rev}}}{\partial T}\right)_p ; \quad C_V \triangleq \left(\frac{\partial Q}{\partial T}\right)_V \quad \text{[J/K]} \quad \text{(closed system)} \tag{A.36}$$

However, as shown in (A.35), we have at constant pressure that the supplied heat equals the change in enthalpy, $dQ^{\text{rev}} = dH$; and at constant volume, we have from (A.31) that the supplied heat equals the change in internal energy, $dQ = dU$. This leads to the following equivalent definitions of heat capacities

$$C_p \triangleq \left(\frac{\partial H}{\partial T}\right)_p ; \quad C_V \triangleq \left(\frac{\partial U}{\partial T}\right)_V \quad \text{[J/K]} \tag{A.37}$$

These definitions are more useful because H and U are state functions, and they also imply that C_p and C_V are state functions. For solids and liquids, V_m is usually small so $\Delta H \approx \Delta U$ and we have $C_p \approx C_V$.

Ideal gas. It can be shown that the internal energy U of an ideal gas is only a function of temperature and composition (that is, independent of volume and pressure). For a system with constant composition, we can then write $dU = C_V dT$. Furthermore, $dH = dU + d(pV)$ where $pV = nRT$ for an ideal gas. For 1 mole of an ideal gas, we then have

$$dH_m = d(U_m + pV_m) = dU_m + RdT = C_{V,m}dT + RdT = \underbrace{(C_{V,m}(T) + R)}_{C_{p,m}(T)} dT$$

This implies that for an ideal gas with constant composition: (1) H is only a function of temperature, and (2) the difference between $C_{p,m}(T)$ and $C_{V,m}(T)$ is constant and equal to R.

A.11 Adiabatic reversible expansion of ideal gas

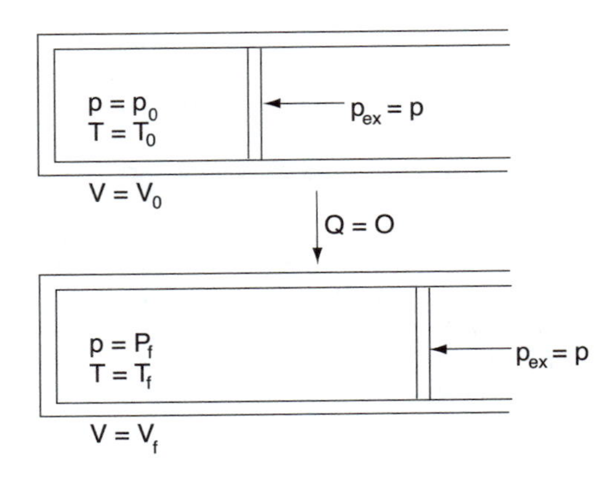

Figure A.9: Adiabatic reversible expansion

We consider, as shown in Figure A.9, an adiabatic ($Q = 0$) reversible expansion of an ideal gas in a closed system, from an initial pressure p_0 to a final pressure p_f. We assume for simplicity that the heat capacity C_p is constant, or, more precisely, that the ratio

$$\gamma \triangleq \frac{C_p}{C_V} \tag{A.38}$$

is constant (independent of T). We then have the following relationship between p and V during this expansion,

$$pV^\gamma = \text{constant} \tag{A.39}$$

or equivalently the following relationship between p and T,

$$\boxed{\frac{T_f}{T_0} = \left(\frac{p_f}{p_0}\right)^{\frac{\gamma-1}{\gamma}} = \left(\frac{p_f}{p_0}\right)^{\frac{R}{C_p}}} \quad \text{(ideal gas)} \tag{A.40}$$

where C_p [J/K mol] is the molar heat capacity. Of course, the ideal gas law also applies, so we also have $\frac{p_0 V_0}{T_0} = \frac{p_f V_f}{T_f} = nR$. (A.39) and (A.40) are not exact, because C_p generally increases with temperature, but nevertheless they are very useful.

Proof. The first law of thermodynamics (A.23) (energy balance) for a closed system gives for an adiabatic process ($Q = 0$) with reversible pV work

$$\Delta U = W^{\text{rev}} = -\int_{V_0}^{V_f} pdV$$

Equivalently, the differential form becomes

$$dU = -pdV \tag{A.41}$$

Let us here consider 1 mol of gas ($n = 1$ mol). For an ideal gas, we have $p = RT/V$, and U is only a function of temperature, that is, $dU = C_V dT$. (A.41) then gives $C_V dT = -pdV$ or

$$C_V dT = -\frac{RT}{V} dV \;\Rightarrow\; \frac{dT}{T} = -\frac{R}{C_V}\frac{dV}{V}$$

If C_V is assumed constant (independent of temperature), we get by integrating from state 0 to f

$$\ln \frac{T_f}{T_0} = -\frac{R}{C_V} \ln \frac{V_f}{V_0} \quad \Rightarrow \frac{T_f}{T_0} = \left(\frac{V_0}{V_f}\right)^{R/C_V}$$

Here, for an ideal gas

$$\frac{V_0}{V_f} = \frac{T_0}{T_f}\frac{p_f}{p_0}$$

and we get

$$\frac{T_f}{T_0} = \left(\frac{p_f}{p_0}\right)^{\frac{R/C_V}{1+R/C_V}} = \left(\frac{p_f}{p_0}\right)^{\frac{R}{C_V+R}} = \left(\frac{p_f}{p_0}\right)^{\frac{R}{C_p}}$$

Note that here

$$\frac{R}{C_p} = \frac{\gamma - 1}{\gamma}$$

and we have derived (A.40) and thereby also (A.39). □

- **Important comment**. (A.39) and (A.40) provide relationships between p, V and T for an adiabatic reversible process in a **closed system**. However, they apply generally to any reversible, adiabatic process, also for **open systems**. This follows because for such process $Q^{\text{rev}} = 0$, and from the definition of entropy in (B.3), we have $\Delta S = Q^{\text{rev}}/T = 0$, so **the entropy is constant**. Since entropy is a state function (which is not obvious, but it can be shown), this implies that (A.39) and (A.40) apply to *any isentropic process.*

Finally, let us calculate the expansion (pV) work W^{rev} for this process in a *closed system.* The energy balance for a closed system with $Q = 0$ gives $W = \Delta U$, and with assumed C_V constant we have

$$W^{\text{rev}} = \Delta U = C_V(T_f - T_0) = C_V T_0 \left(\left(\frac{p_f}{p_0}\right)^{R/C_p} - 1\right) \tag{A.42}$$

Example A.10 Expansion of closed system. *Consider $n = 1$ mol of an ideal gas at $T_0 = 400$ K and $p_0 = 10$ bar and let it expand reversibly to $p_f = 1$ bar in a closed system. Calculate the performed work and the supplied heat for (a) an isothermal process and (b) an adiabatic process (see Figure A.10). Data: $C_{V,m} = 25$ J/mol K can be assumed constant.*

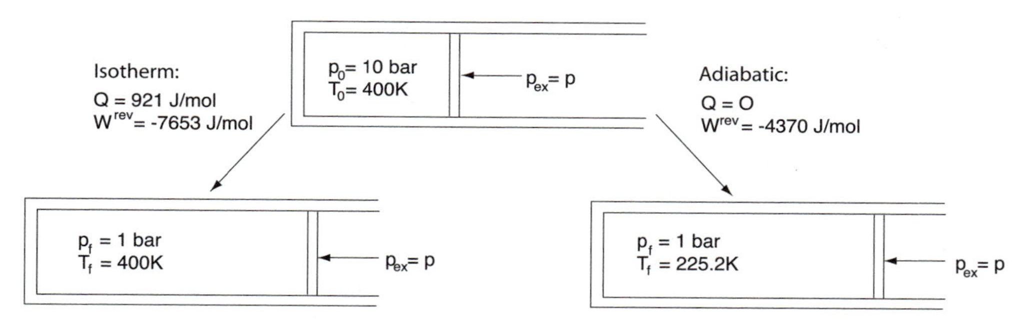

Figure A.10: Reversible expansion in closed system: (a) Isothermal. (b) Adiabatic

(a) For the reversible **isothermal** *process, we have $T_f = T_0 = 400$ K. From (A.29), the performed work is*

$$(-W^{\text{rev}}) = -RT_0 \ln \frac{p_f}{p_0} = 8.31 \cdot 400 \cdot \ln 10 = 7653 \text{ J/mol}$$

For an ideal gas at constant temperature, we have $\Delta U = 0$ and from the energy balance the supplied heat equals the performed work

$$Q = (-W) = 7653 \quad \text{J/mol}$$

(b) For the reversible **adiabatic** *process, we have $Q = 0$ and the final temperature is from (A.40)*

$$T_f = T_0 \left(\frac{p_f}{p_0}\right)^{\frac{\gamma-1}{\gamma}} = 400 \cdot \left(\frac{1}{10}\right)^{0.249} = 225.2 \text{ K}$$

The performed work during the expansion is, from (A.42),

$$(-W^{\text{rev}}) = C_V T_0 \left(1 - \frac{T_f}{T_0}\right) = 25 \cdot 400 \left(1 - \frac{225.2}{400}\right) = 4370 \text{ J/mol}$$

The performed work is significantly smaller than for the isothermal process because the gas is cooled during the expansion and thereby has a smaller final volume.

For an expansion process, where we want to extract (perform) work, we therefore conclude that it is *favorable to supply heat* (isothermal process). Conversely, *for compression it is favorable with cooling* in order to reduce the supplied compression work (note that this also applies to continuous processes).

Example A.11 Maximum temperature variation with elevation. *The outdoor temperature is generally lower at higher elevations. Show that the (maximum) temperature drop per 100 m at 273K is about 1.0 K.*

Solution. *In ExampleA.2 (page 331), we derived the barometric formula for the air pressure's dependency on elevation (height above sea level) is,*

$$\frac{p}{p_0} = \exp\left(-\frac{Mg}{RT_0}h\right)$$

This expression was derived at constant temperature T_0, but we now want to use it to calculate the temperature change with elevation. This seems a bit odd, but it is acceptable if the elevation difference (and thereby the pressure and temperature change) is sufficiently small. From (A.40), we have for a reversible adiabatic expansion

$$\frac{T}{T_0} = \left(\frac{p}{p_0}\right)^{\frac{\gamma-1}{\gamma}}$$

which inserted into the barometric formula gives

$$\frac{T}{T_0} = \exp\left(-\frac{Mg}{RT_0}h\right)^{\frac{\gamma-1}{\gamma}} = \exp\left(-\frac{Mg}{RT_0}\frac{\gamma-1}{\gamma}h\right)$$

With $h = 100$ m and $T_0 = 273.0$ K we get

$$T = T_0 \cdot \exp\left(-\frac{Mg}{RT_0}\frac{\gamma-1}{\gamma}h\right) = 273 \cdot \exp\left(-\frac{29 \cdot 10^{-3} \cdot 9.81}{8.31 \cdot 273}\frac{1.4-1}{1.4}100\right) = 272.0K$$

That is, the temperature drop is about 1.0 K per 100 m.

Comment: *This is the maximum temperature drop we can expect. If it exceeds this value (about 1 K), then we get an unstable situation where the upper layer is too heavy, so that we get a large exchange of mass between the upper cold and lower hot layers. Therefore, the actual temperature drop is usually smaller; typically about 0.5 K per 100 m, and we have a stable situation with the air layers at rest.*

A.12 Pressure independence of U and H for ideal gas: Joule's experiment

Ideal gas. For an ideal gas with constant composition, internal energy and enthalpy are, as mentioned, only a function of temperature, that is, they are independent of pressure. We will now describe an experiment that was historically used to derive this. The example is mostly intended as an eye-opener for those with an extra interest.

Example A.12 Joule's experiment. *Joule placed two metal containers in a water bath, as shown in Figure A.11. One container (tank 1) was filled with gas (air) at* $p_0 = 22$ *atm while the other (tank 2) was empty (vacuum; evacuated). Prior to the experiment, the temperature was* T_0 *everywhere (in the water bath and in the containers). He then opened a valve and let the air from tank 1 flow to the other container (tank 2) until the pressures were equal in both containers. He did not observe any temperature change in the water bath following the experiment. If the experiment was accurately performed, we can then conclude that internal energy is independent of pressure.*

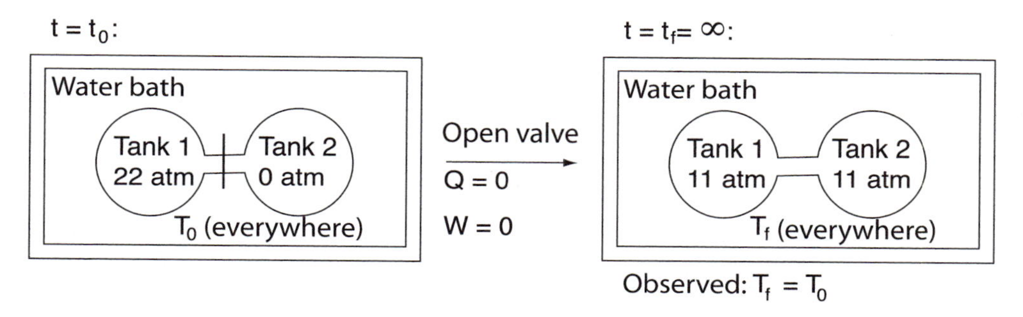

Figure A.11: Joule's experiment

To analyze the experiment in more detail, we define the two containers as the system, and the water bath as the surroundings. It is then obvious that $W = 0$ *since the system's volume (the total volume of the two containers) is constant, and Joule's observation that the temperature was unchanged in the bath implies that* $Q = 0$*. From the energy balance* $\Delta U = U_f - U_0 = Q + W$*, we then have that the internal energy of gas is unchanged*

$$U_f = U_0$$

and since the temperature, despite the pressure changes, was unchanged ($T_f = T_0$)*, we conclude that* U *for the gas is a function of temperature only (i.e.,* U *is independent of pressure).*

Note that even though the energy is constant in this experiment, the energy has been "degraded" because the energy that originally was "concentrated" in tank 1 (at high pressure) has been distributed to both tanks (at lower pressure). A degradation of this kind corresponds to increased entropy and takes place for all natural (spontaneous) processes.

This shows that internal energy is independent of pressure for an ideal gas (strictly speaking, the assumption of ideal gas is only valid at low pressure, but Joule's measurements were too inaccurate to detect the small temperature change he really should have observed). The same applies for enthalpy since $H = U + pV = U + nRT$ for ideal gas; so if U is independent of p then H is also independent of p.

Example A.13 Joule's experiment (continued): What happens before the temperature equilibrium is reached? *The following is a treat for those who are particularly interested. The analysis above considered what happens when time t_f goes to infinity such that temperature equilibrium is reached. But before equilibrium is reached, there will be temporary temperature changes: When the valve is opened, the temperature T_1 in tank 1 drops because of the work the gas in tank 1 performs on tank 2. More specifically, T_1 initially is a function of pressure p_1 according to the equation $T_1/T_0 = (p_1/p_0)^{\frac{\gamma-1}{\gamma}}$; see (A.40) and Example C.1 on page 393. (Later, the temperature rises back to T_0 because of the heat transfer between tank 1 and the surrounding water bath). On the other hand, the temperature in tank 2 will immediately rise to γT_0 after the valve is opened, see (4.21), but it will then drop because the temperature in tank 1 drops (and it will later drop further to T_0 because of the heat transfer to the surrounding water bath). As time goes to infinity, the temperature will, as shown above, return to its original temperature T_0. In summary, temporary changes take place which could give temporary changes in the temperature of the water bath (depending on how good the heat transfer is from each of the two tanks).*

Numerical example. *Let us take a look at a specific case where we assume that the heat transfer between the tanks and the water bath is very slow (for example, because the tanks are insulated). We assume that the two tanks are equal in size, $T_0 = 300$ K, $p_0 = 22$ atm and $\gamma = 1.4$. At the end of the first part of the experiment (which here refers to the time before the heat transfer to the water bath has any effect), we have mechanical equilibrium, that is, $p_1 = p_2$. But what is the pressure? The energy balance $U_f - U_0 = 0$ for the two combined tanks, assuming constant C_V, gives*

$$n_1 C_V (T_1 - T_0) + n_2 C_V (T_2 - T_0) = 0 \quad \Rightarrow \quad n_1 T_1 + n_2 T_2 = n_0 T_0$$

which using the ideal gas law gives $p_1 V_1 + p_2 V_2 = p_0 V_0$, and with $V_1 = V_2 = V_0$, we derive that $p_1 = p_2 = p_0/2 = 11$ atm (as expected). The temperature in tank 1 initially drops from $T_0 = 300$ K to $T_1 = (11/22)^{0.4/1.4}\ T_0 = 0.820\ T_0 = 246.1$ K. The temperature in tank 2 rises initially from T_0 to $\gamma T_0 = 420$ K, but then drops because of the temperature drop in tank 1. From the mass balance and the ideal gas law, we have at any given time that

$$n_1 + n_2 = n_0 \quad \Rightarrow \quad \frac{p_1}{T_1} + \frac{p_2}{T_2} = \frac{p_0}{T_0}$$

At the end of the first part of the experiment (before heat transfer to the water bath has any effect), we have $p_1 = p_2 = p_0/2$ which gives

$$\frac{1}{T_1} + \frac{1}{T_2} = \frac{2}{T_0}$$

and with $T_0 = 300K$ and $T_1 = 246.1K$, we find that $T_2 = 384.1K$. Also note that $n_1/n_2 = T_2/T_1 = 1.56 \neq 1$ at this time, which is before the heat transfer to the water bath has started. The heat transfer will eventually result in a temperature equilibrium where the temperature is 300 K everywhere. This will be accompanied by additional mass moving from tank 1 to tank 2, so that we eventually have $n_1/n_2 = 1$.

A.13 Calculation of enthalpy

Here, we consider the enthalpy's dependency on temperature, pressure and composition $H(T, p, n_i)$. We begin with the dependency on composition.

A.13.1 Composition dependency

For **ideal mixtures** (both for liquid and gas), the enthalpy of the mixture is the sum of the contributions from the pure components, that is, at a given T and p:

$$H = nH_m = \sum_i n_i H^*_{m,i} = n \sum_i x_i H^*_{m,i} \quad [J] \tag{A.43}$$

Here, H_m [J/mol] is the molar enthalpy, $H^*_{m,i}$ [J/mol] is the molar enthalpy for pure component i at T and p, $n = \sum_i n_i$ is total number of moles, n_i [mol] is the amount of component i in the mixture and $x_i = n_i/n$ is the mole fraction [mol i/mol]. The same applies on weight basis (where n is the mass in kg).

For non-ideal mixtures, we need to include the contribution from heat of mixing (see page 359).

A.13.2 Temperature dependency

For **ideal gases** with constant composition, the enthalpy is only a function of temperature, $dH = C_p dT$, that is, the enthalpy is independent of pressure. This applies also to most **solids and liquids**. If we know the enthalpy in state 1, then the enthalpy in state 2 is

$$\boxed{H(T_2) = H(T_1) + \int_{T_1}^{T_2} C_p(T)dT} \tag{A.44}$$

Temperature-varying heat capacity. For more accurate numerical calculations, empirical correlations are used for $C_p(T)$, for example, a polynomial form

$$C_p(T) = A + BT + CT^2 + DT^3 \quad \text{[J/mol K]} \tag{A.45}$$

Integration gives

$$H(T_2) - H(T_1) = \int_{T_1}^{T_2} C_p(T)dT =$$

$$A(T_2 - T_1) + \frac{B}{2}(T_2^2 - T_1^2) + \frac{C}{3}(T_2^3 - T_1^3) + \frac{D}{4}(T_2^4 - T_1^4) \tag{A.46}$$

Average heat capacity. It can be practical to define an average $\bar{C}_p$ over a certain temperature interval, for example from T_1 to T_2,

$$\bar{C}_p \triangleq \frac{\int_{T_1}^{T_2} C_p(T)dT}{T_2 - T_1} \tag{A.47}$$

and we have that

$$H(T_2) = H(T_1) + \bar{C}_p(T_2 - T_1) \tag{A.48}$$

Constant heat capacity. For the case where C_p can be assumed independent of temperature, we have $\bar{C}_p = C_p$, and

$$H(T_2) = H(T_1) + C_p(T_2 - T_1) \tag{A.49}$$

Linear heat capacity. For the case where the heat capacity can be approximated by a linear temperature function, that is, $C_p(T) = A + BT$, the mean heat capacity equals the arithmetic mean value,

$$\bar{C}_P = \frac{C_p(T_1) + C_p(T_2)}{2} \tag{A.50}$$

Proof: $C_p(T) = A + BT$ inserted into (A.44) gives $H(T_2) - H(T_1) = A(T_2 - T_1) + \frac{B}{2}(T_2^2 - T_1^2)$. But here we have that $(T_2^2 - T_1^2) = (T_2 - T_1)(T_2 + T_1)$ and we get $H(T_2) - H(T_1) = \frac{(A+BT_1)+(A+BT_2)}{2}(T_2 - T_1) = \frac{C_p(T_1)+C_p(T_2)}{2}(T_2 - T_1)$. □

Example A.14 Heating of a gas stream. *$m = 2$ kg/s of an ideal gas is heated from $T_{\text{in}} = 290$ K to $T_{\text{out}} = 620$ K. The specific heat capacity is*

$$c_p(T) = 862 + 0.43T \quad [\text{J/kg K}]$$

where the temperature T is in K. How much heat must be supplied and what is the mean heat capacity?

The energy balance for a steady-state continuous process gives that the supplied heat is equal to the enthalpy change

$$Q = H_{\text{out}} - H_{\text{in}} = m \int_{T_{\text{in}}}^{T_{\text{out}}} c_p(T)dT = m\bar{c}_p(T_{\text{out}} - T_{\text{in}})$$

Since the heat capacity is linear in temperature, we can use the mean value

$$\bar{c}_p = 862 + 0.43\frac{290 + 620}{2} = 1058 \ J/K \ kg$$

and we get

$$Q = 2kg/s \cdot 1058J/K \ kg \cdot (620K - 290K) = 698kJ/s$$

Mixtures. Note that the value for C_p (and for A, B, C, D) given above is for the mixture. For an ideal mixture, C_p is easily obtained from pure component data, and we have that (here on molar basis)

$$C_p = \sum_i x_i C^*_{p,i} \quad [\text{J/mol K}] \tag{A.51}$$

where x_i is the mole fraction and $C^*_{p,i}$ [J/mol K] is the heat capacity for pure component i. Data for pure components, for example, on polynomial form

$$C^*_{p,i}(T) = A_i + B_iT + C_iT^2 + D_iT^3 \quad [\text{J/mol K}]$$

are found in the literature.[9] Instead of calculating $C^*_{p,i}(T)$ for each component and then obtaining the mean C_p using (A.51), it is often more practical to first compute the mean coefficients

$$A = \sum_i x_iA_i, \ B = \sum_i x_iB_i, \ C = \sum_i x_iC_i, \ D = \sum_i x_iD_i \tag{A.52}$$

and then calculate $C_p(T)$ for the mixture using (A.45).

For ideal mixtures, it is also easy to determine C_p for a mixture of several streams as the mean of the C_p's for the individual streams, that is, we can use the above equations, but let i represent stream number (instead of component). This follows trivially from evaluating the streams as the sum of their components.

A.13.3 Pressure dependency

Ideal gas. For ideal gas, the enthalpy is independent of pressure, see page 353, and we have

$$H'(T_1) = H'(T_0) + \int_{T_0}^{T_1} C'_{p,m}(T)dT$$

where C'_p is the ideal gas heat capacity which is independent of pressure.

[9] A good source for physical data is Poling, Prausnitz and O'Connell, *The properties of gases and liquids*, 5th Ed., McGraw-Hill (2001).

For **real gases** (and to a certain degree, for liquids and solids), enthalpy is a function of pressure. For gases, we generally use ideal gas as the reference state (ideal gas is indicated with $'$ in the following). Since all real gases can be described as ideal gas as $p \rightarrow 0$, we get

$$H(T_1, p_1) = H'(T_1) + \underbrace{\int_0^{p_1} \mu_T(T_1, p) dp}_{=0 \text{ for ideal gas}} \quad \left[\frac{\text{J}}{\text{mol}}\right] \tag{A.53}$$

where

$$\mu_T \triangleq \left(\frac{\partial H}{\partial p}\right)_T \quad [\text{J/mol bar}] \tag{A.54}$$

is called the isothermal Joule-Thompson-coefficient. From thermodynamic identities, it can be shown that [10]

$$\mu_T = \left(\frac{\partial H}{\partial p}\right)_T = V - T\left(\frac{\partial V}{\partial T}\right)_p \tag{A.55}$$

where $\left(\frac{\partial V}{\partial T}\right)_p$ can be determined from an equation of state, e.g., the SRK equation. For ideal gases, it is easy to show that $\mu_T = 0$, and μ_T is also small for most liquids and solids because the volume is relatively smaller for condensed phases. It can also be shown that[11]

$$\mu_T = -C_{p,m}\mu \quad [\text{J/mol bar}]$$

where $\mu = (\partial T/\partial p)_H$ [K/bar] is the (normal) **Joule-Thompson coefficient.**

In practice, we don't use μ_T for calculating the pressure dependency of enthalpy. Rather, for hand calculations for pure components, we use thermodynamic diagrams (as given for methane, ammonia and water in appendix). More generally, the integral of μ_T in (A.53) is calculated numerically using an equation of state.

A.14 Thermochemistry

Thermochemistry is the evaluation of enthalpy changes for standard processes such as evaporation, mixing and reaction. It is very important to note that enthalpy is a state function. This implies that the enthalpy change can be evaluated as the sum of idealized subprocesses that take the system from the initial to the final state. Typical subprocesses are:

1. **Change in temperature at constant pressure (same phase),** $\Delta_T H = H(T_2, p_0) - H(T_1, p_0) = \int_{T_1}^{T_2} C_p(T) dT$
2. **Change in pressure at constant temperature (same phase),** $\Delta_p H = H(T_0, p_2) - H(T_0, p_1)$ ($= 0$ for ideal gas)

[10] From the exact differential for enthalpy, $dH = TdS + Vdp$, see (B.68), we derive $(\partial H/\partial p)_T = V + T(\partial S/\partial p)_T$ where from the Maxwell relation $(\partial S/\partial p)_T = -(\partial V/\partial T)_p$, and (A.55) follows.

[11] Write the total differential for enthalpy as (constant composition)

$$dH = \left(\frac{\partial H}{\partial T}\right)_p dT + \left(\frac{\partial H}{\partial p}\right)_T dp$$

Set $dH = 0$ and we find for 1 mole

$$\mu = \left(\frac{\partial T}{\partial p}\right)_H = -\left(\frac{\partial H}{\partial p}\right)_T / \left(\frac{\partial H}{\partial T}\right)_p = -\mu_T / C_{p,m}$$

3. **Change in phase (*transition*) at constant temperature,** $\Delta_{\mathrm{trs}}H$, for example

$$\begin{aligned} \text{Evaporation (reverse of condensation)}: &\quad \Delta_{\mathrm{vap}}H \text{ for } l \to g \\ \text{Fusion} = \text{melting (reverse of freezing)}: &\quad \Delta_{\mathrm{fus}}H \text{ for } s \to l \\ \text{Sublimation (reverse of deposition)}: &\quad \Delta_{\mathrm{sub}}H \text{ for } s \to g \end{aligned}$$

Note here that since the enthalpy is a state function, we have

$$\Delta_{\mathrm{sub}}H = \Delta_{\mathrm{fus}}H + \Delta_{\mathrm{vap}}H$$

4. **Mixing of two or more streams (substances) at constant temperature and pressure,** $\Delta_{\mathrm{mix}}H$ ($= 0$ for ideal gas)
5. **Chemical reaction at constant temperature and pressure,** $\Delta_r H$

The total enthalpy change is, as mentioned, the sum of the individual processes since enthalpy is a state function. Generally, there are many ways (paths) to go between two given states but the enthalpy change is always the same. The so-called Hess' law (page 362) is a special case of enthalpy being a state function applied to chemical reactions.

A.14.1 Heat of vaporization

The heat of vaporization $\Delta_{\mathrm{vap}}H(T)$ is the enthalpy to go from liquid to gas at a given temperature T.

$$\Delta_{\mathrm{vap}}H(T) = H(g,T) - H(l,T) \quad [J/mol; J/kg] \tag{A.56}$$

The pressure is the saturation pressure at temperature T. The heat of vaporization is used for pure components, but for mixtures, the term is not well defined and should be avoided. This is because the composition of the liquid phase for mixtures differs from the composition of the vapor phase.

$\Delta_{\mathrm{vap}}H$ is a function of temperature and decreases to 0 at the critical point. Let us illustrate this with an example. Make sure you understand this!

Example A.15 Heat of vaporization *Calculate the heat (enthalpy) of vaporization at temperature T when the heat of vaporization at temperature T_0 is known and constant heat capacities are assumed for gas and liquid.*

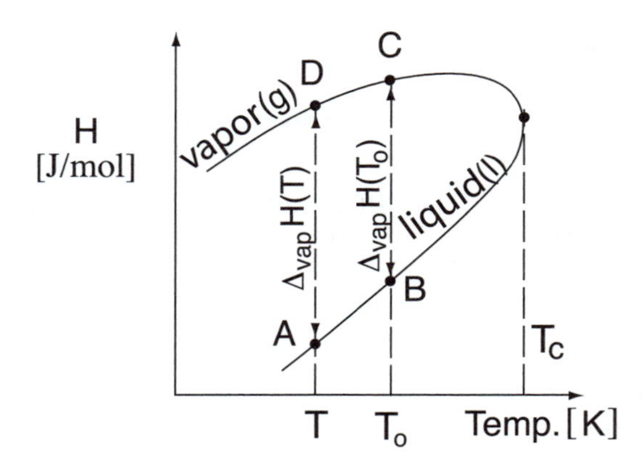

Figure A.12: The heat of vaporization is a function of temperature

Solution. *We want to go from liquid at temperature T to gas at temperature T (from A to D on Figure A.12). Enthalpy is a state function and we can therefore find this as the sum of the following three subprocesses:*

1. *A to B: Heat the liquid from T to T_0. $\Delta_1 H = C_p(l)(T_0 - T)$ [J/mol] or [J/kg].*
2. *B to C: Evaporate the liquid at T_0. $\Delta_2 H = \Delta_{\rm vap} H(T_0)$ [J/mol] or [J/kg]*
3. *C to D: Cool the gas from T_0 to T. $\Delta_3 H = C_p(g)(T - T_0)$ [J/mol] or [J/kg]*

That is, the heat of vaporization at temperature T is

$$\Delta_{\rm vap} H(T) = \Delta_1 H + \Delta_2 H + \Delta_3 H = \Delta_{\rm vap} H(T_0) + (C_p(l) - C_p(g))(T_0 - T)$$

Note that we generally have $C_p(l) > C_p(g)$ so that the heat of vaporization decreases with increasing temperature. Note that we have neglected the influence of pressure for subprocesses 1 and 3. This is acceptable as long as we are not too close to the critical point. (At the critical point there is no longer any difference between gas and liquid and the heat of vaporization is 0).

Exercise A.5 * Heat of vaporization for water. *Given for water: $C_p(l) = 75.4$ J/mol K and $C_p(g) = 33.6$ J/mole K (both assumed constant) and $\Delta_{\rm vap} H = 40.6$ kJ/mol at 100 °C and 1 atm. (a) Use the result from Example A.15 to calculate the heat of vaporization for water at 25 °C.*
(b) The critical temperature for water is 647.3 K. Compare this with the value of T that gives $\Delta_{\rm vap} H(T) = 0$ using the simplified expression where we assume constant heat capacities.

A.14.2 Heat of mixing

Consider mixing of two or more streams with same temperature T. The enthalpy is constant during the mixing process, but the temperature of the mixture, $T_{\rm mix}$, will generally differ from T. The heat of mixing $\Delta_{\rm mix} H$ (the heat of solution, the enthalpy of mixing) is defined as *the heat that must be supplied in order to maintain constant temperature T*. The heat of mixing is usually given compared to the pure components, that is, we define the heat of mixing as (at a given temperature T and given pressure p, here given on mole basis)

$$\Delta_{\rm mix} H(T) = H_m(T) - \sum_i x_i H^*_{m,i}(T) \quad [\text{J/mol}] \tag{A.57}$$

where H_m [J/mol] is the molar enthalpy of the mixture, $H^*_{m,i}$ [J/mol] is the molar enthalpy of pure component i, and x_i is the mole fraction of component i in the mixture.

Example A.16 Nitric acid. *Calculate the heat of mixing [kJ/kg solution] when 40 weight% nitric acid is produced from the pure components (HNO_3 and H_2O). Data: see page 417.*

Solution. *We mix 0.4 kg HNO_3 with 0.6 kg water in order to produce 1 kg 40% nitric acid. With molar mass 63 g/mol and 18 g/mol, this corresponds to 400/63= 6.35 mol HNO_3 and 600/18=33.33 mol H_2O. The number of moles $H_2 0$ per mol HNO_3 (non-diluted acid) is then $n = 33.33/6.35 = 5.25$. From the formula on page 417, we then have that the heat of mixing is $\frac{(-37.5)\cdot 5.25}{5.25+1.74} = -28.2$ kJ/mol HNO_3, that is,*

$$\Delta_{\rm mix} H = -28.2 \frac{\text{kJ}}{\text{mol HNO}_3} \cdot 6.35 \frac{\text{mol HNO}_3}{\text{kg solution}} = -179 \ \frac{\text{kJ}}{\text{kg solution}}$$

A.14.3 Heat of reaction

Let a stoichiometric mixture at temperature T react completely at constant pressure p. The heat of reaction (reaction enthalpy) $\Delta_r H(T,p)$ for the given chemical reaction is defined as *the heat that must be supplied to keep constant reaction temperature* T, when reactants and products are assumed to be in their standard form.

For an **endothermic** reaction, the heat of reaction is positive, that is, we must supply heat in order to keep the temperature constant. For an **exothermic** reaction, the heat of reaction is negative, that is, we must cool (remove heat) in order to keep the temperature constant.

The heat of reaction is usually given per mole reacted according to the given reaction equation $0 = \sum_i \nu_i A_i$, where ν_i is the stoichiometric coefficient.

The reactants and products may be in many states, and we here consider the *standard* heat of reaction, where the pressure is $p^\ominus = 1$ bar and the reactants and products are in their standard state as pure components. We then have

$$\Delta_\mathrm{r} H^\ominus(T) = \sum_i \nu_i H_m^\ominus(i) = \sum_{i=\mathrm{products}} \nu_i H_m^\ominus(i) - \sum_{i=\mathrm{reactants}} |\nu_i| H_m^\ominus(i) \quad [\mathrm{J/mol}] \tag{A.58}$$

where $H_m^\ominus(i)(T)$ [J/mol] is the standard enthalpy of pure component i. The subscript m is used in order to clearly show that it is a molar quantity, but is often omitted. The temperature T is often chosen to be 298.15 K (25°C).

To calculate the standard enthalpy $H_m^\ominus$, we usually choose the elements in their standard state (e.g., O_2 (g), H_2 (g), N_2 (g), C (s) (graphite), etc.) as the reference. We then have that $H_m^\ominus(i) = \Delta_\mathrm{f} H^\ominus(i)$ [J/mol] is the standard enthalpy of formation of pure component i from its elements at temperature T. However, there are also other possible reference states, for example, the combustion products and we then have $H_m^\ominus(i) = \Delta_\mathrm{c} H^\ominus(i)$ [J/mol] which is the standard enthalpy of combustion (see page 362). Since enthalpy is a state function, the value of the heat of reaction $\Delta_\mathrm{r} H^\ominus$ is of course the same independent of the choice of reference state.

- The heat of reaction is often given per mol (e.g., in kJ/mol), and we can ask the question "per mol of what ?" The answer is that it is the stoichiometric coefficients of the given reaction that tells how many moles of each component that react.

Example A.17 *Consider the reaction*

$$CO(g) + 2H_2(g) = CH_3OH(g)$$

which can be written in the standard form $0 = \sum_i \nu_i A_i$ *with stoichiometric coefficients*

$$\nu(C0) = -1; \quad \nu_(H_2) = -2; \quad \nu(CH_3OH) = 1$$

From (A.58), the heat of reaction is the enthalpy of the products minus the enthalpy of the reactants with the given stoichiometric coefficients, that is,

$$\Delta_r H = H^\ominus(CH_3OH) - H^\ominus(CO) - 2H^\ominus(H_2)$$

With the elements as the reference, this gives

$$\Delta_\mathrm{r} H^\ominus = \Delta_\mathrm{f} H^\ominus(CH_3OH) - \Delta_\mathrm{f} H^\ominus(CO) - 2\Delta_\mathrm{f} H^\ominus(H_2)$$

At 298.15 K, we get with data given on page 416:

$$\Delta_\mathrm{r} H^\ominus(298) = (-200.66) - (-110.53) - 2 \cdot 0 = -90.13 \text{ kJ/mol}$$

In summary, we have

$$CO(g) + 2H_2(g) = CH_3OH(g); \quad \Delta_r H^\ominus(298) = -90.13 \text{ kJ/mol}$$

or

$$\Delta_r H^\ominus(298) = \frac{-90.13 \text{ kJ}}{1 \text{ mol CO reacted}} = \frac{-90.13 \text{ kJ}}{2 \text{ mol H}_2 \text{ reacted}} = \frac{-90.13 \text{ kJ}}{1 \text{ mol CH}_3\text{OH formed}}$$

Note, however, that the heat of reaction is half of this value if only 1 mol (rather than 2 mol) of H_2 reacts. That is, we have

$$0.5\ CO(g) + H_2(g) = 0.5\ CH_3OH(g); \quad \Delta_r H^\ominus(298) = -45.06 \text{ kJ/mol}$$

The heat of reaction is a function of temperature (but not of pressure since pressure is 1 bar at standard conditions). If we have data for the reaction at temperature T_0 (e.g., $T_0 = 298.15$ K), then the heat of reaction at temperature T is

$$\Delta_r H^\ominus(T) = \Delta_r H^\ominus(T_0) + \int_{T_0}^{T} \Delta_r C_p^\ominus(T) dT \tag{A.59}$$

where $\Delta_r C_p^\ominus(T) = \sum_i \nu_i C_p^\ominus(i, T)$ is the change in heat capacity during the reaction. (A.59) is derived by calculating the standard enthalpy for each component at temperature T and pressure $p^\ominus = 1$bar, see (A.44)

$$H^\ominus(i, T) = H^\ominus(i, T_0) + \int_{T_0}^{T} C_p^\ominus(i, T) dT$$

and inserting this into $\Delta_r H^\ominus(T) = \sum_i \nu_i H_m^\ominus(i, T)$. If the heat capacities are assumed constant (independent of temperature), then (A.59) simplifies to

$$\Delta_r H^\ominus(T) = \Delta_r H^\ominus(T_0) + \Delta_r C_p^\ominus \cdot (T - T_0) \tag{A.60}$$

Example A.18 *Consider again the reaction*

$$CO(g) + 2H_2(g) = CH_3OH(g)$$

What is the standard heat of reaction at 270 °C? We have $\Delta_r H^\ominus(298) = -90.13$ kJ/mole and the pure components' heat capacities at 298.15 K are

$$C_p^\ominus(CO)(g) = 29.14 \ \frac{\text{J}}{\text{mol K}}; \ C_p^\ominus(H_2)(g) = 28.82 \ \frac{\text{J}}{\text{mol K}}; \ C_p^\ominus(CH_3OH)(g) = 43.89 \ \frac{\text{J}}{\text{mol K}}$$

and we have

$$\Delta_r C_p^\ominus(298) = 43.89 - (-29.14) - 2 \cdot 28.82 = -42.89 \ \frac{\text{J}}{\text{mole K}}$$

If we assume that $\Delta_r C_p^\ominus$ is independent of temperature, then we get from (A.60)

$$\Delta_r H^\ominus(T) = \Delta_r H^\ominus(298) + \Delta_r C_p^\ominus \cdot (T - 298.15)$$

$$= -90130 \frac{\text{J}}{\text{mol}} - 42.89 \frac{\text{J}}{\text{mol K}} (270 - 25) K = -100638 \ \frac{\text{Jr}}{\text{mol}}$$

that is, $\Delta_r H^\ominus(270^o C) = -100.64$ kJ/mol. For more accurate calculations, one should use data for the heat capacity as a function of temperature and integrate (A.59).

We can also calculate the reaction enthalpy by adding the reaction enthalpies of individual reactions that sum up to give the total reaction. This is the so-called **Hess' law**, which is best illustrated by an example.

Example A.19 Hess' law. *We want to find the enthalpy of reaction at 298.15 K for the gas phase reaction where ethylene and water form ethanol,*

$$\text{Reaction 1}: \quad C_2H_4 + H_2O = C_2H_5OH$$

by using data for the following individual reactions (everything in gas phase and everything at 298.15 K):

$$\begin{aligned} &\text{Reaction 2}: && C_2H_5OH + 3O_2 = 2CO_2 + 3H_2O; \quad \Delta_r H_2^\ominus = -1233.59 \text{kJ/mol} \\ &\text{Reaction 3}: && C_2H_4 + 3O_2 = 2CO_2 + 2H_2O; \quad \Delta_r H_3^\ominus = -1322.12 \text{kJ/mol} \end{aligned}$$

We have that

$$\text{Reaction 1} \quad = \quad \text{Reaction 3} \quad - \quad \text{Reaction 2}$$

and from Hess' law (see proof below), the heat of reaction for reaction 1 is that of reaction 3 minus that of reaction 2, i.e.,

$$\Delta_r H_1^\ominus = \Delta_r H_3^\ominus - \Delta_r H_2^\ominus = -1322.12 - (-1233.59) = -88.53 \text{ kJ/mol}$$

It is easy to prove Hess' law. We have for reaction 1:

$$\Delta_r H_1 = H(C_2H_5OH) - H(C_2H_4) - H(H_2O)$$

Similar expressions apply for reactions 2 and 3, and we get

$$\begin{aligned} \Delta_r H_3 - \Delta_r H_2 &= 2H(CO_2) + 2H(H_2O) - H(C_2H_4) - 3H(O_2) \\ &- [2H(CO_2) + 3H(H_2O) - H(C_2H_5OH) - 3H(O_2)] \\ &= H(C_2H_5OH) - H(C_2H_4) - H(H_2O) \end{aligned}$$

which equals $\Delta_r H_1$ *and Hess' law is proven.*

A.14.4 Heat of combustion

In combustion reactions, the feed mixture (*fuel*) reacts with oxygen to produce the combustion products CO_2, SO_2, N_2 and H_2O. By complete combustion, we mean that the maximum amount of oxygen is consumed and also that CO_2 is generated rather then CO. The composition of the combustion product (flue gas) can be given on "wet" basis (when we include H_2O =water) or on "dry" basis (without water). Examples of complete combustion reactions are

$$\begin{gathered} C + 2O_2 = CO_2 \\ CH_4 + 2O_2 = CO_2 + 2H_2O \\ C_nH_m + \left(n + \frac{m}{4}\right) O_2 = nCO_2 + \frac{m}{2} H_2O \\ H_2S + 1.5O_2 = SO_2 + H_2O \end{gathered}$$

The *negative* of the standard heat of combustion $\Delta_c H^\ominus$ ([J/mol] or [J/kg]), also known as the **heating value**, can be found in the literature for many substances. Note that "per mol" or "per kg" here refers to 1 mol or 1 kg of the substance. More precisely, the heat of combustion is defined for the complete combustion of the substance in its natural state (for example, methane (CH_4) as a gas, and methanol (CH_3OH) as a liquid) to the combustion products in their natural state, that is, to CO_2 (g), SO_2 (g) and N_2 (g) (all at 298.15 K). For the water product there are two choices:

1. The natural state of water at 298.15 K and 1 bar is liquid, and the corresponding heat of combustion with H_2O (l) as the product is known as the **higher heat of combustion (gross heating value, GHV)**. This is the value (usually in [kcal/100 g] or in [kJ/100 g]) given on food labels. Specifications for the heating value of natural gas for domestic use are also given in terms of the higher heating value (GHV), and the specification is typically that the GHV should be between 38 MJ/Sm^3 and 42 MJ/Sm^3. Note here that methane has GHV=37.7 MJ/Sm^3 and ethane has GHV=66.0 MJ/Sm^3, so the GHV specification indirectly puts quite strong specifications on the composition.
2. In many cases the combustion product is a gas, and the corresponding heat of combustion with H_2O (g) as the product is known as the **lower heat of combustion (net heating value)**. This value is lower than the gross heating value because one does not make use of the heat released when water condenses.
   The (higher) heat of combustion (GHV) for most hydrocarbons (including gas and oil) is about 48000 kJ/kg (or about 1150 kcal/100 g), whereas the lower heat of combustion (to H_2O (g)) is about 45000 kJ/kg.
   The lower heat of combustion is commonly used when giving the heating value of liquid and solid fuels. For example, the energy unit "1 ton oil equivalent" (toe) corresponds to the lower heating value of 1 ton of crude oil, which is approximately 42 GJ (see page 12).

For most purposes (except for buying and selling) the difference between the two heating values is not very important. The difference depends on the amount of hydrogen in the fuel and the difference varies from 11% for methane, to about 5% for heavier hydrocarbons, and down to no difference for carbon (C) or coal with no hydrogen.

The (standard) heat of combustion can be used for calculating the (standard) heat of formation, $\Delta_f H^\ominus$ (see the example below). The following data for the standard heat of formation at 298.15 K are useful for this:

$$\begin{array}{rrl} O_2(g): & 0 & \text{kJ/mol} \\ N_2(g): & 0 & \text{kJ/mol} \\ CO_2(g): & -393.51 & \text{kJ/mol} \\ H_2O(g): & -241.82 & \text{kJ/mol}; \quad H_2O(l): \quad -285.83 \quad \text{kJ/mol} \\ SO_2(g): & -296.83 & \text{kJ/mol} \end{array}$$

Example A.20 Heat of formation from heat of combustion. *We want to find the heat (enthalpy) of formation for amino ethane ($C_2H_5NH_2$) liquid. In an old reference (*CRC Handbook *from 1977–78), we find that the (higher) heat of combustion for ethyl amine (l) at 25 ^oC is 409.5 kcal/mol, which converted to SI units is* $\Delta_c H^\ominus = 409.4 kcal/mol \cdot 4.184 kJ/kcal = 1713.3$ *kJ/mol. The combustion reaction for 1 mol of the substance is*

$$C_2H_5NH_2(l) + 3.75O_2(g) = 2CO_2(g) + 3.5H_2O(l) + 0.5N_2(g)$$

Using the heat of formation data given above, the standard heat of reaction for this combustion reaction is

$$\Delta_r H^\ominus = 2(-393.51) + 3.5(-285.83) + 0.5 \cdot 0 - 3.75 \cdot 0 - x = -1713.3 \quad [\text{kJ/mol}]$$

(note the negative sign on the right hand side since the combustion reaction is exothermic). Here, x is the unknown heat of formation for amino ethane (l) at 298.15 K, and we find $x = -74.1$ kJ/mol.

Comment. *In the reference book* SI Chemical Data (2002), *we find data for the gas (rather than for the liquid). Specifically, it is given that the heat of formation for amino ethane (g) is -46 kJ/mol and that the heat of vaporization is 32 kJ/mol (both at 298.15 K). This gives that the heat of formation for amino ethane (l) is $-46 - 32 = -78$ kJ/mol, which is a bit different from the value -74.1 kJ/mol found above. (I would assume that* SI Chemical Data *is more reliable, since it is newer, but I am not sure).*

Exercise A.6* *The (higher) heat of combustion for methyl formate ($HCOOCH_3$) (l) to H_2O (l) at 25 °C is, in an old reference book, given as 234.1 kcal/mol. (a) Calculate the standard heat of formation for methyl formate. (b) Which value do you find in a reference book (e.g., in* SI Chemical Data*)?*

Exercise A.7 *Find the heat of combustion for some foodstuff and check if this is consistent with the calculated (theoretical) values.*

A.15 Alternative reference states

Although enthalpy is a state function and the change in enthalpy between two states is uniquely given, it is important to realize that the "absolute" numerical value of the enthalpy is *not* a unique quantity. Thus, if we specify an (absolute) enthalpy $H(T,p,n_i)$ in a given state (for example for a stream) then this must always be specified relative to a defined *reference state*, which must be the same for all streams we consider.

A.15.1 Elements as reference ("absolute" enthalpy)

The "safest" (in the sense that it can always be used as a common reference state) is to use the elements at $p^\ominus = 1$ bar and 298.15 K as the standard reference state, that is, we set $H = 0$ for the elements in their standard (natural) state at 1 bar and 298.15 K. The "absolute" enthalpy $H(T,p,n_i)$ is then the enthalpy change when we (1) take the elements and react them to components of interest (heat of formation $\Delta_f H^\ominus$); (2) convert to the actual phase ($\Delta_{trs} H$), (3) heat from 298.15 K to the actual temperature T, then (4) produce the actual mixture (heat of mixing), and finally (5) compress the mixture from 1 bar to the actual pressure p. We then have that the "absolute" enthalpy is

$$H = \Delta_1 H_1 + \Delta_2 H + \Delta_3 H + \Delta_4 H + \Delta_5 H$$

In more detail, the individual terms are (on a molar basis):

(1) chemical energy of formation (form the actual components in their standard state at 1 bar and 298 K from the elements in their standard state at 1 bar and 298 K)

$$\Delta_1 H = \Delta_f H^\ominus(298) = \sum_i n_i \Delta_f H_m^\ominus(i, 298)$$

where n_i [mol] is the amount of component i and $\Delta_f H_m^\ominus(i, 298)$ [J/mol] (f for *formation*) is the enthalpy for formation of component i in its standard state (phase).

\+ **(2) energy for phase change** ("latent heat") (if the actual mixture has a different phase from the "natural" of one or more components)

$$\Delta_2 H = \Delta_{trs} H^\ominus = \sum_{\text{component with different phase}} n_i \Delta_{trs} H^\ominus(i)$$

where $\Delta_{trs} H_m^\ominus(i)$ [J/mol] (trs for *transition*) is the energy for phase change for component i from its standard phase to the mixture phase at 298 K. (This is, for example, a positive number if the component is in gas phase while the standard state is a solid).

+ **(3) thermal energy** ("sensitive heat"). This is for heating the components at 1 bar from 298 K to T:

$$\Delta_3 H = H^{\ominus}(T) - H^{\ominus}(298) = \sum_i n_i \int_{298}^{T} C_{p,m}(i,T)dT$$

+ **(4) heat of mixing** (sometimes called the heat of solution) for forming the actual mixture from its pure components at temperature T,

$$\Delta_4 H = \Delta_{\text{mix}} H = n\Delta_{\text{mix}} H_m$$

The heat of mixing is 0 for ideal liquid mixtures and for ideal gases. For real gases, the heat of mixture is usually included in the term $\Delta_5 H$ for pressure correction and is computed using an equation of state.

+ **(5) pressure correction** (going from $p^{\ominus} = 1$ bar to p at temperature T). From (A.55), we have

$$\Delta_5 H = \Delta_p H = n \int_{1\text{ bar}}^{p} \left[V_m - T\left(\frac{\partial V_m}{\partial T}\right)_p \right] dp$$

where V_m [m^3/mol] is the molar volume of the actual mixture. The term $\Delta_p H$ is 0 for ideal gases (where $V_m = RT/p$), and is close to 0 for liquids and solids since they have a very small molar volume. For real gases, the pressure correction H_5 may be important, and can, as just mentioned, be obtained from an equation of state.

Comments.

1. This is one way of going from elements to the actual mixture, but since enthalpy is a state function, that is, only dependent on initial state (here: the elements at 298 K and 1 bar) and the final state (here: the final mixture at T and p), we get the same value for H if we go other ways (for example, first heating the elements and then forming the components etc.).
2. We can always choose the elements as the reference state because all streams can be formed from this state. Another common reference state is the combustion products, CO_2, H_2O, SO_2 etc. (which gives the standard heat of combustion, $\Delta_c H^{\ominus}$).
3. Use of "absolute" enthalpies in the energy balance is generally recommended, and is discussed on page 105.

To summarize, the absolute total enthalpy H for a stream with *the elements at 1 bar and 298 K as the reference* is the sum of the enthalpy changes in all these subprocesses, that is,

$$H = \Delta_f H^{\ominus} + \underbrace{\Delta_{\text{trs}} H^{\ominus} + \underbrace{(H^{\ominus}(T) - H^{\ominus}(298)) + \Delta_{\text{mix}} H + \Delta_p H}_{H^2}}_{H^1} \quad [J] \tag{A.61}$$

In practical calculations, the use of an equation of state takes care of the three terms related to phase change, heat of mixing and pressure correction, whereas the thermal energy term $H^{\ominus}(T)$ is calculated from C_p' data for ideal gases.

A.15.2 Other reference states for enthalpy

For hand calculations, we often evaluate the enthalpy differences directly ("method 2," see page 106) and it is in fact unnecessary to define reference state. For numerical calculations using a computer, it is recommended that you always use the elements at 298 K and 1 bar as a reference because you are then sure to avoid errors. Nevertheless, other reference states are sometimes used for practical calculations.

No reactions. With no chemical reactions, the terms related to heat of formation ($\Delta_f H^\ominus$) will drop out when we calculate enthalpy changes. There is then little point in "carrying" these terms, in part because their numerical value is often large. We then, instead, specify the absolute enthalpy as $H = H^1$, see equation (A.61), that is, with the *components* in their standard state as the reference.

No phase change. If, in addition, all of the streams have the same phase, then the terms for the latent heat ($\Delta_{trs} H^\ominus$) will drop out of the energy balance. There is then little point in "carrying" this term, and we may specify the absolute enthalpy as $H = H^2$, that is, with *the components at 1 bar and 298 K in the phase that the streams have as reference.*

Other cases. As you may already have understood, there is an unlimited number of possible choices for the reference state. For example, we can choose a different reference temperature than 298 K, and we can even choose different reference temperatures for different components (for example, this may be practical for calculating distillation columns, where the enthalpy is often set to zero for a component as liquid at its boiling point at the column pressure). Nevertheless, it always necessary to have a *common reference state* which all mixtures and streams in the process can be formed from.

A.15.3 Internal energy

Above we showed how to calculate the enthalpy H. Given the value for H, we find the internal energy from

$$U = H - pV$$

where V is the system volume. Note that we usually choose $H = 0$ in the reference state (for example, the elements at 298.15 K and 1 bar), and since $pV > 0$, we will then have $U < 0$ (not equal to zero) in the reference state.

A.15.4 Gibbs energy

The Gibbs energy G is defined as

$$G = H - TS \tag{A.62}$$

and can therefore be calculated once we know the enthalpy H and the entropy S (the entropy S is defined further in Appendix B). H and S are both state functions, and thereby G is also a state function. If we choose $H = 0$ and $S = 0$ in the reference state (for example, the elements at 298.15 K and 1 bar), we also get $G = 0$ in the reference state.

Example A.21 *In the reference book* SI Chemical Data, *we find the following values:*

	$\Delta_f H^\ominus(298)$ [kJ/mol]	$\Delta_f G^\ominus(298)$ [kJ/mol]	$S^\ominus(298)$ [J/mol K]
$H_2(g)$	0	0	131
$O_2(g)$	0	0	205
$H_2O(g)$	−242	−229	189

Note that data for both $\Delta_f G^\ominus(298)$ and $S^\ominus(298)$ are included, which we will show is not really necessary. $S^\ominus(i, 298)$ [J/mol K] is the "absolute" entropy of the component with a perfect crystal at 0 K as the reference. Alternatively, we can from the given absolute entropies, calculate the entropy of formation, with the elements as ideal gas at 298.15 K as reference.

For example, for H_2O *the formation reaction is* $H_2 + \frac{1}{2}O_2 = H_2O$ *and we get:*

$$\Delta_f S^{\ominus}(298) = \sum_i \nu_i S^{\ominus}(i, 298) = (-1) \cdot 131 + (-0.5) \cdot 205 + 189 = -44.5 \text{ J/mol K}$$

From the given value for $\Delta_f H^{\ominus}(298)$*, we can then calculate* $\Delta_f G^{\ominus}(298)$*:*

$$\begin{aligned}\Delta_f G^{\ominus}(298) &= \Delta_f H^{\ominus}(298) - \Delta_f S^{\ominus}(298) \cdot 298.15\ [K] \\ &= (-242 \cdot 10^3 + 44.5 \cdot 298.15) \text{ J/mol} = -228.7 \cdot 10^3 \text{ J/mol}\end{aligned}$$

which is consistent with the given value of -229 *J/mol. This shows that the listing of* $\Delta_f G^{\ominus}$ *in* SI Chemical Data *is not really necessary, as it can be computed from the given values for* $\Delta_f H^{\ominus}$ *and* $S^{\ominus}$.

APPENDIX B

More thermodynamics: Entropy and equilibrium

In order to proceed to the second law of thermodynamics, we need the state function entropy. This is a measure of the disorder in a given state, or more precisely for the probability of the state. The second law of thermodynamics states that the total entropy increases for all natural processes, and reaches its maximum at the equilibrium state. From this, we can derive quantitative conditions for equilibrium. We can derive how much work it is theoretically possible to extract from heat. The reader should go quickly through this appendix to gain an overview, and then return to the individual topics as needed later.

B.1 Entropy and the second law of thermodynamics

B.1.1 Introduction

We know the following from experience:

- Heat is transferred from high temperature to low temperature (but not *vice versa*);
- A bouncing ball will eventually come to rest (but a ball at rest will not suddenly start bouncing around);
- If we mix salt and sugar in a bowl, we will eventually have a homogeneous mixture (it is highly unlikely that by stirring we will end up with an ordered state where pure sugar and pure salt end up in separate parts of the bowl);
- Methane reacts with oxygen in the air to form CO_2 and water (but water and CO_2 in the air will not suddenly combine to form methane).

All of these processes (and all other natural or spontaneous processes) are *irreversible*. If we show a film, we can use our experience about natural processes to tell whether the film is run forwards or backwards.

Is there a common quantitative measure that explains the direction of all these processes? Yes, we can look at the *entropy* S. Loosely speaking, the entropy is a measure of the *disorder* which, according to the second law of thermodynamics, must increase for all natural processes. The reason is that the probability of an ordered state is very low. Another statement of the second law is then that all natural processes move towards a more probable state, that is, the entropy increases.

$$\text{Increased entropy } S_{\text{total}} \quad \Leftrightarrow \quad \text{More probable}$$

Note that it is the total entropy in the system plus the surroundings, $S_{\text{total}} = S + S_{\text{sur}}$, that must increase according to the second law of thermodynamics.

The system's entropy S is a state variable, so let us first recall what a state variable is.

> *A* **state variable** *is a variable (function; property) that only depends on the system's state (typically, specified by pressure, temperature and composition). The value of a state function is independent of how the system arrived at its state.*

Note that we here by the term "state" mean the macroscopic state, as observed from macroscopic properties, such as pressure, temperature and volume.

B.2 Definition of entropy

There are two views that may be used for defining entropy; the microscopic and macroscopic (thermodynamic) views. We start with the microscopic view, which is the most intuitive and emphasizes the fundamental character of entropy. It may be used to accurately compute the entropy of simple systems using statistical mechanics. However, for our purposes, where we deal with macroscopic quantities such as temperature, pressure and volume, the macroscopic view is more useful. Fortunately, the two views give the same value for the entropy.

B.2.1 Definition 1 of entropy: Microscopic

The probability of a given (macro)state is determined by the number of microstates corresponding to this macrostate. Quantitatively, the entropy of a given (macro)state is defined as

$$S = k \ln \Omega \tag{B.1}$$

where

$k = 1.380658 \cdot 10^{-23}$ *J/K* $= R/N_A =$ *Boltzmann's constant (gas constant per molecule)*

$\Omega =$ *number of microstates corresponding to a given (macro)state (including both position and energy of the molecules).*

Example B.1 *Consider a box with 4 molecules: two A-molecules (A_1 and A_2) and two B-molecules (B_1 and B_2). What is the likelihood of having two molecules of the same kind at one side of the box? To compute this, we consider the order of the molecules from the left to the right inside the box. Each of the 4 molecules may be first, and for each of these 3 molecules may be second, and for each of these 2 molecules may be third, giving a total of $4 \cdot 3 \cdot 2 = 24$ microstates. These are*

1.$A_1A_2B_1B_2$ 2.$A_1A_2B_2B_1$ 3.$A_1B_1A_2B_2$ 4.$A_1B_1B_2A_2$ 5.$A_1B_2A_2B_1$ 6.$A_1B_2B_1A_2$

7.$A_2A_1B_1B_2$ 8.$A_2A_1B_2B_1$ 9.$A_2B_1A_1B_2$ 10.$A_2B_1B_2A_1$ 11.$A_2B_2A_1B_1$ 12.$A_2B_2B_1A_1$

13.$B_1A_1A_2B_2$ 14.$B_1A_1B_2A_2$ 15.$B_1A_2A_1B_2$ 16.$B_1A_2B_2A_1$ 17.$B_1B_2A_1A_2$ 18.$B_1B_2A_2A_1$

19.$B_2A_1A_2B_1$ 20.$B_2A_1B_1A_2$ 21.$B_2A_2A_1B_1$ 22.$B_2A_2B_1A_1$ 23.$B_2B_1A_1A_2$ 24.$B_2B_1A_2A_1$

There are two macrostates:

Macrostate 1. *Ordered state with molecules of the same kind on each side of the box. 8 microstates (no.s 1, 2, 7, 8, 17, 18, 23, 24) correspond to this macrostate, so its entropy is $S_1 = k \ln \Omega = k \ln 8 = 2.08k$.*

Macrostate 2. *Unordered state with molecules of different kinds on each side of the box. 16 microstates correspond to this macrostate, so its entropy is* $S_2 = k \ln \Omega = k \ln 16 = 2.77k$.

As expected, the "unordered" macrostate 2 is the most likely one and has the highest entropy. Now, this box had only 4 molecules, so the difference in entropy was quite small, but as the number of molecules increases, the likelihood of an unordered state becomes increasingly larger.

B.2.2 Definition 2 of entropy: Macroscopic (thermodynamic)

The macroscopic definition of entropy may be stated as follows:

> *There exists a state function called* **entropy** S. *The change in* S *between two states can be found by considering an (idealized) reversible process between the two states, and we have*
>
> $$\Delta S = S_2 - S_1 = \int_1^2 \frac{dQ^{\text{rev}}}{T} \tag{B.2}$$
>
> *or for a small (differential) change*
>
> $$\boxed{dS = \frac{\delta Q^{\text{rev}}}{T}} \tag{B.3}$$
>
> *Here* Q^{rev} *is the heat supplied in the reversible process and* T *is the system's temperature.*

This definition is not very intuitive, so let us at least establish that is reasonable:

1. We already know from the microscopic definition that there is a state function S related to "the degree of disorder."
2. It seems reasonable that the disorder (S) increases when we supply heat (since heat is "disorganized energy transfer," see page 345).
3. It seems reasonable that we have to divide the supplied heat by the system temperature T, because the change in disorder is larger when T is small.
4. For a reversible process, work is *not* included in the calculation of the entropy change in (B.3). This seems reasonable since work is "organized energy transfer," and thereby does not change the degree of disorder.

Let us further try to establish that item 3 is reasonable with two illustrative examples.

Example B.2 Disorder in lecture room. *Consider a lecture room with students where everyone is sitting down ("the crowd has a low temperature"). The degree of disorder ("the entropy") will then rise sharply if one of the students gets up ("heat is added") and starts walking about in the room. Then consider a break where most of the students are moving around ("the crowd has a high temperature"). The degree of disorder is now so large that it is hardly noticeable if one more student gets up and starts walking about ("the change in entropy is small even though the same amount of "heat" is added").*

Example B.3 Disorder in a gas. *For an ideal gas, we know from kinetic gas theory that the temperature is directly related to the kinetic energy of the molecules. At low temperature, the gas molecules move slowly and the degree of disorder will increase sharply if the wall is heated such that it vibrates more and gives the gas molecules that collide with it a real "kick." On the other hand, the degree of disorder increases considerably less if we supply the same amount of heat to a gas mixture where the temperature is already high.*

In Section B.3 (page 373) on the Carnot process it is proved that entropy is a state function for an ideal gas.

Note that entropy is defined in (B.3) by considering an (idealized) reversible process, but it can also be used for (real) irreversible process because entropy is a state function: *The entropy change for an irreversible process can be found by considering the heat supply Q^{rev} in an idealized reversible process between the same two states.*

Note that we here use the term reversible. The closely related terms "lossless" and "frictionless" are often used to denote that the mechanical subprocesses are reversible.

B.2.3 The second law of thermodynamics

The second law of thermodynamics. *For any real process, the total entropy change for a system and its surroundings is positive. The total entropy change is zero for a reversible process. Mathematically:*

$$\boxed{\Delta S_{\text{total}} = \Delta S + \Delta S_{\text{sur}} \geq 0} \tag{B.4}$$

where ΔS is the change in the system's entropy and ΔS_{sur} is the change in the surrounding's entropy. $\Delta S_{\text{total}} = 0$ for a reversible process.

The second law is here formulated as a postulate, but all observations so far confirm that it is true. There exists many equivalent, and certainly more obvious, versions of the second law. One is the following:

> *Heat cannot spontaneously flow from a material at lower temperature to a material at higher temperature.*

Another version is *Lord Kelvin's postulate* (1851)

> *It is impossible to extract heat from a body with uniform temperature and convert it completely to work without at the same time changing the state of another body.*

(The word "completely" is, strictly speaking, redundant because if there are no changes in the state of any bodies then we must have, from the first law of thermodynamics, that all heat is converted to work). Kelvin's postulate was based on experiences from heat engines and seems reasonable.

From Kelvin's postulate, we can (i) derive that entropy is a state function, (ii) that changes are given by $dS = dQ^{\text{rev}}/T$, and (iii) that that total entropy is constant for a reversible process. These derivations, which go back to the work of Clausius in the period 1850–1865 (he also introduced the term entropy), are straightforward but still relatively involved (they cover pages 26–38 in the excellent book of Denbigh[1]), and this is why many, like us, prefer to start directly by postulating the entropy function.

Computing the entropy. The total entropy change (ΔS_{total}) is the sum of the entropy changes in the system (ΔS) and the surroundings (ΔS_{sur}), which may be obtained as follows:

- ΔS. The entropy change for the *system* is a state function and can be calculated from (B.2), $\Delta S = \int \frac{dQ^{\text{rev}}}{T}$ where T is the system's temperature and Q^{rev} is found by considering an (idealized) reversible process between the system's initial and final states. In Section B.4, we use a reversible closed system to derive general expression for how S depends on pressure, temperature and composition.
- ΔS_{sur}. In principle, the entropy change for the surroundings can also be calculated from (B.3), that is, $dS_{\text{sur}} = \delta Q^{\text{rev}}_{\text{sur}}/T_{\text{sur}}$. However, we must make sure that unrelated irreversibilities in other parts of the surroundings don't "help" achieving "impossible"

[1] K. Denbigh, *The principles of chemical equilibrium*, Cambridge Press, 4th Ed, 1981.

changes in our system. For example, if we try to separate a mixture of ethanol and water (the system), then it doesn't help that someone somewhere else in another part of the world (the surroundings) is mixing a drink, thereby making an irreversible mixture of ethanol and water. The entropy change for the surroundings is therefore calculated by assuming that all internal changes in the surroundings are *reversible*, that is, we therefore always assume that $dQ_{\mathrm{sur}}^{\mathrm{rev}} = -dQ$, where dQ is the actual heat supplied to the process from the surroundings. In summary, we have that

$$\Delta S_{\mathrm{sur}} = \int \frac{-dQ}{T_{\mathrm{sur}}} \tag{B.5}$$

The second law of thermodynamics for a process can then be written

$$\Delta S_{\mathrm{total}} = \Delta S + \int \frac{-dQ}{T_{\mathrm{sur}}} \geq 0 \tag{B.6}$$

where $-Q$ is (the actual) heat transferred from the system to the surroundings, and T_{sur} is the surrounding's temperature. We may have several kinds of surroundings, for example a cold reservoir and a hot *reservoir*.[2] Often, we further assume that the heat transfer to the surroundings is reversible such that $T_{\mathrm{sur}} = T$, where T is the system's temperature, but this is not a requirement and depends on the particular process.

Note that from (B.6) that it is possible to have processes where *the system's entropy* decreases (that is, $\Delta S < 0$) provided heat is transferred to the surroundings (that is, Q is negative) such that the surrounding's entropy increases even more. On the other hand, for an adiabatic process ($Q = 0$), the system's entropy must always increase, that is, $\Delta S \geq 0$. Furthermore, if the adiabatic process is reversible, we must have $\Delta S = 0$, that is, the entropy is constant (isentropic process).

B.3 Carnot cycle for ideal gas

We here consider the hypothetical[3] **Carnot cycle** for an ideal gas that operates between a heat reservoir at temperature T_H and a cold reservoir at temperature T_C (see Figure 8.2, page 199). It is a closed system which, in each cycle, passes through four steps of isothermal and adiabatic processes, as illustrated in Figure B.1. We assume that all four processes in the system are reversible. We will now use the first law of thermodynamics (= energy balance) for a closed system, which applies to each step,

$$\Delta U = Q + W$$

together with our knowledge about an ideal gas (including that U is a function of temperature only, so $\Delta U = 0$ for an isothermal process), to prove that

1. Entropy is a state function for an ideal gas.
2. The second law of thermodynamics is correct for an ideal gas.
3. There exists a cyclic process that can achieve the Carnot factor $(1 - T_C/T_H)$ for conversion of heat to work.

For the two **isothermal steps**, we have $\Delta U = 0$, so from the first law

$$W = -Q$$

[2] A reservoir is a body with infinite heat capacity or a pure fluid that condenses/boils at a constant temperature.

[3] The Carnot cycle is a heat engine which can be realized in practice, but it is hypothetical in the sense that it is not used in practice (for economic reasons).

Here, the reversible expansion (pV) work for an ideal gas (see (A.29)) is

$$W = RT \ln \frac{p_f}{p_0}$$

For the two **adiabatic steps** we have $Q = 0$, and for an ideal gas where for simplicity we assume constant heat capacity, the work is then

$$W = \Delta U = C_V \Delta T$$

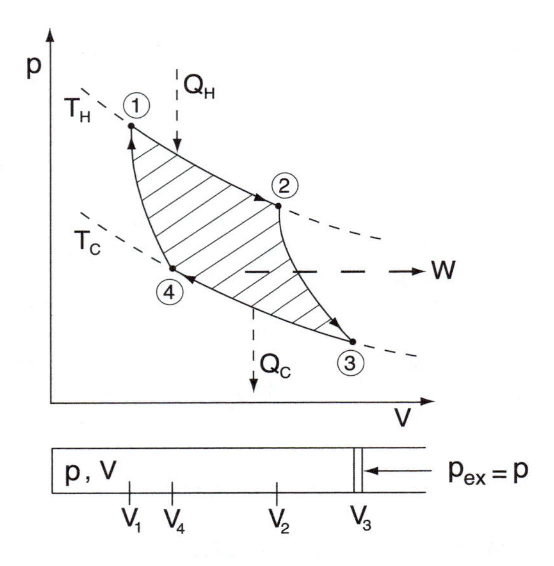

Figure B.1: Carnot cycle

In summary, the work for the four steps of the **Carnot cycle** for an ideal gas is:

1 → 2: Isothermal reversible expansion at T_H,

$$W_{12} = RT_H \ln \frac{p_2}{p_1} = -|Q_H| \quad \text{(Work is performed)}$$

2 → 3: Adiabatic reversible expansion from T_H to T_C,

$$W_{23} = C_V(T_C - T_H) \quad \text{(Work is performed)}$$

3 → 4: Isothermal reversible compression at T_H,

$$W_{34} = RT_C \ln \frac{p_4}{p_3} = |Q_C| \quad \text{(Work is supplied)}$$

4 → 1: Adiabatic reversible compression from T_C to T_H,

$$W_{41} = C_V(T_H - T_C) \quad \text{(Work is supplied)}$$

For a full cycle, the net supplied work in the four steps is

$$W = W_{12} + W_{23} + W_{34} + W_{41}$$

Here, the works in the two adiabatic steps cancel since $W_{23} = -W_{41}$. We further have for an ideal gas, from (A.40), that the following applies for the two adiabatic reversible steps

$$\frac{p_1}{p_2} = \frac{p_4}{p_3}$$

The total performed work is then

$$(-W) = |W| = RT_H \ln \frac{p_1}{p_2} + RT_C \ln \frac{p_3}{p_4} = R(T_H - T_C) \ln \frac{p_1}{p_2}$$

(we use the absolute sign to show clearly that $(-W)$ is positive, which means that net work is performed). Furthermore, $|Q_H| = RT_H \ln \frac{p_1}{p_2}$ and we derive

$$\frac{|W|}{|Q_H|} = \frac{T_H - T_C}{T_C} = 1 - \frac{T_C}{T_H}$$

which is known as the **Carnot factor** for conversion of heat Q_H to work $|W|$ in a reversible cycle operating with a hot and cold reservoir. As shown in Chapter 8.3 this result can also be derived using the second law of thermodynamics and the definition of entropy (see (8.4)), but here we derived it for an ideal gas without using the second law nor the term entropy.

Let us also calculate the entropy changes from the definition in (B.3). We find for the four steps

$$\Delta S_{12} = \frac{|Q_H|}{T_H}, \ \Delta S_{23} = 0, \ \Delta S_{34} = -\frac{|Q_C|}{T_C}, \ \Delta S_{41} = 0$$

and we get

$$\Delta S = \frac{|Q_H|}{T_H} - \frac{|Q_C|}{T_C} = 0$$

This means that the entropy change is zero in a cyclic process and, consequently, we have shown that entropy indeed is a state function for an ideal gas.

Comment. The net performed work for a full cycle is given by the shaded area in the pV-diagram in Figure B.1. This is interesting to know, but this fact was not used in the above derivation. Note that the fact also holds for a cyclic process involving a real gas.

Example B.4 Check that entropy is a state function for ideal gas. *Let us consider a closed system with 1 mol ideal gas and assume constant heat capacity $C_p = 29.1$ J/mol K (which for an ideal gas gives $C_V = C_p - R = 20.8$ J/mol K and $\gamma = C_p/C_V = 1.4$). In the initial state (1), we have $T_1 = 300$ K and $p_1 = 1$ bar. In the final state (2), we have $p_2 = 5$ bar and the temperature $T_2 = 475K$ is achieved by a reversible adiabatic compression (see process A below).*

We consider three different reversible processes that take the system from states 1 and 2:

Process A. *Reversible adiabatic compression from the initial state (T_1, p_1) to the final state (T_2, p_2). From (A.40), we have that (note that we are not using the term entropy):*

$$T_2 = T_1 \left(\frac{p_2}{p_1}\right)^{\frac{\gamma-1}{\gamma}} \tag{B.7}$$

Inserting numerical values gives $T_2 = 475$ K.

Process B. *This process consists of two reversible subprocesses:*

(a) *Reversible isothermal compression from p_1 to p_2 at T_1, followed by*

(b) *Heating from T_1 to T_2 at constant pressure p_2.*

Process C. *This process resembles process B, but the order of the subprocesses is switched:*

(a) *Heating from T_1 to T_2 at constant pressure p_1, followed by*

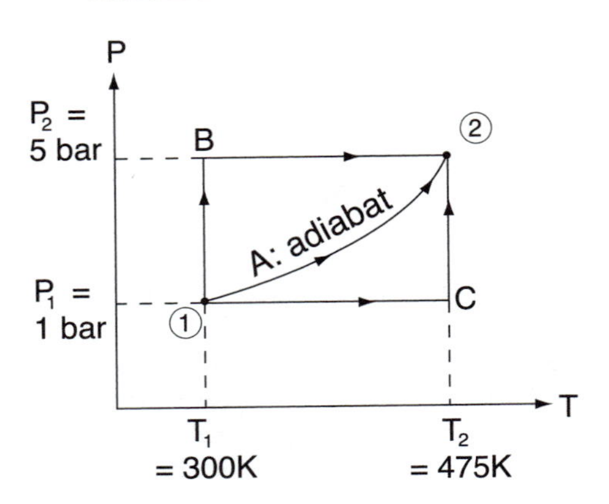

Figure B.2: Internal energy is a state function

(b) *Reversible isothermal compression from p_1 to p_2 at temperature T_2.*

Task:

1. *Calculate the net supplied work and heat for the three processes.*
2. *Calculate the entropy changes and show that entropy is a state function.*

Solution. *1. First, we calculate supplied work and heat for the three processes. From the first law of thermodynamics for a closed system, we have*

$$\Delta U = Q + W$$

where ΔU is the same for the three processes since internal energy is a state function. For an ideal gas, internal energy is only a function of the temperature, and we find

$$\Delta U = U_2 - U_1 = C_V(T_2 - T_1) = 20.8 \cdot (475 - 300) = 3640 \quad [J/mol]$$

In words, the sum of the supplied energy Q and work W must equal $\Delta U = 3640$ [J/mol] for all three processes.

Process A *is adiabatic and we have $Q_A = 0$. The supplied work is then, from the first law, given by*

$$W_A = \Delta U = 3640 \quad [J/mol]$$

For **process B**, *there is first (a) an isothermal expansion at T_1, where for ideal gas $\Delta U = 0$, and we get $Q_a^{\mathrm{rev}} = -W_a^{\mathrm{rev}} = RT_1 \ln(p_2/p_1)$. This is followed (b) by an isobaric heating, where we get $Q_b = C_p(T_2 - T_1)$. The total supplied heat is*

$$Q_B^{\mathrm{rev}} = \underbrace{-RT_1 \ln \frac{p_2}{p_1}}_{Q_{Ba}} + \underbrace{C_p(T_2 - T_1)}_{Q_{Bb}} = -4012 + 5093 = 1081 \quad [J/mol]$$

The work that must be supplied is then from the first law of thermodynamics

$$W_B = \Delta U - Q_B = 3640 - 1081 = 2559 \quad [J/mol]$$

We need to supply less work than for process A because the compression occurs at a lower temperature.

Similarly, we have for **process C**

$$Q_C = C_p(T_2 - T_1) - RT_2 \ln \frac{p_2}{p_1} = 5093 - 6353 = -1260 \quad [J/mol]$$

The compression occurs at a high temperature, which is unfavorable, so the supplied work is larger:

$$W_C = \Delta U - Q_C = 3640 + 1260 = 4900 \quad [J/mol]$$

The results are summarized as follows

	Q^{rev}	W^{rev}	ΔU	ΔS
Process A	0	3640	3640	0
Process B	1081	2559	3640	0
Process C	−1260	4900	3640	0

2. Let us finally show that ΔS is the same (in fact, it is zero) for all three processes, as it should be if it is a state function. The three processes are all reversible so we first note that Q^{rev} is <u>*not*</u> *a state function. Let us now compute $\Delta S = \int dQ^{\text{rev}}/T$ (see page 371) for our three reversible processes.*

Process A *is adiabatic with $Q^{rmrev} = 0$ so we have $\Delta S_A = 0$.*

Process B *is an isothermal expansion from p_1 to p_2 (with $dQ^{\text{rev}} = -dW^{\text{rev}} = pdV = -RTdp/p$) followed by an isobaric heating from T_1 to T_2 (with $dQ^{\text{rev}} = C_p dT$). Using the definition of entropy, $\Delta S = \int dQ^{\text{rev}}/T$, and integrating then gives (for details see Section B.4):*

$$\Delta S_B = -R \ln \frac{p_2}{p_1} + C_p \ln \frac{T_2}{T_1} \tag{B.8}$$

By inserting T_2/T_1 from (B.7) and using $\frac{\gamma-1}{\gamma} = \frac{R}{C_p}$ we derive that

$$\Delta S_B = 0$$

Finally, for **process C**, *we have*

$$\Delta S_B = C_p \ln \frac{T_2}{T_1} - R \ln \frac{p_2}{p_1} = 0$$

Consequently, we find for this example that $\Delta S_A = \Delta S_B = \Delta S_C$, which confirms that entropy is a state function. Our derivation was for an ideal gas, but this fact holds for any material.

B.4 Calculation of the system's entropy

Here, we derive expressions for the system's entropy change ΔS for three important processes:

1. Phase transition
2. Temperature change
3. Pressure change

The derivations are for reversible processes in a closed system, but since entropy is a state function, the result applies to any process between the same states, including an irreversible continuous process. Before we continue let us first briefly review what we know about a reversible closed system.

Reversible closed system. The energy balance for a closed system is $U_2 - U_1 = Q + W$ where, for a reversible closed process, $W = -\int_1^2 pdV$. For a small (differential) change, the energy balance is then $dU = \delta Q^{\text{rev}} - pdV$ or

$$\delta Q^{\text{rev}} = dU + pdV \tag{B.9}$$

The enthalpy is $H = U + pV$, so if pressure is constant we have

$$\text{Constant pressure}: \quad \delta Q^{\text{rev}} = dH \tag{B.10}$$

B.4.1 Entropy change for phase transition

When a liquid evaporates, the degree of disorder increases so we expect that the entropy increases. The same applies similarly for a melting substance. Here, we derive the entropy change for phase transitions (see page 358) at *constant pressure*. From (B.10) or (A.35) we then have $Q^{\rm rev} = \Delta_{\rm trs}H$ [J/mol], and from (B.3) the entropy change for the phase transition is

$$\boxed{\Delta_{\rm trs}S = \frac{\Delta_{\rm trs}H}{T_{\rm trs}}} \tag{B.11}$$

This applies to pure substances where $T_{\rm trs}$ is constant during the phase transition. As expected, the entropy increases for phase transitions where heat must be supplied ($\Delta_{\rm trs}H > 0$), such as melting (fusion) and evaporation. Note that there is equilibrium between the two phases during the phase transition and the temperature $T_{\rm trs}$ is determined by the given pressure. For example, the temperature is 0^oC when ice melts to water at 1 atm, and the temperature is 100^oC when water evaporates at 1 atm.

As a rule of thumb, the entropy of vaporization (evaporation) at the normal boiling point (1 atm) is typically

$$\textbf{Trouton's rule}: \quad \Delta_{\rm vap}S \approx 85 \text{ J/K mol}$$

This is because the increase in disorder during the transfer from liquid to gas at 1 atm is similar for all substances.

Example B.5 *For n-butane, the normal boiling point (1 atm) is $T_b = T_{\rm vap} = 272.7$ K (we usually use the symbol T_b for the boiling point), and the heat of vaporization at this temperature is $\Delta_{\rm vap}H = 22.41$ kJ/mol. The entropy of vaporization is then*

$$\Delta_{\rm vap}S = \frac{\Delta_{\rm vap}H}{T_b} = \frac{22.41 \cdot 10^3}{272.7} = 82.17 \ J/mol\ K$$

which is reasonably consistent with Trouton's rule.

The largest deviations from Trouton's rule are expected for substances with abnormal structural organizations, especially in the liquid phase. This is the case for water, where strong hydrogen bonding results in "more order" in the liquid phase than normal, and as a result the increase in disorder during evaporation is larger than normal. This is confirmed by computing the entropy of vaporization for water at its normal boiling point, which is $\Delta_{\rm vap}S = \Delta_{\rm vap}H/T_b = (40.68 \cdot 10^3 \text{ J/mol})/(373.15 \text{ K}) = 109.0$ J/mol K.

Example B.6 Entropy of melting (fusion). *For water, the heat of melting for the phase transition between ice (s) and water (l) is $\Delta H_{\rm fus} = 6.01$ kJ/mol. The entropy change when ice is melting is then $\Delta S_{\rm fus} = 6.01$ kJ/mol / 273.15 K = 22.00 J/mol K, which is as expexted positive since the liquid phase is less ordered. However, in winter water freezes spontaneously (the opposite of melting) when the temperature drops below 0 ^{o}C, and the system's entropy* decreases, *$\Delta S = -22.00$ J/mol K. How can this be compatible with the second law of thermodynamics?*

The explanation is that the entropy of the surroundings (the air) increases even more. The air must necessarily be a little colder in order to make the water freeze, and if we assume that the air temperature is -1^oC, the entropy change for the air (the surroundings) according to (B.5) is $\Delta S_{\rm sur} = 6.01$ kJ/(mol ice) / 272.15 K = 22.08 J/(mol ice) K. The total entropy change is then $\Delta S_{\rm total} = \Delta S + \Delta S_{\rm sur} = (-22.00 + 22.08)$ J/mol K = +0.08 J/mol K which as expected is positive.

B.4.2 Entropy change for temperature change

When temperature increases (heating), the molecules move faster and we expect the entropy to increase. Let us consider heating at *constant pressure* without phase transition. For a small temperature change dT, the supplied heat amount (in a closed system) is then $\delta Q^{\text{rev}} = dH = C_p dT$ [J/mol] (see and (B.10) and (A.36)) and from (B.3), the entropy change is

$$dS = C_p \frac{dT}{T} \quad \text{(constant } p\text{)}$$

which integrated gives

$$\boxed{S(T_2) - S(T_1) = \int_{T_1}^{T_2} C_p(T) \frac{dT}{T}} \quad \text{(constant } p\text{)} \tag{B.12}$$

This is indeed positive since $C_p > 0$. For the simplified case with *constant heat capacity* (independent of the temperature), we get

$$\Delta S = S(T_2) - S(T_1) = C_p \ln \frac{T_2}{T_1} \quad \text{[J/molK]} \quad \text{(constant } p \text{ and } C_p\text{)} \tag{B.13}$$

B.4.3 Entropy change for pressure change for ideal gas

Let us consider an increase of pressure (compression) at constant temperature. The volume then becomes smaller such that the molecules have less space, and we expect that there is "more order" such that entropy decreases. Let us confirm this for **ideal gas** where internal energy is only a function of temperature, that is, $dU = 0$ when T is constant. The energy balance (B.9) for a closed system gives $\delta Q^{\text{rev}} = pdV$ and thus the entropy change is

$$dS = \frac{\delta Q^{\text{rev}}}{T} = \frac{pdV}{T}$$

For 1 mol of an ideal gas, we have $V = RT/p$ which at constant T gives $dV = -(RT/p^2)dp$ which inserted gives $dS = -Rdp/p$. Integrated, we get

$$\Delta S = S(p_2) - S(p_1) = -R \ln \frac{p_2}{p_1} \quad \text{[J/mol]} \quad \text{(ideal gas; constant } T\text{)} \tag{B.14}$$

In summary, from (B.13) and (B.14), the entropy change for 1 mol of ideal gas (with constant composition) when going from an arbitrary state 1 to state 2 is

$$S(T_2, p_2) - S(T_1, p_1) = \int_{T_1}^{T_2} C_p(T) \frac{dT}{T} - R \ln \frac{p_2}{p_1} \quad \text{[J/mol K]} \quad \text{ideal gas} \tag{B.15}$$

B.4.4 Isentropic expansion of ideal gas

Here, we use the above results to rederive the result (6.8) that we previously derived for an adiabatic expansion of an ideal gas. Assuming constant heat capacity, we get from (B.15)

$$\Delta S = S(T_2, p_2) - S(T_1, p_1) = C_p \ln \frac{T_2}{T_1} - R \ln \frac{p_2}{p_1} \tag{B.16}$$

For an *isentropic* process, we have $\Delta S = 0$ and we derive

$$C_p \ln \frac{T_2}{T_1} = R \ln \frac{p_2}{p_1}$$

This gives

$$\frac{T_2}{T_1} = \left(\frac{p_2}{p_1}\right)^{R/C_p} \tag{B.17}$$

which, since $R/C_p = \frac{\gamma-1}{\gamma}$, is identical to expression (6.8) that we have used many times. However, whereas the "original" derivation on page 350 assumed an adiabatic process, we have now shown that it holds for any isentropic process involving an ideal gas with constant C_P.

B.4.5 Standard entropy change for chemical reaction

Here, we consider the *standard* entropy change for a chemical reaction. This is not the actual entropy change for the reaction, because the important mixing entropy is not included. Consider a reaction $0 = \Sigma_i \nu_i A_i$, where as usual the stoichiometric coefficient ν_i is negative for a reactant and positive for a product. The standard entropy change of reaction is the difference between the entropy of the products as pure components in their standard state and the entropy of the reactants as pure components in their standard state, that is

$$\Delta_r S^\ominus(T) = \sum_{i=\text{components}} \nu_i S_m^\ominus(i) = \sum_{i=\text{products}} \nu_i S_m^\ominus(i) - \sum_{i=\text{reactants}} |\nu_i| S_m^\ominus(i) \quad [\text{J/mol K}] \tag{B.18}$$

This is similar to the definition of the standard heat (enthalpy) of reaction, see (A.58). However, unlike enthalpy, it is possible to assign an absolute value for the entropy $S_m^\ominus(i)$ of each component. This is because from the third law of thermodynamics there exists a perfectly ordered state (corresponding to only one microstate) where the entropy is zero. The "absolute" value of the entropy ($S_m^\ominus(i)$) relative to this perfect state is tabulated in many reference books. However, in this book we are not really interested in absolute entropies, but rather in changes. We may therefore alternatively use data for enthalpy and Gibbs energy of formation ($\Delta_f H^\ominus(i), \Delta_f G^\ominus(i)$) to obtain the standard enthalpy and Gibbs energy of reaction ($\Delta_r H^\ominus, \Delta_r H^\ominus$), and from this the standard entropy of reaction,

$$\Delta_r S^\ominus = \frac{\Delta_r H^\ominus - \Delta_r G^\ominus}{T} \tag{B.19}$$

Example B.7 *We want to find the standard entropy change at 298.15 K for the gas-phase reaction*

$$CO_2 + 3H_2 = CH_3OH + H_2O$$

The following data at 298.15 K are found in the reference book SI Chemical Data*:*

	$\Delta_f H^\ominus$ (kJ/mol)	$\Delta_f G^\ominus$ (kJ/mol)	$S^\ominus$ (kJ/mol K)
CO_2	−394	−394	214
H_2	0	0	131
CH_3OH	−210	−163	240
H_2O	−242	−229	189

Using (B.18) we find

$$\Delta_r S^\ominus = S^\ominus(CH_3OH) + S^\ominus(H_2O) - S^\ominus(CO_2) - 3S^\ominus(H_2) = -178\text{J/K mol}$$

Alternatively, we find from the above data that $\Delta_r S^\ominus = -49$ kJ/mol and $\Delta_r G^\ominus = 2$ kJ/mol and using (B.19) we get

$$\Delta_r S^\ominus = \frac{(-49-2)\cdot 10^3\text{J/mol K}}{298.15\text{K}} = -171\text{J/K mol}$$

The two methods should give the same result, but for this example the deviation between the two values, −178 kJ/mol K and −171 kJ/mol K, is large. The deviation is mainly because of round-off errors because SI Chemical Data *uses no decimals. If we use the more accurate component data given on page 416, we find* $\Delta_r H^\ominus = -48.97$ *kJ/mol and* $\Delta_r G^\ominus = 3.83$ *kJ/mol which gives* $\Delta_r S^\ominus = -177$ *J/mol. Thus, based on this example, it seems better to base the calculation of* $\Delta_r S^\ominus$ *on data for the absolute entropy* $S^\ominus$*.*

B.5 Mixtures (variable composition)

Consider a process where, from the pure components at T and p, we produce a mixture at the same T and p. The mole fraction of each component after mixing is x_i. During such a mixing process, there is generally a change in the properties. In order to quantify them, let us introduce the following notation (here for volume V, but the same notation applies to any extensive variable, such as H, S and G):

- V_i^*: molar volume of pure component i [m^3/mol]
- $\bar{V}_i$ (or just V_i): partial molar volume = molar volume of component i in the mixture [m^3/mol]
- $V = \sum_i n_i \bar{V}_i$: total molar volume of the mixtures [m^3]
- $\Delta_{\text{mix}} V = V - \sum_i n_i V_i^* = \sum_i n_i(\bar{V}_i - V_i^*)$: mixing volume (volume change during mixing of 1 mol) [m^3]

Note that everything is at a given T and p. Similarly, the enthalpy and entropy of mixing are:

$$\Delta_{\text{mix}} H \triangleq H - \sum_i n_i H_i^* = \sum_i n_i(\bar{H}_i - H_i^*) \quad \text{[J]} \tag{B.20}$$

$$\Delta_{\text{mix}} S \triangleq S - \sum_i n_i S_i^* = \sum_i n_i(\bar{S}_i - S_i^*) \quad \text{[J/K]} \tag{B.21}$$

For an ideal mixture (gas or liquid), both the mixing volume and heat of mixing are zero,

$$\bar{V}_i = V_i^* \quad \Rightarrow \quad \Delta_{\text{mix}} V = 0 \tag{B.22}$$

$$\bar{H}_i = H_i^* \quad \Rightarrow \quad \Delta_{\text{mix}} H = 0 \tag{B.23}$$

That is, for an ideal mixture, the volume and enthalpy of a component in the mixture is the same as the volume and enthalpy of the pure component at the same pressure and temperature.[4]

But what about the entropy of mixing? Based on experience, mixing is clearly an irreversible process (no one has ever observed that air magically separates itself such that there is oxygen in one part of the room and nitrogen in the other). Thus, the entropy of mixing is clearly positive – even for an ideal mixture:

$$\Delta_{\text{mix}} S > 0 \tag{B.24}$$

We want to derive the mixing entropy for an ideal mixture by considering the case of **ideal gas**. In ideal gas each component (and molecule) in the mixture behaves as if no other components were present. A consequence of this is (**Gibbs' theorem**):

> For an ideal gas, the entropy of a component i in *the mixture* (at its partial pressure $p_i = x_i p$) equals the entropy of a *pure component* i at pressure p_i, that is, $\bar{S}_i(p) = S_i^*(p_i)$.

[4] For real mixtures, the mixing volume and heat of mixing are not zero. For example, we found on page 55 that $\Delta_{\text{mix}} V$ is negative when we mix water and ethanol, that is, the volume is reduced. Further, we found on page 113 that $\Delta_{\text{mix}} H$ is negative when we mix water in acid, that is, heat is released. For other non-ideal mixtures the values of $\Delta_{\text{mix}} V$ and $\Delta_{\text{mix}} H$ may be positive.

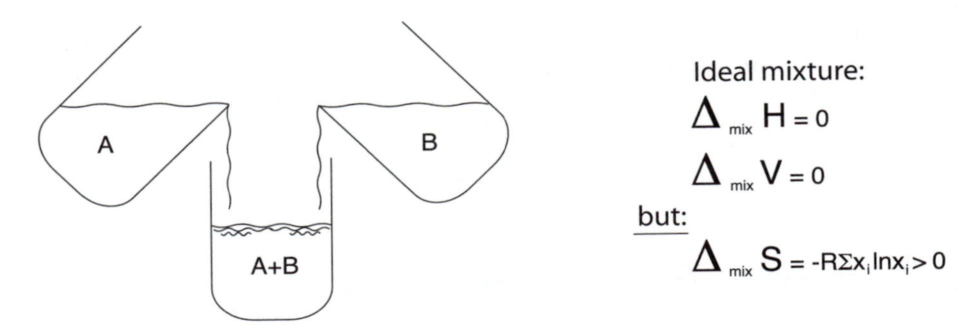

Figure B.3: Mixture is an irreversible process

B.5.1 Osmotic pressure

A consequence of Gibbs' theorem is that for an ideal (reversible) membrane that only lets component A through, the pressure p_A^* on the side with pure A is equal to the partial pressure of A in the mixture on the other side, that is,

$$p_A^* = x_A p$$

This implies that there will be a pressure difference over the membrane, known as the **osmotic pressure** Π, which for an ideal membrane and ideal gas is

$$\Pi = p - p_A^* = (1 - x_A)p$$

Thus, for a binary mixture, $\Pi = x_B p$, where B is the component that does not go through the membrane. Introducing the concentration $c_B = n_B/V = x_B n/V$ [mol/m^3] of molecules that do not pass through the membrane, we derive

$$\Pi = c_B pV/n = c_B RT \qquad [\text{N/m}^2] \tag{B.25}$$

This equation is here derived for a gas membrane. An identical equation was derived by van't Hoff [5] for a liquid phase membrane. The equation can be used to determine the molar mass of large molecules, by dissolving a known mass (in gram) of the molecule in the liquid, and measuring the resulting osmotic pressure difference over a membrane with the pure liquid on the other side.

Example B.8 *Let us calculate the osmotic pressure of seawater at 288K (15° C). We assume a salinity of 3.5% l or 35 kg/m^3, and assume that the only salt is NaCl with a molar mass of 58.4 kg/kmol. The molar concentration of NaCl is then* 35 *kg/m^3/*58.4 *kg/kmol* = *0.60 kmol/m^3* = 600 *mol/m^3. However, when NaCl dissolves in water it splits into Na^+ and Cl^-, so the concentration of dissolved molecules in seawater is twice as high, that is, $c_B = 2 \cdot 600$ mol/m^3. With a liquid phase membrane that only lets water pass, the osmotic pressure is then from van't Hoff's equation (B.25)*

$$\Pi = c_B RT = 2 \cdot 600 \cdot 8.31 \cdot 288 \text{ N/m}^2 = 2872000 \text{ N/m}^2 = 28.7 \text{ bar}$$

[5] Jacobus van't Hoff (1852-1911) received the first Nobel prize in chemistry in 1901 for his work on liquid solutions, including osmotic pressure.

B.5.2 Entropy of mixing for ideal mixture

From (B.14), we know the entropy's dependency on pressure for an ideal gas, and we then get from Gibbs' theorem that the entropy for 1 mol of component i in the mixture is:

$$\bar{S}_i(p) = S_i^*(p_i) = S_i^*(p) - R\ln\frac{p_i}{p} = S_i^*(p) - R\ln x_i \quad \text{[J/mol K]} \tag{B.26}$$

where $x_i = p_i/p$ is the mole fraction of component i in the mixture and $S_i^*(p)$ is the entropy of pure component i at pressure p. The entropy for an ideal gas mixture at T and p is then

$$S(T,p) = \sum_i n_i\bar{S}_i(T,p) = \sum_i n_i S_i^*(T,p) \underbrace{- \sum_i n_i R\ln x_i}_{\Delta_{\text{mix}}S} \quad \text{[J/K]} \tag{B.27}$$

We have derived that the mixing entropy for an ideal mixture is

$$\boxed{\Delta_{\text{mix}}S = -R\sum_i n_i\ln x_i = -nR\sum_i x_i\ln x_i \quad \text{[J/K]}} \quad \text{(constant } T \text{ and } p) \tag{B.28}$$

Since $\ln x_i$ is negative, (B.28) gives as expected that the entropy increases when we mix pure components. (B.28) applies similarly for an ideal liquid mixture (*ideal solution*)[6].

Example B.9 *For an equimolar two-component ideal mixture (with $x_1 = x_2 = 0.5$), we have that*

$$\Delta_{\text{mix}}S = -R\cdot 2\cdot 0.5\cdot\ln 0.5 = 0.69R \quad \text{[J/mol K]}$$

Similarly, for an equimolar five-component mixture (with $x_i = 0.2$),

$$\Delta_{\text{mix}}S = -R\cdot 5\cdot 0.2\cdot\ln 0.2 = 1.61R \quad \text{[J/mol K]}$$

and for an equimolar N_c-component mixture (with $x_i = 1/N_c$),

$$\Delta_{\text{mix}}S = N_c\left(-R\frac{1}{N_c}ln\frac{1}{N_c}\right) = R\ln N_c \quad \text{[J/mol K]}$$

Note that $\Delta_{\text{mix}}S$ in theory can become arbitrarily large if the number of components N_c is large, but because we take the logarithm the increase is in practice small. For example, an equimolar mixture with 1 million components has $\Delta_{\text{mix}}S = 13.8R$ [J/mol K], which is only a factor 20 larger than for a two-component mixture.

Comment. Equation (B.28) for the ideal mixing entropy is of fundamental importance, and is in fact the basis for all reaction and vapor/liquid equilibrium calculations.

B.6 Equilibrium

We mentioned that the total entropy (in the universe) always increases. More generally, we have that **the entropy in an isolated system always increases** (an isolated system is a closed system without exchange of heat and work with the surroundings, and the universe can be considered as a special case of an isolated system). But in an isolated system, it is limited how much the entropy can increase, and we will eventually as time goes to infinity reach an **equilibrium state**:

[6] Inspired by thermodynamics, communications technology also use the term entropy, defined similarly to (B.28) but with x_i replaced by probability. Information is the opposite of entropy, and during communication the information content always decreases, that is the entropy increases.

Left to itself, an isolated system will reach an equilibrium state where the entropy has a maximum.

In a system at equilibrium, we know from experience that temperature and pressure are uniform and that the composition is also uniform within each phase. Chemical equilibrium is also reached, although in practice some reactions may run so slowly that we can neglect them. These observations can be derived mathematically by seeking the state with maximum entropy.

Example B.10 Temperature equilibrium. *We consider an isolated system that consists of two bodies; one body (for example a hot bathtub) with temperature* T_1 *and another body (for example air) with temperature* T_2*, and we have initially* $T_1 > T_2$*. The total entropy is* $S\text{total} = S_1 + S_2$*. If we transfer an amount of heat* $|\delta Q|$ *from the bathtub to the air, the entropy in the bathtub will decrease by* $dS_1 = -|\delta Q|/T_1$*, while the entropy in the air will increase by* $dS_2 = |\delta Q|/T_2$*. As long as* $T_1 > T_2$*, we will have* $|dS_2| > |dS_1|$ *and the total entropy change* $dS\text{total} = dS_1 + dS_2$ *is positive. At equilibrium, entropy has reached its maximum and we have for any change that* $dS_{\text{total}} = 0$*. From this we derive the equilibrium condition* $T_1 = T_2$*, which is consistent with our observations.*

B.6.1 Equilibrium at constant temperature and pressure: Gibbs energy

We can always determine the equilibrium state by maximizing the total entropy S_{total} (which is not a state variable). However, we are often interested in finding equilibrium state at a given value of T and p, and it then turns out that the equilibrium condition is equivalent to minimizing the Gibbs[7] energy G of the system (which is a state variable!). Here, we derive this equilibrium condition for a closed system, but since G is a state function, the result is generally applicable. We start from the second law of thermodynamics, which states that for all real processes we must have that

$$\Delta S + \Delta S_{\text{sur}} \geq 0$$

where ΔS is a state variable. We assume that the interaction with the surroundings is reversible, which in particular implies that $T_{\text{sur}} = T$. Since temperature is assumed constant, we then have from (B.5)

$$\Delta S_{\text{sur}} = -\frac{Q}{T_{\text{sur}}} = -\frac{Q}{T}$$

Since pressure p is constant, we further have $Q = \Delta H$, which applies to a closed system with reversible pV work, see (A.35)). The second law of thermodynamics, then gives that all real processes at constant T and p must satisfy $\Delta S - \Delta H/T \geq 0$ or (by multiplying both sides by T)

$$\Delta H - T\Delta S \leq 0 \tag{B.29}$$

Now, the Gibbs energy of the system is defined as $G = H - TS$(which is a state function), and since T is constant, we see from (B.29) that for all natural processes we must have

$$\Delta G \leq 0 \quad \text{(constant } T \text{ and } p\text{)} \tag{B.30}$$

[7] It was the American physicist Josiah Willard Gibbs (1839–1903) at Yale University who, at the end of the 19th century, laid the basis for modern chemical thermodynamics. Even though he was awarded the first doctoral degree (PhD) in the US in engineering, Gibbs was a theoretician who worked alone and had little contact with practical engineering problems. Nonetheless, his work has later had a deep practical impact.

In other words, all natural processes will, at a given T and p, proceed in a direction that reduces G. Eventually, we reach an (internal) equilibrium state where *the system's* Gibbs energy G has reached its minimum, and we must have for any perturbation (change Δ) at the equilibrium state that

$$\Delta G = 0 \quad \text{(equilibrium at constant } T \text{ and } p) \tag{B.31}$$

or, more precisely, in differential form:

$$(dG)_{T,p} = 0 \tag{B.32}$$

B.6.2 Chemical equilibrium

We consider chemical equilibrium at given temperature and pressure. Assume that the reactants A and B react to the products C and D following the general reaction equation:

$$\boxed{0 = \sum_i \nu_i A_i = \nu_A A + \nu_B B + \nu_C C + \nu_D D} \tag{B.33}$$

where ν_i is the stoichiometric coefficient for component i, which is negative for reactants. We now apply the equilibrium condition (B.31) to the chemical reaction (B.33) and find

$$\boxed{\Delta_r G = 0} \quad \text{(constant } T \text{ and } p) \tag{B.34}$$

where $\Delta_r G$ [J/mol] is the change in Gibbs energy when 1 mol reacts according to the given reaction equation.

Derivation of (B.34): Let ξ [mol] be the extent of reaction. At a given T and p, the equilibrium condition (B.32) becomes $(dG)_{T,p} = \Delta_r G \cdot d\xi = 0$. Since this must hold for any value of $d\xi$, we get $\Delta_r G(T,p) = 0$. □

In (B.34) we have that

$$\Delta_r G = \sum_i \nu_i \bar{G}_i \quad \text{[J/mol]} \tag{B.35}$$

where $\bar{G}_i = \bar{G}_i(T,p)$ is the (partial) Gibbs energy for component i in the mixture. Here, $\bar{G}_i$ is also called the *chemical potential* and is given the symbol μ_i,

$$\mu_i \triangleq \bar{G}_i \triangleq \left(\frac{\partial G}{\partial n_i}\right)_{T,p,n_j} \tag{B.36}$$

We can further write $\mu_i = \bar{G}_i(T,p) = \bar{H}_i(T,p) - T\bar{S}_i(T,p)$. The change in Gibbs energy for the reaction is then (at given T and p)

$$\Delta_r G = \sum_i \nu_i \bar{G}_i = \underbrace{\sum_i \nu_i \bar{H}_i}_{\Delta_r H} - T \underbrace{\sum_i \nu_i \bar{S}_i}_{\Delta_r S} \quad \text{[J/mol]} \tag{B.37}$$

For an **ideal gas**, we have from (B.23) and (B.26) that

$$\bar{H}_i(T,p) = H_i^*(T,p^\ominus) \tag{B.38}$$

$$\bar{S}_i(T,p) = S_i^*(T,p^\ominus) - R\ln(p_i/p^\ominus) \tag{B.39}$$

Because we are considering chemical reactions, it is practical to use the elements as the reference state. We then have

$$
\begin{aligned}
H_i^*(T,p^\ominus) &= H_i^\ominus(T) = \text{standard enthalpy of formation at temperature } T \\
S_i^*(T,p^\ominus) &= S_i^\ominus(T) = \text{ standard entropy of formation at temperature } T
\end{aligned}
$$

We then get that the chemical potential for an ideal gas can be written as

$$\mu_i \triangleq \bar{G}_i(T,p) = G_i^\ominus(T) + RT\ln(p_i/p^\ominus) \tag{B.40}$$

where $G^\ominus{}_i(T) = H^\ominus{}_i(T) - TS^\ominus{}_i(T)$ is the standard Gibbs energy of formation for component i.

For real gases, liquids and solids, we can more generally in (B.40) introduce the **activity** a_i relative to a defined standard state at T and 1 bar. For gases, the standard state is ideal gas. We can then write the chemical potential as

$$\mu_i \triangleq \bar{G}_i(T,p) \triangleq G_i^\ominus(T) + RT\ln a_i \tag{B.41}$$

From (B.35) and (B.41), we then have for the reaction (B.33)

$$\Delta_r G = \sum_i \nu_i \bar{G}_i = \sum_i \left(\nu_i G_i^\ominus(T) + \nu_i RT\ln a_i\right)$$

$$= \Delta_r G^\ominus + RT\left(\nu_A \ln a_A + \nu_B \ln a_B + \nu_C \ln a_C + \nu_D \ln a_D\right)$$

For logarithms it holds that

$$n\ln a = \ln a^n \quad \text{and} \quad \ln a^n + \ln b^m = \ln(a^n b^m)$$

so this can be written

$$\Delta_r G = \Delta_r G^\ominus + RT\prod_i a_i^{\nu_i} = \Delta_r G^\ominus + RT\ln Q \tag{B.42}$$

where $\Delta_r G^\ominus$ is the standard change in Gibbs energy for the reaction at T and $p^\ominus = 1$ bar, and Q is defined by

$$Q \triangleq \prod_i a_i^{\nu_i} \tag{B.43}$$

At chemical equilibrium, we have from (B.34) $\Delta_r G(T,p) = 0$ and we derive the equilibrium condition:

$$\Delta_r G = \Delta_r G^\ominus + RT\ln Q = 0 \quad (\text{given } T \text{ and } p) \tag{B.44}$$

We now introduce the **equilibrium constant** K defined by

$$\boxed{\ln K(T) \triangleq -\frac{\Delta_r G^\ominus(T)}{RT}} \tag{B.45}$$

The equilibrium condition for the chemical reaction (B.33) is then $\ln Q = \ln K$ or $K = Q$ or

$$\boxed{K = \prod_i a_i^{\nu_i} = \frac{a_C^{\nu_C} a_D^{\nu_D}}{a_A^{-\nu_A} a_B^{-\nu_B}} \quad (\text{given } T, p)} \tag{B.46}$$

where we, in the final equality, have assumed that A and B are reactants, and C and D are products.

- a_i is here the **activity** relative to the component's standard state used when calculating the equilibrium constant K. Note that we, for each component, are free to choose the standard state. For example, for gas phase reactions, we usually choose the ideal gas as the standard state for all components (even though some of the components may be liquid at 298.15 K). The activity is generally a function of composition, T and p.
- For gas components, the standard state is ideal gas at T and $p^{\ominus} = 1$ bar.

  — For *ideal gas*, we then have, as shown in (B.40), that the activity of component i is

$$a_i = \frac{p_i}{p^{\ominus}} = y_i \frac{p}{p^{\ominus}} \tag{B.47}$$

  where p_i is the partial pressure, $p^{\ominus} = 1$ bar and y_i is the mole fraction in the mixture. That is, the activity is directly proportional to the partial pressure and is independent of temperature.

  — For **real gases**, we replace the partial pressure p_i by its **fugacity** f_i [bar] and write $a_i = f_i/p^{\ominus}$. Alternatively, we write $a_i = \phi_i p_i/p^{\ominus}$ where ϕ_i is the **fugacity coefficient**. For an ideal gas, $\phi_i = 1$.

- For liquid components, the standard state is usually pure liquid at T and 1 bar and we often write

$$a_i = \gamma_i x_i \tag{B.48}$$

  where γ_i is the activity coefficient and x_i is the mole fraction in the mixture. For a pure component we have $\gamma_i = 1$, and for an ideal liquid mixture we also have $\gamma_i = 1$ (strictly speaking, this holds only if we neglect the effect of pressure on the liquid phase activity, or alternatively, we must in (B.48) include a liquid pressure correction term, similar to the Poynting factor on page 188).
- The standard Gibbs energy for the reaction is usually obtained from

$$\Delta_r G^{\ominus}(T) = \sum_i \nu_i \Delta_f G^{\ominus}(i, T) \quad [J/mol]$$

  where $\Delta_f G^{\ominus}(i, T)$ [J/mol] is the Gibbs energy of formation for component i (relative to the elements in their standard state at 298.15 K and 1 bar), and ν_i is the stoichiometric coefficient for component i in the reaction. In reference books (see page 415), we find values for the Gibbs energy of formation specified at 298 K, so we can easily obtain the standard Gibbs energy of reaction at 298 K,

$$\Delta_r G^{\ominus}(298) = \sum_i \nu_i \Delta_f G^{\ominus}(i, 298)$$

  In order to obtain $\Delta_r G^{\ominus}(T)$, and thereby the equilibrium constant at other temperatures, we can use

$$\Delta_r H^{\ominus}(T) = \Delta_r H^{\ominus}(298) + \int_{298.15}^{T} \Delta_r C_p^{\ominus}(T) dT \tag{B.49}$$

$$\Delta_r S^{\ominus}(T) = \Delta_r S^{\ominus}(298) + \int_{298.15}^{T} \frac{\Delta_r C_p^{\ominus}(T)}{T} dT \tag{B.50}$$

$$\Delta_r G^{\ominus}(T) = \Delta_r H^{\ominus}(T) - T\Delta_r S^{\ominus}(T) \quad [J/mol] \tag{B.51}$$

  where $\Delta_r C_p^{\ominus}(T) = \sum_i \nu_i C_p^{\ominus}(i, T)$ is the change in heat capacity for the reaction.

A little more on the temperature dependency of the equilibrium constant

The equilibrium constant $K(T)$ is a function of the temperature, and in order to study this closer, we start with the defin/Althtion of the equilibrium constant:

$$\ln K(T) = -\frac{\Delta_r G^{\ominus}(T)}{RT} = -\frac{\Delta_r H^{\ominus}(T)}{RT} + \frac{\Delta_r S^{\ominus}(T)}{R} \tag{B.52}$$

If we differentiate (B.52), we get **van't Hoff's equation**[8]

$$\frac{d \ln K}{dT} = \frac{\Delta_r H^{\ominus}(T)}{RT^2} \tag{B.53}$$

or

$$\frac{d \ln K}{d(1/T)} = \frac{-\Delta_r H^{\ominus}(T)}{R} \tag{B.54}$$

that is, when we plot $\ln K$ as a function of $1/T$ the slope is $-\Delta_r H^{\ominus}(T)/R$. The plot is a straight line if $\Delta_r H^{\ominus}(T)$ is constant.

For practical calculations, four different assumptions are commonly used to compute the equilibrium constant's temperature dependency:

1. *Very rough assumption:* Assume that K is independent of T (this is only OK if the heat of reaction, $\Delta_r H^{\ominus}$, is close to 0).
2. *Less rough assumption:* Assume that $\Delta_r H^{\ominus}$ is independent of the temperature. Integration of (B.53) then gives the "integrated form" of van't Hoff's equation

$$\ln \frac{K(T)}{K(T_0)} = -\frac{\Delta_r H^{\ominus}}{R}\left(\frac{1}{T} - \frac{1}{T_0}\right) \tag{B.55}$$

3. *Minor assumption:* Assume that the heat capacities are constant, or more exactly that the change in heat capacity for the reaction, $\Delta_r C_p^{\ominus} = \sum_i \nu_i C_p^{\ominus}(i)$, is independent of temperature. With this assumption, (B.49) and (B.50) are simplified to

$$\Delta_r H^{\ominus}(T) = \Delta_r H^{\ominus}(T_0) + \Delta_r C_p^{\ominus} \cdot (T - T_0) \tag{B.56}$$

$$\Delta_r S^{\ominus}(T) = \Delta_r S^{\ominus}(T_0) + \Delta_r C_p^{\ominus} \cdot \ln \frac{T}{T_0} \tag{B.57}$$

and we obtain $\Delta_r G^{\ominus}(T) = \Delta_r H^{\ominus}(T) - T\Delta_r S^{\ominus}(T)$ and from this $K(T) = \exp\{-\Delta_r G^{\ominus}(T)/RT\}$.

4. *No assumption:* Use data for the heat capacity $C_p^{\ominus}(T)$ from a reference book[9]. For example, it may be in the form

$$cp_i(T) = A_i + B_i T + C_i T^2 + D_i T^3$$

Integration of equations (A.44) and (B.12) then gives the enthalpy and the entropy for each component at temperature T,

$$\Delta_f H_i^{\ominus}(T) = \Delta_f H_i^{\ominus}(T_0) + A_i(T - T_0) + \frac{B_i}{2}(T^2 - T_0^2) + \frac{C_i}{3}(T^3 - T_0^3) + \frac{D_i}{4}(T^4 - T_0^4) \tag{B.58}$$

$$\Delta_f S_i^{\ominus}(T) = \Delta_f S_i^{\ominus}(T_0) + A_i \ln \frac{T}{T_0} + B_i(T - T_0) + \frac{C_i}{2}(T^2 - T_0^2) + \frac{D_i}{3}(T^3 - T_0^3) \tag{B.59}$$

[8] The derivation of (B.53), where we differentiate (B.52), presupposes constant $\Delta_r H^{\ominus}$ and $\Delta_r S^{\ominus}$ (independent of temperature). However, a more complicated derivation, where we use the Gibbs-Helmholtz equation, see (B.70), gives that van't Hoff's equation (B.53) is exact, that is, it applies also when $\Delta_r H^{\ominus}(T)$ and $\Delta_r S^{\ominus}(T)$ are not constant.

[9] A good source for ideal gas-heat capacities is: Poling, Prausnitz and O'Connell, *The properties of gases and liquids*, 5th Ed., McGraw-Hill (2001)

From this, we can obtain for the reaction

$$\Delta_r H^\ominus(T) = \sum_i \nu_i \Delta_f H_i^\ominus(T)$$

$$\Delta_r S^\ominus(T) = \sum_i \nu_i \Delta_f S_i^\ominus(T)$$

and $\Delta_r G^\ominus(T) = \Delta_r H^\ominus(T) - T\Delta_r S^\ominus(T)$, and from this find the exact value of $K(T) = \exp\{-\Delta_r G^\ominus(T)/RT\}$. Alternatively, we get the same result by using $\Delta_r C_p^\ominus(T) = \Delta_r A + \Delta_r B \cdot T + \Delta_r C \cdot T^2 + \Delta_r D \cdot T^3$ and integrating equations (B.49) and (B.50).

For comparison of the above methods see the ammonia example on page 176.

Example B.11 *We want to find the equilibrium constant for the combustion reaction*

$$CO + \frac{1}{2}O_2 = CO_2; \quad K = \frac{(p_{CO_2}/p^\ominus)}{(p_{CO}/p^\ominus)(p_{O_2}/p^\ominus)^{0.5}}$$

where $p^\ominus = 1$ *bar and we assume ideal gas. From the ideal gas component data on page 416, we find* $\Delta_r G^\ominus(298) = -282.98\text{kJ/mol}, \Delta_r G^\ominus(298) = -257.19\text{kJ/mol}$ *and* $\Delta_r C_p^\ominus(298 = -6.71\text{J/K mol}$. *This gives* $\Delta_r S^\ominus(298) = (\Delta_r H^\ominus(298) - \Delta_r G^\ominus(298))/298.15K = -86.57$ *J/K mol. Using (B.56) and (B.57), we can then obtain the equlibrium constant* $K(T)$ *as a function of temperature,*

T [K]	500	1000	1500	2000	2500	3000
$K(T)$	$1.04E25$	$1.22E10$	$1.12E5$	317	9.0	0.81

```
% Computed using MATLAB:
  dh0=-282.98E3; ds0=-86.57; dcp0=-6.71; T0=298.15;
  T=500; dh=dh0+dcp0*(T-T0); ds=ds0+dcp0*log(T/T0); dg=dh-T*ds; K=exp(-dg/(8.31*T))
```

We note that the equilibrium is strongly shifted towards CO_2 *at low temperatures, which is expected since the reaction is exothermic. In practice, combustion of fuels at low temperature may give a high* CO *concentration (incomplete combustion), but this is because of kinetics; adding a catalyst (as used for cars) or adding a secondary combustion chamber with higher temperature (as used in modern wood stoves) may give complete combustion (*CO_2*) as expected from equilibrium thermodynamics.*

B.7 The fundamental equation of thermodynamics and total differentials

In this book, we have stayed away from most of the intricate thermodynamic relationships of partial derivatives, but let us end this Appendix with a few useful relationships.

The starting point for these seemingly mysterious relationships is that a thermodynamic system at internal equilibrium has $2 + N_c$ independent variables (see page 332), where N_c is the number of components. For example, we can specify T, p + the amounts of each of the N_c components. These specifications define the state in the system, and we can then calculate from them all other state variables (for example V, U, S, etc.). The main reason for the seemingly myriad of formulas is that there are many other choices, in addition to T and p, for specifying the two degrees of freedom. The "natural" (canonical, most suitable) variables depend on which thermodynamic function we consider.

Let us start with internal energy, where the "natural" canonical variables turn out to be S and V (+ the molar amounts), that is, we write

$$U = U(S, V, n_1, n_2, \dots n_c) \tag{B.60}$$

We are interested in changes in U and express the total differential

$$dU = \left(\frac{\partial U}{\partial S}\right)_{V,n_1,\dots,n_c} dS + \left(\frac{\partial U}{\partial V}\right)_{S,n_1,\dots,n_c} dV + \sum_{i=1}^{c}\left(\frac{\partial U}{\partial n_i}\right)_{S,V,n_{k(k\neq i)}} dn_i \tag{B.61}$$

The first law of thermodynamics for a closed system gives $dU = dQ + dW$, where for a reversible process $dQ = TdS$ and $dW = -pdV$. This is at constant composition, and by adding the contributions for changes in the number of moles, which can be expressed by the chemical potential μ_i, we get the so-called **fundamental equation** of thermodynamics:

$$dU = TdS - pdV + \sum_i \mu_i dn_i \tag{B.62}$$

This is here derived for a reversible process in a closed system, but since U is a state function, (B.62) applies for all systems at internal equilibrium. A comparison of (B.61) and (B.62) then gives the following exact relationships between temperature, pressure and chemical potential and the partial derivative of U:

$$\text{Temperature}: \quad T = \left(\frac{\partial U}{\partial S}\right)_{V,n_1,\dots,n_c} \tag{B.63}$$

$$\text{Pressure}: \quad p = -\left(\frac{\partial U}{\partial V}\right)_{S,n_1,\dots,n_c} \tag{B.64}$$

$$\text{Chemical potential}: \quad \mu_i = \left(\frac{\partial U}{\partial n_i}\right)_{S,V,n_{k(k\neq i)}} \tag{B.65}$$

These are actually the thermodynamic definitions of pressure and temperature. Fortunately, they are consistent with our "ordinary" definitions of pressure and temperature from everyday life.

In the expression (B.62) for internal energy, the independent variables are S and V, while we, for practical calculations, often want to specify p and T. Is there any state variable or combination that has p and T as its "natural" variables? Yes, this is the Gibbs energy defined by

$$G = U + pV - TS$$

In order to show this, we differentiate this equation and get $dG = dU + pdV + Vdp - TdS - SdT$ which, when combined with (B.62), gives

$$dG = -SdT + Vdp + \sum_i \mu_i dn_i \tag{B.66}$$

which is the total differential of G in the variables T, p and mole amounts. By differentiating the function $G(T, p, n_i)$, we find that

$$S = -\left(\frac{\partial G}{\partial T}\right)_{p,n_1,\dots,n_c} \;;\quad V = \left(\frac{\partial G}{\partial p}\right)_{T,n_1,\dots,n_c} \;;\quad \mu_i = \left(\frac{\partial G}{\partial n_i}\right)_{T,p,n_{k(k\neq i)}} \tag{B.67}$$

where we note that the last equation is consistent with our definition in (B.36) of the chemical potential μ_i as the partial Gibbs energy. Accordingly, for H the "natural" variables are S and p and we have

$$dH = TdS + Vdp + \sum_i \mu_i dn_i \tag{B.68}$$

and

$$T = \left(\frac{\partial H}{\partial S}\right)_{p,n_1,\ldots,n_c} \quad ; \quad V = \left(\frac{\partial H}{\partial p}\right)_{S,n_1,\ldots,n_c} \tag{B.69}$$

With different combinations and introduction of other variables, as heat capacity, we can derive many other relationships between the different partial derivatives, which probably seems more or less mysterious to most people, but which actually have a rather simple basis.

Example B.12 *As an example of one of these "mysterious" relationships, let us derive the Gibbs-Helmholtz equation which states that the temperature dependency of the function G/T is determined by the enthalpy (assuming constant composition):*

$$\left(\frac{\partial(G/T)}{\partial T}\right)_p = -\frac{H}{T^2} \tag{B.70}$$

Proof: *The left side can be written as $\frac{1}{T}(\partial G/\partial T)_p - \frac{G}{T^2}$, that is, we want to show that $(\partial G/\partial T)_p = (G-H)/T$. This is simple: We have $(G-H)/T = -S$ which, according to (B.67), is identical to $(\partial G/\partial T)_p$.*

APPENDIX C

Differential balances: Examples

Here, we consider some more difficult examples where we must solve differential balances.

C.1 Emptying of gas tank

Example C.1 Emptying of gas tank. *This is Example 4.3 shown in Figure 4.5(b) (page 105): A container with gas at 10 bar and 300 K (state 0) is emptied such that the final pressure is 1 bar (state f). Calculate the corresponding final temperature when we assume adiabatic conditions and constant $\gamma = C_p/C_V = 1.4$. Note that the temperature inside the tank will decrease because of the displacement (pv-work) performed on the surroundings by the exiting gas.*

We have previously given up on solving this exercise, but it can be solved using the differential energy balance. Let the volume of the container be V. The number of moles in the container is n [mol] which varies with time. The outflow is $\dot{n}_{\text{out}}$ [mol/s]. The dynamic material balance (11.5) with no chemical reaction is

$$\frac{dn}{dt} = \dot{n}_{\text{in}} - \dot{n}_{\text{out}} \quad [\text{mol/s}]$$

Since the inflow is zero this gives

$$\frac{dn}{dt} = -\dot{n}_{\text{out}} \quad [mol/s]$$

Next, consider the dynamic energy balance (11.11),

$$\frac{dU}{dt} = \dot{H}_{\text{in}} - \dot{H}_{\text{out}} + \dot{Q} + \dot{W}_s - p_{\text{ex}}\frac{dV}{dt} \quad [\text{J/s}]$$

Note that it is the internal energy, and not the enthalpy, that enters on the left hand side. This is important for gases. The volume of the container is constant so $dV/dt = 0$. The system is adiabatic so $\dot{Q} = 0$. No mechanical work is supplied so $\dot{W}_s = 0$. There is no inflow so $\dot{H}_{\text{in}} = 0$. The energy balance then simplifies to

$$\frac{dU}{dt} = -\dot{H}_{\text{out}} \quad [J/s]$$

We introduce the molar enthalpy from the identity

$$U = nU_m = n(H_m - pV_m) \quad [J]$$

where $pV_m = RT$ (ideal gas). Accordingly, we have

$$\dot{H}_{\text{out}} = \dot{n}_{\text{out}} H_m$$

where we have assumed perfect mixing such that at any given time $H_{m,\text{out}} = H_m$. *Inserted into the energy balance, we now get*

$$\frac{d}{dt}[n(H_m - RT)] = -\dot{n}_{\text{out}} H_m$$

which upon insertion of the material balance gives

$$n\frac{dH_m}{dt} - nRT\frac{dT}{dt} = -RT\dot{n}_{\text{out}}$$

For an ideal gas, enthalpy depends only on temperature and we have $dH_m = C_{p,m}dT$, *and we get*

$$\frac{dT}{dt} = -\frac{R}{C_{p,m} - R}\,T\,\frac{\dot{n}_{\text{out}}}{n} = -(\gamma - 1)\,T\,\frac{\dot{n}_{\text{out}}}{n}$$

where $\gamma = C_{p,m}/C_{V,m}$. *We are actually not interested in the time here, but rather in finding the relationship between temperature* T *and the amount of gas* n *left in the tank. We can eliminate* dt *by introducing* $\dot{n}_{\text{out}}dt = -dn$ *from the material balance, and we derive*[1]

$$\frac{dT}{T} = (\gamma - 1)\frac{dn}{n} \tag{C.1}$$

Integration from state 0 to state f *gives*

$$\ln(T_f/T_0) = (\gamma - 1)\ln(n_f/n_0)$$

or

$$\frac{T_f}{T_0} = \left(\frac{n_f}{n_0}\right)^{\gamma-1}$$

Since the gas is ideal, we finally find

$$\frac{n_f}{n_0} = \frac{p_f}{p_0}\frac{T_0}{T_f} \quad \Rightarrow \quad \frac{T_f}{T_0} = \left(\frac{p_f}{p_0}\right)^{\frac{\gamma-1}{\gamma}} \tag{C.2}$$

Together with $p_f/p_0 = 0.1$, $T_0 = 300K$ *and* $(\gamma - 1)/\gamma = 0.286$, *this gives*

$$T_f = 0.1^{0.286} \cdot 300K = 155.38\ K$$

Note that emptying the tank gives a large drop in temperature from 300 K to 155.38 K. Correspondingly, filling a tank gives a temperature increase as shown in Example 4.2 (page 104), but note that the governing equations for the two cases are different.

Comment: *By comparing (C.2) with (B.17) we note that the state change for the process follows that of an* **isentropic** *process. This is a bit surprising because the overall process is clearly irreversible.*

C.2 Logarithmic mean temperature difference

Here, we derive the expression for the logarithmic mean temperature difference for an ideal countercurrent heat exchanger by (1) formulating differential energy balances for the two sides of the heat exchanger, (2) combining them and (3) integrating the resulting differential equation in $T_h - T_c$. Note that we assume constant heat capacities for the two sides.

[1] An alternative and simpler derivation of (C.1) is: Start from the energy balance of differential form $dU = -dH_{\text{out}} = H_m dn$ where we note that dn is negative. Introducing $dU = d(U_m n) = U_m dn + n dU_m$ gives $n dU_m = (H_m - U_m)dn = RTdn$ which, with $dU_m = C_{V,m}dT$, gives (C.1).

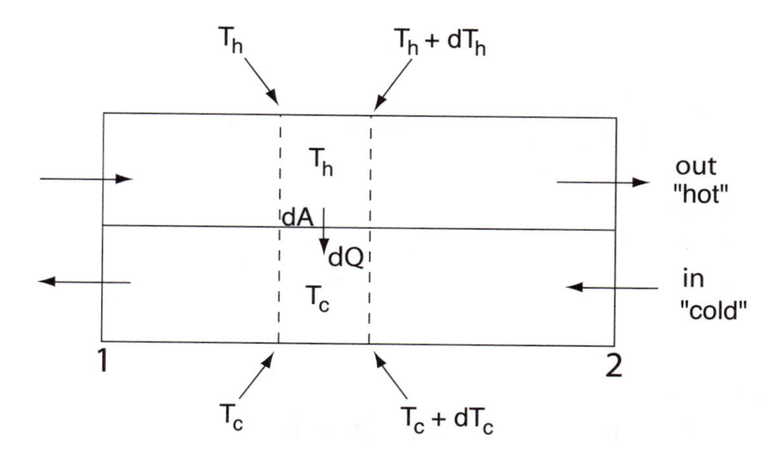

Figure C.1: Energy balance over small (differential) part of countercurrent heat exchanger

(1) We consider a small (differential) part of the heat exchanger with area dA, see Figure C.1. The heat transferred is

$$dQ = (T_h - T_c)UdA \tag{C.3}$$

where $T_h - T_c$ is the "local" temperature difference between the hot and cold sides. Next, differential energy balances for the cold and hot side give

$$m_c C_{p,c}(T_c - (T_c + dT_c)) = m_c C_{p,c}(-dT_c) = dQ \tag{C.4}$$

$$m_h C_{p,h}((T_h + dT_h) - T_h) = m_h C_{p,h} dT_h = -dQ \tag{C.5}$$

where the negative sign for $(-dT_c)$ in (C.4) appears because the flow on the cold side is reverse when we go from point 1 to 2 (see the figure).

(2) Multiplying and then adding the equations, (C.4)$\cdot m_h C_{p,h}+$ (C.5)$\cdot m_c C_{p,c}$, we derive

$$m_h C_{p,h} m_c C_{p,c}(dT_h - dT_c) = (m_h C_{p,h} - m_c C_{p,c}) \underbrace{(T_h - T_c)UdA}_{dQ}$$

$$\Rightarrow \frac{d(T_h - T_c)}{T_h - T_c} = \frac{m_h C_{p,h} - m_c C_{p,c}}{m_h C_{p,h} m_c C_{p,c}} UdA = -RUdA$$

where we have introduced

$$R = \frac{m_c C_{p,c} - m_h C_{p,h}}{m_h C_{p,h} m_c C_{p,c}}$$

(3) We integrate through the heat exchanger (from 1 to 2) assuming constant heat capacities

$$\ln \frac{(T_h - T_c)_2}{(T_h - T_c)_1} = \ln \frac{\Delta T_2}{\Delta T_1} = -RUA \quad \Rightarrow \quad \ln \frac{\Delta T_1}{\Delta T_2} = RUA \tag{C.6}$$

The energy balance of the whole heat exchanger

$$m_h C_{p,h}(T_{h,\text{in}} - T_{h,\text{out}}) = m_c C_{p,c}(T_{c,\text{out}} - T_{c,\text{in}}) = Q$$

gives

$$\frac{Q}{m_h C_{p,h}} = T_{h,\text{in}} - T_{h,\text{out}}; \quad \frac{Q}{m_c C_{p,c}} = T_{c,\text{out}} - T_{c,\text{in}}$$

Subtracting yields

$$\underbrace{Q\left(\frac{1}{m_h C_{p,h}} - \frac{1}{m_c C_{p,c}}\right)}_{R} = \underbrace{(T_{h,\text{in}} - T_{c,\text{out}})}_{\Delta T_1} - \underbrace{(T_{h,\text{out}} - T_{c,\text{in}})}_{\Delta T_2} \tag{C.7}$$

which combined with (C.6) gives

$$Q = \frac{\Delta T_1 - \Delta T_2}{R} = \frac{\Delta T_1 - \Delta T_2}{\ln \frac{\Delta T_1}{\Delta T_2}} UA \tag{C.8}$$

□

C.3 Batch (Rayleigh) distillation

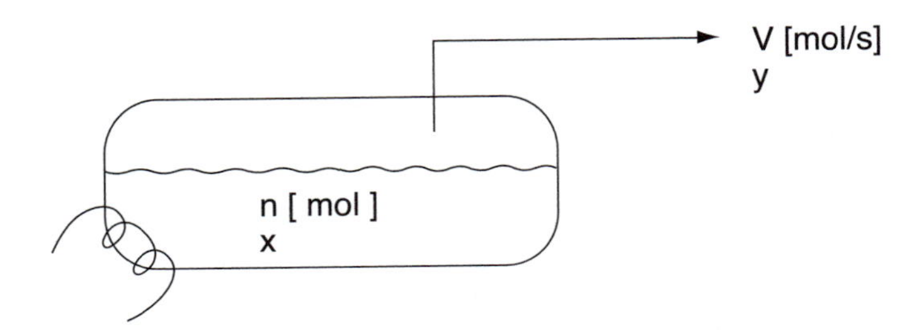

Figure C.2: Single-stage batch (Rayleigh) distillation

A single-stage batch (or more exactly, semi-batch since the product is continuously removed) distillation process is shown in figure C.3. This is often called *Rayleigh distillation* or open distillation. What happens is simply that we boil a mixture, condense the vapor and collect the condensate. We get a separation of the components, since the lightest components are enriched in the vapor (gas) phase.

Assume that the amount of liquid in the apparatus at any time is n [mol], and that we continuously remove a vapor flow $V = Q/\Delta_{\text{vap}}H$ [mol/s] by supplying heat. Let us consider a component with mole fraction x in the liquid phase and mole fraction y in the vapor phase. The total mass balance for the tank gives when we neglect the holdup of gas:

$$\frac{dn}{dt} = -V \quad [mol/s] \tag{C.9}$$

Similarly, the component mass balance becomes

$$\frac{d(xn)}{dt} = -yV \tag{C.10}$$

$$\Rightarrow n\frac{dx}{dt} + x\frac{dn}{dt} = -yV$$

Inserting dn/dt from (C.9) gives

$$n\frac{dx}{dt} = -V(y - x) \tag{C.11}$$

For a "light" component, $y - x$ is positive so dx/dt negative, that is, x (mole fraction of the component in the liquid) decreases over time.

(C.11) can be a bit difficult to solve since n depends on time. However, we are often not interested in x as a function of time, but rather in x as a function of n. A common "trick" is then to eliminate dt by dividing (C.11) by (C.9), which gives

$$n\frac{dx}{dn} = y - x$$

which can be rearranged to

$$\frac{dn}{n} = \frac{dx}{y - x} \tag{C.12}$$

which is known as the *Rayleigh equation.* The equation can be integrated if we know y as function of x from the vapor/liquid equilibrium.

APPENDIX D

Summary of the whole book

Here, we summarize the most important concepts and formulas from the book (you may think there are too many formulas, so having them in one place may be helpful).

The most important basis for process engineering calculations is to choose a control volume and formulate the relevant balance equations for:

- total mass [kg]
- mole component [mole A]
- energy [J] (if we are interested in temperature changes)
- momentum or mechanical energy (if we are interested in pressure changes)

Some important points that you should know:

1. A stream (and the state of a system) is generally specified by $N_c + 2$ independent pieces of information (for example, N_c component masses, temperature and pressure). N_c is the number of components.
2. For any control volume, we can generally formulate N_c+2 independent balances. They are: N_c independent mass balances (for example, 1 total mass balance and $N_c - 1$ independent component balances), 1 energy balance and 1 momentum balance or mechanical energy balance.
3. The energy balance for a **closed system**, where we neglect changes in kinetic and potential energy (the first law of thermodynamics), is

$$\Delta U = Q + W \tag{D.1}$$

   where $\Delta U = U_f - U_0$ is the change in the system's internal energy from *time* t_0 to t_f. Here, we use the convention that Q is the supplied heat and W is the supplied work (from the surroundings to the system).
4. Internal energy U [J/mol] is a state function, that is, a function of composition, temperature and pressure. For an *ideal gas*, internal energy is only a function of temperature.
5. Enthalpy $H = U + pV$ [J/mol] is also a state function, that is, a function of composition, temperature and pressure. For an *ideal gas*, enthalpy is only a function of temperature.
6. Heat Q and work W are *not* state functions, but the *sum* is uniquely determined by the initial and final states, see (D.1).
7. The heat capacity is defined as the amount of heat that must be supplied to a closed system in order to increase its temperature by 1 K in a reversible process (that is, the same amount of heat is liberated when cooling by 1 K). For heating under *constant volume*, we then define

$$C_V \triangleq \left(\frac{\partial Q^{\rm rev}}{\partial T}\right)_V \quad [J/K] \quad \text{(closed system)}$$

For a closed system with constant volume, we get from the energy balance (first law; $dU = \delta Q + \delta W$) that $dU = \delta Q$, which gives the following alternative (equivalent) definition

$$C_V = \left(\frac{\partial U}{\partial T}\right)_V \quad [J/K]$$

For heating under *constant pressure*, we accordingly define

$$C_p \triangleq \left(\frac{\partial Q^{\text{rev}}}{\partial T}\right)_p \quad [J/K] \quad \text{(closed system)}$$

For a closed system with constant pressure, we get from the energy balance $dU = \partial Q - pdV$, i.e., $dH = dU + pdV = \partial Q$ which gives the following alternative (equivalent) definition

$$C_p = \left(\frac{\partial H}{\partial T}\right)_p \quad [J/K]$$

8. For an ideal gas, the following applies on a molar basis,

$$C_p = C_V + R \quad [\text{J/mol, K}]$$

(This follows because for 1 mol of ideal gas, we have $d(pV) = d(RT) = RdT$).

9. The work W can be written as $W = W_{\Delta V} + W_s$, where $W_{\Delta V}$ is the expansion (pV) work related to volume changes and W_s is useful mechanical work (shaft work). For a closed system, the work for a small volume change is $\delta W_{\Delta V} = -p_{\text{ex}} dV$ [J], and we get

$$W_{\Delta V} = -\int_{V_0}^{V_f} p_{\text{ex}} dV$$

where p_{ex} is the surrounding's pressure. For a reversible process, we have $p_{\text{ex}} = p$.

10. For an **open system** (continuous process), the following generalized version of the first law of thermodynamics (energy balance) in (D.1) applies:

$$E_f - E_0 = E_{\text{in}} - E_{\text{out}} + Q + W \tag{D.2}$$

where $W = W_{\Delta V} + W_s$ is the supplied work and $E = U + E_K + E_P + \cdots$ is the system's total energy. In many cases, it is sufficient to consider changes in internal energy and we have $E = U$. Note that in $W_{\Delta V}$ we also include the flow work that the streams perform as they enter or exit the system, and we have

$$W_{\Delta V} = p_{\text{in}} V_{\text{in}} - p_{\text{out}} V_{\text{out}} - \int_{V_0}^{V_f} p_{\text{ex}} dV$$

Introducing the definition of enthalpy,

$$H_{\text{in}} = U_{\text{in}} + p_{\text{in}} V_{\text{in}}$$

$$H_{\text{out}} = U_{\text{out}} + p_{\text{out}} V_{\text{out}}$$

the energy balance for an open system becomes (for the case where assume $E = U$)

$$U_f - U_0 = H_{\text{in}} - H_{\text{out}} + Q + W_s - \int_{V_0}^{V_f} p_{\text{ex}} dV \tag{D.3}$$

11. For an open system at steady-state (continuous process with no accumulation), we have $U_0 = U_f$ and $V_0 = V_f$, and the energy balance becomes

$$0 = H_{\text{in}} - H_{\text{out}} + Q + W_s$$

which is a **very important** result for process engineering calculations! By writing $\Delta H = H_{\text{out}} - H_{\text{in}}$, it can be written in the form

$$\Delta H = Q + W_s \qquad \text{(D.4)}$$

where you should note that the "Δ" represents the difference between in and out flows (whereas the "Δ" in (D.1) represents the difference between the initial and final times).

12. For an open system (continuous process), the reversible shaft work related to a small *pressure change* is

$$\partial W_s^{\text{rev}} = V dp \quad [J]$$

(Do not confuse it with $\partial W^{\text{rev}} = -pdV$, which is the reversible pV-work for *volume changes.*) Integrating from state 1 to state 2 gives

$$W_s^{\text{rev}} = \int_{p_1}^{p_2} V dp \quad [J]$$

or on mass basis (introducing $m = \rho V$ where the mass flow m is assumed constant)

$$\frac{W_s^{\text{rev}}}{m} = \int_{p_1}^{p_2} \frac{dp}{\rho} \quad [J/kg]$$

13. The above equations are used to calculate the compression or expansion work for continuous processes. For 1 mol of ideal gas, we have $V = RT/p$ and derive

$$dW_s^{\text{rev}} = RT\frac{dp}{p} \quad \text{[J/mol]} \qquad \text{(ideal gas)}$$

14. For *isothermal* reversible compression (which requires cooling if the pressure increases), we then get by integration that the shaft work is

$$W_s^{\text{rev}} = RT \ln \frac{p_2}{p_1} \quad \text{[J/mol]} \qquad \text{(ideal gas)}$$

15. For *adiabatic* reversible compression, we get a temperature increase. If we assume that $\gamma = C_p/C_V$ is constant, we can derive $pV^\gamma =$ constant, or equivalently (see page 350)

$$\frac{T_2}{T_1} = \left(\frac{p_2}{p_1}\right)^{\frac{\gamma-1}{\gamma}} = \left(\frac{p_2}{p_1}\right)^{R/C_{p,m}} \qquad \text{(ideal gas with constant } C_p\text{)}$$

The corresponding shaft work for the compression is from the energy balance

$$W_s = C_p(T_2 - T_1) \quad \text{[J/mole]} \qquad \text{(ideal gas with constant C}_\text{p}\text{)}$$

Note that for an adiabatic reversible process, the entropy is constant, and we can also derive the expression for T_2/T_1 from this starting point (see page 380).

16. Above we neglected changes in other energy forms, such as kinetic and potential energy, but if this assumption is incorrect, we can always include them by replacing the internal energy U in the equations above by the total energy $E = U + E_P + E_K + \cdots$. Here the potential energy for a mass m is

$$E_P = mgz \quad [J]$$

and the kinetic energy is

$$E_K = \alpha \cdot m \frac{v^2}{2}$$

where we define v [m/s] $= \frac{q[\text{m}^3/\text{s}]}{A[\text{m}^2]}$ and the factor α corrects for the velocity profile not being flat (we have $\alpha \approx 1$ for the most common case of turbulent flow, while $\alpha = 2$ for laminar pipe flow).

With potential and kinetic energy included, the steady state energy balance (D.4) then becomes:

$$H_2 + m\alpha_2 \frac{v_2^2}{2} + mgz_2 = H_1 + m\alpha_1 \frac{v_1^2}{2} + mgz_1 + Q + W_s \quad [J] \tag{D.5}$$

(1 is "inflow" and 2 is "outflow"). This can, for example, be used to calculate the temperature drop if a gas accelerates to a higher velocity in a choke valve. Thus, we use the energy balance with kinetic energy included if we want to calculate the effect of velocity changes on temperature.

17. In a heat exchanger, heat is transferred through a wall from a hot to a cold stream. The heat transferred is proportional to the area and the temperature difference, that is, we get

$$Q = UA\Delta T$$

where U [W/m^2,K] is the thermal conduction number. Here the temperature difference ΔT varies through the heat exchanger, but if we assume constant heat capacity, we can, for ideal countercurrent flow (and also for cocurrent flow), use the logarithmic mean of the temperature differences at the two ends (1 and 2),

$$\Delta T_{lm} = \frac{\Delta T_1 - \Delta T_2}{\ln \frac{\Delta T_1}{\Delta T_2}}$$

18. Entropy is a state function, and changes in its value can be obtained by computing $\Delta S = S_2 - S_1 = \int_1^2 dS = \int_1^2 \delta Q^{\text{rev}}/T$ for a (imaginary) reversible process between the same states.
19. For an ideal gas, we have

$$S(T_2, p_2) = S(T_1, p_1) + \int_{T_1}^{T_2} C_p \frac{dT}{T} - R \ln \frac{p_2}{p_1} \quad \text{[J/mol K]} \quad \text{(ideal gas)}$$

Note that the entropy for an ideal gas is a function of pressure (while the enthalpy for an ideal gas is independent of pressure).

20. For an ideal mixture, the entropy of mixture (which is always positive) is

$$\Delta_{\text{mix}} S = -R \sum_i x_i \ln x_i \quad \text{[J/mole K]} \quad \text{(ideal mixture)}$$

21. The entropy and the enthalpy for an ideal gas stream is then calculated from the following formulas

$$S(T,p) = \underbrace{\sum_i n_i S^*_{m,i}(T_0, p_0)}_{S^*(T_0,p_0)} + \int_T^{T_0} C_p(T) \frac{dT}{T} - nR \ln \frac{p}{p_0} - R \sum_i n_i \ln x_i \quad [J/K]$$

$$H(T,p) = \underbrace{\sum_i n_i H^*_{m,i}(T_0, p_0)}_{H^*(T_0,p_0)} + \int_T^{T_0} C_p(T) T \quad [J]$$

where $C_p[J/K] = \sum_i n_i C_{p,m}(i)$ is the mixture heat capacity. Usually, the elements at $T_0 = 298.15$ K and $p_0 = 1$ are chosen as reference, and we have

$$S^*(T_0, p_0) = \sum_i n_i \Delta_f S_m^{\ominus}(i, 298)$$

$$H^*(T_0, p_0) = \sum_i n_i \Delta_f H_m^{\ominus}(i, 298)$$

The same formulas also apply for an ideal liquid mixture, except that the entropy's pressure dependency is omitted (that is, we do not include the term $-nR \ln \frac{p}{p_0}$).

22. The second law of thermodynamics states that the total entropy $S_{\text{total}} = S + S_{\text{sur}}$ increases for all natural process. For a (imaginary) reversible process, the increase is zero. S is the system's entropy and S_{sur} is the surrounding's entropy.
23. At equilibrium, the total entropy S_{total} reaches its maximum.
24. At a given p and T, this is equivalent to the system's Gibbs energy $G = H - TS$ reaching a minimum, that is, $(dG)_{T,p} = 0$ (see page 384).
25. For chemical equilibrium at T and p, this can be expressed as $\Delta_r G = 0$, and introducing the "equilibrium constant"

$$\ln K \triangleq -\frac{\Delta_r G^{\ominus}}{RT}$$

the equilibrium condition for the reaction

$$0 = \nu_A A + \nu_B B + \nu_C C + \nu_D D = 0$$

(where A and B are reactants, so $\nu_A < 0$ and$\nu_B < 0$) can be written in the form

$$K = \prod_i a_i^{\nu_i} = \frac{a_C^{\nu_C} a_D^{\nu_D}}{a_A^{|\nu_A|} a_B^{|\nu_B|}} \quad \text{(given } T, p\text{)}$$

Here a_i is the activity of component i relative to its standard state used when obtaining $\Delta_r G^{\ominus}$ (the change in standard Gibbs energy for the reaction at T) and K. For an ideal gas, the activity is $a_i = p_i/p^{\ominus}$ where $p_i = y_i p$ is the partial pressure and $p^{\ominus} = 1$ bar.

26. The following expressions can be used to find $\Delta_r G^{\ominus}(T)$, and the equilibrium constant $K(T)$, from data at 298 K when the heat capacity is assumed constant:

$$\Delta_r H^{\ominus}(T) = \Delta_r H^{\ominus}(298) + \Delta_r C_p^{\ominus} \cdot (T - 298)$$

$$\Delta_r S^{\ominus}(T) = \Delta_r S^{\ominus}(298) + \Delta_r C_p^{\ominus} \cdot \ln \frac{T}{298}$$

$$\Delta_r G^{\ominus}(T) = \Delta_r H^{\ominus}(T) - T\Delta_r S^{\ominus}(T)$$

$$K(T) = e^{-\frac{\Delta_r G^{\ominus}(T)}{RT}}$$

(see the example for the ammonia synthesis on page 176).

27. Assume that we have available heat at temperature T_H and cooling at temperature T_C, and extract the difference as work, $|W| = |Q_H| - |Q_C|$. Then for a reversible process

$$\Delta S_{\text{total}} = -\frac{|Q_H|}{T_H} + \frac{|Q_C|}{T_C} = 0$$

and we derive that the maximum work is given by the Carnot "efficiency" (factor), that is,

$$\frac{|W|_{\max}}{|Q_H|} = 1 - \frac{T_C}{T_H}$$

28. In the **mechanical energy balance**, we concentrate on the mechanical energy terms in the energy balance. If we have no friction, then mechanical energy can be reversibly converted from one form to another. For a steady-state process, the mechanical energy balance can then be written as (1 is here "in" while 2 is "out").

$$m\alpha_2 \frac{v_2^2}{2} + mgz_2 + m\int_{p_1}^{p_2} \frac{dp}{\rho} + \Phi = m\alpha_1 \frac{v_1^2}{2} + mgz_1 + W_s \quad [J; J/s]$$

where Φ [J] is the friction loss, which is always positive. For the case with no friction, no shaft work ($W_s = 0$) and constant density ρ (incompressible flow), this gives the Bernoulli equation

$$p + \rho gz + \rho \frac{v^2}{2} = \text{const}$$

The mechanical energy balance is used when we want to calculate pressure changes.

APPENDIX E

Additional problems

E.1 Test exam

EXAM IN PROCESS ENGINEERING

Time: 0900 - 1500

Allowed: Approved calculator. Mathematical formula collection. *SI Chemical Data.* (No printed or handwritten texts allowed).

For all problems: State clearly additional assumptions. Explain all answers.

Problem 1

500 g CO_2 (s) (dry ice) is left in a 2 liter container at 35 ^oC.

(a) What is the final pressure in the container, provided it does not explode?

(b) Assume that the container explodes at a pressure of 70 bar. How much work does the gas perform on the surroundings (air) when the container explodes? The surroundings have the same temperature as the container, so assume an isothermal process. The surrounding's pressure is 1 bar.

Data: Constants for CO_2 in van der Waals equation of state: a = 3.640 L^2 bar mol^{-2}, $b = 0.04267$ L mol^{-1} (Here L means liter). Molar mass for CO_2 is 44 g mol^{-1}.

Problem 2

A steam reformer produces synthesis gas (a mixture of CO, CO_2 and H_2) from methane by the following reactions

$$CH_4 + H_2O = CO + 3H_2; \quad \Delta_r H^{\ominus}(1148K) = 246\frac{kJ}{mole}, \ \Delta_r G^{\ominus}(1148K) = -66\frac{kJ}{mole}$$

$$CO + H_2O = CO_2 + H_2; \quad \Delta_r H^{\ominus}(1148K) = -38\frac{kJ}{mole}, \ \Delta_r G^{\ominus}(1148K) = 0\frac{kJ}{mole}$$

where the last is known as the shift reaction.

(a) Introduce the extent of reaction ξ_j for the reactions and formulate the mass balances (you can denote the feed with n^0 and the product with n).

(b) The feed to the steam reformer is 10000 mol/s and contains 30 mol% methane and 70 mol% steam (H_2O). Compute the composition of the product when it is assumed that 80% of the methane is converted and 770 mol/s CO_2 is produced.

(c) Formulate the energy balance. The feed is at 300 ^oC and the product at 875 ^oC. How much heat must be supplied to the reactor?

(d) What is the pressure in the reactor if it is assumed that the first reaction is in equilibrium? How far is the shift reaction from equilibrium?

(e) How does the conversion of methane change if the pressure drops?

Problem 3

Components A and B react in liquid phase according to the reaction

$$A + B \rightarrow 2P$$

The reaction rate is

$$r_A = -k \cdot c_A \cdot c_B \quad [kmol/m^3, s]$$

where the reaction rate constant k follows Arrhenius equation,

$$k = Ae^{-\frac{E}{RT}}$$

The activation energy is $E = 16.67$ kJ/mol. At 50 °C, $k = 0.166$ [m^3/kmol s].

The reaction takes place in a continuous stirred tank reactor (CSTR) with constant volume $V = 0.3$ m^3. The total volumetric feed velocity is 3.6 m^3/h and the reactants A and B are fed in the stoichiometric ratio, $c_{A0} = c_{B0} = 0.5$ kmol/m^3. The liquid density is assumed constant.

(a) What is the reaction rate constant at 60 °C ?

(b) Formulate the mass balance for the system (derive the design equation for the reactor).

(c) What is the conversion if the reactor operates isothermally at 60 °C ?

Problem 4

2 kg/s of a product with 2 weight% potassium (K) and 3 weight% phosphorus (P) should be produced. In the first step, stream 1 (which contains 2 weight% K and water) is mixed with stream 2 (which contains 2 weight% P and water), and in the second step, pure water is removed (stream 3) by evaporation. Find the value of the three streams.

Problem 5

You have available water at 100 °C and 1 bar that should be used to produce 1 kg/s of superheated steam at 5 bar and 300 °C in a continuous process. The following two alternatives are considered

1. (i) Evaporation at 1 bar, (ii) adiabatic lossless compression from 1 bar to 5 bar, (iii) heating of the steam until 300 °C.
2. (i) Pumping from 1 bar to 5 bar, (ii) heating of the liquid to the boiling point (152 °C) and evaporation at 5 bar, (iii) heating of the steam until 300 °C.

(a) Sketch a flow sheet for the two alternatives.

(b) Find the supplied heat and work in each step for the two alternatives.

(c) Compare the sum of the supplied work and heat for the two alternatives. Why might one choose, in practice, the alternative with pumping?

Data. You can assume ideal gas and use the following data. The heat capacity for water is $C_p(l) = 4.18$ kJ/kg K (liquid) and $C_p(g) = 1.87$ kJ/kg K (vapor). The heat of vaporization is 2260 kJ/kg at 100 °C/1bar and 2140 kJ/kg at 152 °C/5bar. For the vapor, $\gamma = C_p/C_V = 1.33$. The density of water is $\rho = 1000$ [kg/m^3].

E.2 Solution

EXAM IN PROCESS ENGINEERING

Suggested solution

Solution to Problem 1

Note that CO_2 is a gas at 35 ^{o}C. The number of moles of gas in the container is

$$n = \frac{m}{M} = \frac{500g}{44g/mol} = 11.36\ mol$$

(a) Inserting the numerical values ($T_1 = 308K$, $V = 2l = 2 \cdot 10^{-3}m^3$) in the van der Waals equation of state gives

$$p = 192.0 - 117.4 = 74.5\ bar$$

(b) The work performed is

$$W = \int_{V_1}^{V_2} p_{\rm ex} dV = p_{\rm ex}(V_2 - V_1)$$

where $p_{\rm ex} = 1bar = 10^5 N/m^2$, $V_1 = 2 \cdot 10^{-3}m^3$ and V_2 is (assuming ideal gas at 1 bar at the final temperature $T_2 = 308K$)

$$V_2 = \frac{nRT_2}{p_{\rm ex}} = 0.291m^3$$

which gives that the performed work is $W = p_{\rm ex}(V_2 - V_1) = 10^5(0.291 - 0.002) = 28900$ J (if we put $V_2 = 0.291m^3$ into the van der Waals equation, we get $p_2 = 0.995$ bar, which is close to 1 bar, so the assumption of ideal gas is OK).

Comment: If the container weighs 10 kg, then this work should be enough to lift the container about 289 m up into the air (using $W = mgh$ and $g = 10\text{m/s}^2$). However, this assumes that the gas leaves the container at 35 ^{o}C (isothermally), which is not realistic. In practice, the temperature falls in the container; for an ideal gas with adiabatic condition, we have $T_2/T_1 = (p_2/p_1)^{\frac{\gamma-1}{\gamma}}$. This means that the gas that leaves the cylinder has less volume and the work for lifting the cylinder into the air is less. However, the total work is the same as given above, and the remaining "useless" work is liberated afterwards, when the gas that left the cylinder is heated to 35°C (but this work will only give rise to a "small breeze" which cannot be used to lift the container).

Solution to Problem 2

$$CH_4 + H_2O = CO + 3H_2; \quad \xi_1\ [mol/s]$$
$$CO + H_2O = CO_2 + H_2; \quad \xi_2\ [mol/s]$$

(a) The component balances give (mol/s):

$$\begin{aligned}
n_{CH_4} &= n^0_{CH_4} - \xi_1 \\
n_{H_2O} &= n^0_{H_2O} - \xi_1 - \xi_2 \\
n_{CO} &= n^0_{CO} + \xi_1 - \xi_2 \\
n_{CO_2} &= n^0_{CO_2} + \xi_2 \\
n_{H_2} &= n^0_{H_2} + 3\xi_1 + \xi_2
\end{aligned}$$

(b) Given $n^0_{tot} = 10000$ mol/s and $n^0_{CH_4} = 3000$ mol/s and $n^0_{H_2O} = 7000$ mol/s. 80% of the methane is converted $\Rightarrow$ $\xi_1 = 0.8 \cdot 3000 = 2400 mol/s$. It generates 770 mol/s

$CO_2 \quad \Rightarrow \xi_2 = 770$ mol/s. We then get [mol/s]:

$$\begin{aligned} n_{CH_4} &= 3000 - 2400 = 600 \\ n_{H_2O} &= 7000 - 2400 - 770 = 3830 \\ n_{CO} &= 2400 - 770 = 1630 \\ n_{CO_2} &= 770 \\ n_{H_2} &= 3 \cdot 2400 + 770 = 7970 \\ n_{tot} &= 14800 \end{aligned}$$

The composition becomes

$$x_{CH_4} = 0.0405, \; x_{H_2O} = 0.2588, \; x_{CO} = 0.1101, \; x_{CO_2} = 0.0520, \; x_{H_2} = 0.5385$$

(c) The energy balance for a continuous steady-state process is $H_{\text{out}} = H_{\text{in}} + Q + W_s$ [J/s]. Here, $W_s = 0$ and we get

$$Q = H_{\text{out}} - H_{\text{in}}$$

To compute $H_{\text{out}} - H_{\text{in}}$, we consider an idealized process that takes the inflow (feed) to the outflow (product):

1. Heating of inflow from 300 °C to 875 °C (we assume constant heat capacity and take data for C_p from the table on page 416):

$$\Delta_1 H = n^0_{CH_4} \cdot C_{p,CH_4} \cdot (875 - 300) + n^0_{H_2O} \cdot C_{p,H_2O} \cdot (875 - 300)$$

$$= 3000 \cdot 35.31 \cdot 575 + 7000 \cdot 33.58 \cdot 575 = 196.1 \cdot 10^6 \text{ J/s} = 196.1 \text{ MW}$$

2. Reaction at 875 °C:

$$\Delta_2 H = \xi_1 \cdot \Delta_r H_1^\ominus + \xi_2 \cdot \Delta_r H_2^\ominus$$

$$= 2400 \cdot 246 \cdot 10^3 + 770 \cdot (-38 \cdot 10^3) = 561.1 \cdot 10^6 \text{ J/s} = 561.1 \text{ MW}$$

We get that the heat that must be supplied is

$$Q = H_{\text{out}} - H_{\text{in}} = \Delta_1 H + \Delta_2 H = 757.2 \; MW$$

(d) Equilibrium constants at 1148 K (875 °C)

$$K_1 = e^{\frac{-\Delta_r G_1^\ominus}{RT}} = 1011$$

$$K_2 = e^{\frac{-\Delta_r G_2^\ominus}{RT}} = 1$$

Reaction 1 is in equilibrium (assume ideal gas)

$$\frac{\frac{p_{CO}}{p^0} \cdot \left(\frac{p_{H_2}}{p^0}\right)^3}{\frac{p_{CH_4}}{p^0} \cdot \frac{p_{H_2O}}{p^0}} = K_1$$

Here, $p^0 = 1$ bar and the partial pressure is $p_i = x_i p$, where p is the total pressure. This gives

$$\frac{x_{CO} \cdot x_{H_2}^3}{x_{CH_4} \cdot x_{H_2O}} \left(\frac{p}{p^0}\right)^2 = K_1$$

Inserting numerical values, we find that the pressure is

$$p = 1 \text{ bar} \cdot \sqrt{\frac{1011}{1.64}} = 24.8 \text{ bar}$$

For the shift reaction, we have

$$Q_2 = \frac{x_{CO_2} \cdot x_{H2}}{x_{CO} \cdot x_{H_2O}} = 0.98$$

which is about the same as the equilibrium constant $K_2 = 1$, i.e., the reaction is in equilibrium.

(e) From Le Chatelier's principle: There is a net generation of molecules (moles) in reaction 1 and it is favorable to have a low pressure, i.e., the conversion of methane will increase by reducing the pressure. (However, the reaction is slower at low pressure and this has the opposite effect). (**Comment.** We can confirm this result quantitatively by assuming equilibrium and using the equilibrium constant found above. We find that the conversion of methane is 84% at 20 bar, 94% at 10 bar and 98% at 5 bar.

Solution to Problem 3

(a) The reaction rate constant k follows the Arrhenius equation

$$\frac{k(T_2)}{k(T_1)} = \frac{e^{-E/RT_2}}{e^{-E/RT_1}}$$

With data $T_1 = 323$K, $T_2 = 333$K, $E/R = 2000K^{-1}$ and $k(T_1) = 0.166$ m^3/kmol s, we get

$$k(T_2) = 0.200\text{m}^3/\text{kmol s}$$

(b) Mass balance for component A (Out = In + Generated) [kmol A/s]:

$$n_A = n_{A0} + r_A V$$

Solving with respect to V gives the "design equation" $V = \frac{n_{A0}-n_A}{-r_A}$. Here, $n_A = \dot{V} c_A$ where $\dot{V}$ $[m^3/s]$ is the volumetric stream. Since we have stoichiometric feed, $c_B = c_A$, the mass balance gives

$$\dot{V} c_A = \dot{V} c_{A0} - k c_A^2 V$$

(c) Inserting numerical values ($\dot{V} = 0.001\ m^3/s, V = 0.3\ m^3, c_{A0} = 0.5\ kmol/m^3$, $k = 0.200$ m^3/kmol s) gives a second-order equation

$$0.06 c_A^2 + 0.001 c_A - 0.0005 = 0$$

which gives $c_A = 0.083\ kmol/m^3$, and the conversion of A is

$$X_A = \frac{n_{A0} - n_A}{n_{A0}} = \frac{c_{A0} - c_A}{c_{A0}} = \frac{0.5 - 0.083}{0.5} = 0.834 \quad (83.4\%)$$

Solution to Problem 4

First, we sketch a simple flow sheet (not shown). Mass balances for the overall process with two inflows (streams 1 and 2) and two outflows (stream 3 and product) [kg/s] give

$$\begin{aligned} \text{Total}: &\quad m_1 + m_2 = m_3 + m && \text{(E.1)} \\ \text{Potassium}: &\quad 0.02 m_1 = 0.02 m && \text{(E.2)} \\ \text{Phosphorus}: &\quad 0.02 m_2 = 0.03 m && \text{(E.3)} \end{aligned}$$

Solution:

$$\begin{aligned} (E.2) \text{ gives}: &\quad m_1 = m = 2kg/s \\ (E.3) \text{ gives}: &\quad m_2 = \frac{0.03}{0.02} m = 3kg/s \\ (E.1) \text{ gives}: &\quad m_3 = m_1 + m_2 - m = 3kg/s \end{aligned}$$

Solution to Problem 5

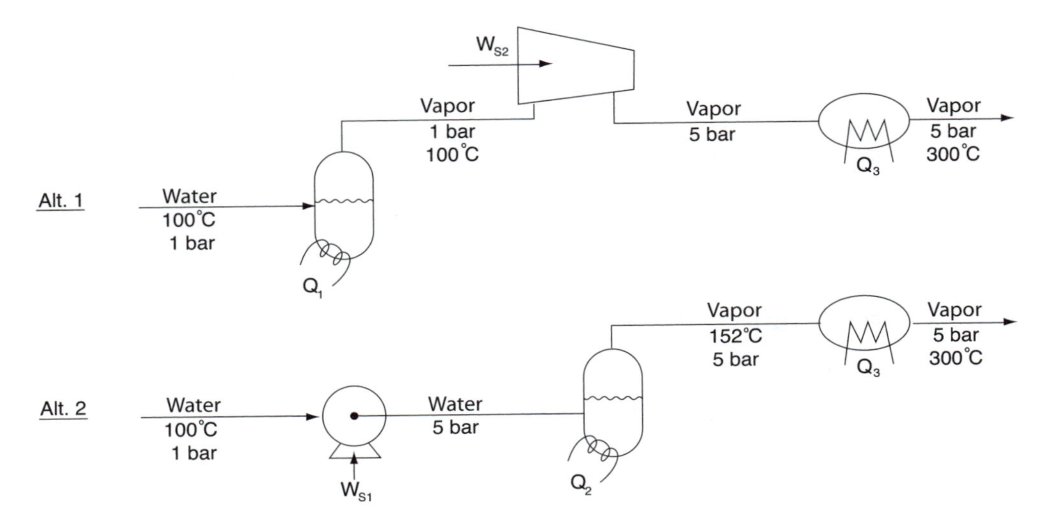

Figure E.1: Alternative processes for producing superheated vapor (exercise 5)

(a) The alternative flow sheets are shown in Figure E.1.

(b) We assume constant heat capacities, $C_p(l) = 4.18$ kJ/kg K and $C_p(g) = 1.87$ kJ/kg K. The mass flow is $m = 1$ kg/s.

Alternative 1.

(i) Evaporation at 1 bar

$$Q_1 = \Delta_{\text{vap}} H(100^oC) \cdot m = 2260\ kJ/s$$

(ii) Adiabatic compression from 1 to 5 bar

$$\frac{T_2}{T_1} = \left(\frac{p_2}{p_1}\right)^{\frac{\gamma-1}{\gamma}} = 5^{0.248} = 1.49$$

which gives $T_2 = 1.49T_1 = 55K = 283$ ^{o}C. Supplied work

$$W_{s2} = C_p(g)(T_2 - T_1) \cdot m = 1.87 \cdot 183 \cdot 1 = \mathbf{342\ kJ/s}$$

(iii) Heat up the gas

$$Q_3 = 1.87 \cdot (300 - 283) \cdot 1 = 32\ kJ/s$$

Sum of the energy supplied in the three steps:

$$Q_1 + W_{s2} + Q_3 = 2260 + 342 + 32 = \mathbf{2634\ kJ/s}$$

Alternative 2.

(i) Pumping from 1 to 5 bar. Pump work

$$W_{s1} = \dot{V} \cdot \Delta p = \frac{1\ kg/s}{1000\ kg/m^3} \cdot (5-1)10^5\ N/m^2 = 400\ J/s = \mathbf{0.4\ kJ/s}$$

(this gives a temperature rise of about 0.1 ^{o}C, which can be neglected).

(ii) Heating of water from 100 ^{o}C to 152 ^{o}C and evaporation at 152 ^{o}C (5bar):

$$Q_2 = 4.18 \cdot 52 \cdot 1 + 2140 \cdot 1 = 2357\ kJ/s$$

(iii) Heating of vapor from 152 ^{o}C to 300 ^{o}C

$$Q_3 = 1.87 \cdot 148 \cdot 1 = 277\ kJ/s$$

Sum of the energy supplied in the three steps:

$$W_{s1} + Q_2 + Q_3 = 0.4 + 2357 + 277 = \mathbf{2634\ kJ/s}$$

(c) The total energy supplied is the same for the two alternatives (2634 kJ/s) because enthalpy is a state function and the total energy balance in both cases is

$$H_{\text{out}} - H_{\text{in}} = Q + W$$

where the in- and out-states are the same for both alternatives. Nevertheless, alternative 2 (with pumping) is chosen because the supplied mechanical work W_s is much smaller (0.4 kW in pump work versus 342 kW in compression work). However, note that we, for the pump alternative, must supply heat for evaporation at a *higher* temperature (152 ^{o}C instead of 100 ^{o}C).

E.3 Some more exercises

Mass balances

Exercise E.1* *A feed gas stream consists of methanol and an inert gas. Methanol is to be removed from this stream by dissolving it into water in an absorption column (see Figure 1.4). The outgoing gas stream is free of methanol and the inert gas is not soluble in water.* **Data.** *The gas feed is 100 mol/s with 7 mol% methanol. The water feed is 500 mol/s.*

(a) Formulate three independent mass balances.

(b) State any assumptions made and find the amount and composition of the two product streams.

Exercise E.2 *3 kg/s of a solution with 20 weight% ethanol at temperature 50 ^{o}C should be produced. You have available three streams: Pure ethanol at 20 ^{o}C, pure water at 20 ^{o}C and pure water at 100 ^{o}C.*

(a) Formulate the mass and energy balances.

(b) How much is needed of each stream? (Data: heat capacity for water is $c_p = 4.2$ kJ/kg,K and for ethanol $c_p = 2.6$ kJ/kg,K. Heat of mixing can be neglected).

Energy balances and compression

Exercise E.3 *(i) Find the changes in U, V, H, S and B when we take 1 mol of ideal gas with C_p=30 J/K,mol from 400K/ 10 bar (state 1) to 300K/ 4 bar (state 2). (ii) For an open system, suggest a continuous steady-state process that takes us from state 1 (inflow) to state 2 (outflow) (there may be several possible processes). What is Q, W_s and $Q + W_s$ (per mol in) for your suggested process? (iii) For a closed system, suggest a process that takes us from state 1 (at start) to state 2 (at the end) (there may be several possible processes). What is Q, W and $Q + W$ for your suggested process? (iv) Is it possible to have a process without cooling (i.e., with $Q = 0$)?*

Exercise E.4 *A gas stream of 30800 kmol/h with temperature 76 ^oC is to be compressed from 99 to 104 bar. The heat capacity can be assumed constant, c_p=31 J/mol,K.*

(a) What is the work and the temperature after a lossless (reversible) adiabatic compression?

(b) What is the real work and the temperature with an adiabatic efficiency of 0.74?

Some exercises with chemical reaction

Exercise E.5 *Formaldehyde ($HCHO$) is produced from methanol (CH_3OH) by combustion over a silver catalyst in a gas phase reactor,*

$$CH_3OH+\frac{1}{2}O_2 = HCHO+H_2O; \quad \Delta_r H^\ominus(973K) = -143\frac{kJ}{mole}, \ \Delta_r G^\ominus(973K) = -218\frac{kJ}{mole}$$

(a) Introduce the extent of reaction ξ for the reaction and formulate the mass balance equations (you can denote the feed with n_0 and the product with n).

(b) Find the product composition when the feed consists of 27 mol% methanol, 36% water, 7% oxygen and 30% nitrogen and we assume complete conversion of the limiting reactant.

(c) Formulate the energy balance. What is the feed temperature if the reactor operates adiabatically and the product temperature is 700 ^oC?

(d) In practice, there is not complete conversion, but the reaction is in equilibrium at 700 ^oC and 1.5 bar. Find the fraction of oxygen in the product.

(e) What is the heat of reaction and equilibrium constant at 600 ^oC?

Exercise E.6* *In a continuous methanol reactor that operates at 100 bar, the following gas phase reactions take place*

$$CO + 2H_2 = CH_3OH$$

$$CO_2 + H_2 = CO + H_2O$$

(the last is the "shift reaction").

(a) Introduce the extent of reaction ξ_j for the reactions and formulate the mass balances (you can denote the feed with n_0 and the product with n).

(b) The feed is 5000 mol/s and consists of $7.0\% CO, 2.0\% CO_2, 75.0\% H_2$ and $16.0\% CH_4$. Find the product composition when 2/3 of the supplied CO and CO_2 reacts to methanol and this generates 40 mol/s of water.

(c) Formulate the energy balance. The feed is at 150 ^oC and the product at 270 ^oC. How much cooling is needed in the reactor? (Use the component data given on page 416.)

(d) What is the product composition if 300 mol/s methanol is produced and the shift reaction is in equilibrium at 270 ^oC? Assume that the equilibrium constant for the shift reaction is $K = 0.016$.

(e) Find the equilibrium constant for the first reaction at 270 ^oC and check how close it is from the equilibrium found in (d).

Exercise E.7* **Combustion of ammonia to NO.** *Nitric acid is commercially produced in a process where ammonia is burned with air (with platinum as catalyst) to generate nitric gases that are cooled and absorbed in water. We consider the first part of the process (see Figure E.2), where we assume that ammonia burns completely following the gas phase reaction:*

$$NH_3 + 1.25O_2 = NO + 1.5H_2O$$

Data: *Assume that the air contains 21% oxygen. The standard heat of reaction is $\Delta_r H^\ominus(298K) = -227$ kJ/mol and $\Delta_r H^\ominus(1213K) = -223$ kJ/mol. Assume constant mean heat capacities:*

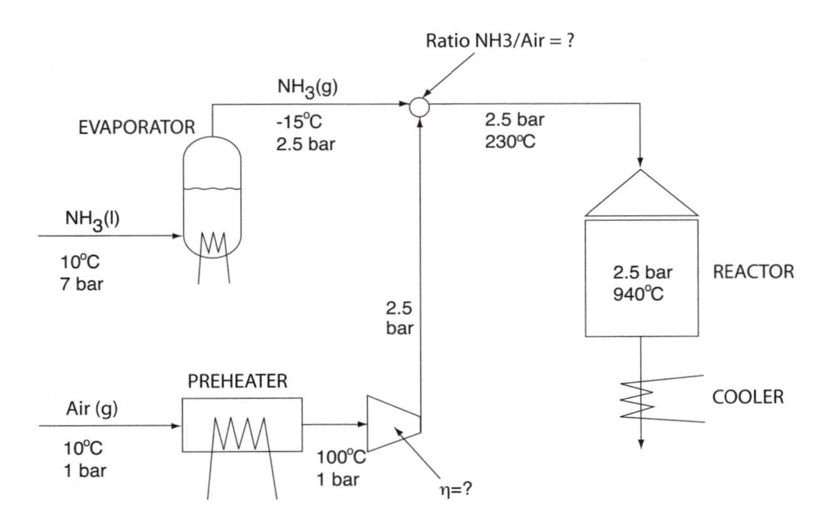

Figure E.2: First step in the production of nitric acid

$C_p(NH_3)(g) = 37$ J/K mol (up to 230 °C); = 54 J/K mol (230 °C to 940 °C)
$C_p(luft)(g) = 29$ J/K mol (up to 230 °C); = 32 J/K mol (230 °C to 940 °C)

(a) The reactor feed is a gas mixture of air and ammonia. Introduce the extent of reaction of the reaction and formulate the reactor mass balances.

(b) Find the composition of the reactor feed and reactor product when we assume an adiabatic reactor.

(c) Find the efficiency of the compressor from the data given in the figure.

(d) We have neglected the formation of NO_2. To see that this is a reasonable assumption, find the ratio between NO_2 and NO, which could be theoretically obtained if the reaction

$$NO + 0.5O_2 = NO_2$$

was in equilibrium at 940 °C and 2.5 bar (use data from the appendix to find the equilibrium constant).

Exercise E.8 Condensation of nitric acid. *We consider here the second step in the process for production of nitric acid; see Figure E.3. A gas stream (stream 1) at 200 °C has the composition (mole percent): 9% NO, 7% O_2, 70% N_2 and 14% H_2O. The stream is cooled to 30 °C such that the resulting condensate (stream 2) is 40% nitric acid (i.e., 40 weight% HNO_3 and the rest H_2O). The pressure in the process is 1 atm.*

Data: *The reaction equation is*

$$2NO(g) + \frac{3}{2}O_2(g) + H_2O(g) \rightarrow 2HNO_3(l)$$

The partial pressure of water (in stream 3) in equilibrium with nitric acid (stream 2) is given in Figure E.3 (note that 1 atm = 760 mm Hg). The partial pressure of HNO_3 can be neglected. Heat of mixing: see page 417 (remaining data can be taken e.g. from SI Chemical Data*).*

(a) Find the amount and composition of the three streams with 100 mol in stream 1 as basis. Find the composition of the gas (stream 3) together with the conversion of NO to HNO_3.

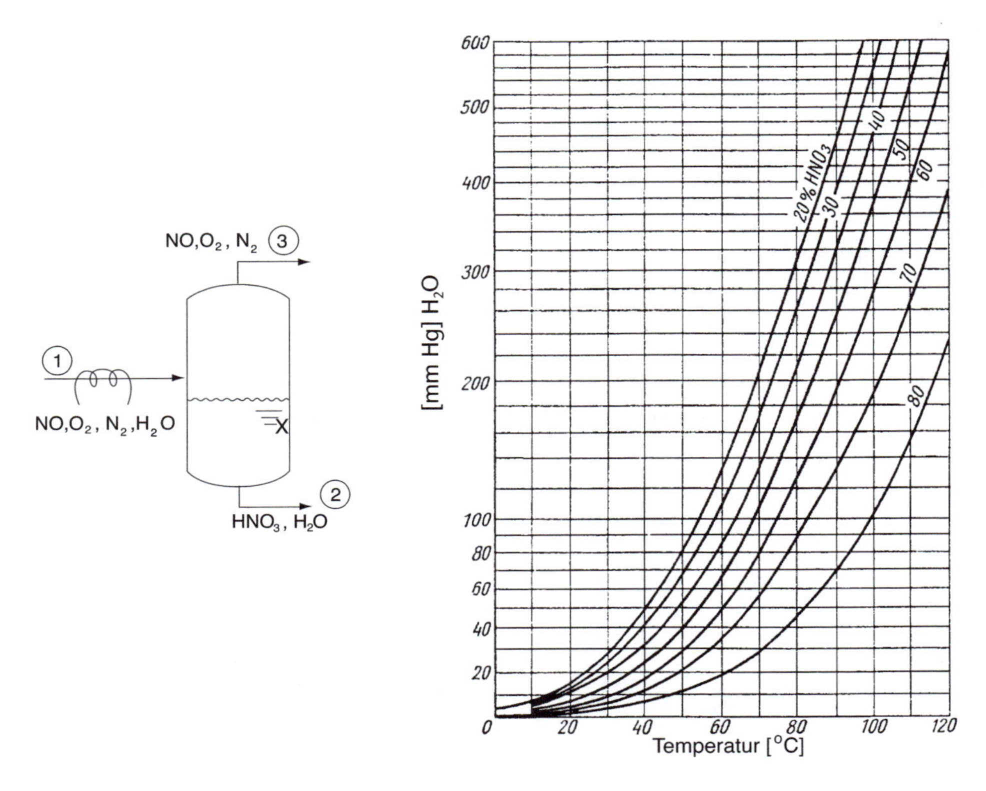

Figure E.3: (a) Process for condensation of nitric acid. (b) Partial pressure of water in nitric acid solution as function of temperature

(b) Find the standard enthalpy of reaction. Is the reaction exothermic?

(c) How can the process be changed to increase the conversion? (Some suggestions: change pressure, change the temperature, recycle, add air, add water.)

(d) Estimate the cooling needed when the feed stream is 100 mol/s (assume constant heat capacity for ideal gas and estimate the heat of mixing).

APPENDIX F

Data

Data are needed for all practical calculations. Some particularly useful numbers are given on page 15, and thermodynamic data for some selected gases are given on the next page (page 416). On page 186, Henry's constant for the solubility of some gases in water is given, and Antoine parameters to compute the vapor pressure of selected components are given on page 190. On page 417, some data for heat of mixing for mixture of acids, bases and salts in water are given. For additional data, see under *data* in the Index.

Furthermore, the following thermodynamic diagrams are included:

- pH diagram for methane (page 418)
- pH diagram for ammonia (page 419)
- HS diagram for water (steam) (page 420)
- c_p for liquids as function of temperature (page 421)
- c_p for gases as function of temperature (page 422)

On the book's home page, you find electronic versions of these thermodynamic diagrams for methane, ammonia and water, and in addition diagrams for:

- air, • CO_2, • ethane, • ethylene,
- propane, • propylene, • n-butane, • Refrigerant R134a

Here are some other sources for physical and thermodynamic data:

- G. Aylward and T. Findlay, *SI Chemical Data*, Wiley, 6th Edition, 2007. This small handbook is used as a reference in many chemistry classes. It is a bit inaccurate since decimals are not included for most thermodynamic quantities. For example, the heat of formation for CO_2 (g) is given as -394 kJ/mol, rather than -394.36 kJ/mol (as given on page 416). Also, the price is a little stiff when compared to its modest size and contents.
- Poling, Prausnitz and O'Connell, *The Properties of Gases and Liquids*, 5th Ed., McGraw-Hill, 2001. This is an excellent book. It contains component data for 468 components, including data for energy of formation and heat capacities as function of temperature for ideal gas. The book also contains methods for estimating data for other components.
- *CRC Handbook of Chemistry and Physics.* It contains a lot of data, but perhaps not so much for a process engineer.
- *Perry's Chemical Engineers' Handbook.* It contains a lot of data of interest for process calculations, but it is mainly a reference work for equipment calculations.
- Built-in data based in commercial simulation programs such as Aspentech, Hysys, Chemcad, Pro-II, etc.
- Many other books and journals, for example *Journal of Chemical Engineering Data.*

Component data for some gases

	M [g/mol]	T_b [K]	$\Delta_{vap}H$ [kJ/mol]	T_c [K]	p_c [bar]	$C_p^{\ominus}$ [J/K mol]	$\Delta_f H^{\ominus}$ [kJ/mol]	$\Delta_f G^{\ominus}$ [kJ/mol]
$CH_4(g)$	16.04	111.7	8.17	190.6	46.0	35.31	−74.81	−50.72
$C_2H_4(g)$	28.05	169.4	13.53	282.3	50.4	43.56	+52.26	+68.15
$C_2H_6(g)$	30.07	184.6	14.70	305.5	48.7	52.63	−84.68	−32.82
$C_3H_8(g)$	44.10	231.0	19.04	369.8	42.5	73.50	−103.85	−23.49
$C_6H_6(g)$	78.12	353.2	30.72	562.0	49.0	136.1	+49.0	+124.3
$CO(g)$	28.01	81.7	6.04	132.9	34.5	29.14	−110.53	−137.17
$CO_2(g)$	44.01	*	*	304.1	73.7	37.11	−393.51	−394.36
$HCHO(g)$	30.03	253.8	24.48	408	66	35.40	−108.57	−102.53
$CH_3OH(g)$	32.04	337.7	35.21	512.6	81.0	43.89	−200.66	−161.96
$C_2H_5OH(g)$	46.07	35.56	351.8	513.9	61.5	65.44	−277.69	−174.78
$H_2(g)$	2.016	20.3	0.89	33.2	13.0	28.82	0	0
$H_2O(g)$	18.015	373.15	40.68	647.1	220.6	33.58	−241.82	−228.57
$O_2(g)$	32.00	90.1	6.82	154.5	50.4	29.36	0	0
$O_3(g)$	48.00	161.8	14.20	261.0	55.7	39.20	+142.7	+163.2
$Ar(g)$	39.95	87.3	6.43	150.9	49.0	20.79	0	0
$Cl_2(g)$	70.91	239.1	20.41	417.0	77.0	33.91	0	0
$HCl(g)$	36.46	188.2	16.15	324.7	83.1	29.12	−92.31	−95.30
$He(g)$	4.003	4.30	0.08	5.19	2.27	20.79	0	0
$N_2(g)$	28.01	77.4	5.58	126.2	33.9	29.13	0	0
$NO(g)$	30.01	121.4	13.78	180.0	64.8	29.84	+90.25	+86.55
$NO_2(g)$	46.01	294.3	16.55	431.4	101	37.20	+33.18	+51.31
$NH_3(g)$	17.03	239.8	23.35	405.4	113.5	35.06	−46.11	−16.41
$H_2S(g)$	34.08	212.8	18.68	373.4	89.6	34.23	−20.63	−33.56
$SO_2(g)$	64.06	263.1	24.94	430.8	78.8	39.87	−296.83	−300.19

* CO_2 has no normal boiling point because it does not exist as liquid for pressures below 5.2 bar (triple point pressure). At 1 atm, CO_2 sublimes (goes directly from solid to vapor) at 195 K.

- The table gives the molar mass (M), normal boiling temperature at 1 atm (T_b), heat of vaporization ($\Delta_{vap}H$) at T_b, critical temperature (T_c), critical pressure (p_c), ideal gas heat capacity $C_p^{\ominus}(298)$, standard enthalpy (heat) of formation $\Delta_f H^{\ominus}(298)$ and standard Gibbs energy of formation $\Delta_f G^{\ominus}(298)$. The latter three are for the standard state as ideal gas at $p^{\ominus} = 1$ bar and 298.15 K.
- Since the data in this table are for gases, $\Delta_f H^{\ominus}$ and $\Delta_f G^{\ominus}$ are the enthalpy and Gibbs energy of reaction for forming the component as ideal gas from the elements in their standard state (everything at 1 bar and 298.15 K). The standard states for the elements in the table are: H_2(g), N_2(g), O_2(g), C(s, graphite) and S(s, rhombic). For example, for methanol, the formation reaction is $C(\text{s, graphite}) + 2H_2(g) + 0.5O_2(g) = CH_3OH(g)$.

Data for heat of mixing[1]

We define for a mixture at given T and p:

- H_i^* [J/mol] – enthalpy of pure component i.
- $\bar{H}_i = \partial H / \partial n_i$ [J/mol] – partial enthalpy for component i in the mixture, where H is the enthalpy of mixing.
- $\bar{H}_i - H_i^*$ [J/mol] – partial heat of mixing for component i.
- $H_m = \sum_i x_i \bar{H}_i$ [J/mol] – molar enthalpy for the mixture.
- $\Delta_{\text{mix}} H = H - H^* = \sum_i x_i (\bar{H}_i - H_i^*)$ [J/mol] – heat of mixing.

We have mostly omitted subscript m that is used to indicate molar quantities.

Partial heat of mixing (solution heat) for producing an infinite diluted solution at 20 ^oC:

Component i	Partial heat of mixing $\bar{H}_i - H_i^*$ [kJ/mol i]
H_2SO_4	−96
HBr, HI	−80
HCl	−75
HNO_3	−31
KOH	−56
$NaOH$	−43
NH_3	−35
$CaCl_2$	−75
K_2CO_3, Na_2CO_3	−25

The above table gives heat of mixing when 1 mol of pure component (undiluted acid or base or pure salt) is mixed with a large amount (n mol) of water, where "large" is roughly $n > 10$ mol. For smaller amounts of water (more concentrated solutions), the following correlations can be used (heat of mixing for 1 mol acid and n mol water):

$$\text{HCl}: \quad (-72.6 + 50.1/\text{n}) \frac{\text{kJ}}{\text{mol HCl}} \quad (1 < \text{n} < 10)$$

$$\text{H}_2\text{SO}_4: \quad \left(\frac{-74.7}{1 + 1.80/\text{n}}\right) \frac{\text{kJ}}{\text{mol H}_2\text{SO}_4} \quad (0 < \text{n} < 10)$$

$$\text{HNO}_3: \quad \left(\frac{-37.5}{1 + 1.74/\text{n}}\right) \frac{\text{kJ}}{\text{mol HNO}_3} \quad (0 < \text{n} < 10)$$

[1] Data from S.D. Beskow, *Technisch-Chemische Berechnungen*, Deutscher Verlag für Grundstoffindustrie, 1962.

1

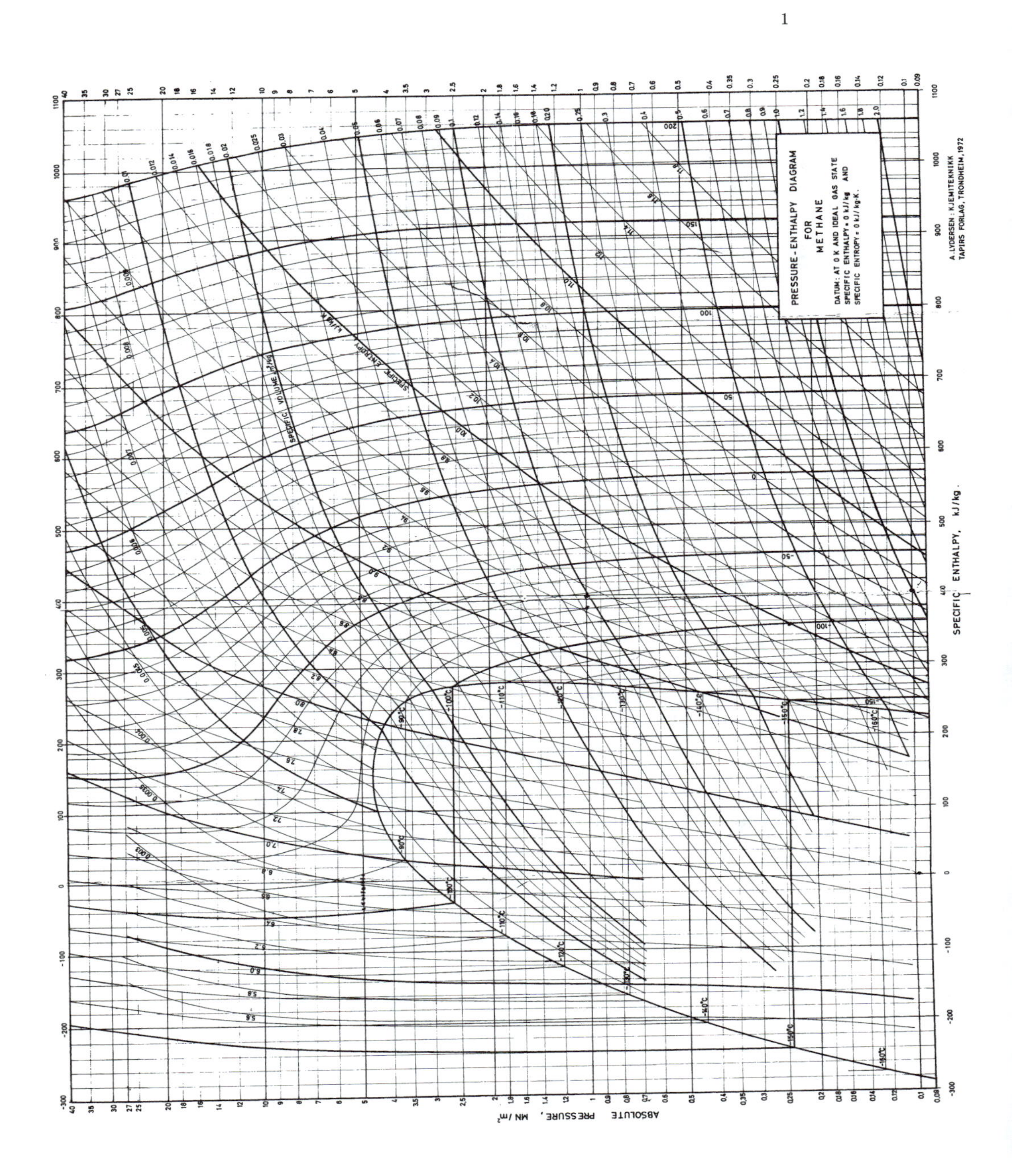

Figure F.1: Pressure enthalpy diagram for methane
From: Aksel Lydersen, *Kjemiteknikk*, Tapir, Trondheim, 1972

1

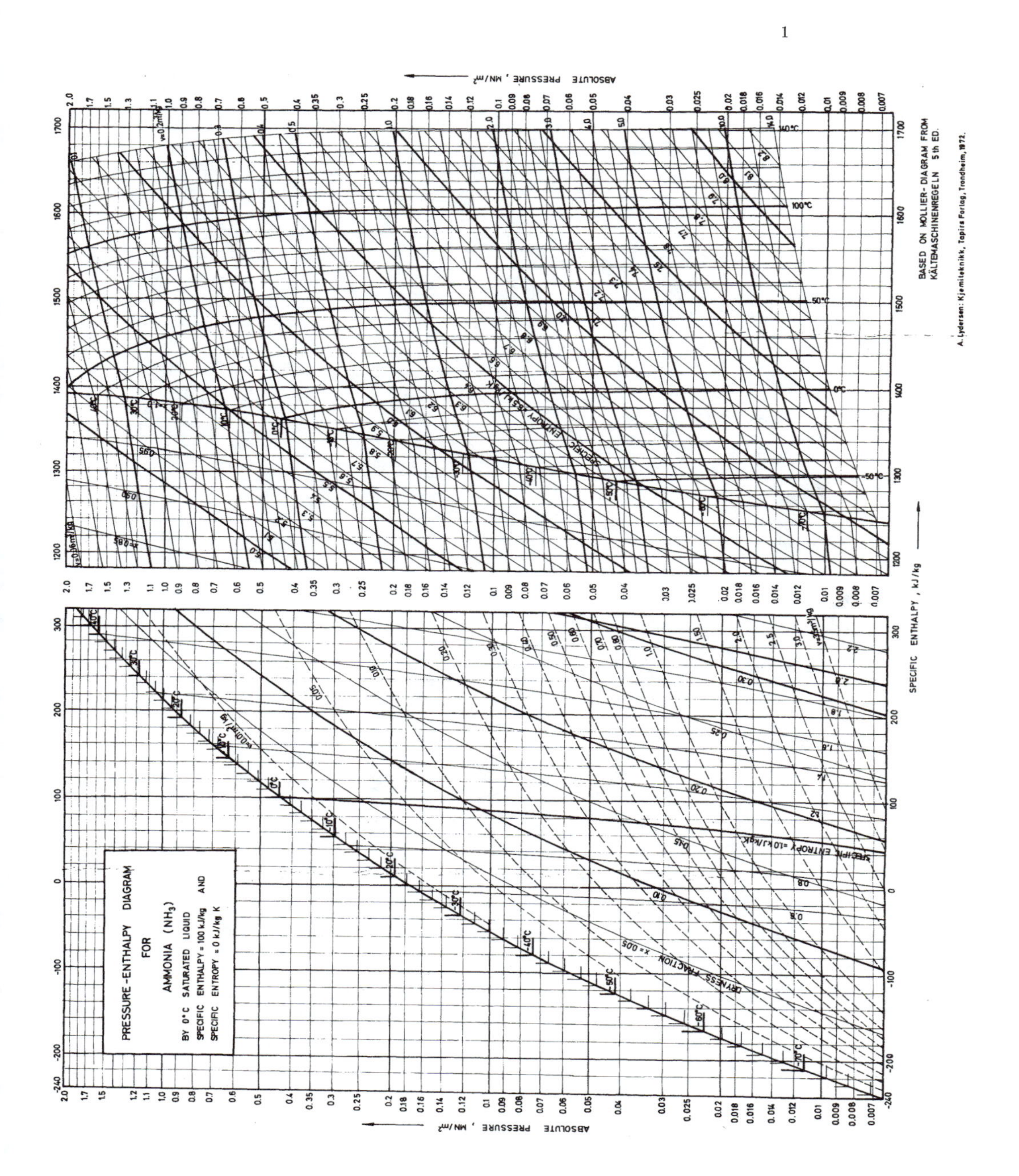

Figure F.2: Pressure enthalpy diagram for ammonia
From: Aksel Lydersen, *Kjemiteknikk*, Tapir, Trondheim, 1972

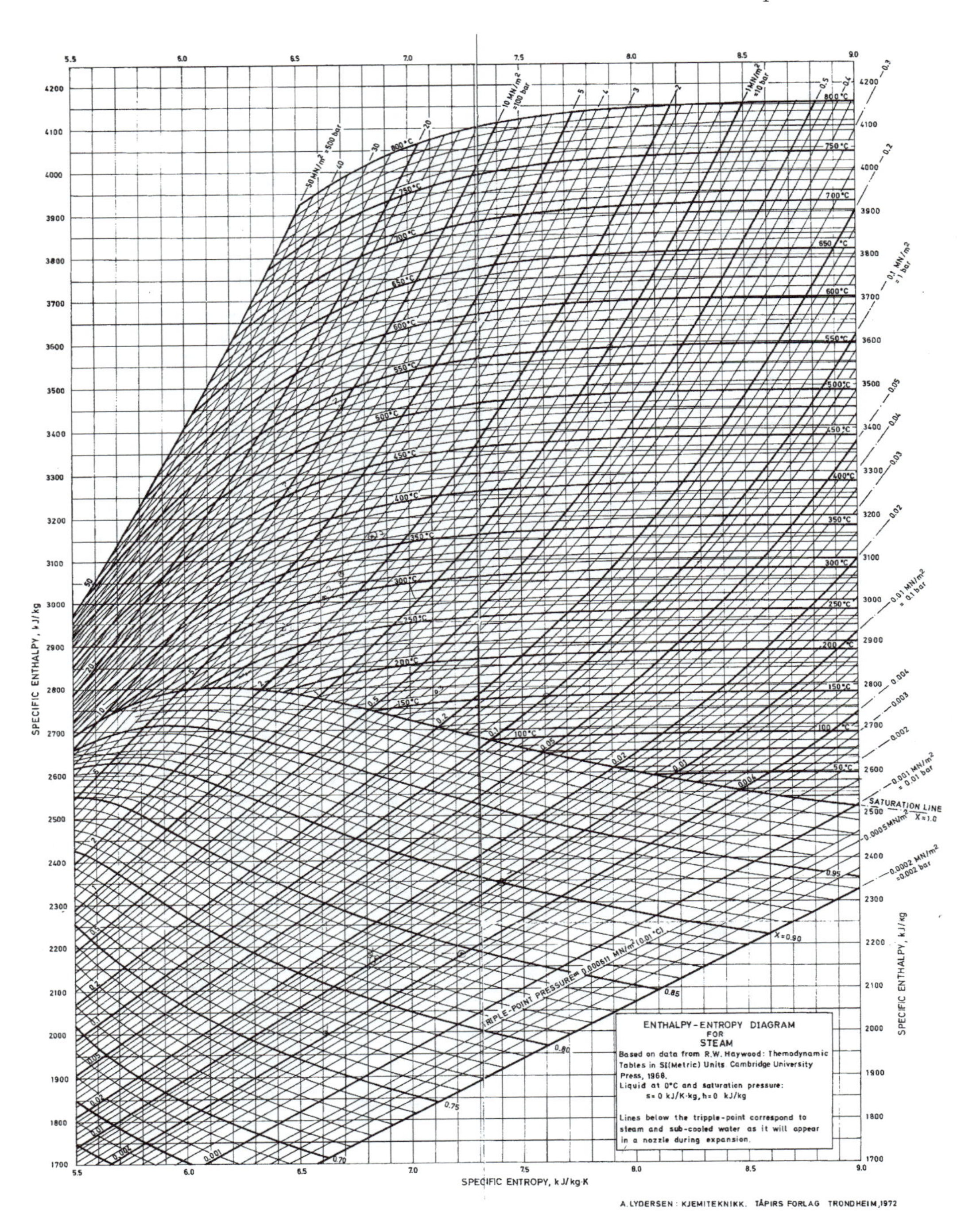

Figure F.3: Enthalpy-entropy diagram for water (steam)
From: Aksel Lydersen, *Kjemiteknikk*, Tapir, Trondheim, 1972

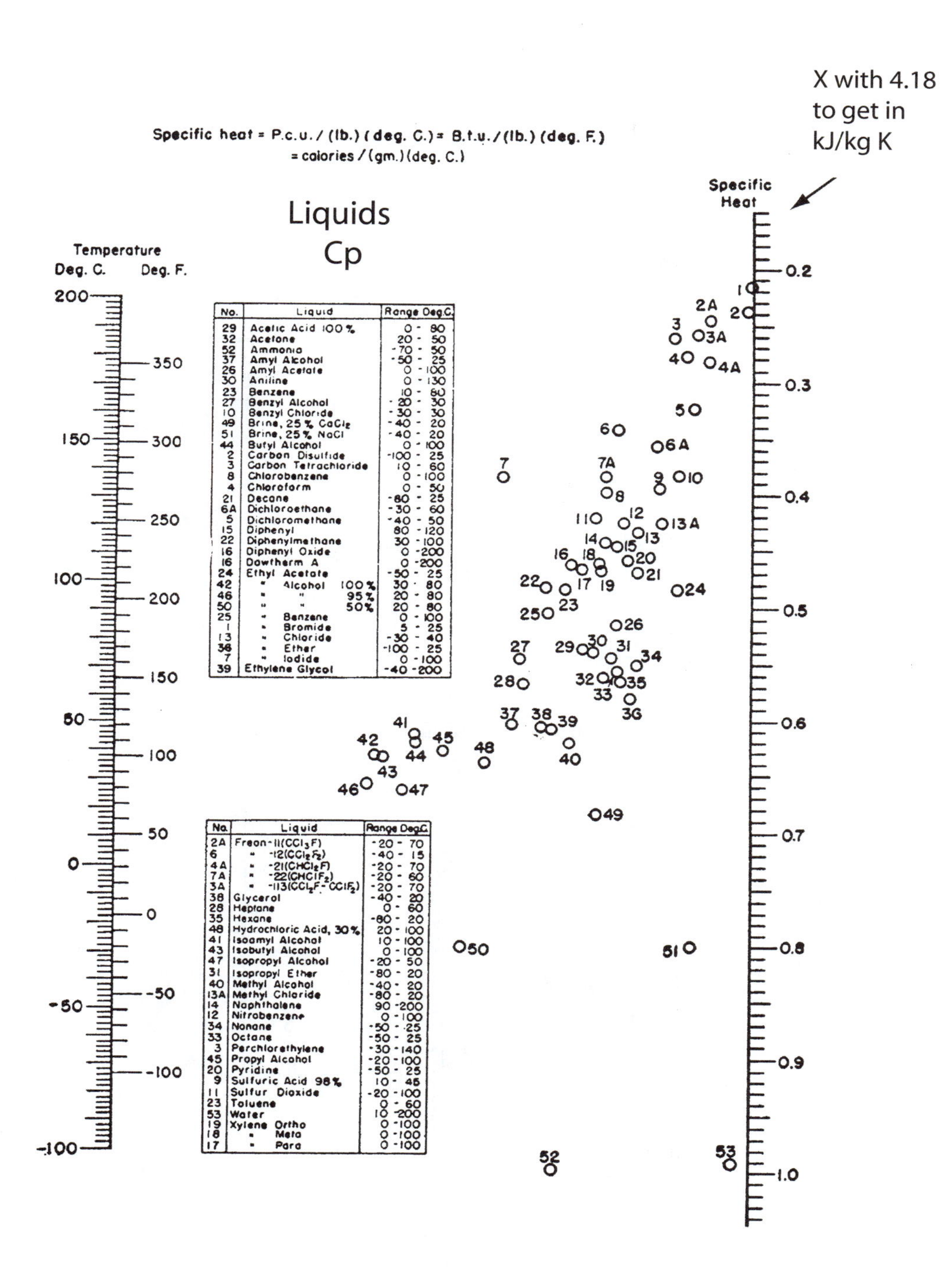

No.	Liquid	Range Deg.C.
29	Acetic Acid 100%	0 - 80
32	Acetone	20 - 50
52	Ammonia	-70 - 50
37	Amyl Alcohol	-50 - 25
26	Amyl Acetate	0 - 100
30	Aniline	0 - 130
23	Benzene	10 - 80
27	Benzyl Alcohol	-20 - 30
10	Benzyl Chloride	-30 - 30
49	Brine, 25% $CaCl_2$	-40 - 20
51	Brine, 25% NaCl	-40 - 20
44	Butyl Alcohol	0 - 100
2	Carbon Disulfide	-100 - 25
3	Carbon Tetrachloride	10 - 60
8	Chlorobenzene	0 - 100
4	Chloroform	0 - 50
21	Decane	-80 - 25
6A	Dichloroethane	-30 - 60
5	Dichloromethane	-40 - 50
15	Diphenyl	80 - 120
22	Diphenylmethane	30 - 100
16	Diphenyl Oxide	0 - 200
16	Dowtherm A	0 - 200
24	Ethyl Acetate	-50 - 25
42	" Alcohol 100%	30 - 80
46	" " 95%	20 - 80
50	" " 50%	20 - 80
25	" Benzene	0 - 100
1	" Bromide	5 - 25
13	" Chloride	-30 - 40
36	" Ether	-100 - 25
7	" Iodide	0 - 100
39	Ethylene Glycol	-40 - 200

No.	Liquid	Range Deg.C.
2A	Freon-11(CCl_3F)	-20 - 70
6	" -12(CCl_2F_2)	-40 - 15
4A	" -21($CHCl_2F$)	-20 - 70
7A	" -22($CHClF_2$)	-20 - 60
3A	" -113($CCl_2F-CClF_2$)	-20 - 70
38	Glycerol	-40 - 20
28	Heptane	0 - 60
35	Hexane	-80 - 20
48	Hydrochloric Acid, 30%	20 - 100
41	Isoamyl Alcohol	10 - 100
43	Isobutyl Alcohol	0 - 100
47	Isopropyl Alcohol	-20 - 50
31	Isopropyl Ether	-80 - 20
40	Methyl Alcohol	-40 - 20
13A	Methyl Chloride	-80 - 20
14	Naphthalene	90 - 200
12	Nitrobenzene	0 - 100
34	Nonane	-50 - 25
33	Octane	-50 - 25
3	Perchlorethylene	-30 - 140
45	Propyl Alcohol	-20 - 100
20	Pyridine	-50 - 25
9	Sulfuric Acid 98%	10 - 45
11	Sulfur Dioxide	-20 - 100
23	Toluene	0 - 60
53	Water	10 - 200
19	Xylene Ortho	0 - 100
18	" Meta	0 - 100
17	" Para	0 - 100

Figure F.4: Heat capacity for liquids as function of temperature
From: *Perry's Chemical Engineers' Handbook*

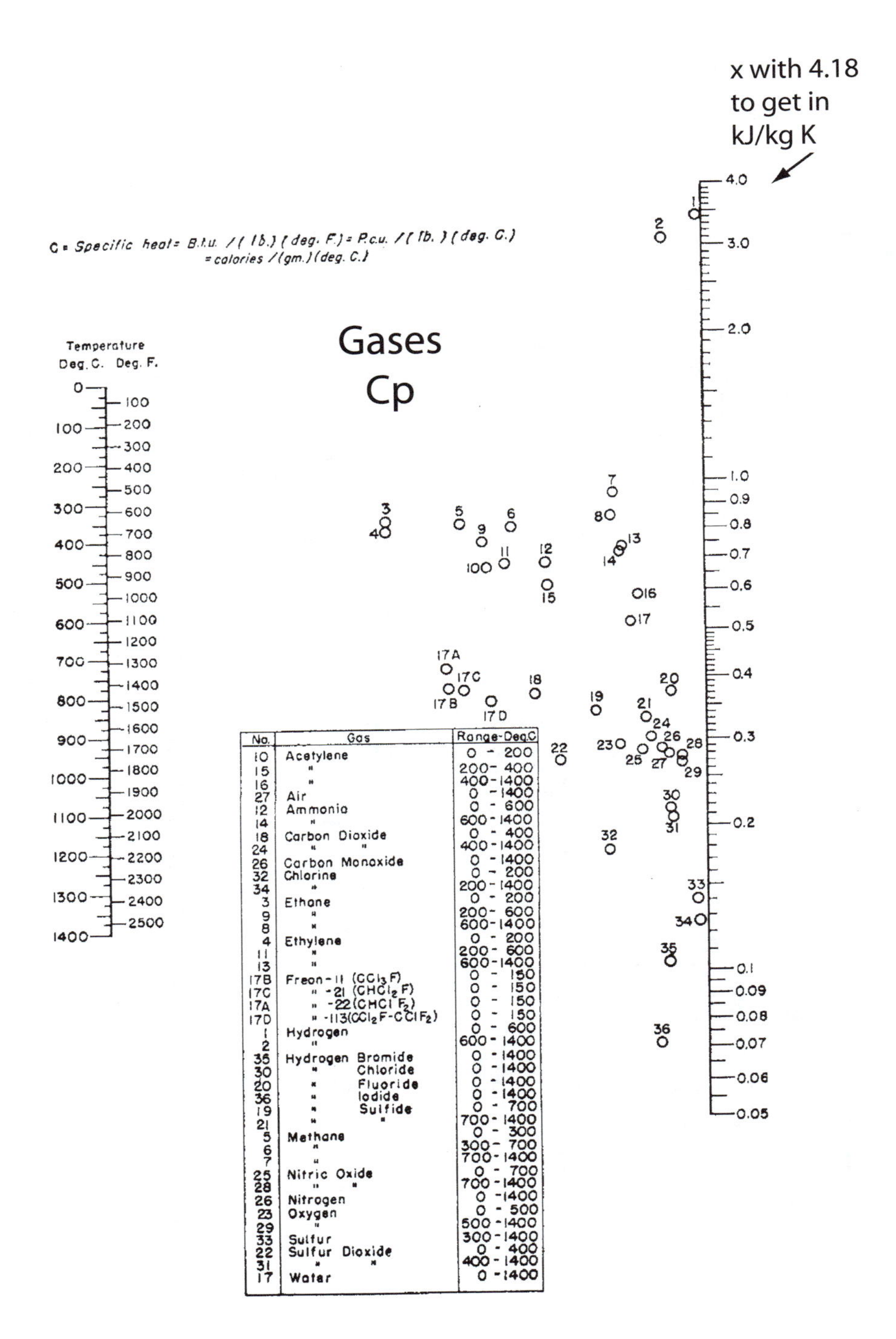

No.	Gas	Range-Deg.C
10	Acetylene	0 - 200
15	"	200 - 400
16	"	400 - 1400
27	Air	0 - 1400
12	Ammonia	0 - 600
14	"	600 - 1400
18	Carbon Dioxide	0 - 400
24	" "	400 - 1400
26	Carbon Monoxide	0 - 1400
32	Chlorine	0 - 200
34	"	200 - 1400
3	Ethane	0 - 200
9	"	200 - 600
8	"	600 - 1400
4	Ethylene	0 - 200
11	"	200 - 600
13	"	600 - 1400
17B	Freon-11 (CCl_3F)	0 - 150
17C	" -21 ($CHCl_2F$)	0 - 150
17A	" -22 ($CHClF_2$)	0 - 150
17D	" -113 (CCl_2F-$CClF_2$)	0 - 150
1	Hydrogen	0 - 600
2	"	600 - 1400
35	Hydrogen Bromide	0 - 1400
30	" Chloride	0 - 1400
20	" Fluoride	0 - 1400
36	" Iodide	0 - 1400
19	" Sulfide	0 - 700
21	" "	700 - 1400
5	Methane	0 - 300
6	"	300 - 700
7	"	700 - 1400
25	Nitric Oxide	0 - 700
28	" "	700 - 1400
26	Nitrogen	0 - 1400
23	Oxygen	0 - 500
29	"	500 - 1400
33	Sulfur	300 - 1400
22	Sulfur Dioxide	0 - 400
31	" "	400 - 1400
17	Water	0 - 1400

Figure F.5: Heat capacity for gases as function of temperature
From: *Perry's Chemical Engineers' Handbook*

APPENDIX G

Solutions to starred exercises

Detailed solutions are available at the book's home page. Solutions to *non-starred* exercises are only available to instructors.

1.1, page 9. 44.70 m/s
1.3, page 13. 12.7 kW
1.6, page 26. $C = C_V \cdot 63.09\text{e} - 6 \cdot \sqrt{14.5\text{e} - 2} = C_V/41625$
1.7, page 31. 27.8 $/GJ
1.8, page 31. 27.8 $/GJ (assuming the lower heating value is 45 MJ/kg)
1.9, page 32. 0.46 $
1.10, page 32. 0.372 kg of oil (with water as a gas product)
1.11, page 32. 0.63 $/l = 2.38 $/gal
1.12, page 32. 0.71 $/GJ (Saudi), 7.63 $/GJ (Europe), 13.25 $/GJ (USA)
1.13, page 32. 0.32 $/Sm^3
1.14, page 32. 1277 kg/s. 70.6 GW.
1.15, page 32. (a) 10.9 billion $/year. (b) 19.8 billion $/year.
1.16, page 32. 1.18 Sm^3.
1.17, page 32. 1.2 E12 l/y, 26636 kg/s, 1.2 E12 J/s = 1.2 TW.
1.18, page 32. 504 EJ/y, $1.4 \cdot 10^{14}$ kWh/y, 16 TW, 2.7 kW/person
1.19, page 33. 2.3 years.
1.20, page 33. 484 000 km^2 (0.3% of the earth's area).
1.21, page 33. 0.78 $/kWh
1.22, page 34. 44 W, 92 W, 218 W
1.23, page 34. 370g.
1.24, page 34. (a) 3600 W. (b) 305W.
1.25, page 34. 2000 W.
1.26, page 34. 111 W
1.27, page 34. 1.7 W
1.28, page 35. 7.9 W.
1.29, page 35. (a) 27°C. (b) 12°C. (c) −13°C.
1.30, page 36. 3.54 kg.
1.31, page 36. 339 W to 678 W.
1.32, page 36. 12 liters.
2.1, page 46. (ii) 20.3 mmol/l
2.3, page 52. 1183 kmol/h.
2.4, page 52. (a) 0.227 g S/l. (b) 0.077 g S/l. (c) Better with several small rinsings.
2.5, page 52. (a) 3. (b) 1.3527 kg/s
2.9, page 55. (a) 94.99 wt-% ethanol, (b) 40.25 g water and 763.98 g ethanol, (c) 42.39 weight% and 48.20 volume-% ethanol, (d) 0.9301 g/l, (e) 1.938 l.
2.10, page 56. See equations in Example 2.16 (page 73).
2.11, page 61. (a) 22.06 mol/s and 529.5 mol/s. (b) 8.59 mol/s and 850.3 mol/s. (c) 10.2

mol/s and 244.6 mol/s.
2.12, page 61. (a) 977.84 mol/s (b)255.7 mol/s
2.13, page 62. 0.0521
2.16, page 64. (a) 0.44 kg/s, (b) 50%, 0.418 kg/s
2.17, page 71. $R_W = 0.41$.
3.1, page 84. (a) methanol, 0.95, (b) 0.2 mol, (c), 0.7895, (d) 0.75/0.95 = 0.7895
3.2, page 85. (b) 2.6%C_3H_8, 48.7%C_3H_6, 48.7%H_2. (c) 9.0065 (d)9.54%.
3.4, page 91. See detailed solution available on book's home page.
3.5, page 91. (a) $N_r = 1$, (c) $N_r = 5$. See detailed solution.
3.6, page 94. 11.5% CO, 4.8% CO_2, 3.9% CH_4, 52.7% H_2, 0.1% N_2, 27.1% H_2O.
4.2, page 111. (a) 0.841, 0.093, 1.566 [kg/s]. (b) 0.670, -0.145, 1.965 [kg/s]
4.4, page 111. 4.2, 0.3, 3.5, 1 [kg/s].
4.5, page 112. 5.6, 0.4, 5.0, 1 [kg/s], that is, 0.1 kg/s oil is lost.
4.7, page 113. (b) 72 K. (c) 326 kW.
4.9, page 118. (a) 105 ^oC (b)2.25 MW.
5.1, page 138. (a) 5.1 kg/s. (b) $UA = 69910$ W/K. (c) 304 ^oC and 7.22 kg/s
5.2, page 139. (a) 3.19 kg/s cooling water. (b) Countercurrent flow: $\Delta T_{\text{lm}} = 13.1$K and $A = 28$ m^2. (c) Cocurrent flow: $\Delta T_{\text{lm}} = 9.9$K and $A = 37$ m^2. (d) Two tube passes: We read $F \approx 0.88$ and find $A = 28$ m$^2/0.88 = 32$m^2.
6.1, page 149. This is mostly an exercise in integration; note that $\int x^n dx = \frac{1}{n+1}x^{n+1}$.
6.2, page 153. 547 kW
6.4, page 153. (a) 41.8 kW, (b) 25.3 kW
6.5, page 158. 3.79 MW. 75.2%.
7.3, page 179. Assuming constant $\Delta_r C_p^\ominus$ gives $K(940^oC) = 0.033$ and the ratio is 0.012.
7.5, page 182. See detailed solution.
7.6, page 182. 2.5 K, which is surprisingly large!
7.8, page 195. (a) 1.1585 bar, (b) 0.0018 bar
7.9, page 195. (a) 294.7 K, (b) 415.0K
7.11, page 195. 34.9 mol/s
7.13, page 196. (a) $x_1 = 0.5525$, (b) Bubble point, $T = 373.3408$K, $y_1 = 0.4473$, (c) Dew point. $T = 346.7518$K, $x_1 = 0.1835$.
8.1, page 202. (a) No, (b) Yes, zero
8.4, page 222. (a) 2.97 MJ, (b) -3.3 MJ
10.3, page 270. (a) 0.643. (b) 0.835.
10.5, page 270. (a) 0.0020 mol/l s. (b) $n_A = n_{A0} - 2kc_A^2 V$. (c) 74.1 m^3. (d) Yes (reaction order positive).
10.6, page 270. (a) $k = 0.0713$ min^{-1} gives $t = 9.72$ min. (b) 51%.
11.2, page 299. $\tau = mc_{pL}/wc_{pV}$
A.1, page 328. (1/18.015 e-3) mol = 55.5 mol
A.2, page 331. (a) T_b[K] = 46.13+3816.44/(11.6703+1.25$\cdot10^{-4}$h[m]). (b) $dT_b/dh = -0.0035$ K/m. (c) 344.8 K (71.7 ^oC).
A.3, page 341. (b) $z = 0.84$, (c) $z = 0.894$, (d) $z = 0.864$.
A.5, page 359. (a) 43.7 kJ/mol, (b) 1344 K
A.6, page 364. (a)-379.2 kJ/mol. (b)-387 kJ/mol.
E.1, page 411. 507 mol/s liquid out
E.6, page 412. (b) 90 mol/s CO, (c) 10.5 MW, (e) 86.5 mol/s CO
E.7, page 412. (b) Feed: 11% NH_3 and 89% air. Product: 11% NO, 16% H_2O, 5% O_2, 68% N_2. (c) 67.9%. (d) $K = 0.033$; $NO_2/NO = 0.012$.

INDEX

- A (area), 132
- a (activity), 386
- a (annum) (1 a = 1 y), 8
- Å (Ångstrøm), 9
- absolute enthalpy, 105, 364
- absolute pressure, 11
- absolute temperature (T), 10
- absorber, 21
- absorption, 21
- acceleration of gravity (g), 2
- acentric factor, 340
- acre-foot (Sigurd's favorite unit ☺), 9
- acrylonitrile, 81, 87
- activation energy (E), 256, 259
- activity (a), 386
- activity coefficient (γ), 188, 387
- adiabatic, 19
- adiabatic expansion, 350
  - derivation pV^γ, 350
  - irreversible (isenthalpic), 171
  - reversible (isentropic), 170
- adiabatic process, 170
- adiabatic temperature change, 112, 120
- air, data, 16
- airplane wing, 241
- algebraic equation, 316
- ammonia, 417
  - chemical equilibrium, 176
  - combustion, 412
  - compression, 159
  - data, 416
  - flash, 194
  - Henry's law, 187
  - kinetics, 258, 259
  - refrigerant, 208
  - synthesis gas, 94
  - thermodynamic diagram, 419
- ammonia plant
  - heat recovery, 229
- analysis, 75
- Antoine parameters (data), 190
- Antoine's equation, 181
- API gravity, 13
- Arrhenius equation, 256, 276
- atm (1 atm = 1.01325 bar), 11, 16
- atmospheric pressure, 11
- atom matrix (A), 89, 329
- atomic balance, 41, 86, 329
- atomic mass (weight), 327
- auditorium
  - balance, 40
  - entropy, 371
- autocatalysis, 260
- automatic control, 322
- availability, *see* exergy
- Avogadro's law, 334
- Avogadro's number (N_A), 327
- axiom, 347
- azeotrope, 183

- B (exergy), 213
- balance
  - degrees of freedom, 66
  - dynamic, 273–325
  - no accumulation, 43
  - over time period, 42
  - per unit of time, 42
  - procedure for, 67
  - solvability, 66
    - quick analysis, 70
  - steady-state process, 43
  - with reaction, 77
- balance equation, 66
  - dynamic, 274
  - general form, 42
- balancing chemical reactions, 328
- bank account, 40
- bar, 11
- barometric formula, 331
  - temperature change, 352
- basis, 66, 67
  - smart, 48
- batch distillation, 396
- batch process, 20, *27*[1]
  - energy balance, 102
  - recycle, 64
  - typical batch size, 28
- batch reactor, 78, 262

[1] Page numbers in *italic* refer to definitions.

bathtub, dynamics, 296
battery, 213
bbl (barrel), 9
BCFD (billion cubic feet per day), 15
Bernoulli's equation, 240–251
biology, 266
bioproteins, 86
blending
  linear mixing rule, 50
blower, 24
boiler, 21
boiling point (T_b), 16, 378
  elevation, 184
  pressure variation, 331
boundary, 19
Boyle's law, 334
BPD (barrels per day), 15
Btu (British thermal unit, 12
bubble point
  df, 190
butane
  i-butane (refrigerant R600a), 209
  n-butane, thermodynamic diagram, 415
bypass, 300

^{o}C (degrees Celsius), 10
C_v (valve coefficient), 26
C_p, C_V, *see* heat capacity
calcium chloride, heat of mixing, 113
calorie, 16
calorie (cal), 11
calorimetry, 347
canonical variable, 332, 389
carbon, 327
carbon dioxide, *see* CO_2
Carnot cycle, 373
Carnot efficiency, *see* Carnot factor
Carnot factor, 200, 211, 375
  log mean, 201, 233
  steam turbine process, 233
catalysis, 259
  car, 259
Celsius (^{o}C), 10
centrifuge, 26
Charles' law, 334
chemical bonding energy, 345
chemical equilibrium, 174, 179, 385
  ammonia synthesis, 176
  methanol reactor, 92
  reaction kinetics, 258
chemical potential (μ), 385, 390
  VLE, 180, 182
chemical reaction, 77
  balancing, 328
  energy balance, 118
    dynamic, 281
choke
  flow measurement, 242
  valve, 26
choke valve, 114
Clapeyron, 180, 343
Clausius-Clapeyron, 181
closed system, 346, 377
  energy balance, 101
CO
  combustion, 389
CO_2
  combustion, 389
  contents in air, 16
  data, 416
  refrigerant, 208, 209
  solubility in water, 186
  sublimation, 416
  thermodynamic diagram, 415
coal, 17
coalescer, 21
cocurrent flow, 133
coefficient of performance (COP), 204
cogeneration, *see* combined cycle
cold reservoir, 199
colligative property, 185
column
  section, 21
combined cycle, 203, 226
combustion, 119, 226, 229, 412
  ammonia, 412
  engine, 203
  equlibrium between CO and CO_2, 389
combustion gas turbine, 203
component balance, 48
  concentration
    dynamic, 277
  with reaction, 78
    dynamic, 276
compressibility factor (z), 337
  generalized diagram, 336
compressible flow in pipe, 247
compression, 144
  intermediate cooling, 151, 152
  isentropic, 157
  isothermal, 152, 156
  real gas, 155, 156
  work, 144
compression ratio, 153
compressor, 22, 70
condensation, 333, 358
condenser, 22, 23
conservation law, 40
conservation principle
  energy, 40
  mass, 40
  momentum, 40

conserved quantity, 40
  balance equation, 43
constitutive equation, 20
continuity equation, 126, 240
continuous process, *28*
  reactor, 79
  steady-state, 43
  typical production rates, 28
continuous stirred tank
  concentration response, 289
  temperature response, 290
    with heat exchange, 291
continuous stirred tank reactor (CSTR), *see* reactor
control, 322
control valve, 26
control volume, 19, 67, 76
conversion (X), 79, *80*
  total, 85
conversion factors, 8–15
  exercises, 31
cooking plate, dynamics, 293
cooling process
  minimum work, 219
countercurrent flow, 133
countercurrent flow principle, 47, 130
critical compressibility (z_c), 337
critical data (T_c, p_c), 416
critical point, *180*, 333, 337, 339
critical pressure (p_c), 180, 333, 337, 339, 342
critical pressure ratio, 128
critical temperature (T_c), *180*, 186, 333, 337, 339, 342
critical volume (V_c), 339
crude oil
  classification, 13
  density (API gravity), 13
  heating value, 12, 363
  price, 31
crystallization, 62
crystallizer, 22
CSTR, *see* reactor
cubic equation of state, 338
cyclic process, 198, 373
  Carnot cycle, 373
  Diesel cycle, 203
  Otto cycle, 203
  Rankine cycle, 203
  Rankine refrigeration cycle, 206
cyclone, 22

d (day), 8
DAE system, 279, 283, 316
Dalton's law, 335
data, 415
  air, 16, 296
  ammonia, pH-diagram, 419
  Antoine parameters, 190
  data for some gases, 416
  heat capacity
    gases, 422
    liquids, 421
  heat of mixing, 417
  Henry's constant, 186
  hydrocarbons, 17
  methane, pH-diagram, 418
  water, 16, 296
    HS-diagram, 420
decanter, 22
degrees of freedom, 66
  equation counting, 68
  quick analysis, 69, 70
delay (θ), 286
demister, 22
density (ρ), 5, 13
deposition, 358
design, 75
  heat exchanger, 131
  versus simulation, 139
deviation variables, 289
dew point, 191
dialysis, 44
Diesel cycle, 203
differential equation, 273
differential-algebraic equation (DAE) set, 316
distillation, 22, 223
  batch, 396
  continuous, dynamics, 313
  entropy, 168
  side stream, 51
dot notation ($\dot{X}$), 5, 274
dry basis, *16*, 362
dryer, 23
dynamic flash, 316
dynamic model, 273
dynamic pressure, 240
dynamics, 273–325
  distillation, 313
  gas tank, 393
  process, 21
  simulation, 303–325
    Euler integration, 303

E, *see* energy
E (exa), 2
economy, 29
efficiency (η), 22, 26, *209*
  Carnot, 200, 211
  compression, 144
  energy, 210
  exergy, 211, 215

heat exchanger, 140
solar cell, 33
thermal, 210
thermodynamic, 209
turbine, 145
Einstein, 41, 97
electric energy, 345
electrochemical work, 98, 344
elementary kinetics, 256
elementary reaction, 257
end time (t_f), 76
endothermic, 360
energy, *344*
chemical, 345
conversion factor, 11
electric, 345
fun exercises, 31
global consumption, 37
internal (U), *see* internal energy
kinetic, 344
latent, 345
potential, 345
pressure correction, 365
rotation, 345
sensitive, 365
surface, 345
thermal, 345, 365
total, 345
energy balance, 95–128
dynamic, 277–283
enthalpy, 100
enthalpy balance, 102
gas tank, 393
general form, 100
isenthalpic, 114
mechanical, 237–251
reaction, 118
reading rule, 102
steady state, 102, 126
with kinetic and potential energy, 125
energy efficiency, 210
energy forms, 96
engine
gasoline, 203
enthalpy (H), 99, 105, *348*, 354, 357
absolute, 105, 364
constant pressure, 348
continuous process, 99
dynamic energy balance, 278
heat of reaction, 360
ideal mixture, 355
of reaction, 360
pressure dependency, 356
Joule's experiment, 353
pressure relief valve, 172
reference state, 364, 365
subprocesses, 106, 122
temperature independency, 355
thermochemistry, 357
vaporization, 358
enthalpy of formation ($\Delta_f H^\ominus$), 219, 360, 364, 416
data, 416
from heat of combustion, 363
enthalpy-entropy (HS) diagram, 156
water, 420
entropy (S), 156, 161–179, *371*
calculation of, 163, 173, 377
disorder, 369
friction, 249
heat exchanger, 167
ideal mixture, 165, 381
maximum (equilibrium), 384
phase transition, 378
state function, 375
enzyme, 260
equation of state, 20, 334–343
entropy, 167
ideal gas, 334
Peng-Robinson (PR), 340
Soave-Redlich-Kwong (SRK), 340
van der Waals, 337
equilibrium, 173, 179, 383
chemical, 174, 385
Gibbs energy, 384
temperature, 384
vapor/liquid, 179
equilibrium constant (K), 386
temperature dependency, 388
equivalent work, 215
ethane, 83, 341
data, 416
thermodynamic diagram, 415
ethylene
thermodynamic diagram, 415
Euler integration, 304
algorithm, 304
MATLAB, 306, 307
mixing tank, 304
evaporation, 358
dynamic energy balance, 280
evaporator, 24
energy balance, 280
temperature response, 299
Excel introduction, 59
exergy, *213*
exothermic, 360
expansion, 144
closed system, 351
isentropic, 157
isothermal, 156
real gas, 118, 156

expansion machine, 25
expansion work, 145
extensive variable, 3, *20*
extent of reaction (ξ)
  methanol reactor, 92
extent of reaction (ξ), 79
extraction, 24

Fahrenheit (°F), 10
fan, 24
  typical power, 13
feedback control, 322
Fick's law, 47
filtration, 24
first law of thermodynamics, 161
fittings, 24
flash, 24, 64, *189*
  UV, 317
  pH, 195
  pT, 193
  bubble point, 190
  calculations, 64, 189–196
  dew point, 191
  dynamic, 316, 317
  Rachford-Rice, 65, 193
flotation, 24
flow in pipe
  gas, 247
  friction pressure drop, 242
flow process, 21
flow sheet, 66, 67, 109
flow system, *see* open system
flow work (W_{flow}), 99
flue gas, 362
fluid ounce (fl.oz.), 10
force (F), 330
  conversion factor, 10
formaldehyde, 412
fossil fuels
  exergy, 219
freezing, 358
freezing point depression, 184, 185
frequency response, 285
friction, 26
  second law, 249
friction factor (f), 243
friction loss (Φ), 239
friction pressure drop
  rule of thumb, 245
friction pressure drop (Δp_f), 240
  flow in pipe, 242
frictionless, 372
  flow, 240
ft (foot, '), 9
fuel, 12, 17
fuel cell, 98, 102, 213, 226, 344
fugacity (f), 342, 387
  component (f_i), 186
fugacity coefficient (ϕ), 178, 342, 387
fusion (melting), 186, 358

G, *see* Gibbs energy
g (acceleration of gravity), 2
g_c (mystical factor in many US books), 10
G (giga), 2
g-mol, 9
gain (k), 286
gal (gallon), 9
gas, *116*
  dynamics, 298
  emptying container, 393
  filling tank, 104
  flow in pipe, 247, 248
  ideal, *see* ideal gas
  volume, 14
gas constant (R), 2, 335
gas power plant, 236
gas turbine, 226
  combustion, 203
gasoline, 32
gasoline engine, 203
gauge pressure (g), 11
Gibbs energy (G), 366, 385
  equilibrium, 384
Gibbs energy of formation ($\Delta_f G^\ominus$), 219, 416
  data, 416
Gibbs phase rule, 332
Gibbs' theorem, 381
Gibbs, Josiah Willard (1839–1903), 384
Gibbs-Helmholtz, 391
global warming, 16
  potential (GWP), 209
gpm (gallons per minute), 15
Gr (Grashof number), 296
Grand Canyon, 36
gravity
  API, 13
  specific (spgr), 13
gross heating value (GHV), 363
  natural gas, 33
Guldberg and Waage, 175, 259

H, *see* enthalpy
h (hour, hr), 8
half time, 269
HAZOP analysis, 273
heat (Q), 329, 345, *345*
heat capacity, 16, *349*
  constant pressure (C_p), 349
  constant volume (C_V), 349
  data
    gases, 422

hydrocarbons, 17
liquids, 421
water, 16
ideal gas, 17, 342, 349
ideal mixture, 356
mean ($\bar{C}_p$), 355
molar ($C_{p,m}$), 4
polynomial form, 355
specific (c_p), 5
heat engine, *198*
Carnot cycle, 373
heat exchanger, 24, 70, 129–142
cocurrent flow, 133, 137
countercurrent flow, 133, 134, 137
efficiency, 140
entropy change, 167
logarithmic mean temperature, 132, 394
pressure drop, 245
shell and tube, 130
two tube-pass, 137
typical numbers, 135
heat of combustion, 362
gross heating value (GHV), 363
higher, 362
hydrocarbons, 17
lower, 362
heat of combustion ($\Delta_c H^\ominus$), 363
heat of formation, *see also* enthalpy of formation
heat of formation ($\Delta_f H^\ominus$), 363
heat of reaction (enthalpy), 360
heat of vaporization ($\Delta_{vap} H$), 16, 281, 358
heat pump, 203
heat reservoir, 199, 373
heat transfer coefficient (h), *see also* overall heat transfer coefficient (U), 136
heating value, *see also* heat of combustion
higher, 363
lower, 219
Henry's law, 186
Hess' law, 358, 362
higher heating value, 363
horsepower (hp), 12
humans
artificial kidney, 44
energy consumption, 34
heart work, 34
heat transfer, 35
sweating, 35
hydrocarbons
data, 17
hydrogen
data, 416
synthesis gas, 94, 405
hydrostatic pressure, 330

ideal gas, 334–335
equation of state (gas law), 20, 334
mass basis, 335
heat capacity, 17
mole volume, 14
ideal mixture, 381
density, 54
entropy, 165, 381–383
vapor/liquid equilibrium, 182
ideal solution, *see* ideal mixture
ideal tank reactor (CSTR), *see* reactor
implicit equation, 191
impulse response, 285
inch ("), 9
incompressible, 126, 154
flow, 240
independent equation, 68
independent reaction, 88
index problem, 321
industrial ecology, 212
inertia, time constant, 287
initial time (t_0), 76
input, 303
integrator dynamics, 297
intensive variable, 3, *20*
internal energy (U), 96, 97, 345, *347*
calculation of, 366
constant volume, 347
interpolation, 54
inventory, 39
inverse response, 311, 312
investment cost, 29
irreversible process, 369
isenthalpic, 21
pressure relief, 114, 171
isentropic, 21, 373, 379
expansion, 170, 350
ideal gas example, 172
pressure relief
ideal gas-example, 172
shaft work, 149
isobaric, 21
isochoric, 21
isothermal, 21
reactor, 260
shaft work, 148
iteration, 139
IUPAC, 2

J (Joule), 2, 11
Joule's experiment, 353
Joule-Thompson coefficient (μ), 357
Joule-Thompson effect, 116, 117
Joule-Thompson valve, 26, 171

K (Kelvin), 2, 10
k (kilo), 2

K value, 188
Kelvin's postulate, 372
key reactant, 81
kg (kilogram), 2, 8
kidney, artificial, 44
kinetic energy, 96, 125, 344
  neglect, 97
kinetics, *see* reaction kinetics
  elementary, 256
KTPA (kilo ton per annum), 15
kWh (kilowatt hour), 12

l (liter), 9
laminar flow, 243, 297
Lang-factor, 30
Langmuir, 257
latent energy, 345
latent heat, 364
law of mass action, 175, 259
lb (pound), 8
lb-mol (strange unit in US books), 9
lb_f (pound force) (strange unit in US books), 10
Le Chatelier's principle, 173
length, conversion factor, 9
Levenspiel plot, 265
lever rule, 54
light, speed of, 41
limiting reactant, 81
linearization, dynamics, 301–303
linearized model, 303
liquid, *116*
LNG (liquefied natural gas), 219
log-mean Carnot factor, 201, 233
logarithmic mean temperature, 132, 394
lossless, 372
lower heating value, 219

M (molar mass), 3, 328
m, *see* mass
M (mega), 2
m (meter), 2, 9
m (milli), 2
M (roman numeral for 1000), 15
mass (m), 40
  conversion factor, 8
mass balance, 47–76
  component, 48, 276
  dynamic, 275
  matrix formulation, 90, 254
  mixing, 49
  mixture, 51
  recycle, 55
  total, 43
  with reaction, 77–94
mass flow (mass rate) ($\dot{m} \equiv w$), 5, 41
mass flow rate ($\dot{m} \equiv w$), 274
material balance, *see* mass balance
MATLAB
  ammonia equilibrium, 177
  DAE system, 319
  distillation, 314
  Euler integration, 307
  fmincon, 321
  isothermal CSTR, 311
  ode15s, 319
  odeeuler.m, 307
  rank, 89
  stirred tank, 306
  three stirred tanks, 309
  vector calculations, 315
matrix, 303
matrix formulation of mass balances, 90, 254
Maxwell, 339
mean velocity (v), 125
mechanical energy, 239
mechanical energy balance, 237–251
melting point depression, 185
membrane unit, 24
  artificial kidney, 44
methane
  combustion, 119
  data, 416
  thermodynamic diagram, 418
methanol, 412
  data, 416
  energy balance, 120
  equilibrium, 92
  mass balance, 92
  synthesis gas, 94
MIGD (million imperial gallons per day), 15
mile, 9
min (minute), 8
mixer, 24, 70
mixing rule
  linear blending, 50
  SRK equation of state, 341
mixing tank
  dynamics, 306
mixture, 381
  energy balance, 107
  heat of mixing ($\Delta_{\text{mix}}H$), 112, 281, 359, 365, 381
  ideal, 381
  mixing entropy ($\Delta_{\text{mix}}S$), 381
  mixing volume ($\Delta_{\text{mix}}V$), 55, 381
  mixing work
    natural gas power plant, 220
  non-ideal, 188
  separation work, 222
    distillation, 223
MM (million), 15

mol, 2, 9
molar enthalpy (H_m), 4
molar flow (mole rate) ($\dot{n} \equiv F$), 5
molar flow rate ($\dot{n} \equiv F$), 274
molar fraction, *see* mole fraction (x_i)
molar heat capacity ($C_{p,m}$), 4
molar mass (M), 3, 328
  data, 416
  mixture, 328
molar quantity (X_m), 4
molar volume (V_m), 4
  ideal gas, 14
mole, 327
mole fraction (x_i), 4
molecular weight, 3, *328*, *see also* molar mass (M)
momentum (mv), 40
momentum balance, *see* mechanical energy balance
  bath tub, 297
Mount Everest, 331
MTPA (million ton per annum), 15
multivariable systems, 303

n (nano), 2
N (Newton), 2, 10
N (normal), 14
natural convection, 296
  air, 296
  water, 296
natural gas, 225
  combustion, 119
  heating value, 33
  heating value (GHV), 33
  sales gas, 33
  sales gas specification, 33
  ton oil equivalent (toe), 33
natural gas power plant, 225
negative feedback, 322
Newton's 2nd law, 41, 238
nitric acid, 359
  production, 412, 413
Nm^3 (Normal cubic meter), 14
non-flow system, *see* closed system
non-ideal mixture, 188
normal boiling point, 378
normal boiling point (T_b), 16
normal cubic meter [Nm^3], 14
notation, *1*
  simplified, 7
Nu (Nusselt number), 296
number of heat transfer units (N_{tu}), 141

oil, 12, 17
open system
  energy balance, 95, 100

osmotic pressure, 221, 382
  seawater, 382
Otto cycle, 203
ounce (oz), 10
output, 285
overall heat transfer coefficient (U), 132, 135, 293, 295
  conversion factor, 135
  typical values, 135

p, *see* pressure
P (peta), 2
p (pico), 2
packings, 22
paper machine, 69
partial molar enthalpy, 281
partial molar quantity ($\bar{X}_i$ or X_i), 381
partial pressure (p_i), 335
Pascal (Pa), 2
Peng-Robinson (PR) equation of state, 340, 342
perpetuum mobile, 162, 239
phase rule, Gibbs, 332
phase transition, 358
  dynamic energy balance, 280
  entropy, 378
physical chemistry, 327–391
PID controller, 322
platinum catalyst, 259, 260
poise (P) (unit for viscosity), 243
polytropic, 148
polytropic process, 149
postulate, 347
potential energy, 96, 125, 345
  neglect, 97
pound (lb), 8
pound force (lb_f) (strange unit in US books), 10
pound mole (lb-mol) (strange unit in US books), 9
power
  conversion factor, 12
Poynting factor, 183, 188
ppb, 4
ppb (parts per billion), 2
ppm, 4
ppm (parts per million), 2
Pr (Prandtl number), 296
PRBS response, 285
pressure
  atmospheric (1 atm)), 11
pressure (p), *330*
  absolute, 11
  barometric formula, 331, 352
  conversion factor, 11
  gauge pressure, 11

standard ($p^\ominus = 1$ bar), 3
thermodynamic, 390
pressure drop
friction (Δp_f), 242
gas, 247
heat exchanger, 245
pressure dynamics, 299
pressure ratio
critical, 128
pressure relief
ideal gas, 114
isenthalpic, 114
real gas, 118
pressure-enthalpy (pH) diagram, 118, 156, 333
ammonia, 419
methane, 115, 418
pressure-volume (pV) work, 99
process, v, 18
batch, 20
continuous, 21
dynamic (non-steady state), 21
reversible, 20, *331*
semi-batch, 20
steady-state, 21
process design, 75
production cost, 31
production rate, 28
propane
thermodynamic diagram, 415
propylene
thermodynamic diagram, 415
psi (pounds per square inch), 11
psia (pounds per square inch absolute), 11
psig (pounds per square inch gauge), 11
pump, 25, 70
pump work (W_s), 154
pumping head, 242
purge, 56
methanol process, 71

Q, *see* heat
Qatar, 33
quench, 25

R (gas constant), 2
R134a (refrigerant), 159, 209
thermodynamic diagram, 415
R600a (isobutane), 209
R717 (ammonia), 208, *see also* ammonia
Rachford-Rice flash equation, 193
radiator, 129
Rankine (R), 10
Rankine cycle, 203, 226
power plant, 227
reverse (refrigeration), 206
Rankine refrigeration cycle, 203, 206
Raoult's law, 182
Rashig-ring, 22
rate, 5
rate of reaction, *see* reaction rate
Rayleigh distillation, 396
Re (Reynolds number), 243
reaction, 77
energy balance, 118
independent, 88
reaction engineering, 253–271
reaction kinetics, 253
chemical equilibrium, 258
reaction mechanism, 256
reaction order, 257, 267
reaction rate (r), *253*, 277
reaction rate constant (k), 256
reactor, 25, 70, 260–271
batch, 78, 262, 299
continuous, 79
continuous stirred tank (CSTR), 262, 277
dynamic energy balance, 281
dynamics isothermal, 311
dynamics, first-order reaction, 299
dynamics, second-order reaction, 313
plug flow (PFR), 264
similarity batch and plug flow, 267
real gas, 335–343
compression, 155, 159
generalized diagram, 336
pressure relief, 118
reboiler, 23
recycle, 55
batch process, 64
mass balance, 62
methanol process, 71
Redlich-Kwong equation, 340
reduced pressure (p_r), 337
reduced temperature (T_r), 337
reference state, 106, 364
elements, 364
other, 365
reflux, 23
refrigerant, 208
CO_2 (R744), 208, 209
ammonia (R717), 208
CFCs, freons (R12, R22), 208
i-butane (R600a), 209
R134a, 159, 209, 415
refrigerator, 129, 203, 205
ammonia cycle, 206
R134a cycle, 159
relative volatility (α), 183
reservoir, 199, 373
residence time, 290, 291
reactor, 267

residence time distribution (RTD), 285
restriction, pressure drop, 241
reversible process, *331*, *372*, 377
  entropy, 371
Reynolds number (Re), 243
rotational energy, 345
rule of thumb
  friction pressure drop, 245
  heat exchanger $\Delta T_{\min}$, 137
  heating value of oil and natural gas, 32
  Trouton's rule for ΔS, 378
  vapor pressure water, 181

s (second), 2, 8
S (standard), 3, 14
salt power, 221
saturated, 16, 116
saturation pressure, 65, 180, *see also* vapor pressure (p^{sat})
scaling, 66, 67
scaling exponent, 29
SCMH (standard cubic meters per hour), 15
scrubber, 25
sea water
  boiling point, 185
  freezing point, 186
second law of thermodynamics, 161
selectivity (ϕ), *82*
semi-batch process, 20
sensitive heat, 365
separator, 25, 70
settler, 25
shaft work (W_s), 98, 344
  adiabatic, 149
  ideal gas, 148
  important example, 150
  isentropic, 149
  isothermal, 148
  real gas, 155
  reversible, 145
shell and tube heat exchanger, 130, 136
shift reaction, 92, 405, 412
SI units, 2, 7
simulation, 75
  dynamic, 303–325
  heat exchanger, 139
  versus design, 139
Sm^3 (standard cubic meter), 14
☺, 9, 93, 210, 250
Soave-Redlich-Kwong (SRK), 178, 187, 188, 340, 357
solar cells, 33
solubility
  Henry's constant, 186
solubility of gas, 186
solute, 184
solvent, 184
specific enthalpy (h), 5
specific gravity (spgr), 13
specific heat capacity (c_p), 5
specific quantity, 5
speed of sound (c_s), 127, 247
split fraction, 70
splitter, 25, 70
spreadsheet introduction, 59
standard ambient temperature (298.15 K), 3
standard cubic meter [Sm^3], 14
standard pressure ($p^{\ominus}$), 3
standard state, 3
standard temperature and pressure (STP), 3
state, 19
state diagram, 109, *see also* thermodynamic diagram
state function, *371*
state variable, 19, *332*
  dynamics, 303
  number of, 332
static pressure, 240
steady-state process, 21, *43*
  balance, 283
  component balance, 78
steady-state value (*), 283
steam
  low pressure, 216
steam, H_2O (g), 16
  steam reformer, 405
  steam turbine, 158, 227
  thermodynamic diagram, 420
  steam engine, 197
step response, 285, 286
  first-order, 289
  higher-order, 308
stirred tank
  dynamics three tanks, 308
stoichiometric coefficient (ν), 79
stoichiometric matrix (N), 90
STP (standard temperature and pressure), 3
stream, 20
stream data, 67
stripping, 25
styrene, 124
subcooled liquid, 206
sublimation, 358
  CO_2, 416
supercritical, *116*
superheated vapor, 228, 232
surface energy, 345
symbols, 1
synthesis, 76
synthesis gas, 94, 405

heat exchanger, 137
system, 19
adiabatic, 19
boundary, 76
closed, 19, 346
isolated, 19
open, 19
system boundary, 19

T, *see* temperature
t, *see* time
T (tera), 2
t (ton), 8
Taylor-series expansion, 301
T_b (normal boiling point), 16
temperature, *329*
absolute (T), 10
Celsius (t), 10
conversion factor, 10
dynamic energy balance, 279
thermodynamic (T), 390
thermal conductivity (k), 136, 296
thermal energy, 345, 365
thermal expansion coefficient (β), 296
thermal power, 197–236
thermochemistry, 357
thermocouple, 293
thermodynamic diagram, 115, 156, 333
HS-diagram, water, 420
pH-diagram, 333
ammonia, 419
ammonia cooling cycle, 207
methane, 418
pV-diagram, 333
other components, 415
thermodynamics, *329*, 327–391
0th law, 161
1st law, 101, 161, 345
2nd law, 161, 173, 372
3rd law, 163
fundamental equation, 389
history of, 197
partial derivatives, 389
third law of thermodynamics, 163
time (t), units for, 8
time constant (τ), 286
bathtub, 298
continuous stirred tank, 290
with heat exchanger, 291
continuous stirred tank reactor (CSTR)
first order reaction, 299
second-order reaction, 313
cooking plate, 293
first-order system, 287
gas tank, 298
outflow sink, 302
thermocouple, 295
time delay (θ), 287
time response, 284–290
toe (ton oil equivalent), 12, 17, 363
ton, 8
long, 8
metric (tonne), 8
short, 8
torr (mm Hg), 11
tower, 21
TPA (ton per annum), 15
TPD (ton per day), 15
trays, 22
triple point, 416
Trouton's rule, 378
turbine, 25, 70, 144
turbulent flow, 243, 297, 302

U, *see* internal energy
U, *see* overall heat transfer coefficient
UNIFAC, 188
unit operations, *21*
units
check , 7
conversion, 8
exercises, 31
SI, 2
unsteady state, *see* dynamic

V, *see* volume
v, *see* velocity
valve, 26
relative capacity coefficient (C_d), 245
choke, 26
energy balance
leak, 126
isenthalpic, 114, 172
Joule-Thompson, 26
pressure drop, 244
valve coefficient (C_v), 26
valve equation, 26, 244
van der Waals equation of state, 155, 337
van der Waals, Johannes (1837–1923), 337
van't Hoff, 176
equation, 388
osmotic pressure, 382
van't Hoff, Jacobus (1852–1911), 382
vapor (= saturated gas), 16, 179
vapor pressure (p^{sat}), *16*, *180*, 182, 340
water, 16, 180, 181
vapor/liquid-equilibrium (VLE), 179–189
distillation, 314
non-ideal, 188
Raoult's law, 182
relative volatility, 183
SRK, 188, 342

UNIFAC, 188
velocity (v), 97, *125*, 126, 127, 239
viscosity, 243
viscosity (μ), 296
volume (V)
conversion factor, 9
gas, 14
standard, 14
volume change work, 99, 346–347
volumetric flow rate ($\dot{V} \equiv q$), 5, 274

W, *see* work
W (Watt), 2
water, 16
boiling point elevation, 185
boiling point variation
elevation, 331
pressure, 182
data, 16, 416
freezing point depression, 185
Henry's law, 186
thermodynamic diagram, 420
vapor pressure, 16, 181, 182
experimental data, 181
rule of thumb, 181
water fall, 96
water turbine, 155
wet basis, 362
window pane, heat loss, 135
wood, 17
work (W), *344*, 345
expansion, 145
flow, 99
forms, 98
from heat, 197–377
lost, 239
sign convention, 3, 98
volume change, 99, 346–347
working fluid, *see also* refrigerant, 208

X, *see* conversion
x_i (mole fraction), 4

y (year), 8
yard, 9
yield (Y), *82*
total, 85